Methods in Enzymology

Volume 154

RECOMBINANT DNA

Part E

METHODS IN ENZYMOLOGY

EDITORS-IN-CHIEF

John N. Abelson   Melvin I. Simon

Methods in Enzymology

Volume 154

Recombinant DNA

Part E

EDITED BY

Ray Wu

SECTION OF BIOCHEMISTRY
MOLECULAR AND CELL BIOLOGY
CORNELL UNIVERSITY
ITHACA, NEW YORK

Lawrence Grossman

DEPARTMENT OF BIOCHEMISTRY
THE JOHNS HOPKINS UNIVERSITY
SCHOOL OF HYGIENE AND PUBLIC HEALTH
BALTIMORE, MARYLAND

ACADEMIC PRESS, INC.
Harcourt Brace Jovanovich, Publishers
San Diego New York Berkeley Boston
London Sydney Tokyo Toronto

ACADEMIC PRESS, INC.
1250 Sixth Avenue, San Diego, California 92101

United Kingdom Edition published by
ACADEMIC PRESS INC. (LONDON) LTD.
24–28 Oval Road, London NW1 7DX

LIBRARY OF CONGRESS CATALOG CARD NUMBER: 54-9110

ISBN 0–12–182055–6 (alk. paper)

PRINTED IN THE UNITED STATES OF AMERICA

87 88 89 90 9 8 7 6 5 4 3 2 1

Table of Contents

Section I. Methods for Cloning cDNA

Section II. Identification of Cloned Genes and Mapping of Genes

Section III. Chemical Synthesis and Analysis of Oligodeoxynucleotides

Section IV. Site-Specific Mutagenesis and Protein Engineering

Contributors to Volume 154

Article numbers are in parentheses following the names of contributors.
Affiliations listed are current.

TOM ALBER (27), *Institute of Molecular Biology, University of Oregon, Eugene, Oregon 97403*

DANNY C. ALEXANDER (3), *Calgene, Inc., Davis, California 95616*

K. ARAI (1), *Department of Molecular Biology, DNAX Research Institute of Molecular and Cellular Biology, Palo Alto, California 94304*

S. L. BEAUCAGE (15), *Department of Genetics, Stanford University, Stanford, California 94305*

HELMUT BLÖCKER (13), *GBF (Gesellschaft für Biotechnologische Forschung mbH), D-3300 Braunschweig, Federal Republic of Germany*

JEF D. BOEKE (10), *Department of Molecular Biology and Genetics, The Johns Hopkins University, School of Medicine, Baltimore, Maryland 21205*

M. BROWNSTEIN (1), *Laboratory of Molecular Genetics, National Institute of Child Health and Human Development, Bethesda, Maryland 20205*

PAUL CARTER (20), *Department of Biomolecular Chemistry, Genentech, Inc., South San Francisco, California 94080*

M. H. CARUTHERS (15), *Department of Chemistry and Biochemistry, University of Colorado, Boulder, Colorado 80309*

PIERRE CHAMBON (14), *Laboratoire de Génétique Moléculaire, LGME/CNRS et U.184/INSERM, Institute de Chimie Biologique, Faculté de Médecine, 67085 Strasbourg Cedex, France*

CHRISTOPHER COLECLOUGH (4), *Basel Institute for Immunology, CH-4005 Basel, Switzerland*

A. D. DARONE (15), *Centocor, Inc., Malvern, Pennsylvania 19355*

RONALD W. DAVIS (7), *Department of Biochemistry, Stanford University School of Medicine, Stanford, California 94305*

D. R. DODDS (15), *Sepracor, Inc., Marlborough, Massachusetts 01752*

STEPHEN ELLEDGE (7), *Department of Biochemistry, Stanford University School of Medicine, Stanford, California 94305*

GERALD R. FINK (10), *Whitehead Institute for Biomedical Research, Cambridge, Massachusetts 02142, and Massachusetts Institute for Technology, Cambridge, Massachusetts 02139*

JOSEPH R. FIRCA (16), *Pandex Laboratories, Inc., Mundelein, Illinois 60060*

E. F. FISHER (15), *AMGen, Inc., Thousand Oaks, California 91360*

RONALD FRANK (13), *GBF (Gesellschaft für Biotechnologische Forschung mbH), D-3300 Braunschweig, Federal Republic of Germany*

HANS-JOACHIM FRITZ (18), *Max-Planck-Institut für Biochemie, Abteilung Zellbiologie, Am Klopferspitz 18, D-8033 Martinsried bei München, Federal Republic of Germany*

MARK R. GRAY (8), *Department of Biological Chemistry, Harvard Medical School, Boston, Massachusetts 02115*

GISELA HEIDECKER (2), *Section of Genetics, Laboratory of Viral Carcinogenesis, National Cancer Institute, Frederick, Maryland 21701*

LEROY HOOD (16), *Division of Biology, California Institute of Technology, Pasadena, California 91125*

SUZANNA J. HORVATH (16), *Division of Biology, California Institute of Technology, Pasadena, California 91125*

P. C. HUANG (22), *Department of Chemistry, The Johns Hopkins University, School of Hygiene and Public Health, Baltimore, Maryland 21205*

Michael W. Hunkapiller (16), *Applied Biosystems, Inc., Foster City, California 94404*

Tim Hunkapiller (16), *Division of Biology, California Institute of Technology, Pasadena, California 91125*

Mittur N. Jagadish (12), *Division of Protein Chemistry, CSIRO, Parkville 3052, Victoria, Australia*

E. T. Kaiser (25), *Laboratory of Bioorganic Chemistry and Biochemistry, The Rockefeller University, New York, New York 10021*

M. Kawaichi (1), *Laboratory of Molecular Genetics, National Institute of Child Health and Human Development, Bethesda, Maryland 20205*

Peter Kollman (23), *Department of Pharmaceutical Chemistry, University of California, San Francisco, San Francisco, California 94143*

Wilfried Kramer (18), *Max-Planck-Institut für Biochemie, Abteilung Zellbiologie, Am Klopferspitz 18, D-8033 Martinsried bei München, Federal Republic of Germany*

Thomas A. Kunkel (19), *National Institute of Environmental Health Sciences, National Institute of Health, Research Triangle Park, North Carolina 27709*

F. Lee (1), *Department of Molecular Biology, DNAX Research Institute of Molecular and Cellular Biology, Palo Alto, California 94304*

Corey Levenson (21), *Department of Chemistry, Cetus Corporation, Emeryville, California 94608*

David F. Mark (21), *Department of Molecular Biology, Cetus Corporation, Emeryville, California 94608*

M. Matteucci (15), *Genentech, Inc., South San Francisco, California 94112*

Hans W. Djurhuus Matthes (14), *Laboratoire de Génétique Moléculaire, LGME/CNRS et U.184/INSERM, Institute de Chimie Biologique, Faculté de Médecine, 67085 Strasbourg Cedex, France*

Brian W. Matthews (27), *Institute of Molecular Biology, University of Oregon, Eugene, Oregon 97403*

C. Robert Matthews (26), *Department of Chemistry, The Pennsylvania State University, University Park, Pennsylvania 16802*

Gail P. Mazzara (8), *Applied Biotechnology, Inc., Cambridge, Massachusetts 02139*

L. J. McBride (15), *Applied Biosystems, Foster City, California 94404*

Joachim Messing (2), *Waksman Institute of Microbiology, Rutgers University, Piscataway, New Jersey 08854*

Andreas Meyerhans (13), *GBF (Gesellschaft für Biotechnologische Forschung mbH), Mascheroder Weg 1, D-3300 Braunschweig, Federal Republic of Germany*

C. Garrett Miyada (6), *Department of Molecular Genetics, Beckman Research Institute of the City of Hope, Duarte, California 91010*

Georges Natsoulis (10), *Department of Molecular Biology and Genetics, The Johns Hopkins University, School of Medicine, Baltimore, Maryland 21205*

John D. Noti (12), *Molecular and Cell Biology, Triton Biosciences, Inc., Alameda, California 94501*

H. Okayama (1), *Laboratory of Cell Biology, National Institute of Mental Health, Bethesda, Maryland 20892*

Richard Pine (22), *Department of Molecular and Cell Biology, The Rockefeller University, New York, New York 10021*

Sándor Pongor (24), *Institute of Enzymology, Hungarian Academy of Sciences, pf 7 Budapest 1502, Hungary, and Boyce Thompson Institute, Cornell University, Ithaca, New York 14853*

Pranhitha Reddy (8), *Department of Biology, Massachusetts Institute of Technology, Cambridge, Massachusetts 02139*

A. A. Reyes (5), *Department of Molecular Genetics, Beckman Research Institute of the City of Hope, Duarte, California 91010*

JOHN D. ROBERTS (19), *National Institute of Environmental Health Sciences, National Institute of Health, Research Triangle Park, North Carolina 27709*

MICHAEL ROSBASH (8), *Department of Biology, Brandeis University, Waltham, Massachusetts 02254*

KONRAD SCHWELLNUS (13), *GBF (Gesellschaft für Biotechnologische Forschung mbH), Mascheroder Weg 1, D-3300 Braunschweig, Federal Republic of Germany*

MICHAEL SMITH (17), *Department of Biochemistry, University of British Columbia, Vancouver, British Columbia*

MICHAEL SNYDER (7), *Department of Biochemistry, Stanford University School of Medicine, Stanford, California 94305*

Z. STABINSKY (15), *Department of Chemistry and Biochemistry, University of Colorado, Boulder, Colorado 80309*

ADRIEN STAUB (14), *Laboratoire de Génétique Moléculaire, LGME/CNRS et U.184/INSERM, Institute de Chimie Biologique, Faculté de Médecine, 67085 Strasbourg Cedex, France*

DOUGLAS SWEETSER (7), *Whitehead Institute for Biomedical Research, Cambridge, Massachusetts 02142*

ALADAR A. SZALAY (12), *Boyce Thompson Institute for Plant Research, Cornell University, Ithaca, New York 14853*

J.-Y. TANG (15), *Shanghai Institute of Biochemistry, Shanghai, Peoples Republic of China*

JOHN W. TAYLOR (25), *Laboratory of Bioorganic Chemistry and Biochemistry, The Rockefeller University, New York, New York 10021*

JOSHUA TRUEHEART (10), *Whitehead Institute for Biomedical Research, Cambridge, Massachusetts 02142, and Massachusetts Institute of Technology, Cambridge, Massachusetts 02139*

R. BRUCE WALLACE (5, 6), *Department of Molecular Genetics, Beckman Research Institute of the City of Hope, Duarte, California 91010*

ALICE WANG (21), *Department of Molecular Biology, Cetus Corporation, Emeryville, California 94608*

GEORGE M. WEINSTOCK (9), *Department of Biochemistry and Molecular Biology, The University of Texas Medical School at Houston, Houston, Texas 77225*

T. YOKOTA (1), *Department of Molecular Biology, DNAX Research Institute of Molecular and Cellular Biology, Palo Alto, California 94304*

RICHARD A. YOUNG (7), *Whitehead Institute for Biomedical Research, Cambridge, Massachusetts 02142, and Department of Biology, Massachusetts Institute of Technology, Cambridge, Massachusetts 02139*

RICHARD A. ZAKOUR (19), *Molecular and Applied Genetics Laboratory, Allied Corporation, Morristown, New Jersey 07960*

MARK J. ZOLLER (17), *Cold Spring Harbor Laboratory, Cold Spring Harbor, New York 11724*

FRANS J. DE BRUIJN (11), *Max-Planck-Institut für Züchtungsforschung, Abteilung Schell, D-5000 Köln 30, Federal Republic of Germany*

W. F. VAN GUNST (23), *Department of Physical Chemistry, University of Groningen, 9747 AG Groningen, The Netherlands*

Preface

Recombinant DNA methods are powerful, revolutionary techniques for at least two reasons. First, they allow the isolation of single genes in large amounts from a pool of thousands or millions of genes. Second, the isolated genes or their regulatory regions can be modified at will and reintroduced into cells for expression at the RNA or protein levels. These attributes allow us to solve complex biological problems and to produce new and better products in the areas of health, agriculture, and industry.

Volumes 153, 154, and 155 supplement Volumes 68, 100, and 101 of *Methods in Enzymology*. During the past few years, many new or improved recombinant DNA methods have appeared, and a number of them are included in these three new volumes. Volume 153 covers methods related to new vectors for cloning DNA and for expression of cloned genes. Volume 154 includes methods for cloning cDNA, identification of cloned genes and mapping of genes, chemical synthesis and analysis of oligodeoxynucleotides, site-specific mutagenesis, and protein engineering. Volume 155 includes the description of several useful new restriction enzymes, details of rapid methods for DNA sequence analysis, and a number of other useful methods.

RAY WU
LAWRENCE GROSSMAN

METHODS IN ENZYMOLOGY

EDITED BY

Sidney P. Colowick and Nathan O. Kaplan

VANDERBILT UNIVERSITY
SCHOOL OF MEDICINE
NASHVILLE, TENNESSEE

DEPARTMENT OF CHEMISTRY
UNIVERSITY OF CALIFORNIA
AT SAN DIEGO
LA JOLLA, CALIFORNIA

I. Preparation and Assay of Enzymes
II. Preparation and Assay of Enzymes
III. Preparation and Assay of Substrates
IV. Special Techniques for the Enzymologist
V. Preparation and Assay of Enzymes
VI. Preparation and Assay of Enzymes (*Continued*)
Preparation and Assay of Substrates
Special Techniques
VII. Cumulative Subject Index

METHODS IN ENZYMOLOGY

EDITORS-IN-CHIEF

Sidney P. Colowick and Nathan O. Kaplan

VOLUME XVIII. Vitamins and Coenzymes (Parts A, B, and C)
Edited by DONALD B. MCCORMICK AND LEMUEL D. WRIGHT

VOLUME XIX. Proteolytic Enzymes
Edited by GERTRUDE E. PERLMANN AND LASZLO LORAND

VOLUME XX. Nucleic Acids and Protein Synthesis (Part C)
Edited by KIVIE MOLDAVE AND LAWRENCE GROSSMAN

VOLUME XXI. Nucleic Acids (Part D)
Edited by LAWRENCE GROSSMAN AND KIVIE MOLDAVE

VOLUME XXII. Enzyme Purification and Related Techniques
Edited by WILLIAM B. JAKOBY

VOLUME XXIII. Photosynthesis (Part A)
Edited by ANTHONY SAN PIETRO

VOLUME XXIV. Photosynthesis and Nitrogen Fixation (Part B)
Edited by ANTHONY SAN PIETRO

VOLUME XXV. Enzyme Structure (Part B)
Edited by C. H. W. HIRS AND SERGE N. TIMASHEFF

VOLUME XXVI. Enzyme Structure (Part C)
Edited by C. H. W. HIRS AND SERGE N. TIMASHEFF

VOLUME XXVII. Enzyme Structure (Part D)
Edited by C. H. W. HIRS AND SERGE N. TIMASHEFF

VOLUME XXVIII. Complex Carbohydrates (Part B)
Edited by VICTOR GINSBURG

VOLUME XXIX. Nucleic Acids and Protein Synthesis (Part E)
Edited by LAWRENCE GROSSMAN AND KIVIE MOLDAVE

VOLUME XXX. Nucleic Acids and Protein Synthesis (Part F)
Edited by KIVIE MOLDAVE AND LAWRENCE GROSSMAN

VOLUME XXXI. Biomembranes (Part A)
Edited by SIDNEY FLEISCHER AND LESTER PACKER

VOLUME XXXII. Biomembranes (Part B)
Edited by SIDNEY FLEISCHER AND LESTER PACKER

VOLUME XXXIII. Cumulative Subject Index Volumes I–XXX
Edited by MARTHA G. DENNIS AND EDWARD A. DENNIS

VOLUME XXXIV. Affinity Techniques (Enzyme Purification: Part B)
Edited by WILLIAM B. JAKOBY AND MEIR WILCHEK

VOLUME XXXV. Lipids (Part B)
Edited by JOHN M. LOWENSTEIN

VOLUME XXXVI. Hormone Action (Part A: Steroid Hormones)
Edited by BERT W. O'MALLEY AND JOEL G. HARDMAN

VOLUME XXXVII. Hormone Action (Part B: Peptide Hormones)
Edited by BERT W. O'MALLEY AND JOEL G. HARDMAN

VOLUME XXXVIII. Hormone Action (Part C: Cyclic Nucleotides)
Edited by JOEL G. HARDMAN AND BERT W. O'MALLEY

VOLUME XXXIX. Hormone Action (Part D: Isolated Cells, Tissues, and Organ Systems)
Edited by JOEL G. HARDMAN AND BERT W. O'MALLEY

VOLUME XL. Hormone Action (Part E: Nuclear Structure and Function)
Edited by BERT W. O'MALLEY AND JOEL G. HARDMAN

VOLUME XLI. Carbohydrate Metabolism (Part B)
Edited by W. A. WOOD

VOLUME XLII. Carbohydrate Metabolism (Part C)
Edited by W. A. WOOD

VOLUME XLIII. Antibiotics
Edited by JOHN H. HASH

VOLUME XLIV. Immobilized Enzymes
Edited by KLAUS MOSBACH

VOLUME XLV. Proteolytic Enzymes (Part B)
Edited by LASZLO LORAND

VOLUME XLVI. Affinity Labeling
Edited by WILLIAM B. JAKOBY AND MEIR WILCHEK

VOLUME XLVII. Enzyme Structure (Part E)
Edited by C. H. W. HIRS AND SERGE N. TIMASHEFF

VOLUME XLVIII. Enzyme Structure (Part F)
Edited by C. H. W. HIRS AND SERGE N. TIMASHEFF

VOLUME XLIX. Enzyme Structure (Part G)
Edited by C. H. W. HIRS AND SERGE N. TIMASHEFF

VOLUME L. Complex Carbohydrates (Part C)
Edited by VICTOR GINSBURG

VOLUME LI. Purine and Pyrimidine Nucleotide Metabolism
Edited by PATRICIA A. HOFFEE AND MARY ELLEN JONES

VOLUME LII. Biomembranes (Part C: Biological Oxidations)
Edited by SIDNEY FLEISCHER AND LESTER PACKER

VOLUME LIII. Biomembranes (Part D: Biological Oxidations)
Edited by SIDNEY FLEISCHER AND LESTER PACKER

VOLUME LIV. Biomembranes (Part E: Biological Oxidations)
Edited by SIDNEY FLEISCHER AND LESTER PACKER

VOLUME LV. Biomembranes (Part F: Bioenergetics)
Edited by SIDNEY FLEISCHER AND LESTER PACKER

VOLUME LVI. Biomembranes (Part G: Bioenergetics)
Edited by SIDNEY FLEISCHER AND LESTER PACKER

VOLUME LVII. Bioluminescence and Chemiluminescence
Edited by MARLENE A. DELUCA

VOLUME LVIII. Cell Culture
Edited by WILLIAM B. JAKOBY AND IRA PASTAN

VOLUME LIX. Nucleic Acids and Protein Synthesis (Part G)
Edited by KIVIE MOLDAVE AND LAWRENCE GROSSMAN

VOLUME LX. Nucleic Acids and Protein Synthesis (Part H)
Edited by KIVIE MOLDAVE AND LAWRENCE GROSSMAN

VOLUME 61. Enzyme Structure (Part H)
Edited by C. H. W. HIRS AND SERGE N. TIMASHEFF

VOLUME 62. Vitamins and Coenzymes (Part D)
Edited by DONALD B. MCCORMICK AND LEMUEL D. WRIGHT

VOLUME 63. Enzyme Kinetics and Mechanism (Part A: Initial Rate and Inhibitor Methods)
Edited by DANIEL L. PURICH

VOLUME 64. Enzyme Kinetics and Mechanism (Part B: Isotopic Probes and Complex Enzyme Systems)
Edited by DANIEL L. PURICH

VOLUME 65. Nucleic Acids (Part I)
Edited by LAWRENCE GROSSMAN AND KIVIE MOLDAVE

VOLUME 66. Vitamins and Coenzymes (Part E)
Edited by DONALD B. MCCORMICK AND LEMUEL D. WRIGHT

VOLUME 67. Vitamins and Coenzymes (Part F)
Edited by DONALD B. MCCORMICK AND LEMUEL D. WRIGHT

VOLUME 68. Recombinant DNA
Edited by RAY WU

VOLUME 69. Photosynthesis and Nitrogen Fixation (Part C)
Edited by ANTHONY SAN PIETRO

VOLUME 70. Immunochemical Techniques (Part A)
Edited by HELEN VAN VUNAKIS AND JOHN J. LANGONE

VOLUME 71. Lipids (Part C)
Edited by JOHN M. LOWENSTEIN

VOLUME 72. Lipids (Part D)
Edited by JOHN M. LOWENSTEIN

VOLUME 73. Immunochemical Techniques (Part B)
Edited by JOHN J. LANGONE AND HELEN VAN VUNAKIS

VOLUME 74. Immunochemical Techniques (Part C)
Edited by JOHN J. LANGONE AND HELEN VAN VUNAKIS

VOLUME 75. Cumulative Subject Index Volumes XXXI, XXXII, XXXIV–LX
Edited by EDWARD A. DENNIS AND MARTHA G. DENNIS

VOLUME 76. Hemoglobins
Edited by ERALDO ANTONINI, LUIGI ROSSI-BERNARDI, AND EMILIA CHIANCONE

VOLUME 77. Detoxication and Drug Metabolism
Edited by WILLIAM B. JAKOBY

VOLUME 78. Interferons (Part A)
Edited by SIDNEY PESTKA

VOLUME 79. Interferons (Part B)
Edited by SIDNEY PESTKA

VOLUME 80. Proteolytic Enzymes (Part C)
Edited by LASZLO LORAND

VOLUME 81. Biomembranes (Part H: Visual Pigments and Purple Membranes, I)
Edited by LESTER PACKER

VOLUME 82. Structural and Contractile Proteins (Part A: Extracellular Matrix)
Edited by LEON W. CUNNINGHAM AND DIXIE W. FREDERIKSEN

VOLUME 83. Complex Carbohydrates (Part D)
Edited by VICTOR GINSBURG

VOLUME 84. Immunochemical Techniques (Part D: Selected Immunoassays)
Edited by JOHN J. LANGONE AND HELEN VAN VUNAKIS

VOLUME 85. Structural and Contractile Proteins (Part B: The Contractile Apparatus and the Cytoskeleton)
Edited by DIXIE W. FREDERIKSEN AND LEON W. CUNNINGHAM

VOLUME 86. Prostaglandins and Arachidonate Metabolites
Edited by WILLIAM E. M. LANDS AND WILLIAM L. SMITH

VOLUME 87. Enzyme Kinetics and Mechanism (Part C: Intermediates, Stereochemistry, and Rate Studies)
Edited by DANIEL L. PURICH

VOLUME 88. Biomembranes (Part I: Visual Pigments and Purple Membranes, II)
Edited by LESTER PACKER

VOLUME 89. Carbohydrate Metabolism (Part D)
Edited by WILLIS A. WOOD

VOLUME 90. Carbohydrate Metabolism (Part E)
Edited by WILLIS A. WOOD

VOLUME 91. Enzyme Structure (Part I)
Edited by C. H. W. HIRS AND SERGE N. TIMASHEFF

VOLUME 92. Immunochemical Techniques (Part E: Monoclonal Antibodies and General Immunoassay Methods)
Edited by JOHN J. LANGONE AND HELEN VAN VUNAKIS

VOLUME 93. Immunochemical Techniques (Part F: Conventional Antibodies, Fc Receptors, and Cytotoxicity)
Edited by JOHN J. LANGONE AND HELEN VAN VUNAKIS

VOLUME 94. Polyamines
Edited by HERBERT TABOR AND CELIA WHITE TABOR

VOLUME 95. Cumulative Subject Index Volumes 61–74, 76–80
Edited by EDWARD A. DENNIS AND MARTHA G. DENNIS

VOLUME 96. Biomembranes [Part J: Membrane Biogenesis: Assembly and Targeting (General Methods; Eukaryotes)]
Edited by SIDNEY FLEISCHER AND BECCA FLEISCHER

VOLUME 97. Biomembranes [Part K: Membrane Biogenesis: Assembly and Targeting (Prokaryotes, Mitochondria, and Chloroplasts)]
Edited by SIDNEY FLEISCHER AND BECCA FLEISCHER

VOLUME 98. Biomembranes (Part L: Membrane Biogenesis: Processing and Recycling)
Edited by SIDNEY FLEISCHER AND BECCA FLEISCHER

VOLUME 99. Hormone Action (Part F: Protein Kinases)
Edited by JACKIE D. CORBIN AND JOEL G. HARDMAN

VOLUME 100. Recombinant DNA (Part B)
Edited by RAY WU, LAWRENCE GROSSMAN, AND KIVIE MOLDAVE

VOLUME 101. Recombinant DNA (Part C)
Edited by RAY WU, LAWRENCE GROSSMAN, AND KIVIE MOLDAVE

VOLUME 102. Hormone Action (Part G: Calmodulin and Calcium-Binding Proteins)
Edited by ANTHONY R. MEANS AND BERT W. O'MALLEY

VOLUME 103. Hormone Action (Part H: Neuroendocrine Peptides)
Edited by P. MICHAEL CONN

VOLUME 104. Enzyme Purification and Related Techniques (Part C)
Edited by WILLIAM B. JAKOBY

VOLUME 105. Oxygen Radicals in Biological Systems
Edited by LESTER PACKER

VOLUME 106. Posttranslational Modifications (Part A)
Edited by FINN WOLD AND KIVIE MOLDAVE

VOLUME 107. Posttranslational Modifications (Part B)
Edited by FINN WOLD AND KIVIE MOLDAVE

VOLUME 108. Immunochemical Techniques (Part G: Separation and Characterization of Lymphoid Cells)
Edited by GIOVANNI DI SABATO, JOHN J. LANGONE, AND HELEN VAN VUNAKIS

VOLUME 109. Hormone Action (Part I: Peptide Hormones)
Edited by LUTZ BIRNBAUMER AND BERT W. O'MALLEY

VOLUME 110. Steroids and Isoprenoids (Part A)
Edited by JOHN H. LAW AND HANS C. RILLING

VOLUME 111. Steroids and Isoprenoids (Part B)
Edited by JOHN H. LAW AND HANS C. RILLING

VOLUME 112. Drug and Enzyme Targeting (Part A)
Edited by KENNETH J. WIDDER AND RALPH GREEN

VOLUME 113. Glutamate, Glutamine, Glutathione, and Related Compounds
Edited by ALTON MEISTER

VOLUME 114. Diffraction Methods for Biological Macromolecules (Part A)
Edited by HAROLD W. WYCKOFF, C. H. W. HIRS, AND SERGE N. TIMASHEFF

VOLUME 115. Diffraction Methods for Biological Macromolecules (Part B)
Edited by HAROLD W. WYCKOFF, C. H. W. HIRS, AND SERGE N. TIMASHEFF

VOLUME 116. Immunochemical Techniques (Part H: Effectors and Mediators of Lymphoid Cell Functions)
Edited by GIOVANNI DI SABATO, JOHN J. LANGONE, AND HELEN VAN VUNAKIS

VOLUME 117. Enzyme Structure (Part J)
Edited by C. H. W. HIRS AND SERGE N. TIMASHEFF

VOLUME 118. Plant Molecular Biology
Edited by ARTHUR WEISSBACH AND HERBERT WEISSBACH

VOLUME 119. Interferons (Part C)
Edited by SIDNEY PESTKA

VOLUME 120. Cumulative Subject Index Volumes 81–94, 96–101

VOLUME 121. Immunochemical Techniques (Part I: Hybridoma Technology and Monoclonal Antibodies)
Edited by JOHN J. LANGONE AND HELEN VAN VUNAKIS

VOLUME 122. Vitamins and Coenzymes (Part G)
Edited by FRANK CHYTIL AND DONALD B. MCCORMICK

VOLUME 123. Vitamins and Coenzymes (Part H)
Edited by FRANK CHYTIL AND DONALD B. MCCORMICK

VOLUME 124. Hormone Action (Part J: Neuroendocrine Peptides)
Edited by P. MICHAEL CONN

VOLUME 125. Biomembranes (Part M: Transport in Bacteria, Mitochondria, and Chloroplasts: General Approaches and Transport Systems)
Edited by SIDNEY FLEISCHER AND BECCA FLEISCHER

VOLUME 126. Biomembranes (Part N: Transport in Bacteria, Mitochondria, and Chloroplasts: Protonmotive Force)
Edited by SIDNEY FLEISCHER AND BECCA FLEISCHER

VOLUME 127. Biomembranes (Part O: Protons and Water: Structure and Translocation)
Edited by LESTER PACKER

VOLUME 128. Plasma Lipoproteins (Part A: Preparation, Structure, and Molecular Biology)
Edited by JERE P. SEGREST AND JOHN J. ALBERS

VOLUME 129. Plasma Lipoproteins (Part B: Characterization, Cell Biology, and Metabolism)
Edited by JOHN J. ALBERS AND JERE P. SEGREST

VOLUME 130. Enzyme Structure (Part K)
Edited by C. H. W. HIRS AND SERGE N. TIMASHEFF

VOLUME 131. Enzyme Structure (Part L)
Edited by C. H. W. HIRS AND SERGE N. TIMASHEFF

VOLUME 132. Immunochemical Techniques (Part J: Phagocytosis and Cell-Mediated Cytotoxicity)
Edited by GIOVANNI DI SABATO AND JOHANNES EVERSE

VOLUME 133. Bioluminescence and Chemiluminescence (Part B)
Edited by MARLENE DELUCA AND WILLIAM D. MCELROY

VOLUME 134. Structural and Contractile Proteins (Part C: The Contractile Apparatus and the Cytoskeleton)
Edited by RICHARD B. VALLEE

VOLUME 135. Immobilized Enzymes and Cells (Part B)
Edited by KLAUS MOSBACH

VOLUME 136. Immobilized Enzymes and Cells (Part C)
Edited by KLAUS MOSBACH

VOLUME 137. Immobilized Enzymes and Cells (Part D) (in preparation)
Edited by KLAUS MOSBACH

VOLUME 138. Complex Carbohydrates (Part E)
Edited by VICTOR GINSBURG

VOLUME 139. Cellular Regulators (Part A: Calcium- and Calmodulin-Binding Proteins)
Edited by ANTHONY R. MEANS AND P. MICHAEL CONN

VOLUME 140. Cumulative Subject Index Volumes 102–119, 121–134 (in preparation)

VOLUME 141. Cellular Regulators (Part B: Calcium and Lipids)
Edited by P. MICHAEL CONN AND ANTHONY R. MEANS

VOLUME 142. Metabolism of Aromatic Amino Acids and Amines
Edited by SEYMOUR KAUFMAN

VOLUME 143. Sulfur and Sulfur Amino Acids
Edited by WILLIAM B. JAKOBY AND OWEN GRIFFITH

VOLUME 144. Structural and Contractile Proteins (Part D: Extracellular Matrix)
Edited by LEON W. CUNNINGHAM

VOLUME 145. Structural and Contractile Proteins (Part E: Extracellular Matrix)
Edited by LEON W. CUNNINGHAM

VOLUME 146. Peptide Growth Factors (Part A)
Edited by DAVID BARNES AND DAVID A. SIRBASKU

VOLUME 147. Peptide Growth Factors (Part B)
Edited by DAVID BARNES AND DAVID A. SIRBASKU

VOLUME 148. Plant Cell Membranes
Edited by LESTER PACKER AND ROLAND DOUCE

VOLUME 149. Drug and Enzyme Targeting (Part B)
Edited by RALPH GREEN AND KENNETH J. WIDDER

VOLUME 150. Immunochemical Techniques (Part K: *In Vitro* Models of B and T Cell Functions and Lymphoid Cell Receptors)
Edited by GIOVANNI DI SABATO

VOLUME 151. Molecular Genetics of Mammalian Cells
Edited by MICHAEL M. GOTTESMAN

VOLUME 152. Guide to Molecular Cloning Techniques
Edited by SHELBY L. BERGER AND ALAN R. KIMMEL

VOLUME 153. Recombinant DNA (Part D)
Edited by RAY WU AND LAWRENCE GROSSMAN

VOLUME 154. Recombinant DNA (Part E)
Edited by RAY WU AND LAWRENCE GROSSMAN

VOLUME 155. Recombinant DNA (Part F) (in preparation)
Edited by RAY WU

VOLUME 156. Biomembranes (Part P: ATP-Driven Pumps and Related Transport: The Na,K-Pump) (in preparation)
Edited by SIDNEY FLEISCHER AND BECCA FLEISCHER

VOLUME 157. Biomembranes (Part Q: ATP-Driven Pumps and Related Transport: Calcium, Proton, and Potassium Pumps) (in preparation)
Edited by SIDNEY FLEISCHER AND BECCA FLEISCHER

VOLUME 158. Metalloproteins (Part A) (in preparation)
Edited by JAMES F. RIORDAN AND BERT L. VALLEE

VOLUME 159. Initiation and Termination of Cyclic Nucleotide Action (in preparation)
Edited by JACKIE D. CORBIN AND ROGER A. JOHNSON

Section I

Methods for Cloning cDNA

[1] High-Efficiency Cloning of Full-Length cDNA; Construction and Screening of cDNA Expression Libraries for Mammalian Cells

By H. OKAYAMA, M. KAWAICHI, M. BROWNSTEIN, F. LEE, T. YOKOTA, and K. ARAI

cDNA cloning constitutes one of the essential steps to isolate and characterize complex eukaryotic genes, and to express them in a wide variety of host cells. Without cloned cDNA, it is extremely difficult to define the introns and exons, the coding and noncoding sequences, and the transcriptional promoter and terminator of genes. Cloning of cDNA, however, is generally far more difficult than any other recombinant DNA work, requiring multiple sequential enzymatic reactions. It involves *in vitro* synthesis of a DNA copy of mRNA, its subsequent conversion to a duplex cDNA, and insertion into an appropriate prokaryotic vector. Due to the intrinsic difficulty of these reactions as well as the inefficiency of the cloning protocols devised, the yield of clones is low and many of clones are truncated.[1]

The cloning method developed by Okayama and Berg[2] circumvents many of these problems, and permits a high yield of full-length cDNA clones regardless of their size.[3–6] The method utilizes two specially engineered plasmid DNA fragments, "vector primer" and "linker DNA." In addition, several specific enzymes are used for efficient synthesis of a duplex DNA copy of mRNA and for efficient insertion of this DNA into a plasmid. Excellent yields of full-length clones and the unidirectional insertion of cDNA into the vector are the result. These features not only facilitate cloning and analysis but are also ideally suited for the expression of functional cDNA.

To take full advantage of the features of this method, Okayama and

[1] A. Efstratiadis and L. Villa-Komaroff, *in* "Genetic Engineering" (J. K. Setlow and A. Hollaender, eds.), Vol. 1, p. 1. Plenum, New York, 1979.

[2] H. Okayama and P. Berg, *Mol. Cell. Biol.* **1,** 161 (1982).

[3] D. H. Maclennan, C. J. Brandl, B. Korczak, and N. M. Green, *Nature (London)* **316,** 696 (1985).

[4] L. C. Kun, A. McClelland, and F. H. Ruddle, *Cell* **37,** 95 (1984).

[5] K. Shigesada, G. R. Stark, J. A. Maley, L. A. Niswander, and J. N. Davidson, *Mol. Cell. Biol.* **5,** 1735 (1985).

[6] S. M. Hollenberg, C. Weinberger, E. S. Ong, G. Cerelli, A. Oro, R. Lebo, E. B. Thompson, M. G. Rosenfeld, and R. M. Evans, *Nature (London)* **318,** 635 (1985).

Berg[7] have modified the original vector. The modified vector, pcD, has had SV40 transcriptional signals introduced into the vector primer and linker DNAs to promote efficient expression of inserted cDNAs in mammalian cells. Construction of cDNA libraries in the pcD expression vector thus permits screening or selection of particular clones on the basis of their expressed function in mammalian cells, in addition to regular screening with hybridization probes.

Expression cloning has proven extremely powerful if appropriate functional assays or genetic complementation selection systems are available.[8–14] In fact, Yokota *et al.*[11,12] and Lee *et al.*[13,14] have recently isolated full-length cDNA clones encoding mouse and human lymphokines without any prior knowledge of their chemical properties, relying entirely on transient expression assays using cultured mammalian cells. Similar modifications have been made to promote the expression of cDNA in yeast, thereby permitting yeast mutant cells to be used as possible complementation hosts.[15,16]

In this chapter, we describe detailed procedures for the construction of full-length cDNA expression libraries and the screening of the libraries for particular clones based on their transient expression in mammalian cells. Methods for library transduction and screening based on stable expression are described in Vol. 151 of *Methods in Enzymology*. If expression cloning is not envisioned, the original vector[2] or one described by others[17] can be used with slight modifications of the procedure described below.

[7] H. Okayama and P. Berg, *Mol. Cell. Biol.* **2,** 280 (1983).

[8] D. H. Joly, H. Okayama, P. Berg, A. C. Esty, D. Filpula, P. Bohlen, G. G. Johnson, J. E. Shivery, T. Hunkapiller, and T. Friedmann, *Proc. Natl. Sci. Acad. U.S.A.* **80,** 477 (1983).

[9] D. Ayusawa, K. Takeishi, S. Kaneda, K. Shimizu, H. Koyama, and T. Seno, *J. Biol. Chem.* **259,** 1436 (1984).

[10] H. Okayama and P. Berg, *Mol. Cell. Biol.* **5,** 1136 (1985).

[11] T. Yokota, F. Lee, D. Rennick, C. Hall, N. Arai, T. Mosmann, G. Nabel, H. Cantor, and K. Arai, *Proc. Natl. Acad. Sci. U.S.A.* **81,** 1070 (1985).

[12] T. Yokota, N. Arai, F. Lee, D. Rennick, T. Mosmann, and K. Arai, *Proc. Natl. Acad. Sci. U.S.A.* **82,** 68 (1985).

[13] F. Lee, T. Yokota, T. Otsuka, L. Gemmell, N. Larson, L. Luh, K. Arai, and D. Rennick, *Proc. Natl. Acad. Sci. U.S.A.* **82,** 4360 (1985).

[14] F. Lee, T. Yokota, T. Otsuka, P. Meyerson, D. Villaret, R. Coffman, T. Mosmann, D. Rennick, N. Roehm, C. Smith, C. Zlotnick, and K. Arai, *Proc. Natl. Acad. Sci. U.S.A.* **83,** 2061 (1986).

[15] G. L. McKnight and B. C. McConaughy, *Proc. Natl. Acad. Sci. U.S.A.* **80,** 4412 (1983).

[16] A. Miyajima, N. Nakayama, I. Miyajima, N. Arai, H. Okayama, and K. Arai, *Nucleic Acids Res.* **12,** 6639 (1984).

[17] D. C. Alexander, T. D. McKnight, and B. G. Williams, *Gene* **31,** 79 (1984).

Methods

Clean, intact mRNA is prepared from cultured cells or tissue by the guanidine thiocyanate method[18] followed by two cycles of oligo(dT)–cellulose column chromatography. The purified mRNA is then reverse transcribed by the avian myeloblastosis enzyme in a reaction primed with the pcD-based vector primer, a plasmid DNA fragment that contains a poly(dT) tail at one end and a *Hin*dIII restriction site near the other end (Figs. 1 and 2).[7] The vector also contains the SV40 poly(A) addition signal downstream of the tail site as well as the pBR322 replication origin and the β-lactamase gene. Reverse transcription results in the synthesis of a cDNA : mRNA hybrid covalently linked to the vector molecule (Fig. 3). This product is tailed with oligo(dC) at its 3′ ends and digested with *Hin*dIII to release an oligo(dC) tail from the vector end and to create a *Hin*dIII cohesive end. The C-tailed cDNA : mRNA hybrid linked to the vector is cyclized by addition of DNA ligase and a pcD-based linker DNA—an oligo(dG)-tailed DNA fragment with a *Hin*dIII cohesive end (this linker contains the SV40 early promoter and the late splice junctions) (Figs. 1 and 2). Finally, the RNA strand is converted to DNA by nick-translation repair catalyzed by *Escherichia coli* DNA polymerase I, RNase H, and DNA ligase. The end product, a closed circular cDNA recombinant, is transfected into a highly competent *E. coli* host to establish a cDNA clone library.

In the steps that have just been enumerated, double-stranded, full-length DNA copies of the original mRNAs are efficiently synthesized and inserted into the vector to form a functional composite gene with the protein coding sequence derived from the cDNA and the transcriptional and RNA processing signals from the SV40 genome. To screen for or select a particular clone on the basis of the function it encodes, the library is acutely transfected or stably transduced into cultured cells. Procedures for stable transduction are described in Chap. [32] of Vol. 151 of *Methods in Enzymology*.

Preparation of mRNA

Successful construction of full-length cDNA libraries depends heavily on the quality of the mRNA preparation. The use of intact, uncontaminated mRNA is essential for generating full-length clones. Messenger RNA prepared by the guanidine thiocyanate method[18] satisfies the above

[18] J. M. Chilgwin, A. E. Przybyla, R. J. MacDonald, and W. J. Rutter, *Biochemistry* **18,** 5294 (1978).

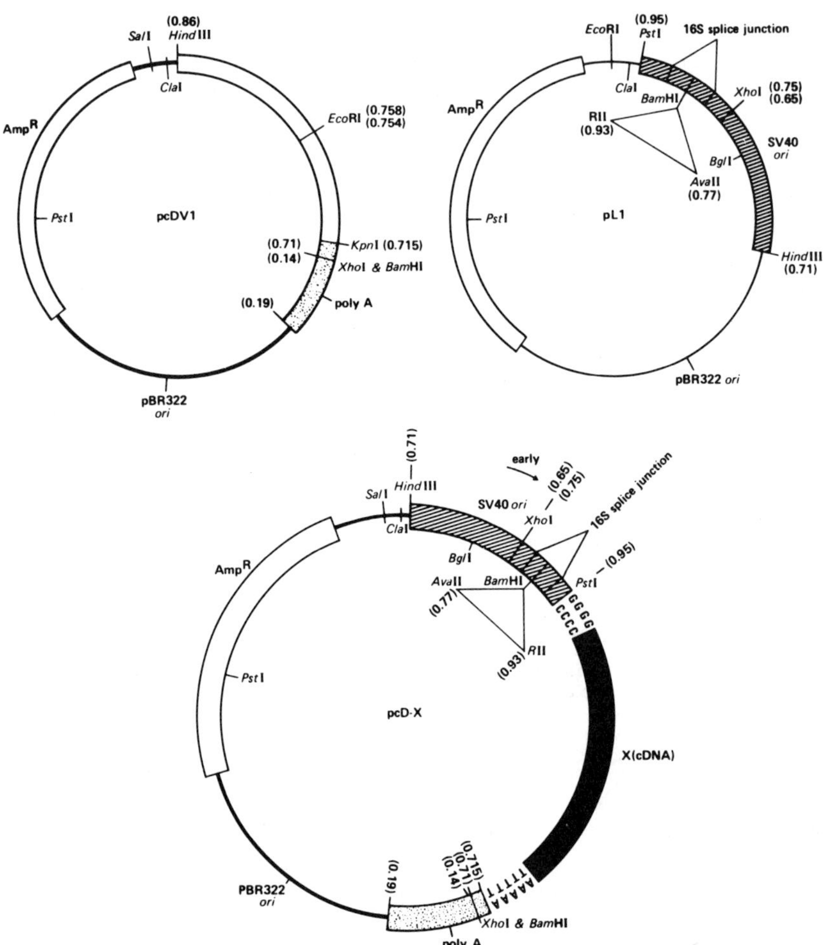

FIG. 1. Structure and component parts of the pcD vector and its precursor plasmids, pcDV1 and pL1. The principal elements of the pcD vector are a segment containing the SV40 replication origin and the early promoter joined to a segment containing the 19 S and 16 S SV40 late splice junctions (hatched area); the various cDNA inserts flanked by dG/dC and dA/dT stretches that connect them to the vector (solid black area); a segment containing the SV40 late polyadenylation signal [poly(A)] (stippled area); and the segment containing the pBR322 β-lactamase gene and the origin of replication (thin and open area). pcDV1 and pL1 provide the pcD-based vector primer and linker DNA, respectively. For the preparation of the vector primer and linker DNA, see Methods sections and Fig. 2.

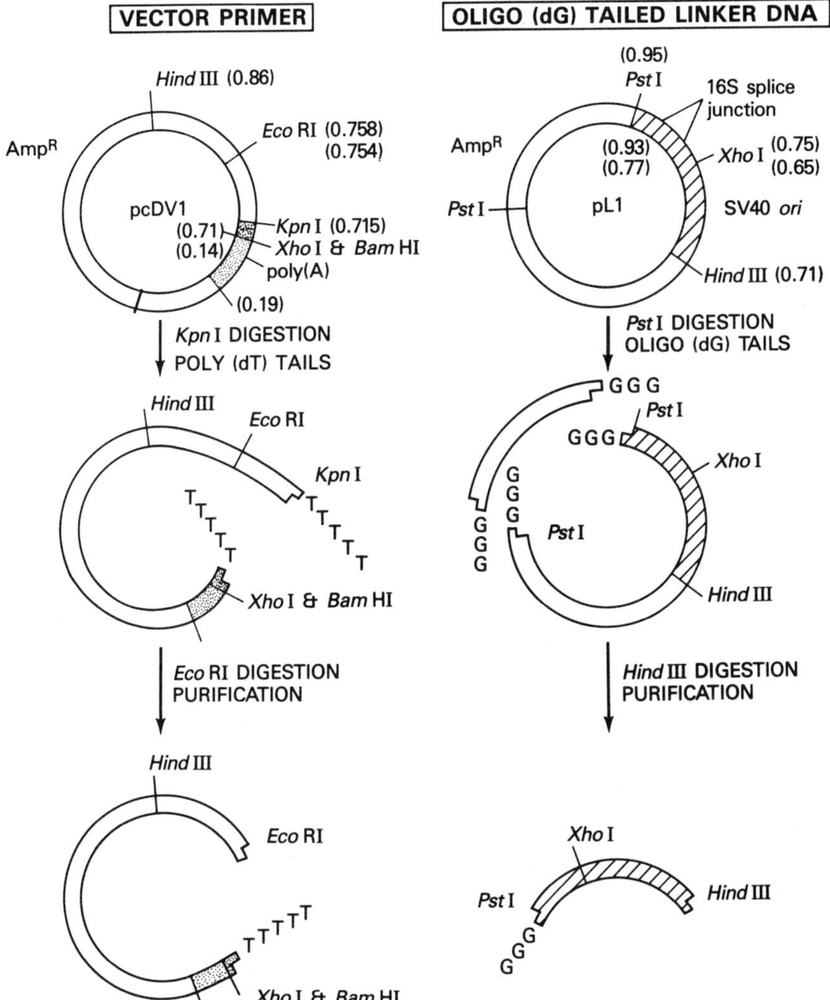

FIG. 2. Preparation of vector primer and linker DNAs.

criteria and is reverse transcribed efficiently. It has successfully been used for cloning a number of cDNAs.[3–14] The method described below is a slight modification of the original method that ensures complete inactivation of RNases through all the steps of RNA isolation.

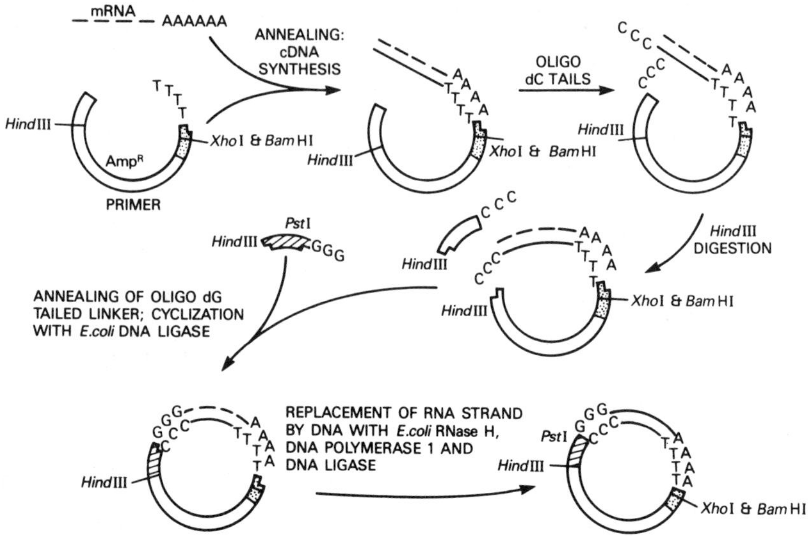

FIG. 3. Enzymatic steps in the construction of pcD–cDNA recombinants. The designations of the DNA segments are as described in Fig. 1. For experimental details and comments, see Methods.

Reagents

All solutions are prepared using autoclaved glassware or sterile disposable plasticware, autoclaved double-distilled water and chemicals of the finest grade. Solutions are sterilized by filtration through Nalgen 0.45 μm Millipore filters and subsequently by autoclaving (except as noted). In general, treatment of solutions with diethyl pyrocarbonate is not recommended since residual diethyl pyrocarbonate may modify the RNA, resulting in a marked reduction in its template activity.

5.5 *M* GTC solution:
 5.5 *M* guanidine thiocyanate (Fluka or Eastman-Kodak), 25 m*M* sodium citrate, 0.5% sodium lauryl sarcosine. After the pH is adjusted to 7.0 with NaOH, the solution is filter-sterilized and stored at 4°. Prior to use, 2-mercaptoethanol is added to a final concentration of 0.2 *M*.

4 *M* GTC solution: 5.5 *M* solution diluted to 4 *M* with sterile distilled water.

CsTFA solution:
 cesium trifluoroacetate (density 1.51 ± 0.01 g/ml), 0.1 *M* ethylenediaminetetraacetic acid (EDTA) (pH 7.0). Prepared with cesium trifluoroacetate (2 g/ml) (CsTFA, Pharmacia) and 0.25 *M* EDTA (pH 7.0).

TE: 10 m*M* Tris–HCl (pH 7.5), 1 m*M* EDTA.
1 *M* NaCl
2 *M* NaCl
1 *M* acetic acid, filter sterilized
Oligo(dT)–cellulose (Collaborative Research), Type 3.
Resins provided by some other supplier may not be useful because mRNA purified on these often has significant template activity for reverse transcription without addition of primer. This is likely to be due to contamination of the mRNA with oligo(dT) leached from the resin.
TE/NaCl: a 1 : 1 mixture of TE and 1 *M* NaCl.
Ethidium bromide stock solution: 10 mg/ml water, stored at 4°.
Yeast tRNA stock solution: 1 mg/ml, dissolved in sterile water.

Procedure

Step 1. Extraction of Total RNA. Approximately 2–4 × 10^8 cells or 1–3 g of tissue are treated with 100 ml of the 5.5 *M* GTC solution. Cultured cells immediately lyse but tissue generally requires homogenization to facilitate lysis. The viscous lysate is transferred to a sterile beaker, and the DNA is sheared by passing the lysate through a 16- to 18-gauge needle attached to a syringe several times until the viscosity decreases. After removal of cell debris by a brief low speed centrifugation, the lysate is gently overlaid onto a 17-ml cushion of CsTFA solution in autoclaved SW28 centrifuge tubes and centrifuged at 25,000 rpm for 24 hr at 15°.

After centrifugation, the upper GTC layer and the DNA band at the interface are removed by aspiration. The tubes are quickly inverted, and their contents are poured into a beaker. Still inverted, they are placed on a paper towel to drain for 5 min, and then the bottom 2 cm of the tube is cut off with a razor blade or scalpel; the remainder is discarded. After the bottom of the tube is removed, the cup that is formed is turned over again and placed on a bed of ice. The RNA pellet is dissolved in a total of 0.4 ml of the 4 *M* GTC solution. After insoluble materials are removed by brief centrifugation in an Eppendorf microfuge, the RNA is precipitated as follows: 10 μl of 1 *M* acetic acid and 300 μl of ethanol are added to the solution and chilled at −20° for at least 3 hr. The RNA is pelleted by centrifugation at 4° for 10 min in a microfuge. The RNA pellet is dissolved in 1 ml of TE, and the insoluble material is removed by centrifugation. One hundred microliters of 2 *M* NaCl and 3 ml of ethanol are added to the solution. The RNA is precipitated by centrifugation after chilling at −20° for several hours. The RNA may be stored as a wet precipitate.

Step 2. Oligo(dT)–Cellulose Column Chromatography. Poly(A)$^+$ RNA is separated from the total RNA by oligo(dT)–cellulose column chromatography.[19] Oligo(dT)–cellulose is suspended in TE, and the fines are removed by decantation. A column 1.5 cm in height is made in an autoclaved Econocolumn (0.6 cm diameter) (Bio-Rad), washed with several column volumes of TE, and equilibrated with TE/NaCl. The RNA pellet is dissolved in 1 ml of TE. It is then incubated at 65° for 5 min, quickly chilled on ice, and 1 ml of 1 *M* NaCl is added. The RNA sample is applied to the column and the flowthrough is applied again. The column is washed with 5 bed volumes of TE/NaCl and eluted with 3 bed volumes of TE. One-half milliliter fractions of TE are collected. The RNA eluted is assayed by the spot test.

Small samples (0.5–3 μl) from each fraction are mixed with 20 μl of 1 μg/ml ethidium bromide (freshly prepared from the stock solution). The mixture is spotted onto a sheet of plastic wrap placed on a UV light box; ethidium bromide bound to RNA in the positive fractions emits a red–orange fluorescence. Fractions containing poly(A)$^+$ RNA are combined, incubated at 65° for 5 min, and chilled on ice. After adding an equal volume of 1 *M* NaCl, the sample is reapplied to the original column that has, in the meantime, been washed with TE and reequilibrated with TE/NaCl. The column is washed and eluted as above. Poly(A)$^+$ RNA eluted from the column is precipitated by adding 0.2 volume of 2 *M* NaCl and 3 volumes of ethanol, chilling on dry ice for 30 min, and centrifuging in a microfuge at 4°. The RNA pellet is dissolved in 20 μl of TE. The RNA concentration is determined by the spot test, as described above, using *E. coli* tRNA solutions of known concentrations as standards (the assay can be used to measure between 100 and 400 ng of RNA). Ethanol is added to a final concentration of 50%, and the solution is stored at −20°. Generally 20–30 μg of poly(A)$^+$ RNA is obtained from 3×10^8 cells or 1 g of tissue. This RNA is more than enough for making a library.

Prior to use, the RNA should be tested. Reverse-transcribe 2 μg of RNA using 0.5 μg of oligo(dT) primer in place of the vector primer under the conditions described in the cDNA Cloning section "Pilot-scale reaction." Calculate the percent conversion to cDNA from the amounts of cDNA synthesized and RNA used [(μg of cDNA synthesized/μg of RNA used) × 100]. Generally 15–20% of the poly(A)$^+$ RNA prepared by this method can be converted to cDNA. If the number is considerably smaller than this, the RNA should not be used.

Comments. Fresh tissue or healthy cells should be used for the preparation of mRNA. Messenger RNA from mycoplasma-infected cells is often partially degraded. The use of such RNA leads to a failure to generate

[19] H. Aviv and P. Leder, *Proc. Natl. Acad. Sci. U.S.A.* **69,** 1408 (1972).

full-length clones. A great deal of caution should be taken to prevent the contamination of glassware and solutions by RNase. Solutions should be freshly prepared each time from stock buffers and solutions that have been guarded against contamination. As long as samples are carefully handled, wearing gloves is not necessary. The use of sodium dodecyl sulfate (SDS) in oligo(dT)–cellulose column chromatography is not recommended since residual SDS may inactivate reverse transcriptase.

Preparation of Vector Primer and Linker DNAs

The pcD-based vector primer and linker DNAs are prepared from pcDV1 and pL1, respectively, using the enzymatic treatments and purification procedures illustrated in Fig. 2. Briefly, pcDV1 is linearized by *Kpn*I digestion, and poly(dT) tails are added to the ends of the linear DNA with terminal transferase. One tail is removed by *Eco*RI digestion. The resulting large fragment is purified by agarose gel electrophoresis and subsequent oligo(dA)–cellulose column chromatography, and used as the vector primer. Untailed or uncut DNA, which produces significant background colonies, is effectively removed by the column purification step.

The best cloning results have been obtained with vector primer having 40 to 60 dT residues per tail. The reaction conditions that allow the addition of poly(dT) tails of this size vary with the lot of transferase and the preparation of DNA. Therefore, optimization of tailing conditions should be established for each preparation with a pilot-scale reaction.

Excessive digestion of DNA with restriction endonucleases should be avoided to minimize nicking of DNA by contaminating nucleases. The ends at nicks serve as effective primers for terminal transferase as well as for reverse transcriptase.[20] The resulting homopolymer tails and branching structures at nicks will reduce cloning efficiency.

Oligo(dG)-tailed linker DNA is prepared by *Pst*I digestion of pL1 DNA. Oligo(dG) tails of 10–15 dG residues are added to the ends. After *Hin*dIII digestion, the tailed fragment that contains the SV40 sequences is purified by agarose gel electrophoresis.

Reagents

*Kpn*I, *Eco*RI, *Pst*I, and *Hin*dIII (New England Biolabs)
Terminal deoxynucleotidyltransferase from calf thymus (Pharmacia)
Oligo(dA)–cellulose (Collaborative Research)
Loading buffer: 1 *M* NaCl, 10 m*M* Tris–HCl (pH 7.5), 1 m*M* EDTA.

[20] T. Nelson and D. Brutlag, this series, Vol. 68, p. 43.

10× *Kpn*I buffer: 60 mM NaCl, 60 mM Tris–HCl (pH 7.5), 60 mM $MgCl_2$, 60 mM 2-mercaptoethanol, 1 mg bovine serum albumin (BSA) (Miles, Pentex crystallized)/ml.

5× *Eco*RI buffer: 0.25 M NaCl, 0.5 M Tris–HCl (pH 7.5), 25 mM $MgCl_2$, 0.5 mg BSA/ml.

10× *Pst*I buffer: 1 M NaCl, 0.1 M Tris–HCl (pH 7.5), 0.1 M $MgCl_2$, 1 mg BSA/ml.

10× *Hin*dIII buffer: 0.5 M NaCl, 0.5 M Tris–HCl (pH 8.0), 0.1 M $MgCl_2$, 1 mg BSA/ml.

10× terminal transferase buffer:

1.4 M sodium cacodylate, 0.3 M Tris base, 10 mM $CoCl_2$. Adjusted to pH 7.6 at room temperature (the pH of the 10-fold diluted solution will be 6.8 at 37°); the buffer is filter-sterilized and stored at 4°.

1 mM dithiothreitol (DTT)

[*methyl*-1′,2′-^{3}H]dTTP and [8-^{3}H]dGTP (New England Nuclear)

NaI solution:

90.8 g of NaI and 1.5 g of Na_2SO_3 are dissolved in a total 100 ml of distilled water. Insoluble impurities are removed by filtering through Whatman No. 1 filter paper, and 0.5 g of Na_2SO_3 is added to the solution to protect the NaI from oxidation.

Glass bead suspension:

200 ml of silica-325 mesh (a powdered flint glass obtainable from ceramic stores) is suspended in 500 ml of water. The suspension is stirred with a magnetic stirrer at room temperature for 1 hr. Coarse particles are allowed to settle for 1 hr, and the supernatant is collected. Fines are spun down in a Sorval centrifuge and resuspended in 200 ml of water. An equal volume of nitric acid is added, and the suspension is heated almost to the boiling point in a chemical hood. The glass beads are sedimented by centrifugation and washed with water until the pH is 5–6. The beads are suspended in a volume of water equal to their own volume and stored at 4°.

Ethanol wash solution: 50% ethanol, 0.1 M NaCl, 10 mM Tris–HCl (pH 7.5), 1 mM EDTA.

TE: 10 mM Tris–HCl (pH 7.5), 1 mM EDTA.

SDS/EDTA stop solution: 5% sodium dodecyl sulfate, 125 mM sodium EDTA (pH 8.0).

BPB/XC: 1% bromophenol blue, 1% xylene cyanol, 50% glycerol.

25× TAE: 121 g of Tris base, 18.6 g of disodium EDTA, 28.7 ml of acetic acid in total 1 liter of water.

Phenol/chloroform: a 1 : 1 (by volume) mixture of water-saturated phenol and chloroform.

Procedure

For a diagram of this procedure, see Fig. 2. To determine optimal conditions, pilot reactions are carried out at each step. In all the procedures described below, submicroliter amounts of enzyme are pipeted with a 1-μl Hamilton syringe connected to an autoclaved Teflon tubing.

Preparation of Vector Primer

Step 1. KpnI Digestion of pcDV1. pcDV1 DNA is prepared by the standard Triton X-100–lysozyme extraction method followed by two cycles of CsCl equilibrium gradient centrifugation.[21]

A small-scale *Kpn*I digestion is carried out at 37° in a 20-μl reaction mixture containing 2 μl of 10× *Kpn*I buffer, 20 μg pcDV1 DNA, and 20 units of *Kpn*I enzyme. One-half microliter aliquots are removed every 20 min up to 2 hr and mixed with 9 μl of TE, 1 μl of SDS/EDTA stop solution, and 1 μl of BPB/XC. The aliquots are then analyzed by agarose gel (1%) electrophoresis in 1× TAE, and the minimum time required for at least 95% digestion of the plasmid is determined.

Large-scale digestion is performed with 500–600 μg of DNA in a proportionally scaled-up reaction mixture for the determined time. The reaction is terminated by adding 0.1 volume of SDS/EDTA stop solution. The mixture is extracted twice with an equal volume of phenol/chloroform. One-tenth volume of 2 *M* NaCl and 2 volumes of ethanol are added to the aqueous solution, and the solution is chilled on dry ice for 20 min and centrifuged at 4° for 15 min in an Eppendorf microfuge. The pellet is dissolved in TE, and the ethanol precipitation is repeated once more. The pellet is dissolved in water (not TE) to make a solution of approximately 4 μg/μl. The DNA concentration is determined by measuring its absorbance at 260 nm. A small sample is analyzed by gel electrophoresis to check the completion of digestion.

Step 2. Poly(dT) Tailing. A pilot tailing reaction is carried out in a 10 μl mixture containing 1 μl of 10× terminal transferase buffer, 1 μl of 1 m*M* DTT, 20 μg of *Kpn*I-digested DNA (19 pmol of DNA ends), 1 μl of 2.5 m*M* [^{3}H]dTTP (250–500 dpm/pmol), and 20 units of terminal transferase. The mixture is warmed to 37° prior to addition of the enzyme. After 5, 10, 15, 20, and 30 min of incubation, 1-μl aliquots are taken with Drummond microcapillary pipettes and mixed with 50 μl of ice-cold TE containing 10 μg of plasmid DNA carrier. The DNA is precipitated by addition of 50 μl of 20% trichloroacetic acid (TCA) to each tube and

[21] L. Katz, D. T. Kingsbury, and D. R. Helinski, *J. Bacteriol.* **114,** 577 (1973).

collected on Whatman GF/C glass filter disks (2.4 cm diameter). The filters are washed 4 times with 10 ml of 10% TCA and rinsed with 10 ml of ethanol. After being dried in a oven, the radioactivity on the filter is measured in a toluene-based scintillation fluid. The average number of dT residues per DNA end is calculated based on the total counts incorporated, the counting efficiency of ^{3}H on a glass filter, the specific activity of the [^{3}H]dTTP used, and the number of DNA ends (19 pmol). Figure 4 shows the typical time course of the tailing reaction. The rate of dT incorporation decreases after 10 min of incubation and levels off at around 15 min. The incubation time that results in the formation of 40–50 dT long tails is determined.

Large-scale tailing is then carried out in a 200-μl reaction mixture containing 400 μg of DNA under the conditions determined by doing the pilot reaction. The reaction is terminated by adding 20 μl of the SDS/EDTA stop solution. After two extractions with phenol/chloroform, the DNA is precipitated by adding 20 μl of 2 *M* NaCl and 400 μl of ethanol as described above. The ethanol precipitation is repeated once more, and the pellet is dissolved in 100 μl of TE.

Step 3. EcoRI Digestion. A miniscale *Eco*RI digestion is performed in a 5-μl reaction volume containing 2 μl of 5× EcoRI buffer, 2.5 μl (8–9 μg) of poly(dT)-tailed DNA, and 10 units of *Eco*RI. The incubation time required for at least 95% digestion of DNA is determined by analyzing the products on agarose gels as described above after incubation times of 30–90 min. The remainder of the DNA (100 μl) is digested in a 200 μl reaction mixture under the conditions determined in the pilot study.

After the reaction is stopped with 20 μl of the SDS/EDTA solution and

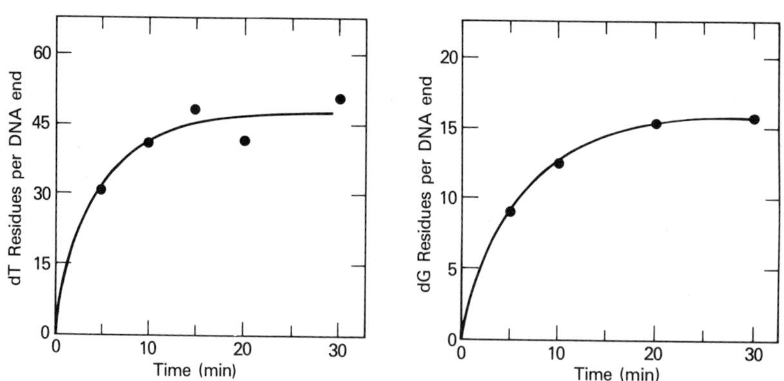

FIG. 4. Time course of poly(dT) (left) and oligo(dG) (right) tailings. Experimental details are described in Methods.

20 μl of BPB/XC, the product is purified by preparative agarose gel (1%, 20 × 25 × 0.6 cm) electrophoresis in 1× TAE buffer. The gel is stained with ethidium bromide (1 μg/ml), and the area of the gel containing the vector primer (the larger DNA fragment) is cut out with a razor blade and sliced into pieces. The gel is weighed and transferred to a plastic bench-top centrifuge tube, to which 2 ml of NaI solution is added for each gram of gel. The tube is placed in a 37° water bath and incubated with occasional vigorous shaking until the gel is completely solubilized. Approximately 0.3 ml of glass bead suspension (0.8 ml/mg DNA) is added, and the solution is cooled on ice and incubated at 4° for 1–2 hrs with gentle rocking. The glass beads, which bind DNA, are recovered by brief centrifugation and washed with 1.5 ml of ice-cold NaI solution once and then twice with 1.5 ml of ice-cold ethanol wash solution. The washed beads are suspended in 1 ml of TE and incubated at 37° for 30 min to dissociate the DNA. The TE is separated from the beads by brief centrifugation in a microfuge; then the beads are extracted once more with 1 ml of TE. Both extracts are pooled, and, after several brief centrifugations to remove residual fine glass particles, the DNA is recovered by ethanol precipitation. This step has 50–80% yield.

Step 4. Oligo(dA)–Cellulose Column Chromatography. Oligo(dA)–cellulose column chromatography is used to remove untailed DNA. All solutions should be RNase free. Oligo(dA)–cellulose is suspended in loading buffer and packed in a column (0.6 cm diameter × 2.5 cm height). The column is washed with several bed volumes of sterile distilled water and equilibrated with the loading buffer at 0–4°. The DNA pellet is dissolved in 1 ml of the loading buffer, cooled on ice and applied to the column. After the column is washed with several bed volumes of the buffer at 0–4°, the bound DNA is eluted with sterile distilled water at room temperature. One milliliter fractions are collected. Small samples are removed from each fraction and the radioactivity is counted in an aqueous scintillation fluid. Radioactive fractions are combined, and the DNA is recovered by ethanol precipitation. The pellet is dissolved in TE to give a solution of about 3 μg/μl. Based on the radioactivity and the amount of DNA recovered, the average length of the poly(dT) tails can be redetermined. The overall yield of vector primer is 30–40%.

The vector primer solution is adjusted to a concentration of 1.2–1.5 μg/μl by adding an equal volume of ethanol, and stored at −20°.

Preparation of Oligo(dG)-Tailed Linker DNA

Step 1. PstI Digestion of pL1. A pilot reaction is performed in a 10-μl solution containing 1 μl of 10× *Pst*I buffer, 5 μg of pL1 DNA, and 5–10 units of *Pst*I. The minimal time for 95% digestion is determined as described above. Two hundred micrograms of pL1 DNA is digested in a 400-μl reaction mixture under the conditions determined. After termination of the reaction with 30 μl of SDS/EDTA solution, the mixture is extracted twice with phenol/chloroform, and the digested DNA is recovered by ethanol precipitation. After being washed with 70% ethanol, the pellet is dissolved in 60 μl of sterile distilled water (not TE). The DNA concentration must accurately be determined by measuring the absorbance at 260 nm.

Step 2. Oligo(dG) Tailing. The optimal conditions for dG tailing are determined in a pilot reaction as above. The reaction mixture (10 μl) contains 1 μl each of 10× terminal transferase buffer, 1 m*M* DTT, and 1 m*M* [^{3}H]dGTP (500–1000 dpm/pmol), 12 μg of the *Pst*I-cut pL1, and 12 units of terminal transferase. The mixture is preincubated for 5–10 min at 37° before adding enzyme. After incubation times ranging from 10 to 60 min, 1-μl samples are taken with Drummond microcapillary pipets, and TCA-insoluble counts are determined as described earlier. From the counts and the amount of DNA used, the size of the dG tails is determined. A typical time course is shown in Fig. 4. Under the conditions used, the dG tailing reaction is self-limiting and stops after 10–15 dG residues are added, perhaps due to the formation of double-strand structures within and/or between tails. As a result of the formation of concatenated DNA by base-pairing between tails, the solution becomes more viscous as the reaction proceeds.

The preparative-scale tailing reaction is carried out with 120 μg of DNA in a 100-μl reaction volume. The reaction is terminated with 10 μl of SDS/EDTA solution. After several extractions with phenol/chloroform, the tailed DNA is recovered by ethanol precipitation. The pellet is washed with 70% ethanol and dissolved in 62 μl of TE (approximately 2 μg DNA/μl).

Step 3. HindIII Digestion. The tailed DNA is digested with *Hin*dIII. After determining the conditions to be used by means of a pilot reaction (10 μl) containing 1 μl of 10× *Hin*dIII buffer, 2 μl (4 μg) of the tailed DNA, and 20 units of *Hin*dIII, the rest of the DNA is digested in a 300-μl reaction mixture. The reaction is stopped, and the product is separated by preparative agarose gel (1.8%) (20 × 25 × 0.6 cm) electrophoresis as described above. The tailed linker fragment (450 bp) is recovered by the glass bead method as described above, except that a somewhat larger amount of glass bead suspension and a longer incubation (12–14 hr) at 4° are required to ensure complete adsorption of the DNA to the beads. The DNA recovered is precipitated with ethanol and dissolved in TE to give a concentration of 0.5–1 μg/μl. The DNA concentration and the precise size of the tails are determined from the absorbance at 260 nm and the radioactivity of the solution. After adjusting the concentration to 0.6–1.0 pmol DNA/μl of TE, an equal volume of ethanol is added, and the linker solution is stored at −20°.

Comments. Contamination of the vector primer and linker DNA solutions by agarose or impurities in the agarose will strongly inhibit the reactions that these DNAs are involved in. DNA recovered by the glass bead method[22] is very clean, and we have not encountered any difficulty in using it. DNA recovered by other methods may not be clean enough. Gel buffers containing borate should not be used; borate inhibits solubilization of agarose gel by NaI. As a result of the formation of high molecular weight DNA by base-pairing between dG-tails, the yield of linker DNA from agarose gels is generally poor.

cDNA Cloning

Steps in the construction of cDNA–plasmid recombinants are illustrated in Fig. 3.

Step 1. cDNA Synthesis

mRNA is reverse-transcribed with the avian myeloblastosis enzyme in a reaction primed with poly(dT)-tailed vector molecules, resulting in the synthesis of cDNA : mRNA hybrids covalently linked to the vector molecules. The use of high concentrations of deoxynucleoside triphosphates (2 m*M* each)[23] is designed to inhibit the enzyme-associated RNase H activity. The latter enzyme activity cleaves off the 5′ end of RNA on the hybrid in an exonucleolytic fashion, thereby inducing the formation of hairpin structures at the single-stranded 3′ ends of the cDNA. Such structures induce cloning artifacts.

Reagents

Poly(A)$^+$ RNA, 10 μg (20% template activity, see section Preparation of mRNA).

Vector primer DNA, 3.2 μg

Avian myelobastosis reverse transcriptase (Seikagaku or Bio-Rad) RNase free.

[22] B. Vogelstein and D. Gillespie, *Proc. Natl. Acad. Sci. U.S.A.* **76,** 615 (1979).

[23] D. L. Kacian and J. C. Mayers, *Proc. Natl. Acad. Sci. U.S.A.* **73,** 2191 (1976).

After being received, the enzyme is aliquoted and stored in liquid nitrogen. Repeated freezing and thawing inactivate the enzyme.

10× reaction buffer:

500 m*M* Tris–HCl (pH 8.5 at 20°), 80 m*M* $MgCl_2$, 300 m*M* KCl, 3 m*M* dithiothreitol (DTT).

Prepared from autoclaved, filtered 1 *M* stock solutions of Tris–HCl and the salts, and filter-sterilized 1 *M* dithiothreitol.

20 m*M* dNTP: a mixture of 20 m*M* dGTP, dATP, and dTTP.

20 m*M* dCTP

Prepared in sterile distilled water and neutralized with 0.1 *N* NaOH (final pH 6–6.5). The pH is monitored by applying small aliquots to narrow-range pH paper with a capillary pipet. The solution is filter-sterilized with a small Millipore filter, and the concentration is determined by measuring absorbance at 260 nm.

20 m*M* [α-^{32}P]dCTP (400–700 cpm/pmol):

Approximately 100 μCi of [α-^{32}P]dCTP (200–5000 Ci/mmol) are dried down under reduced pressure and dissolved in 15 μl of 20 m*M* dCTP.

10% trichloroacetic acid (TCA)

0.25 *M* EDTA (pH 8.0), filter-sterilized and autoclaved after preparation.

10% SDS

Phenol/chloroform

4 *M* ammonium acetate

Procedure

Pilot-scale reaction. Prior to large-scale cDNA synthesis, a small-scale reaction should always be performed to test the mRNA and other reagents. The reaction is carried out as described below but in a 15-μl total reaction volume. Under these conditions, the number of effective mRNA template molecules is about twice the number of primer molecules used (generally only 15–20% of the RNA molecules in the final preparation are effective as templates). After a 30-min incubation, a 1-μl aliquot is removed with a Drummond glass capillary pipet and added to 20 μl of 10% TCA. TCA-insoluble species are collected on a 0.45 μm Millipore filter, type HA, and the radioactivity is measured. From the radioactive counts, the specific activity of the ^{32}P-dCTP, and the amount of vector primer (1 μg = 0.5 pmol as primer) used, the average size of the cDNA can be estimated assuming that 100% of the primer is utilized for the synthesis. Generally the average size of the cDNA falls between 1.0 and 1.2 kb (120–150 pmol dCTP incorporated/μg vector primer). The production of long cDNA (4–6 kb) can be confirmed by phenol extraction and ethanol precipitation of an aliquot of the reaction mixture followed by electrophoresis of the product on denaturing agarose gels (0.7–1%) and autoradiography.

If the estimated size of the cDNA is considerably shorter than expected or if the production of long cDNA is not detected, check the mRNA, the reverse transcriptase, and the vector primer by using globin mRNA (commercially available), reverse transcriptase from another lot or source, and/or oligo(dT) primer as controls.

Large-scale cDNA synthesis. Approximately 6–7 μg of mRNA (20% template activity, see Preparation of mRNA) is ethanol-precipitated in a small Eppendorf microfuge tube (never let the RNA dry completely), and dissolved in 16 μl of 5 m*M* Tris–HCl (pH 7.5). The solution is heated at 65° for 5 min and transferred to 37°. The reaction is immediately initiated by addition of 3 μl each of 20 m*M* dNTP and 20 m*M* ^{32}P-dCTP, 2 μl of vector primer (1.2 μg/μl), 3 μl of 10× reaction mixture, and 3 μl of reverse transcriptase (15 U/μl). After a 30-min incubation, the mixture is terminated with 3 μl of 0.25 *M* EDTA (pH 8.0) and 1.5 μl of 10% SDS. At the end of reaction, a 1-μl aliquot is taken with a Drummond glass capillary pipet and precipitated with 10% TCA, and the radioactivity incorporated is determined as above to monitor the cDNA synthesis. The reaction mixture is extracted with 30 μl of phenol/chloroform twice. The aqueous phase is transferred to another tube, and 35 μl of 4 *M* ammonium acetate and 140 μl of ethanol are added. The solution is chilled on dry ice for 15 min and then warmed to room temperature with occasional vortexing to dissolve free deoxynucleoside triphosphates that have precipitated. As it warms up, the solution clears. The cDNA product is then precipitated by centrifugation in an Eppendorf microfuge for 15 min at 4°. The pellet is dissolved in 35 μl of TE and reprecipitated with ethanol as above. This ethanol precipitation step may be repeated once more to ensure complete removal of free deoxynucleoside triphosphates. Finally the pellet is rinsed with ethanol prior to the next step. The yield of product after three ethanol precipitations is 70–80%.

Comments. The use of clean, intact mRNA, fresh RNase-free reverse transcriptase, and a well-prepared vector primer is essential for the efficient synthesis of long cDNA. Inclusion of RNase inhibitors in the reaction mixture is of little value as long as clean enzyme is used, and may have an adverse effect.

Step 2. C-Tailing

Tails 10–15 dC residues long are added to the 3′ ends of the cDNA and vector by calf thymus terminal transferase. RNA is not a substrate for this enzyme. Addition of poly(A) to the reaction mixture prevents preferential tailing of unutilized vector molecules, thereby minimizing cloning of vector molecules with no inserts.

Reagents

Terminal deoxynucleotidyltransferase from calf thymus:
Pharmacia, minimal nuclease grade. Stored in liquid nitrogen; once thawed, the preparation should be stored at $-20°$ but for no longer than 1 month. The enzyme is unstable.

10× reaction buffer:
See Preparation of Vector Primer and Linker DNAs.

2m*M* [α-^{32}P]dCTP (4000–7000 cpm/pmol):
Approximately 100 μCi of [α-^{32}P]dCTP (200–5000 Ci/mmol) is dried down under reduced pressure and dissolved in 15 μl of 2 m*M* dCTP.

1 m*M* dithiothreitol (DTT)

Poly(A) (Miles), 0.15 μg/μl:
Prepared with sterile water. The average size of the chain is 5–6 S.

Procedure. The pellet is dissolved in 12 μl of water, and 2 μl each of 10× reaction buffer, poly(A) (0.15 μg/μl), and 1 m*M* DTT and 0.6 μl of 2 m*M* ^{32}P-dCTP are added. After vortexing, the solution is preincubated at 37° for 5 min. A 1-μl aliquot is taken with a Drummond microcapillary pipet and precipitated in 20 μl of 10% TCA as described earlier. The reaction is started by the addition of 2 μl of terminal transferase (10–15 units/μl) and lasts for 5 min. Just before stopping the reaction, another 1-μl aliquot is taken and precipitated with TCA as above. The reaction is terminated with 2 μl of 0.25 *M* EDTA (pH 8.0) and 1 μl of 10% SDS. The mixture is extracted with 20 μl of phenol/chloroform, the aqueous phase is collected, and the product is ethanol-precipitated twice in the presence of 2 *M* ammonium acetate as in Step 1. The pellet is rinsed with ethanol.

The average size of C-tails formed, where *A* is the total counts of cDNA formed in the RT reaction (in cpm), *B* the TCA-insoluble counts in the first aliquot (in cpm), *C* the TCA-insoluble counts in the second aliquot (in cpm), *D* the specific activity of the 2 m*M* ^{32}P-dCTP (in cpm/pmol), and 18 × *B*/*A* the recovery of cDNA synthesized or vector primer, is calculated as follows:

$$\text{Average size of C-tails} = \frac{(C - B) \times A}{2 \times D \times B \times 1.2 \text{ pmol (vector primer)}}$$

Formation of C-tails with an average length of 8–15 dC residues should be aimed at. If the length of the tail is much shorter, the binding of the G-tailed linker (see below) to the C-tailed cDNA will not be very strong, and cyclization will be inefficient. If the C-tail is too long (>20–25), expression in mammalian cells will be adversely affected.

To set up tailing conditions for your own lot of enzyme, a pilot reaction should be carried out first at one-half scale with the product of Step 1. Changes in incubation time and/or enzyme amount may be necessary.

Step 3. HindIII Restriction Endonuclease Digestion

Cleavage with *Hin*dIII removes the unwanted C-tail from the vector end and creates a sticky end for the G-tailed linker DNA to bind. *Hin*dIII does not cleave DNA : RNA hybrids efficiently, and we have not had any problem in cloning cDNA containing multiple *Hin*dIII sites.[5]

Using the reaction conditions described below, the *Hin*dIII cleavage is often relatively poor (30–50% digestion). This may be due to inhibition by the large amounts of free mRNA or by salts carried over from the previous step. For the best results, the use of a clean, fresh, active enzyme is necessary. Even with an excess of enzyme it is difficult to obtain complete digestion; that is not necessary in any case and should not be attempted because it will surely lead to a loss of some cDNA clones containing *Hin*dIII sites. The extent of digestion can be roughly estimated by looking at the small fragment (about 500 bp) cleaved off from the vector end on 1.5% agarose gels.

Reagents

*Hin*dIII restriction endonuclease (New England Biolabs), 20 units/μl
10× reaction buffer:
500 mM NaCl, 500 mM Tris–HCl (pH 8.0), 100 mM $MgCl_2$, 1 mg bovine serum albumin (BSA) (Miles, Pentex, crystallized)/ml. Prepared from sterilized stock solutions.

Procedure. The pellet is dissolved in 26 μl of sterilized water, and 3 μl of 10× reaction mixture is added. After brief vortexing, 0.7 μl of *Hin*dIII (20 U/μl) is added to the tube, and it is placed in a 37° water bath for 1 hr. The reaction is stopped with 3 μl of 0.25 M EDTA (pH 8.0) and 1.5 μl of 10% SDS. The mixture is extracted twice with phenol/chloroform, and the product is precipitated twice with ethanol in the presence of ammonium acetate as described above. The pellet is rinsed with ethanol, dissolved in 10 μl of TE, and stored at −20° after addition of 10 μl ethanol (total 20 μl) to prevent freezing. The product is stable for several years under these conditions.

Comments. *Hin*dIII enzyme preparations that have been stored at −20° for more than 2–3 months may not cleave the C-tailed cDNA : mRNA–vector though they will restrict pure plasmids. Use fresh enzyme.

Step 4. Oligo(dG)-Tailed Linker DNA-Mediated Cyclization and Repair of RNA Strand

The *Hin*dIII-cut, C-tailed cDNA : mRNA–vector is cyclized by DNA ligase and the oligo(dG)-tailed linker DNA that bridges the C-tail and the *Hin*dIII end of the hybrid–vector. The RNA strand is then replaced by DNA in a nick-translation repair reaction: RNase H introduces nicks in the RNA, DNA polymerase I nick-translates utilizing the nicks as priming sites, and ligase seals all the nicks. The end products are closed circular cDNA–vector recombinants.

Reagents

*Hin*dIII-cut, C-tailed cDNA : mRNA–vector (Step 3)
dG-tailed linker DNA (0.3 pmol/μl)
5× hybridization buffer:
50 m*M* Tris–HCl (pH 7.5), 5 m*M* EDTA, 500 m*M* NaCl. Stored at −20°.
5× ligase buffer:
100 m*M* Tris–HCl (pH 7.5), 20 m*M* $MgCl_2$, 50 m*M* $(NH_4)_2SO_4$, 500 m*M* KCl, 250 μg BSA/ml. Stored at −20°.
2 m*M* dNTP: mixture of dATP, dGTP, and dTTP (2 m*M* each).
2 m*M* dCTP
10 m*M* βNAD
E. coli DNA ligase (Pharmacia), nuclease free
E. coli DNA polymerase (Boehringer Mannheim), nuclease free
E. coli RNase H (Pharmacia), nuclease free
Contamination of the *E. coli* enzyme preparations by endonuclease specific for double- or single-stranded DNA can be detected by digestion of supercoiled pBR322 or single-stranded ϕX174 DNAs under the conditions specified below for each enzyme followed by analysis of their degradation by agarose gel electrophoresis.

Procedure. One microliter of the *Hin*dIII-digested, C-tailed mRNA : cDNA–vector solution (50% ethanol/TE) is added to a 1.5-ml Eppendorf tube along with 0.08 pmol of oligo(dG)-tailed linker DNA (about 1.5-fold excess over C-tailed ends), 2 μl of 5× hybridization buffer, and enough water to yield a final volume of 10 μl. The tube is placed in a 65° water bath for 5 min, then placed in a 43° bath for 30 min and transferred to a bed of ice. Eighteen microliters of 5× ligase buffer, 70.7 μl of water, and 1 μl of 10 mM NAD are added to the tube, which is then incubated on ice for 10 min. After the addition of 0.6 μg of DNA ligase, the tube is gently vortexed and incubated overnight in a 12° water bath. (The ligase is fairly labile and must be handled with care. It is reasonably stable when stored at −20°.)

After the cyclization step, the RNA strand must be replaced by DNA. The following are added to the tube: 2 μl each of 2 mM dNTP and 2 mM dCTP, 0.5 μl of 10 mM NAD, 0.3 μg of DNA ligase, 0.25 μg of DNA polymerase, and 0.1 unit of RNase H. The mixture is gently vortexed and incubated for 1 hr at 12°, then 1 hr at room temperature. The product is frozen at −20°.

Comments. It is imperative to use pure and active enzymes for this step. Nicking of the cDNA strand by nucleases before the RNA strand is replaced by DNA will completely destroy the recombinant. Before attempting a large-scale transfection, the cyclized product should be tested in a small-scale transfection. One microliter of the product should yield 1–3×10^4 colonies (this product is only 5–10% as active in transfecting cells as a corresponding amount of intact pBR322).

Preparation of Competent DH1 Cells

To make large cDNA libraries with the expenditure of reasonable amounts of mRNA, highly competent cells are required. We routinely prepare competent DH1 cells with transfection efficiencies of 3×10^8 to 10^9 colonies per microgram pBR322 DNA. The method described below is a modification of Hanahan's procedure.[24] It is somewhat simpler than the original, quite reliable, and can be used to prepare the large quantity of DH1 cells needed to construct big libraries.

Reagents

Double-distilled water or Milli Q water (not regular deionized water) should be used to prepare SOB, SOC, and FTB reagents.

SOB medium:

2% Bactotryptone (Difco), 0.5% yeast extract (Difco), 10 mM NaCl, 2.5 mM KCl, 10 mM $MgCl_2$, 10 mM $MgSO_4$.

All of the components except the magnesium salts are mixed and autoclaved. The solution is cooled, made 20 mM in Mg^{2+} with a 2 M Mg^{2+} stock solution (1 M $MgCl_2$ + 1 M $MgSO_4$), and filter-sterilized.

[24] D. H. Hanahan, *J. Mol. Biol.* **166,** 557 (1983).

Freeze–thaw buffer (FTB):

10 m*M* CH_3COOK, 45 m*M* $MnCl_2$, 10 m*M* $CaCl_2$, 3 m*M* hexamine cobalt chloride, 10 m*M* KCl, 10% glycerol.

Adjusted to pH 6.4 with 0.1 *M* HCl, filter-sterilized, and stored at 4°.

Dimethyl sulfoxide (DMSO):

MCB, spectrograde. A fresh bottle of DMSO should be used for best results. Alternatively, the content of a fresh bottle of DMSO can be aliquoted, stored at −20°, and thawed just before use.

Procedure. Frozen stock DH1 cells are thawed, streaked on an LB-broth agar plate, and cultured overnight at 37°. About 10–12 large colonies are transferred to 1 liter of SOB medium in a 4-liter flask, and grown to an OD_{600} of 0.5 at 20–25°, with vigorous shaking of the flask (200–250 rpm). The flask is removed from the incubator and placed on ice for 10 min. The culture is transferred to two 500-ml Sorvall centrifuge bottles and spun at 3000 *g* for 10 min at 4°. The pellet is resuspended in 330 ml of ice-cold FTB, incubated in an ice bath for 10 min, and spun down as above. The cell pellet is gently resuspended in 80 ml of FTB, and DMSO is added with gentle swirling to a final concentration of 7%. After incubating in an ice bath for 10 min, between 0.5 and 1 ml of the cell suspension is aliquoted into Nunc tissue culture cell freezer tubes and immediately placed in liquid nitrogen.

The frozen competent cells can be stored in liquid nitrogen for 1–2 months without a significant loss of competency. Prior to use, each preparation of competent cells should be assayed using standard plasmids, such as pBR322.

Transfection of DH1 Cells

Described below is a typical protocol for establishing a cDNA library containing 1–2 $\times 10^6$ clones in *E. coli*. Depending on the size of the library desired, one should scale the reaction up or down.

Reagents

SOC medium: SOB with 20 m*M* glucose.

Prepare a 2 *M* filter-sterilized glucose stock, add the glucose after making complete SOB medium, filter-sterilize, and store at room temperature.

Procedure. Four milliliters of competent cells are thawed at room temperature and placed in an ice bath as soon as thawing is complete. A maximum of 60 μl of the cyclized cDNA plasmid is added to the 4 ml of cells. The cells are then incubated in an ice bath for 30 min. Four hundred microliters of the transfected cell suspension is dispensed into each of 20 Falcon 2059 tubes in a bed of ice. They are then incubated in a 43° water bath for 90 sec and transferred to an ice bath. After 1.6 ml of SOC is added, the tubes are placed in a 37° incubator for 1 hr and shaken vigorously. The above transfection steps are repeated once more with another 4 ml of competent cells until a total 120 μl of the cyclized product is finally used. The entire suspension of transfected cells is then combined and transferred to 1 liter of L broth containing 50 μg/ml ampicillin. To determine the number of independent clones generated, a 0.1–0.2 ml aliquot is removed, mixed with 2.5 ml of L-broth soft agar at 43°, and plated on LB-broth agar containing ampicillin. Colonies are counted after an overnight incubation. The rest of the culture is grown to confluency at 37° and aliquoted in Nunc tubes. The tubes are stored at −70° or in liquid nitrogen after addition of DMSO (7%).

Comments. Pilot transfections with cyclized plasmid should always be undertaken before attempting to make a large library. A big water bath should be used to do large-scale transfections or else the water temperature will fall. We commonly observe a decrease of 50% in transfection efficiency from that predicted by pilot assays when we do large-scale transfections as above. The final cyclized product (Step 4, above) is only 5–10% as potent in transfecting cell as intact pBR322. (MC1061 *recA* cells also give a transformation efficiency comparable to DH1 cells.)

Transient Expression Screening of a cDNA Library

This screening method[10] relies on transient expression of cDNA clones in mammalian cells. The approach does not require any prior knowledge of the protein product itself. One only needs a specific biological or enzymatic assay for the presence of the protein in the cells or medium. For the techniques to work, the activity sought must be attributable to a single gene product.

In addition to the library of cDNA clones to be transfected, there are two components necessary for transient expression screening. First, one needs an appropriate recipient cell line. Because the pcD vector carries the SV40 early promoter and origin of replication, COS cells have been used as hosts. These cells contain an origin-defective SV40 genome and constitutively produce T antigen, a viral gene product needed to direct DNA replication initiating at the SV40 origin.[25] These cells are capable of greatly amplifying the number of DNA molecules taken up by the trans-

[25] Y. Gluzman, *Cell* **23,** 175 (1981).

fected cells, increasing the amount of gene product synthesized. The second requirement is an efficient method for introducing DNA into the COS cells. Based on previous studies, DEAE-dextran has been used to effectively introduce DNA into recipient cells. Immediately after introducing DNA into the cells they are treated for 3 hr with chloroquine. This has been found to stimulate levels of expression 5- to 10-fold.[26] Secreted products accumulate in the cell supernatants, and these are harvested 72 hr after the transfection and assayed. Using this procedure, Yokota *et al.*[12] and Lee *et al.*[13,14] have directly isolated full-length cDNA clones for mouse interleukin 2, human granulocyte–macrophage colony-stimulating factor, and mouse B-cell-stimulating factor-1 from concanavalin A-activated T-cell cDNA libraries.

Expression cloning of transiently expressed cDNAs can be used successfully provided that a sensitive assay is available and that the clones of interest are present in reasonable abundance (>0.01%). This procedure should be adaptable to screening protocols involving immunological detection of either intracellular or cell surface gene products. If one is screening for rare cDNA clones (<0.01%), prior enrichment of the library to be screened will probably prove necessary.

Reagents

COS cells[25]:
The cells are grown on tissue culture plates in Dulbecco modified Eagle's medium (DME) supplemented with 10% fetal calf serum (FCS), L-glutamine (2 m*M*), and antibiotics (penicillin/streptomycin). They are passed every 3 days by splitting 1 : 10 using a solution of trypsin–EDTA (see below).

Growth medium: DME containing 10% FCS, L-glutamine, penicillin/streptomycin.

Tris-buffered serum-free medium (Tris–SFM): DME buffered with 50 m*M* Tris–HCl (pH 7.4).

Collecting medium: DME containing 4% FCS, L-glutamine, penicillin/streptomycin.

DEAE–dextran stock solution:
20 mg/ml (Pharmacia) in sterile water. Do not autoclave or filter.

Chloroquine stock solution:
10 m*M* in 50 m*M* Tris–HCl (pH 7.4), 140 m*M* NaCl. Filter-sterilize.

Procedure. The main library is split into sublibraries containing an appropriate number of cDNA clones (50–100), and plasmid DNA is pre-

[26] H. Luthman and G. Magnusson, *Nucleic Acids Res.* **11,** 1295 (1983).

pared from the sublibraries. (The size of the sublibraries depends on the sensitivity of the assay for the desired gene product.)

Recipient cells are prepared from confluent plates of COS cells. The medium is removed by suction, and the plates are washed once with phosphate-buffered saline (PBS). One milliliter of trypsin–EDTA is added to each plate, and the plates are incubated at 37° for about 5–10 min to allow the cells to detach. Nine milliliters of DME with 10% FCS is added to each plate, and the cells are released and resuspended by gentle pipeting. Cells are then replated at 10^6 cells/plate in 10 ml of growth medium and incubated at 37° overnight. The medium is removed, and the plates are washed twice with Tris–SFM or PBS. Four milliliters of Tris–SFM containing 80 μl of DEAE–dextran (20 mg/ml) and 10–50 μg of plasmid DNA is added to each plate followed by incubation at 37° for 4 hr. The medium is removed, and the plates are washed with Tris–SFM or PBS. Five milliliters of DME containing 2% FCS, L-glutamine, and 100 μM chloroquine are added to each plate. After incubation at 37° for 3 hr, the plates are washed twice with Tris–SFM or PBS, fed with 4 ml of collecting medium, and incubated at 37°. The supernatants are harvested after 72 hr for assay.

Trouble Shooting Chart

Problems in library construction	Possible causes
1. Low yield of colonies	A. cDNA synthesis
	i. Heavily nicked vector primer—Prepare with clean enzymes
	B. C-tailing
	i. Contamination of dNTP from previous step—Try three ethanol precipitations
	ii. Unstable or inactive enzyme
	iii. Endonuclease contamination in enzyme
	iv. Incomplete phenol extraction or incomplete removal of phenol or ethanol after cDNA synthesis
	C. *Hin*dIII digestion
	i. Inactive or unstable enzyme
	ii. Endonuclease contamination in enzyme
	iii. Incomplete phenol extraction or incomplete removal of phenol or ethanol after C-tailing
	D. Cyclization and repair
	i. Inactive ligase or polymerase
	ii. Nuclease contamination in one of the three enzyme
	iii. Linker DNA contaminated with agarose
	E. Transfection
	i. Incompetent DH1 cells

(*continued*)

Trouble Shooting Chart (*continued*)

Problems in library construction	Possible causes
2. Few plasmids contain inserts	A. cDNA synthesis i. mRNA with low template activity a. Contamination by ribosomal RNA—Use high flow rate for loading a sample and washing column in oligo(dT)–cellulose chromatography b. mRNA contaminated by impurities that inhibit reverse transcriptase—Prepare new one ii. mRNA contaminated by oligo(dT)—Use highest quality oligo(dT)–cellulose; extensively wash column before use iii. Vector primer contaminated by oligo(dA)—Use highest quality oligo(dA)–cellulose; extensively wash column before use iv. Vector primer has T-tails that are too long or too short v. Inactive or unstable reverse transcriptase B. C-tailing i. Degraded poly(A)
3. Inserts are short	A. cDNA synthesis i. mRNA degraded or contaminated with impurities—Prepare new mRNA ii. Inactive or RNase-contaminated reverse transcriptase B. C-tailing i. Endonuclease-contaminated terminal transferase

[2] A Method for Cloning Full-Length cDNA in Plasmid Vectors

By GISELA HEIDECKER and JOACHIM MESSING

The first step in the molecular analysis of many genetic systems is the isolation and cloning of a cDNA copy of an mRNA expressed by the gene of interest. The reasons for this step are mainly that even rare mRNAs in a mammalian cell, for instance, represent one molecule in at most 10^5, a definite improvement in most cases over the genomic situation where a single copy gene is contained within 10^9 base pairs. The increased odds

METHODS IN ENZYMOLOGY, VOL. 154

facilitate the identification of the sought-after sequence, which in the early stages of a study are often done by sib selection through assays like *in vitro* translation of mRNAs hybrid selected from pooled clones.[1] Furthermore, as mRNA is associated with its protein product during translation, it can be enriched by precipitation of polysomes with an appropriate antibody.[2] An alternative is to identify the cDNA clones immunologically by inserting them into a vector that directs the expression of the cDNA in bacteria, an approach often feasible only with cDNAs made from processed mRNA, as only they are colinear with the protein.[3,4]

Cloned cDNAs are very useful in several regards. Even short clones provide molecular probes, the tools to facilitate the isolation of homologous genomic clones and additional cDNA clones. The sequence of full-length cDNA clones allows one to deduce the amino acid sequence of the encoded protein.[5] Full-length cDNA clones have been invaluable in the analysis of the organization and regulation of eukaryotic genes in many hybridization experiments.[6,7] Various methods for the cloning of cDNA have been developed since the discovery of reverse transcriptase, the enzyme that made it all possible.[8-10] The method we present here combines high efficiency with relative simplicity.

Principles

The principal steps of our cDNA cloning procedure are outlined in Fig. 1. The 3′ ends of a linearized pUC plasmid are extended with thymidine residues to an average tail length of 50 nucleotides. The oligo(dT) tails are annealed to the poly(A) tails of mRNA and are used to prime cDNA synthesis, thus already covalently linking the cDNA to the vector DNA. The plasmid–cDNA molecules are extended with oligo(dG) tails, alkali denatured, and sized on alkaline sucrose gradients. Besides provid-

[1] B. M. Paterson, B. E. Roberts, and E. L. Kuff, *Proc. Natl. Acad. Sci. U.S.A.* **74,** 4370 (1977).

[2] R. Palacios, R. D. Palmiter, and R. T. Schimke, *J. Biol. Chem.* **247,** 2316 (1972).

[3] A. J. Korman, P. J. Knudsen, J. F. Kaufman, and J. L. Strominger, *Proc. Natl. Acad. Sci. U.S.A.* **79,** 1849 (1982).

[4] P. H. Seeburg, J. Shine, J. A. Martial, J. D. Baxter, and H. M. Goodman, *Nature (London)* **270,** 486 (1977).

[5] A. Efstratiadis, F. C. Kafatos, and T. Maniatis, *Cell* **10,** 279 (1977).

[6] R. Breathnach, J. L. Mandel, and P. Chambon, *Nature (London)* **270,** 314 (1977).

[7] C. Brack, M. Hirama, R. Lenhard-Schuler, and S. Tonegawa, *Cell* **15,** 1 (1978).

[8] H. Temin and S. Mizutani, *Nature (London)* **226,** 1211 (1970).

[9] A. Efstratiadis and L. Villa-Komaroff, *in* "Genetic Engineering" (J. K. Setlow and A. Hollaender, eds.), Vol. 1, pp. 1–14. Plenum, New York, 1979.

[10] H. Okayama and P. Berg, *Mol. Cell. Biol.* **2,** 161 (1982).

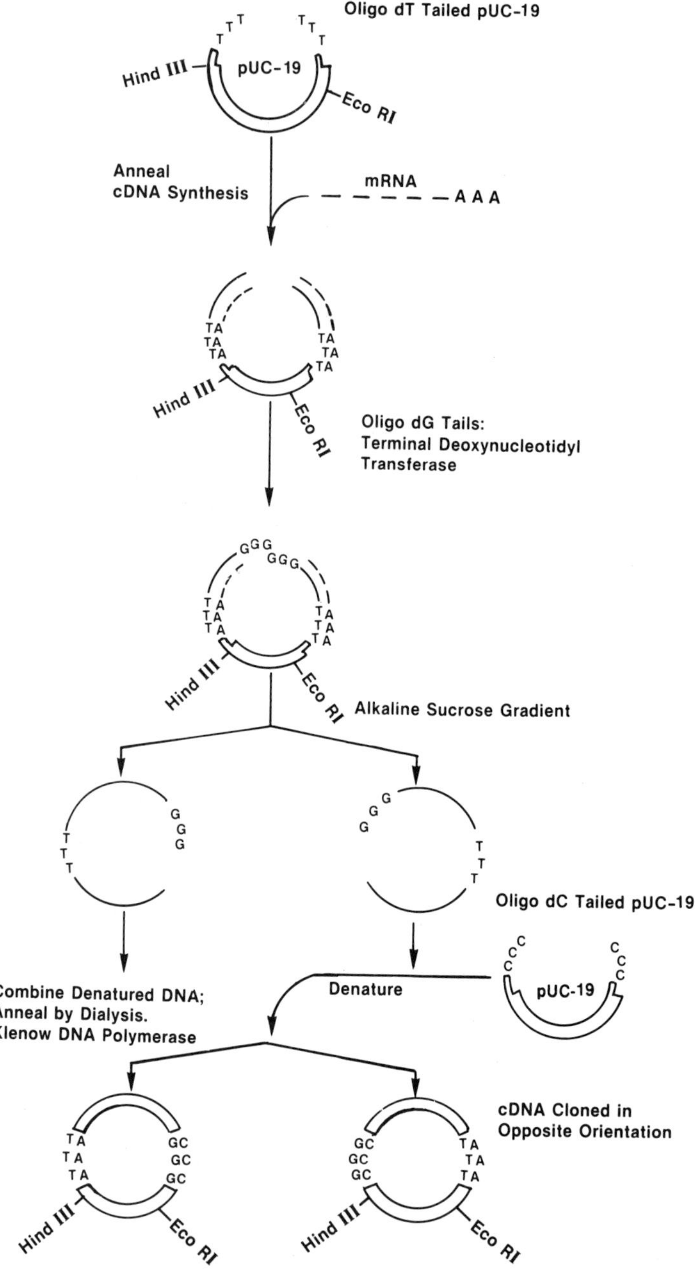

FIG. 1. Flow sheet of the cDNA cloning procedure. Explanations are given in the text. The *Hin*dIII and *Eco*RI sites flank the polylinker region in pUC19.[23]

ing a size selection this step also removes the RNA and denatures the vector. Prior to neutralization and renaturation an excess of oligo(dC)-tailed pUC plasmid is added. The vector strands are reannealed under dilute conditions to favor circularization of the plasmid over concatemerization. The oligo(dC) tail of the second strand vector serves as primer for synthesis of the second strand of the cDNA insert after which the recombinant plasmids are transfected into *Escherichia coli.*

Experimental Procedures

Materials

All restriction enzymes were purchased from either New England Biolabs or Bethesda Research Laboratories (BRL). MLV reverse transcriptase, placental RNase inhibitor (RNasin), and large fragment of *E. coli* DNA polymerase I were also from BRL, AMV reverse transcriptase was obtained from Life Sciences Inc., terminal deoxynucleotidyltransferase (TdT) was from PL-Biochemicals/Pharmacia, who also supplied the deoxynucleotides. Radioactive nucleotides were purchased from New England Nuclear. Competent cells were supplied by BRL or Vector Cloning Systems.

Tailing of Vector DNAs

The integrity of the plasmids used in this scheme is of great importance. Nicks in the backbone of the vector are substrate for terminal transferase and can also greatly diminish the efficiency of the procedure due to network formation during the reannealing step. Great care should thus be taken to start with a preparation of supercoiled plasmid and to use only enzymes that are devoid of nicking activity.

Five hundred micrograms of pUC plasmid DNA is digested with 500 U of *Pst*I in 1 ml of 50 m*M* Tris (pH 7.5), 50 m*M* NaCl, 10 m*M* $MgCl_2$, 1 m*M* dithiothreitol (DTT), and 50 μg/ml BSA for 3 hr at 37°. As undigested plasmid will cause a high background of transformants, the linearized molecules should be purified by banding on a CsCl/ethidium bromide gradient. After removal of the ethidium bromide by *n*-butanol extraction and the CsCl by dialysis against TE (10 m*M* Tris, pH 7.5, 1 m*M* EDTA), the DNA is concentrated by ethanol precipitation and air dried.

The easiest way to control the tail length is through the molar ratio of nucleotides to plasmid ends as the K_m of TdT is very low. We use a 60-fold excess of dTTP or dCTP over ends to generate the dT- and dC-tailed vector molecules. Tailing with dGTP is relatively independent of the nucleotide concentration, as dG homopolymer tails of about 20 nucleotides assume a secondary structure that prevents further extension.[11] We normally use dCTP for tailing the second strand vector and set up our plasmid tailing reactions as follows. The ethanol-precipitated plasmid DNA is resuspended in 0.5 ml 0.3 *M* potassium cacodylate and divided into two equal aliquots. (The cacodylate is prepared by titering a 2 *M* solution of cacodylic acid with KOH to pH 7.0 and adjusting the final concentration to 1 *M*.) To one aliquot we add 20 μl of 1 m*M* dTTP and to the other 20 μl of 1 m*M* dCTP. The other components are the same for the two reactions and consist of the following: 2 μl of 0.1 *M* DTT, 1 μl of BSA 50 mg/ml, 10 μCi of dCTP of >100 Ci/mmol (any nucleotide can be used at this point as long as the specific activity is high enough to allow sufficient incorporation of radiolabel without contributing significantly to the amount of nucleotides), 37.5 μl of 10 m*M* $CoCl_2$, and 150 units of terminal TdT. Adjust the final volume to 375 μl and incubate 45 min at 37°. To tail the second strand vector with dGTP the same conditions can be used, replacing dGTP for dCTP in the procedure. Before terminating the reaction by phenol extraction of the DNA, the tails should be analyzed and the reaction mixture stored at −20°.

We routinely check the tailing reactions in two respects: (1) the amount of label incorporated into the backbone of the plasmid molecule and (2) the average length of the tails. Heat inactivate 10 μl of the tailing reactions, then digest 2 μl with *Hae*II in a final volume of 10 μl and analyze the restriction pattern on a 2% agarose gel in TBE (90 m*M* Tris base, 90 m*M* boric acid, 2 m*M* EDTA). For comparison also run 1 μg of pUC plasmid cut with *Hae*II and *Pst*I. As can be seen in Fig. 2a, the lowest two bands of the tailed plasmid DNA shift up when compared to the untailed DNA. The autoradiograph of the gel shows that the vast majority of the radiolabel resides in these two smallest fragments while the other fragments are not noticeably labeled, indicating that no tailing has occurred at nicks during the *Pst*I digest or the tailing reaction itself. The mobilities of the tailed fragments are aberrant when compared to standards. This reflects the partial single strandedness of the fragments and provides only an estimate of the tail lengths.

To determine the exact lengths it is necessary to digest another 2-μl aliquot of the heat-inactivated tailing reaction with *Eco*RI (or any other enzyme that cuts in the cloning sites). This digest is performed in a volume of 5 μl with 5 units of enzyme for 1 hr. Then 10 μl of denaturing dye (50 m*M* EDTA in formamide with 0.02% bromophenol blue) is added, and the sample is boiled and analyzed on an 8% polyacrylamide gel containing

[11] G. Deng and R. Wu, *Nucleic Acids Res.* **9**, 4172 (1982).

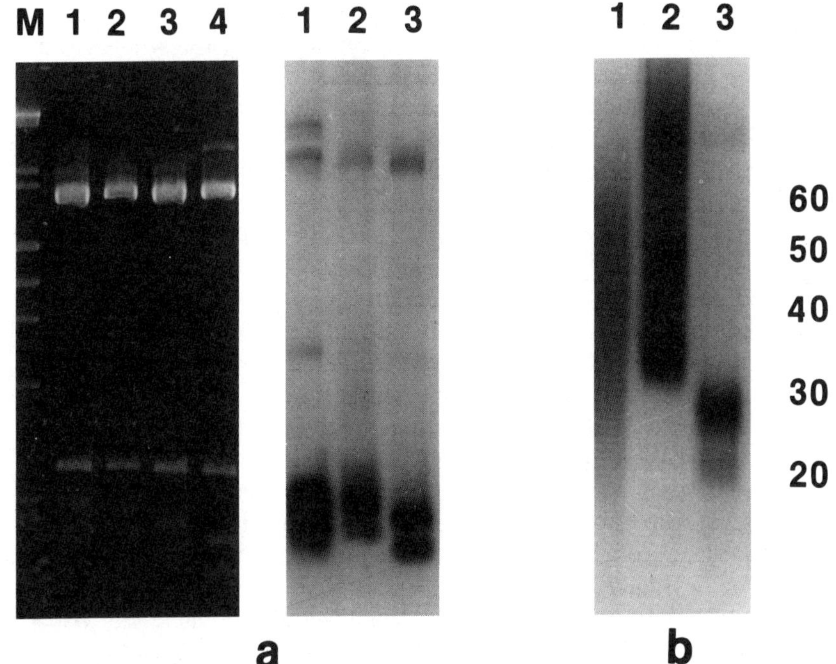

FIG. 2. Analysis of the tailing reaction by gel electrophoresis and autoradiography of labeled DNA. (a) Ethidium bromide stain (left) and autoradiograph (right) of *Hae*II-digested pUC19 DNA which had been tailed with oligo(dT) (lane 1), oligo(dC) (lane 2), and oligo(dG) (lane 3) at the *Pst*I site and separated on a 2% agarose gel electrophoresed in TBE. The tailing conditions were those detailed in the text. Lane 4 on the ethidium bromide stain shows the *Hae*II restriction pattern for untailed pUC19. Lane M contains *Hin*dIII-digested λ DNA and *Hae*III-digested ϕX174 DNA as size markers. Note that the faint bands in the autoradiograph represent partial digest products which are only slightly visible in the stained gel. (b) Autoradiograph of a denaturing acrylamide gel used to separate *Eco*RII-digested pUC19 DNA which had been tailed as in (a).

8 *M* urea in TBE. The autoradiograph of this gel (Fig. 2b) shows a ladder of bands extending around the average tail length. The position in the gel can be determined most accurately by running a DNA sequencing reaction on the same gel. The length of the T-tails should range between no less than 20 nucleotides and no more than 100, while the C-tails can be shorter by 5 nucleotides, but should not be longer. As is apparent in Fig. 2b, the T-tails show a wider range in their length than the C-tails. We do not know the reason for this.

Tailing reactions which result in tails which are much longer or shorter than expected are mostly due to two causes: the concentration of either the plasmid DNA or the nucleotide triphosphate were different from the one assumed. If the tails are too short add more nucleotides, fresh TdT, and continue the incubation. If the tails are too long, the whole reaction has to be repeated. It may be advisable to perform a pilot experiment with one-tenth the amount of plasmid in one-tenth the reaction volume.

After the tailing reaction, the plasmid DNA is extracted with phenol/chloroform and ethanol precipitated. The pellet is washed with 70% ethanol, air dried and dissolved in 250 μl of TE.

cDNA Synthesis

We found that a 2- to 4-fold molar excess of mRNA over tailed plasmid ends will saturate the system and result in over 90% of the tails priming cDNA synthesis. A reaction starting with 2 μg of plasmid DNA and 8 μg of poly(A) mRNA or 40 μg of total RNA will normally generate enough clones for several representative libraries.

The reaction conditions for first strand synthesis are as follows: Dry down 10 μCi [α-^{32}P]dCTP of about 500 Ci/mmol in an Eppendorf tube. Add 2 μl of pUCdT DNA (1 mg/ml), 2.5 μl of 10× buffer (500 m*M* Tris–HCl, pH 8.0, 700 m*M* KCl, 30 m*M* $MgCl_2$, 50 m*M* DTT, 5 mg/ml BSA), 1 μl of RNasin 20 U/μl, 1 μl of dXTPs (12.5 m*M* for each nucleotide triphosphate), and no more than 17.5 μl of RNA, adjust to a volume of 24 μl with dH_2O (distilled), mix, and start the reaction with 1 μl MLV reverse transcriptase (200,000 units/ml) at 42°. We found that Moloney MLV reverse transcriptase works marginally better than AMV reverse transcriptase. It is, however, important to follow the buffer conditions that are recommended by the supplier (BRL). The Mg^{2+} concentration is very critical. The placental RNase inhibitor is included in the reaction as a precaution and should be added to the buffer before the RNA. The volume of the reaction should not be increased as the time to reach $C_0t/2$ for the tails is already about 10 min.

The reaction is terminated after 2 hr by adding 75 μl of TE buffer. The DNA is extracted with phenol/chloroform and ethanol precipitated at room temperature 3 times from a volume of 100 μl in the presence of 2 *M* CH_3COONH_4. The repeated ethanol precipations will remove the unincorporated nucleotides. The supernatant of the last precipitate should contain no more than a trace of the radioactivity that was included in the cDNA synthesis reaction. The precipitate is thoroughly air dried and dissolved in 25 μl of dH_2O.

The cDNA reaction can be analyzed in various ways. We usually separate 5–10% of the reaction on a 1% agarose gel in 2.2 *M* formaldehyde, 40 m*M* MOPS pH 7.0, 10 m*M* NaAc, and 1 m*M* EDTA. Two and one-half microliters of the sample is boiled in 7.5 μl of denaturing dye

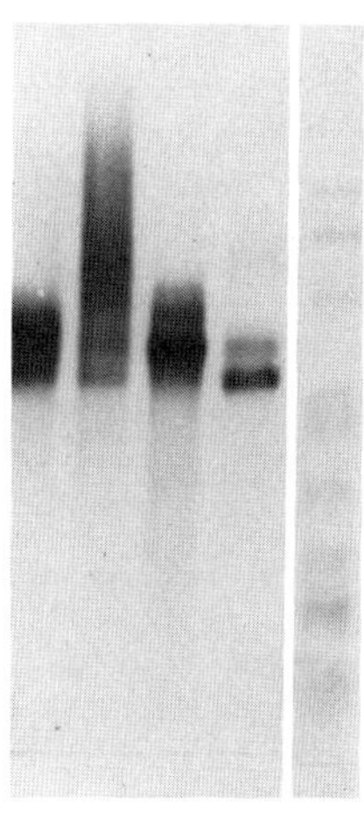

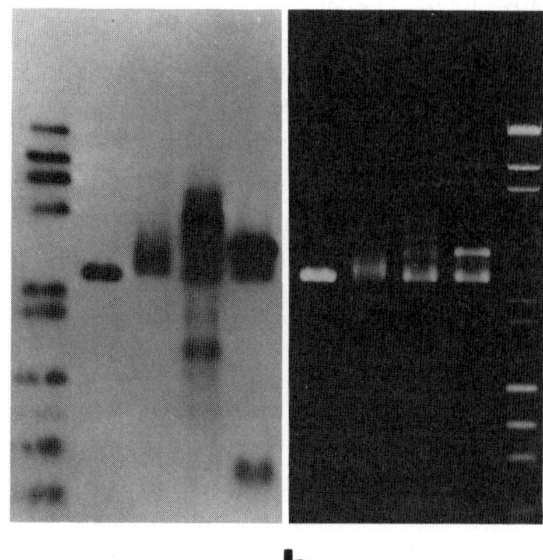

a b

FIG. 3. Analysis of cDNA synthesis by gel electrophoresis and autoradiography of labeled DNA. The RNAs used in the cDNA synthesis were total RNA isolated from CrFk cells (lane 1), viral SRV-1 RNA (lane 2), globin RNA (lane 3), and no RNA (lane 4). Lane M shows the same marker DNAs as Fig. 2. (a) DNAs separated under denaturing conditions on a 1.2% agarose gel containing 2.2 *M* formaldehyde. (b) Samples after RNase treatment and *Eco*RI digest. The ethidium bromide-stained gel (1% agarose in TBE) is on the right; the autoradiograph of the gel is on the left.

solution and electrophoresed for 3–4 hr at 10 V/cm. An autoradiograph of the dried gel should show the radioactivity as a smear shifted up when compared to the vector which is run in parallel (see Fig. 3a). A homogeneous RNA will result in a sharp band as in the case of globin. However, for large RNAs only a fraction of the transcripts will be full length, especially if no precautions are taken to overcome strong stops which are probably caused by secondary structure in the template. An example of this can be seen in lanes 2 in Fig. 3. The inhibiting influence of the secondary structure becomes especially noticeable if poor quality reverse transcriptase is used. To reduce the amount of secondary structure we routinely heat the RNA to 60° followed by quenching on ice prior to the addition to the reaction mix.

Another way of analyzing what fraction of the tails has primed cDNA synthesis and how long the cDNAs are is to digest an aliquot with *Eco*RI in the presence of RNase. The RNase removes all the RNA not paired to DNA, thus ensuring that the molecules will separate according to their double-stranded length. After separation on a 0.8% agarose gel in TBE, the DNA can be visualized with ethidium bromide. Again an equal amount of vector DNA should be run for comparison. Figure 3b, right panel, shows such a gel. In the globin cDNA reaction about 50% of the plasmid material has shifted up in the gel, most of it to a position in agreement with a full-length transcript of globin mRNA having been added to the vector. This reaction was performed with globin and vector ends in equimolar amounts. By changing the ratio to 3/1 mRNA/vector most of the ends will prime cDNA synthesis as can be seen in lane 1, which shows the cDNA of a reaction with 40 μg of total RNA. Although most of the cDNA copies in this reaction are rather short, almost none of the DNA bands at the position occupied by the unreacted plasmid.

Second Tailing Reaction

The choice of nucleotide to be used in the second tailing reaction is somewhat arbitrary, though, of course, determined by the tails added to the second strand vector. Tailing with dCTP in the presence of Mg^{2+} or Co^{2+} will favor extension of protruding 3′ ends and thus enrich for full-length cDNA clones, while adding dG-tails with Mn^{2+} as cofactor is more efficient with any kind of end and should thus increase the overall yield.[11] In our experience the differences in either respect were minor. Using dGTP has the advantage that the tail length does not depend on the ratio of nucleotides to ends in the reaction and, consequently, a large excess of dGTP can be added without increasing the tail length beyond 20 nucleotides. The molar concentration of ends at this stage is sometimes hard to estimate because short reverse transcripts are generated during the cDNA synthesis, possibly due to priming by short RNA fragments acting as random primers. At this point we do not know whether these fragments are present in our RNA preparations or are produced during cDNA synthesis by the RNase H activity of reverse transcriptase.

The dG-tailing reaction is set up as follows: 20 μl of cDNA–plasmid DNA in dH_2O, 10 μl of 1 *M* potassium cacodylate (pH 7.0), 1 μl of 5 m*M* dGTP, 1 μl of 50 m*M* DTT, 1 μl of BSA (5 mg/ml), 20 U TdT, and dH_2O to a final volume of 45 μl are mixed, and the reaction is started with 5 μl 10 m*M* $MnCl_2$ and incubated 20 min at 37°. The conditions for tailing with dCTP are the following: 20 μl of cDNA–plasmid DNA in dH_2O, 10 μl of 1 *M* potassium cacodylate, 1 μl of 0.2 m*M* dCTP, 1 μl of BSA (5 mg/ml), 20 U TdT, and dH_2O to a final volume of 45 μl are mixed, and the reaction is started with 5 μl 10 m*M* $CoCl_2$ and incubated for 45 min at 37°.

Adding the cofactor last in either reaction prevents the precipitation of CoS which may appear as a black solid. The precipitate may be due to degradation of the DTT which is added to this reaction or carried over from the reverse transcriptase reaction. If a precipitate appears the DNA need to be extracted and ethanol precipitated again, and the reaction is set up as before. This will ensure the proper cofactor concentration in the reaction mix. After the reaction the DNA is extracted with phenol/chloroform, ethanol precipitated, and dissolved in 50 μl TE.

Alkaline Sucrose Gradient

The cDNA–plasmid molecules are denatured and size selected on an alkaline sucrose gradient. This step will also remove the RNA and most of the (at this point tailed) short DNA molecules mentioned above. We use an SW40 rotor and 5–30% sucrose gradients made up in 1 *M* NaCl, 0.2 *M* NaOH, and 1 m*M* EDTA in polypropylene tubes. (*Warning:* do not use "ultraclear" tubes as they do not withstand the alkaline conditions for the length of the run.) The sample is adjusted to 0.2 *M* NaOH and layered on the gradient which is centrifuged at 37,000 rpm for 20 hr at 4°. The gradient is fractionated into about 35 fractions of 0.3 ml. The gradient profile is determined by the Cerenkov counts in each fraction. A gel electrophoresis of selected fractions, running some pUCdT in parallel, determines which fractions to pool. Figure 4 shows the profiles of two gradients which were used to separate the cDNA–plasmid molecules of the reactions analyzed in Fig. 3, lanes 2 and 3.

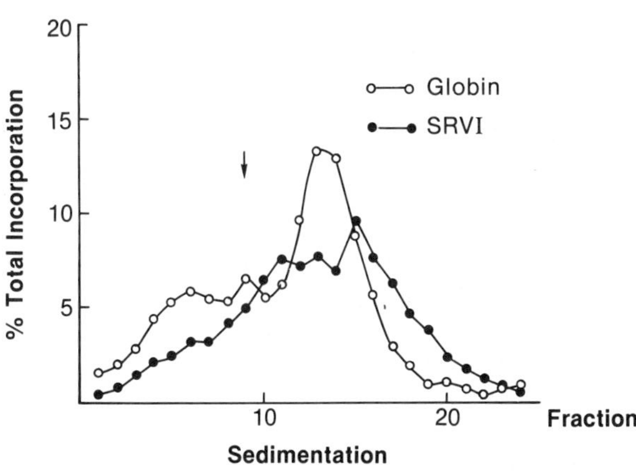

FIG. 4. Preparative alkaline sucrose gradient centrifugation of cDNA–plasmid molecules. After cDNA synthesis and the addition of oligo(dG), the DNA solution is made alkaline by adjusting the solution to 0.2 *M* NaOH and layered on a 5–30% sucrose gradient in 1 *M* NaCl, 0.2 *M* NaOH, and 2 m*M* EDTA. The gradient is centrifuged with a SW40 rotor at 37,000 rpm for 20 hr at 4°. Sedimentation is from left to right. The profiles of two cDNAs derived from SRV-1 and globin RNA are shown. These samples were also analyzed in Fig. 3, lanes 2 and 3. The arrow marks the position of linear vector DNA.

Renaturation of the Plasmid

Based on the fraction of the total plasmid pooled, the amount of DNA can be estimated; e.g., if all the plasmids primed cDNA synthesis and only the longer molecules (which contain 50% of the total counts) are pooled, probably 25% of the plasmid molecules that entered into the cDNA synthesis reaction are present. Based on this estimate we add a 5- to 10-fold excess of second strand plasmid to the pooled material which, thus, automatically denatures the DNA. The sample is then dialyzed against a 1000-fold excess of 10 m*M* Tris (pH 7.5), 100 m*M* NaCl, and 1 m*M* EDTA (TEN) in the cold to remove the NaOH, sucrose, and excess NaCl. Renaturation is achieved by transferring the dialysis tube to 50 volumes of 30% formamide in TEN at 37° for 24 hr. These conditions will promote faithful and quantitative reannealing of the plasmid backbone. The formamide is removed by dialysis against TEN at 4°. This will also ensure that even short tails will hybridize. The concentration of plasmid DNA should not exceed 3 μg/ml to avoid concatemerization over the ends, which would especially affect plasmids carrying long cDNA inserts.[12] After dialysis the DNA is concentrated by ethanol precipitation and dissolved in 25 μl TE.

Second Strand Synthesis

The second strand is primed by the homopolymer tail of the second strand plasmid using the large fragment of *E. coli* DNA polymerase. Add 5 μl of 10× polymerase buffer (500 m*M* Tris, pH 7.5, 100 m*M* $MgCl_2$, 20 m*M* DTT, 1 mg/ml BSA), 5 μl of nucleotide mix (2.5 m*M* for each nucleotide), 10 U of large fragment, and dH_2O to 50 μl. Incubate for 30 min at 14°, and then for 30 min at room temperature. Phenol/chloroform extract and ethanol precipitate the DNA and dissolve it in 50 μl of TE.

Transformation of 2–5 μl should result in approximately 10,000 independent colonies. We use commercially available competent cells and have found that *E. coli* DH1 or its derivatives give the highest efficiency.[13]

Comments

We have applied the procedure given above to many cDNAs. Our goal, as a rule, was to obtain full-length copies of the mRNAs in order to define regulatory and processing signals of the genes of interest, which in our case were the storage protein genes in corn.[14,15] For instance, compar-

[12] A. Dugaiczyk, H. W. Boyer, and H. M. Goodman, *J. Mol. Biol.* **96,** 171 (1979).

[13] D. Hanahan, *J. Mol. Biol.* **166,** 557 (1983).

[14] G. Heidecker and J. Messing, *Nucleic Acids Res.* **11,** 4891 (1983).

[15] G. Heidecker and J. Messing, *Annu. Rev. Plant Physiol.* **37,** 439 (1986).

ison of full-length cDNA clones to the genes of these proteins allowed us to identify the transcriptional start sites, translational signals in the leader sequence, and different usage of polyadenylation signals in the 3′ sequence of the mRNA. This analysis was not possible before since libraries from maize endosperm mRNA that have been prepared earlier were missing significant portions of the 5′ end and even sometimes parts of the 3′ end.[16,17]

Obviously, this is only one of many procedures for the construction of cDNA libraries, and already various modifications come to mind which might increase its usefulness depending on the requirements of the experiment. We would like to use this Comments section to discuss several variations of our protocol.

Our libraries contain cDNA clones in either orientation with respect to the promoter of the β-galactosidase gene on the pUC plasmid. This means that one-sixth of all clones could potentially be expressed as a fusion protein to the very amino-terminal part of β-galactosidase and screened for by immunoblotting. Such a procedure has been described for pUC plasmids.[18] It would be easy to prevent the synthesis of either orientation by removing one of the tails before synthesis of the first strand. As an example, if, as described here, pUC19 T-tailed at the *Pst*I site is used, cleavage with *Sph*I or *Hin*dIII would remove the tail which would prime synthesis of the antisense strand (in respect to the *lac* operon). Still, clones may not be expressed since full-length copies of mRNA might contain in-frame termination codons in the untranslated leader sequence.

However, predetermining the orientation of the cloned cDNA in the plasmid may be used advantageously for the generation of differential libraries if the newly developed plasmids pUC118 or pUC119 are used as vector instead of pUC18 or pUC19. (These plasmids are described by Vieira and Messing in Chap. [1] of Vol. 153 in this series.) Suffice it to say that these plasmids carry the origin of replication and the packaging signals of bacteriophage M13 and thus will produce defective phage particles containing single-stranded plasmid DNA upon infection with an appropriate helper phage like KO7. pUC119 plasmids containing cDNA cloned in its antisense orientation with respect to the *lac* operon would produce defective phage with single-stranded DNA that is complementary to the mRNA. This single-stranded DNA can be hybridized to mRNA that was immobilized onto an appropriate matrix. If the RNA was isolated from a

[16] B. Burr, F. A. Burr, T. P. St. John, M. Thomas, and R. D. Davis, *J. Mol. Biol.* **154,** 33 (1982).

[17] K. Pedersen, J. Devereux, D. R. Wilson, E. Sheldon, and B. A. Larkins, *Cell* **29,** 1015 (1982).

[18] D. M. Helfman, J. R. Feramisco, J. C. Fiddes, G. P. Thomas, and S. H. Hughes, *Proc. Natl. Acad. Sci. U.S.A.* **80,** 31 (1983).

tissue or developmental stage different from the one copied into the cDNA library, then plasmid molecules carrying sequences not represented in the immobilized mRNA are left in solution. These can be transfected into *E. coli* without further treatment as the transformation efficiency of single-stranded plasmids is only 3-fold lower than that of double-stranded plasmids.

Constructing the cDNA library in these "single-stranded plasmid" vectors is also advisable if the library is to be screened with a synthetic oligonucleotide. Colonies which also harbor the helper phage extrude defective phage particles that can be blotted onto nitrocellulose filters. The single-stranded viral DNA is the only DNA available for hybridization to the oligonucleotide as the cellular double-stranded DNA stays in its native form. This significantly decreases the background normally observed when screening with a radiolabeled oligonucleotide.

The use of the single-stranded plasmids also facilitates the generation of templates for DNA sequencing with a universal primer,[19–21] for site-directed mutagenesis,[22,23] and for the generation of second strand probes.[24] A new efficient system for packaging single-stranded plasmid is described in Chap. [1] by Vieira and Messing in Vol. 153 of this series.

On the other hand, one might try to construct cDNA libraries directly in M13 RF instead of plasmid vectors with an M13 origin. It is, however, not easy to propagate very complex libraries since phage with larger inserts or those that are unstable for other reasons may be lost during the amplification of the cDNA library. Although this is not so pronounced in plasmid cDNA libraries, we still found that those also have to be propagated with care. In general, the libraries should only be grown in suspension immediately after the original transformation of host bacteria to a density of about 10^8 cells/ml. The bacteria are then concentrated by centrifugation and frozen at $-70°$ in media containing 15% glycerol. The libraries should be replica plated for screening a cDNA clone. Any clone should be restreaked a few times for further purification. Total library DNA should be isolated from bacteria grown on plates rather than from liquid cultures.

One last aspect we would like to address is the efficiency of cDNA cloning by this method in particular, as well as cloning efficiency in general. Efficiency becomes a major consideration when a gene is expressed

[19] G. Heidecker, J. Messing, and B. Gronenborn, *Gene* **10,** 69 (1980).
[20] J. Messing, this series, Vol. 101, p. 20.
[21] C. Yanisch-Perron, J. Vieira, and J. Messing, *Gene* **33,** 103 (1985).
[22] M. Zoller and M. Smith, this series, Vol. 100, p. 468.
[23] J. Norrander, T. Kempe, and J. Messing, *Gene* **26,** 101 (1983).
[24] N.-T. Hu and J. Messing, *Gene* **17,** 271 (1982).

at low levels in a tissue of limited availability. Various methods and vectors beside the one we present here have been developed to cope with such a situation, among them are highly effective means of introducing the recombinant vectors into the bacterial host cell. Bacteriophage λ-derived vectors allow *in vitro* packaging of the DNA and infection of the host. This method was until recently about 3 orders of magnitude more efficient in introducing DNA into *E. coli* cells on a mole per mole basis than the uptake of naked plasmid DNA following a $CaCl_2$ treatment of the bacteria. The transformation method developed by Hanahan[13] has narrowed this gap to a factor of 10, with efficiencies of 10^5 to 10^6 transformed bacterial colonies per nanogram of plasmid DNA.

With these efficiencies it should be theoretically possible to obtain a representative cDNA library starting with minute, i.e., nanogram amounts of mRNA and plasmid DNA. However, the highest yield we ever obtained following our protocol was about 10^6 colonies starting with 1 μg of globin mRNA and plasmid DNA each, and the efficiency dropped significantly if less material was used to begin with. Obviously, the biochemical and technical aspects of the procedure are the limiting factor, i.e., how little material can be handled and pipetted without losing it to adhesion to reaction tubes, in the supernatant of ethanol precipitation, in the interphase after phenol extractions, etc. In summary, we do no advise scaling down this method much beyond 1 μg of plasmid and the equivalent amount of mRNA.

Finally, we would like to point out that all ingredients for this procedure including the tailed pUC plasmids and the competent cells that give rise to the high transformation efficiencies are commercially available.

[3] An Efficient Vector–Primer cDNA Cloning System

By DANNY C. ALEXANDER

Introduction

The vector–primer cDNA cloning method described by Okayama and Berg[1] represents a major advance in cDNA cloning technology. The high efficiency and proportion of full-length clones make the method very useful, especially to those researchers studying genes that are not highly

[1] H. Okayama and P. Berg, *Mol. Cell. Biol.* **2,** 161 (1982).

expressed. Very large cDNA libraries can be constructed in plasmids, allowing the isolation of cDNAs for rare mRNAs or the isolation of complex sets of mRNAs sharing some biological significance.

In a previous publication[2] we reported some modifications in the Okayama–Berg method that made the system more reliable and versatile in our hands. Two steps, including a restriction enzyme digestion and an agarose gel purification, were eliminated, resulting in a more active vector–primer. Polylinkers were also engineered into the final construct to simplify analysis and subcloning of inserts. Concurrent improvements in *Escherichia coli* transformation have increased the efficiency of the system even more, allowing greater than 10^6 independent cDNA clones to be recovered from a single 1 pmol-scale reaction containing 2–5 μg of mRNA.

The protocols described here are based primarily on the Okayama–Berg method. They are intended for workers with minimal experience in recombinant DNA methods; it is assumed that careful laboratory technique will be practiced. The cloning example described here was chosen because of continuity of data and because its publication does not interfere with results to be published elsewhere by myself or collaborators; it is by no means the best result obtained with the system.

Principle of the Method

The dimer–primer method allows vector-primed cDNA synthesis from both ends of a T-tailed plasmid vector. Because the starting plasmid approximates a head-to-tail dimer, both cDNAs are attached to a functional plasmid bearing a selectable marker gene. As represented in Fig. 1, cDNA synthesis results in DNA copies of two mRNA molecules, each covalently attached to one end of the dimer–primer. These DNA strands then receive homopolymer G-tails, followed by digestion of the dimer at the "hinge" with *Bam*HI to create two independent cDNA–mRNA–vector complexes. These complexes have a *Bam*HI sticky end at one end, and a G-tail at the other. A linker piece, having a *Bam*HI sticky end at one end and a C-tail at the other, is added. The G-tracts and C-tracts anneal at 42°, forming linear complexes. The mixture is diluted and cooled to a temperature at which the four-base *Bam*HI sticky ends can anneal—conditions which highly favor self-closure—then ligation is performed to covalently close the complexes. Repair is completed by nicking the RNA–DNA duplex with RNase H and replacing the RNA with DNA

[2] D. C. Alexander, T. D. McKnight, and B. G. Williams, *Gene* **31,** 79 (1984).

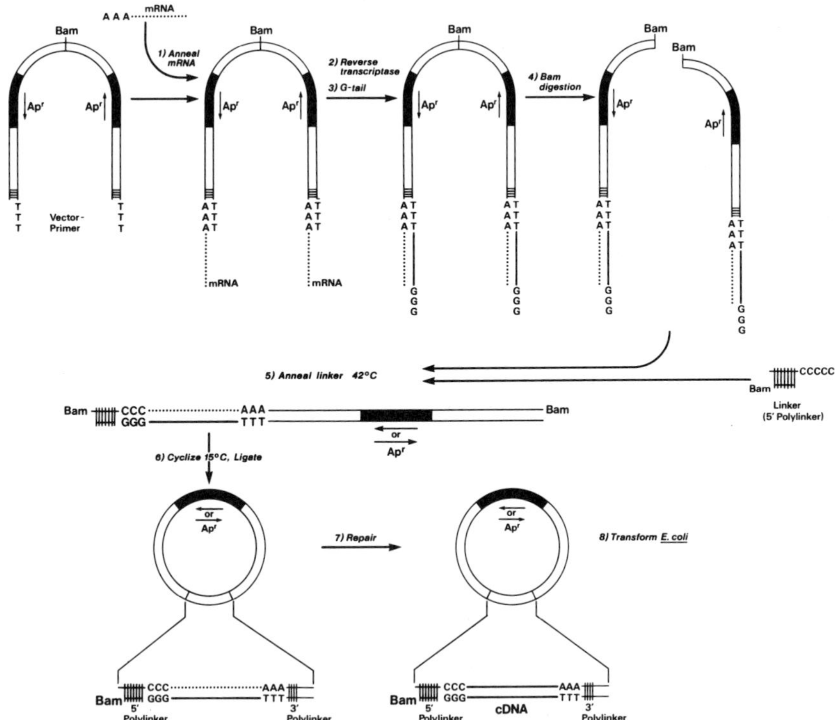

FIG. 1. Schematic of the dimer–primer cDNA cloning method. The term "3′ polylinker" refers to the multiple restriction site segment that is adjacent to the A–T tracts, or the end of the cDNA representing the 3′ sequence of the mRNA; "5′ polylinker" refers to the multiple restriction site segment adjacent to the G–C tract, or the end of the cDNA representing the 5′ sequence of the mRNA.

using DNA polymerase I. Nicks are closed with T4 DNA ligase, and the double-stranded closed circular molecules are transferred into an *Escherichia coli* host. Transformants are selected by virtue of the amplicillin-resistance gene borne by the plasmid vector.

The primer ends of the dimer–primer have been engineered to contain multiple restriction sites directly adjacent to the T-tail. Likewise the linker piece, which is attached to the opposite end of the cDNA insert, contains these same restriction sites, as well as additional sites. Both of these polylinkers are derived from the λCharon 34 polylinker.[2,3] The final

[3] W. A. M. Loenen and F. R. Blattner, *Gene* **26,** 171 (1983).

cDNA constructs thus contain a seven-site polylinker followed by a G–C tract, then the end of the cDNA representing the 5′ sequences of the mRNA; at the opposite end of the cDNA is an A–T tract followed by a four-site polylinker. The order of these components is always the same, but half the clones will contain the vector molecule in one orientation and half in the other.

Materials

Strains

Escherichia coli K12, strain HB101 and strain MM294, were obtained from the American Type Culture Collection. Strains TB-1 and DH5α were purchased from Bethesda Research Laboratories.[4]

Reagents

Oligo(dA)–cellulose was obtained from Collaborative Research. RNasin (placental ribonuclease inhibitor) was from Promega Biotec. Restriction endonuclease *Sst*I and cloned Maloney Murine Leukemia Virus (M-MLV) reverse transcriptase were from Bethesda Research Laboratories; all other enzymes were from Boehringer Mannheim Biochemicals. Radioactive deoxynucleotides were purchased from New England Nuclear or Amersham. Cesium chloride was from KBI.

Media

Analytical plating is done on 15 × 100 mm plates containing L agar (10 g Bactotryptone, 5 g yeast extract, 5 g NaCl, 17 g Bactoagar per liter) and 50 μg/ml ampicillin. Preparative plating is done on 15 × 150 mm plates with the same medium except that the agar concentration is increased to 20 g/liter. The resulting stronger substrate is advantageous when scraping the preparative plates (see below).

2×L broth is prepared by dissolving 20 g Bactotryptone, 10 g yeast extract, and 1 g of NaCl in 1 liter of water, then adjusting the pH to 7 with 1 *M* NaOH (if necessary). After autoclaving (20 min), 10 ml of 20% glucose is added per liter of broth. Preparation of M9 medium is as described by Messing,[5] except that it is supplemented with casamino acids to 0.4% (w/v).

[4] Bethesda Research Laboratories, *Focus* **6,** 4 and 7 (1984).

[5] J. Messing, this series, Vol. 101, p. 20.

General Methods

Plasmid Preparations

The plasmids pARC5 and pARC7, harbored in HB101, are isolated by a modification of the method described by Katz *et al.*,[6] including two bandings in CsCl gradients. Plasmids are extracted extensively in CsCl-saturated isoamyl alcohol or water-saturated *n*-butanol in order to remove ethidum bromide. Alkaline lysis methods are avoided because they were found to cause appreciable contamination by small DNA fragments, which interfere with terminal deoxynucleotidyltransferase reactions in later steps.

Large-Scale Preparation of Competent Cells

Glassware and plasticware are rinsed extensively with distilled water to remove all traces of detergents. TB-1 cells are innoculated into 25 ml of 2×L broth and grown overnight at 30°. Each of four 2.8-liter Fernbach flasks, containing 500 ml 2×L broth prewarmed to 30°, are inoculated with 5 ml of the overnight growth. The flasks are shaken at 150–200 rpm, 30°, until the A_{600} reaches 0.45–0.55. The cells are chilled in ice water for 2 hr and then centrifuged in two 1-liter polypropylene bottles at 2500 *g* (maximum) for 15–20 min, 4°. Cells in each bottle are triturated with a pipet in a small volume (10–20 ml) of ice-cold 100 m*M* $CaCl_2$, 70 m*M* $MnCl_2$, 40 m*M* NaOAc, pH 5.5 (made fresh from the three salts and filter sterilized), and then diluted to 500 ml with the same solution. After 45 min incubation in ice water the cells are centrifuged at 1800 *g* (maximum) for 10 min and then very gently suspended in 50 ml (each bottle) of the ice-cold solution. The cells are pooled, and then 80% glycerol is added drop-wise with gently swirling to give a final concentration of 15% (v/v). Cells are aliquoted into 1.5 ml microfuge tubes on ice, most into aliquots of 1 ml, some into aliquots of 0.2 ml, and placed directly into −80° for storage.

Preparation of Vector–Primer and Linker

Schematic representations of the primer and linker preparation are shown in Figs. 2 and 3, respectively.

SstI Digests

Preparative digests of pARC5 and pARC7 are done under identical conditions. The reaction mixtures contain 10 m*M* Tris–HCl (pH 7.5), 10

[6] L. Katz, D. Kingsbury, and D. R. Helinski, *J. Bacteriol.* **114,** 577 (1973).

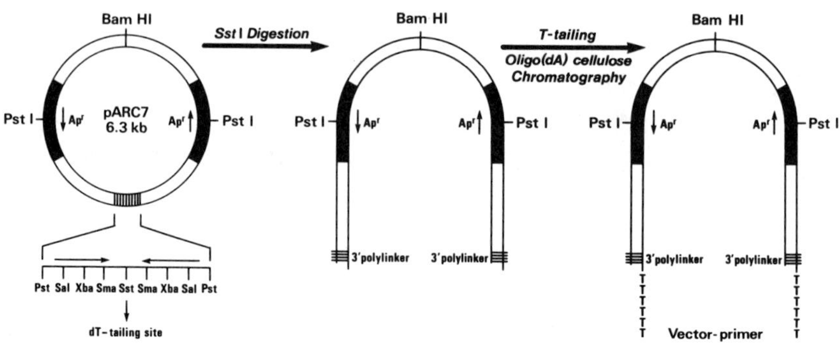

FIG. 2. Schematic of the dimer–primer preparation.

mM $MgCl_2$, 90 mM NaCl, 250 μg/ml plasmid DNA, and 1,000 units/ml *Sst*I (care is taken to keep the glycerol concentration below 5%). Following incubation at 37° for 2 hr the mixture is extracted with 1 volume of phenol : $CHCl_3$ (1 : 1) and then with 1 volume of $CHCl_3$. The aqueous

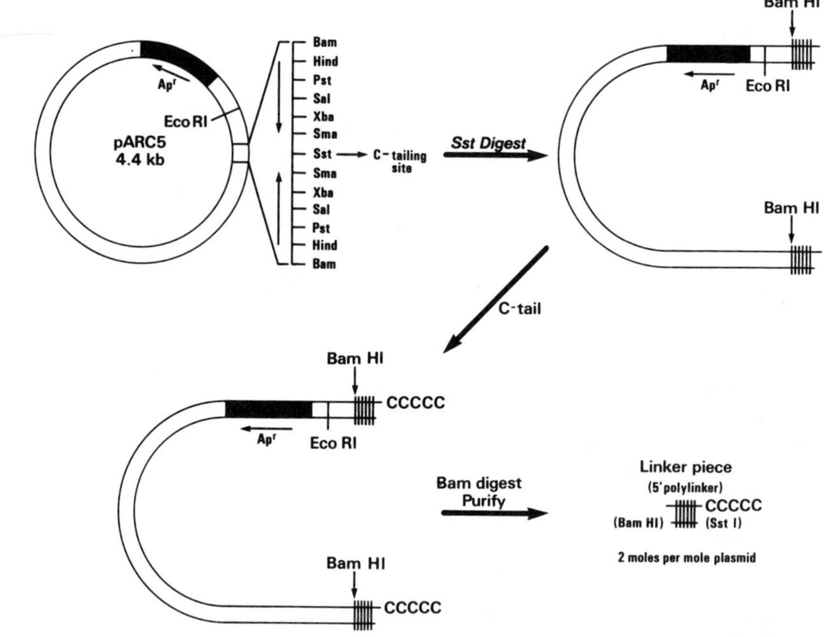

FIG. 3. Schematic of the linker preparation. Note that each mole of pARC5 yields 2 mol of the linker piece.

phase is made 2 *M* in NH_4OAc and precipitated with 2.5 volumes of ethanol. Following centrifugation at 12,000 *g* (maximum) for 10 min the DNA is redissolved in TE (10 m*M* Tris–HCl, pH 8, 1 m*M* EDTA), and the absorbance at 260 nm is determined. When large batches of plasmid are digested and purified, aliquots of 200 μg are precipitated with ethanol as above and stored at −20°.

T-Tailing of Vector–Primer

A 200-μg aliquot of *Sst*I-digested pARC7 is recovered by centrifugation, dried briefly under reduced pressure to remove excess ethanol, and the reaction mixture is assembled by the addition of

1. 114 μl of water to slowly dissolve the DNA
2. 10 μl of 20 m*M* $CoCl_2$
3. 20 μl of 2.5 m*M* deoxythymidine triphosphate (dTTP)
4. 40 μl of 5× CT buffer (0.7 *M* sodium cacodylate, 0.15 *M* Tris–HCl, pH 6.8)
5. 1 μl of [α-^{32}P]dTTP (10 μCi), then the mix is prewarmed to 37°
6. 15 μl of terminal deoxynucleotidyltransferase (TdT, 16 units/μl)

This is a final concentration of approximately 1200 units/ml of the enzyme and approximately 500 pmol of vector ends per ml. The mixture is incubated at 37° for 3.5–4.0 min, and the reaction is stopped by the addition of 2 volumes of TES (10 m*M* Tris–HCl, pH 7.2, 10 m*M* EDTA, 0.2% SDS).

DNA is recovered as follows: The mix (0.6 ml) is extracted twice with 1 volume of phenol : $CHCl_3$ (1 : 1). The two portions of resulting organic phase are sequentially back-extracted with a single 0.3-ml portion of TE, and then this TE is pooled with the main aqueous phase. The aqueous phase is extracted once with 1 volume of $CHCl_3$; 4 *M* NaOAc is added to a concentration of 0.2 *M*, and then 2.5 volumes of ethanol is added. Following centrifugation in a microfuge for 15 min the pellet is washed with 80% ethanol, drained well, redissolved in 200 μl of TE, made 2 *M* in NH_4OAc, and precipitated with 2.5 volumes of ethanol.

Oligo(dA)–Cellulose Chromatography

In order to remove undigested and/or untailed plasmid the ^{32}P-labeled vector is subjected to oligo(dA)–cellulose chromatography. Two hundred milligrams (dry weight) of oligo(dA)–cellulose is used to prepare the column. A 10-ml Econocolumn (Bio-Rad) is convenient for setting up the apparatus with a peristaltic pump in such a way that the effluent can be cycled back onto the column bed. In a cold room (4°) the matrix is washed with 5 ml portions of column buffer (10 m*M* Tris–HCl, pH 7.5, 1 m*M*

EDTA, 1 *M* NaCl) and then water. It is then equilibrated with 10–15 ml of column buffer.

The DNA is dissolved in 200 μl of TE and then made 1 *M* in NaCl by the addition of 50 μl of 5 *M* NaCl. The volume of the sample is kept to a minimum to give a high concentration of reacting ends and to allow maximum time of contact with the matrix. The sample is applied to the column and then recirculated slowly through the bed for 1–2 hr. The column is washed with 15–20 ml column buffer and then brought out of the cold and allowed to warm to room temperature. The bound DNA is eluted with warm (55°) water, and 0.4-ml fractions are collected. The peak fractions are determined by counting 5 μl of each fraction in a scintillation counter. Peak fractions are pooled, made 1 *M* in NaCl, and reapplied to the washed and equilibrated column. (If the volume is excessive at this point the DNA can be concentrated by ethanol precipitation. Generally, if the volume is under 2 ml it can be reapplied directly.) The procedures for washing and sample recovery are repeated and the final pool is precipitated with ethanol. Following centrifugation (15 min in a microfuge) the sample is dissolved in 200 μl of TE, and a portion is used to determine the absorbance at 260 nm. This portion is recovered and pooled with the bulk of the sample, then the vector–primer is aliquoted into 0.5 ml microfuge tubes (usually 2 μg each), made 2 *M* in NH_4OAc, precipitated with ethanol, and stored at −80°.

C-Tailing of Linker Plasmid (see Addendum)

Fifty picomoles (100 pmol ends, 147 μg) of *Sst*I-digested pARC5 (in ethanol) is centrifuged and dried briefly to remove excess ethanol. The reaction is assembled by adding

1. 65 μl water to slowly dissolve the DNA
2. 5 μl 20 m*M* $CoCl_2$
3. 2 μl 10 m*M* dCTP
4. 20 μl 5× CT buffer
5. 1 μl [α-^{32}P]dCTP (10 μCi)
6. 7 μl terminal deoxynucleotidyltransferase (16 units/μl)

The mix is incubated at 22° for 5.0 min. The reaction is stopped by the addition of 2 volumes of TES, and the DNA is extracted and precipitated as described above for the T-tailed vector, with the volumes decreased proportionately.

The resulting DNA is digested for 2 hr at 37° with 500 units of *Bam*HI in a volume of 350 μl containing 6 m*M* Tris–HCl (pH 7.9), 6 m*M* $MgCl_2$, 150 m*M* NaCl. The mix is then made 0.2 *M* in sodium by the addition of NaOAc and precipitated with ethanol.

Gel-Purification of Linker Piece

The *Bam*HI-digested DNA is dissolved in 100 μl of TE, 25 μl of a 5× dye mixture added, and the sample is subjected to native polyacrylamide gel electrophoresis. The 14 × 17 × 0.15 cm gel contains 12% acrylamide : bisacrylamide (20 : 1) and TBE (89 m*M* Tris, 89 m*M* boric acid, 2 m*M* EDTA) and is prerun at 150 V for 1 hr. The sample is loaded into a single 3 cm well and electrophoresed at 200 V until the bromophenol blue dye is near the bottom. The gel is stained briefly in ethidium bromide (1 μg/ml), and the smear just above the 50-bp range is excised and recovered by electroelution (see Fig. 4C, below). Following reduction of the sample volume by evaporation and ethanol precipitation, the linker is further purified using a NACS tip (BRL), or NENSORB 20 column (NEN), according to the manufacturer's specifications. The final sample is dissolved in 100 μl of TE and stored tightly sealed at 4°.

Determination of Homopolymer Tail Lengths

It is recommended that the tail lengths be determined before the final purifications of vector [on oligo(dA)–cellulose] and linker (on polyacrylamide gels). In the protocols described above the homopolymer tails are labeled to a low specific activity by the incorporation of radioactive deoxynucleotides, and thus can be detected directly by autoradiography following gel electrophoresis. If direct comparison to untailed plasmid is desired, the *Sst*I-digested plasmids are end-labeled by the incorporation of radioactive dideoxyadenosine triphosphate or cordycepin according to the supplier's protocol. A portion of each labeled plasmid is digested with *Pst*I to release the homopolymer tails from the plasmid (note *Pst*I site in polylinker) and then subjected to electrophoresis on an 8% polyacrylamide–urea gel such as described by Sanger and Coulson[7] for DNA sequencing. Bands are detected by autoradiography using an intensifying screen. The cut site for *Pst*I is 34 base pairs from the cut site (i.e., tail addition site) of *Sst*I, so that the absolute size of the fragment should be 34 plus the length of the homopolymer tail.

cDNA Cloning

mRNA

Polyadenylated RNAs from many sources, purified by several methods, have been used for cloning in this system. Potential problems result-

[7] F. Sanger and A. R. Coulson, *FEBS Lett.* **87,** 107 (1978).

ing from impurities in the mRNA purification are covered in more detail in the Results and Discussion. Generally, the mRNA is purified by two passes on an affinity matrix such as oligo(dT)–cellulose, and stored in small aliquots as an ethanol precipitate at −80°. Storage of RNA frozen in solution for more than a few days, even at −80°, is avoided. mRNA is never stored lyophilized at any temperature.

First Strand Synthesis

The first strand synthesis, as in most of the reactions in the cDNA cloning protocol, is done in a 0.5 ml microfuge tube in order to facilitate the separation of phases during the organic extractions. The standard reaction is given for 1 pmol of vector ends (2 μg of T-tailed pARC7).

The conditions given here are for synthesis using M-MLV reverse transcriptase and are based on the supplier's recommendations. Conditions for synthesis with AMV reverse transcriptase are given in a previous publication.[2] In neither case have they been determined to be optimal conditions for reverse transcription of a particular mRNA in this system. The mRNA (2–5 μg) pellet is dried briefly under reduced pressure to remove ethanol, as is the T-tailed vector–primer pellet (2 μg in a 0.5 ml microfuge tube). The RNA pellet is dissolved in 22 μl of water, and the vector is dissolved in 5 μl of water. The RNA is transferred to the vector tube and the mixed nucleic acids are heated to 65° for two min. After briefly cooling the mix on ice, the following additions are made:

1. 2 μl RNasin
2. 10 μl of 5× buffer salts (250 m*M* Tris–HCl, pH 8.3, 375 m*M* KCl, 15 m*M* $MgCl_2$, 50 m*M* dithiothreitol)
3. 2.5 μl dNTP mix (10 m*M* in each dNTP)
4. 1 μl of 5 mg/ml bovine serum albumin (nuclease free)
5. 2.5 μl of 1 mg/ml actinomycin D
6. 2 μl [α-^{32}P]dCTP (20 μCi)
7. 3 μl M-MLV reverse transcriptase (600 units)

The mix is incubated for 30–60 min at 37°, then an additional 1 μl of M-MLV reverse transcriptase is added and the incubation continued for 15–30 min.

After the synthesis the DNA is purified as follows: the mix is made 20 m*M* in EDTA by the addition of 2 μl of a 0.5 *M* stock solution; 1 volume (50 μl) of phenol : $CHCl_3$ (1 : 1) is added, and the mix is vortexed vigorously. The aqueous phase is removed, and the organic phase is back-extracted with 50 μl of TE, then aqueous phases are pooled, and the mix is extracted with 100 μl of $CHCl_3$. The aqueous phase removed to a 1.5-ml microfuge tube; 100 μl of 4 *M* NH_4OAc is added followed by 0.5 ml of ethanol. After 15 min on dry ice the mixture is warmed to room temperature with gentle shaking[1] to dissolve unincorporated nucleotides which would interfere with the G-tailing reaction. The mixture is centrifuged in the microfuge for 15 min, and the ethanol carefully removed. Most of the unincorporated radioactivity is in the ethanol phase at this point. The pellet is washed with 80% ethanol and then dissolved in 20 μl of TE. Twenty microliters of 4 *M* NH_4OAc is added, followed by 100 μl of ethanol. The DNA is cooled on dry ice, warmed with shaking as above, centrifuged, and washed. The ethanol supernatant at this point generally contains less than half the remaining radioactivity, as estimated with a Geiger counter, and the counts in the DNA pellet represent almost all incorporated DNA. The DNA is dissolved in 10 μl of water for the next step.

Simple Evaluation of First Strand Synthesis

Before proceeding, the first strand synthesis may be evaluated by observing a portion of the products on a simple agarose gel. One microliter of the first strand products (equivalent to 0.2 μg of vector) is removed and digested for 1–2 hr with *Bam*HI in a 10-μl reaction. The remainder of the DNA may be ethanol precipitated or used directly. The digest is electrophoresed on a 1% agarose minigel containing TAE (40 m*M* Tris–acetate, pH 7.5, 2 m*M* EDTA). Following staining with ethidium bromide (1 μg/ml) the cDNA synthesis can be observed as a smear beginning at 3.1 kb (see Results and Discussion).

Addition of G-Tails

The G-tailing reaction is done in a 20 μl final volume. To the remaining 9 μl (or pelleted DNA redissolved in 9 μl) of first strand products is added

1. 1 μl of 0.2 mg/ml poly(rA)
2. 1 μl of 20 m*M* $CoCl_2$
3. 1 μl of 20 m*M* dGTP
4. 4 μl of 5× CT buffer
5. 3 μl of water
6. 1 μl of terminal deoxynucleotidyltransferase (10 units)

The mix is incubated at 37° for 15 min, diluted to a volume of 50 μl with TE, and then the DNA is purified as described above for the first strand synthesis products, except that only a single ethanol precipitation is performed.

BamHI Digestion

To the ethanol pellet from the G-tailing reaction is added

1. 20.5 μl of water to dissolve the DNA
2. 2.5 μl of 10× *Bam*HI salts (60 m*M* Tris–HCl, pH 7.9, 60 m*M* $MgCl_2$, 1.5 *M* NaCl)
3. 2 μl of *Bam*HI (15–20 units)

The mixture is incubated for 1–2 hr at 37°, and then diluted to 50 μl with TE and purified as described above, again with a single ethanol precipitation. The final pellet is dissolved in 10 μl of TE and stored tightly capped at 4°. For long-term storage, the sample is stored as an ethanol precipitate.

Cyclization and Repair

For analytical purposes portions containing 10% of the preparation are cyclized and repaired with linker versus without linker. The proper amount of the linker preparation is determined empirically as described in the next section.

To 1 μl of the *Bam*-digested cDNA–vector is added

1. 3 μl of water
2. 1 μl of the linker dilution (or 1 μl additional water for the no-linker control)
3. 5 μl of 2× cyclization salts (20 m*M* Tris–HCl, pH 7.5, 2 m*M* EDTA, 200 m*M* NaCl)

The mixture is heated to 65° for 5 min, briefly centrifuged to recover any condensation at the top of the tube, and incubated at 42° for 30–60 min followed by the addition of 90 μl of ligation mixture. The ligation mixture (e.g., for 10 reactions of this size, or one complete library) is made by combining the following (on ice):

1. 100 μl 10× T4 ligase salts (660 m*M* Tris–HCl, pH 7.6, 50 m*M* $MgCl_2$)
2. 50 μl 0.1 *M* DTT
3. 100 μl of 1 mg/ml bovine serum albumin
4. 20 μl of 50 m*M* ATP
5. 100 μl of 10 m*M* spermidine trihydrochloride
6. Water to a final volume of 900 μl

The 100 μl reaction is cooled on ice, and 1 μl (1–3 units) of T4 DNA ligase is added, followed by incubation at 12–15° overnight.

The following morning the mix is supplemented with

1. 2 μl dNTP mixture (2.5 mM each dNTP)
2. 2 μl 50 mM ATP
3. 1 μl additional T4 ligase
4. 0.3–1 μl RNase H (0.3–1 unit)
5. 0.5–1 μl DNA polymerase I (2.5–5 units)

For convenience these components can be premixed on ice just before addition to the reactions. The supplemented reactions are incubated for 1 hr at 14° and then for another hour at room temperature, then diluted 10-fold with cold TE and stored on ice. For longer term storage the mixtures can be frozen, but the efficiency may drop by 10–20%.

Titration of Linker Preparation

Since there is no simple way to measure the concentration of the isolated linker, the optimum amount for cyclization is determined empirically. The first time a linker preparation is used a duplicate cDNA reaction (or a reaction with a standard mRNA) is done, since most of the 1 pmol reaction is used for this determination.

Aliquots of the cDNA–vector preparation are annealed as described above using a 2-fold dilution series of the linker preparation. The maximum possible concentration of the linker is 1 pmol/μl (assuming 100% recovery), and for convenience the most concentrated annealing reaction in the series contains 1 μl of the linker (a theoretical 10-fold excess). Dilutions are made for 0.5, 0.25, 0.125, 0.06, and 0 μl of linker per reaction. The reactions are completed as described above, and cloning efficiencies are determined as described in the following sections. The resulting optimum linker concentration is used for preparative platings as long as a cDNA reaction contains roughly the same amount of mRNA as was used in the test reaction (see Results and Discussion).

Transformation and Analytical Plating

A 200-μl aliquot of competent TB-1 cells is thawed on ice and used immediately for transformation. One hundred microliters of the diluted repaired cDNA is added, and the mix is incubated in an ice water mixture for 30 min. The tube is mixed by gentle inversion and placed in a 37° water bath for 5 min. The entire sample is gently drawn into a 1 ml Pipetman and placed into 3.7 ml of 2×L broth that has been prewarmed to 37°. The cells are incubated for 100 min at 37° with shaking. Aliquots are then plated by spreading on L plates containing 50 μg/ml ampicillin. In practice a 100-μl aliquot of the 4 ml outgrowth usually gives an easily countable density. Larger samples may be plated by briefly centrifuging the cells and resuspending in a smaller volume of broth. Good preparations of vector and linker normally yield plating efficiencies 10- to 50-fold higher than background (sample ligated in the absence of linker). A 1 ng or 10 ng sample of pBR322 in 100 μl of TE is used to transform 200 μl of cells as a control for the transformation procedure.

Preparative Plating

Using the analytical plating results as a guide, the remaining repaired cDNAs, or a portion thereof, is plated on 150-mm plates at a density of 2.5–5 $\times$ 10^4 colonies per plate. For the preparative transformation 0.5-ml aliquots of the repaired cDNAs are added to 1-ml aliquots of thawed competent TB-1 cells in a 1.5-ml microfuge tube. The mix is incubated on ice as before and the heat shock at 37° is extended to 6 min. The cells are then pooled into an appropriate volume (i.e., proportional to the analytical plating) of 2×L broth prewarmed to 37° in an Erlenmeyer or Fernbach flask. It is essential to have a large air-to-liquid ratio for the outgrowth, which is done with shaking at 200 rpm for 100 min. The cells are then chilled on ice and centrifuged at 3500 *g* (maximum) for 5 min. Cells are resuspended in a volume of cold L broth that will yield the desired density when spread on large plates. Each plate will easily accept 200 μl, and drier plates will accept more.

After overnight incubation at 37° the colonies are scraped from the plates using a 10–15 ml aliquot of L broth and a glass spreader. A turntable is helpful for this task. It is also useful to scrape multiple plates with the same aliquot of broth by transferring it to a second or third plate before collection. After all the cells are collected the mixture is triturated or vortexed extensively to insure that no clumps of cells remain. Multiple freezer stocks are made (1–2 ml aliquots in 10% DMSO, −80°), and the cell titer and A_{600} are determined.

In order to preserve the library as a collection of plasmids, standard plasmid preparations are done in supplemented M9 medium (2–4 liters) as described in General Methods, except that the medium is inoculated such that $A_{600} = 0.1$. This somewhat dense inoculum is intended to minimize the number of doublings before chloramphenicol amplification, thus decreasing the chances of altered clone distribution in the population. The plasmid library is stored at 4° in TE containing 10 m*M* NaCl.

Results and Discussion

Vector and Linker Preparations

The preparative digestion of 1 mg of pARC7 yielded 824 μg of linear plasmid. A 200-μg aliquot was used for the addition of homopolymer tails of thymidine residues. The tail lengths were determined as described in the previous sections. The average T-tail length in this preparation was 65, with the bulk of the molecules falling in the 55–75 range (Fig. 4A). The bands should appear as a tight group around the average, not widely and evenly distributed. Wide distributions can result from low concentrations of nucleotide or enzyme. (The multiple bands observed in the untailed plasmid lane are an artifact of the [^{32}P]dideoxyadenosine triphosphate

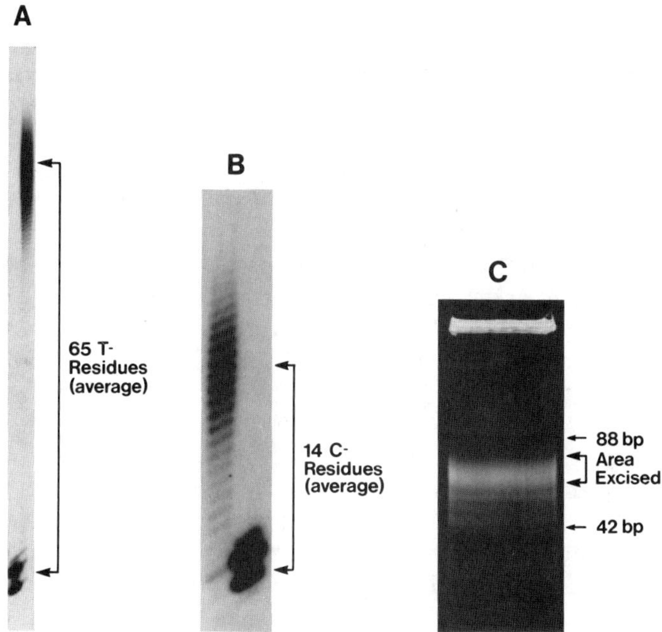

FIG. 4. Homopolymer tail analysis and preparative gel of C-tailed linker. A and B are autoradiograms of 8% polyacrylamide–urea gels. (A) Thymidine homopolymer tract released from T-tailed pARC7 by *Pst*I digestion. In the adjacent lane is a sample of untailed pARC7 that has been end-labeled by the addition of [α-^{32}P]dideoxyadenosine triphosphate and then digested with *Pst*I. (The multiple band pattern is an artifact of the labeling procedure.) (B) Cytidine homopolymer tract released from C-tailed pARC5 by *Pst*I digestion. In the adjacent lane is an end-labeled sample of pARC5 digested with *Pst*I. (C) 12% polyacrylamide preparative gel (TBE) with *Bam*HI-digested C-tailed pARC5. The 1.5-mm-thick gel was stained with ethidium bromide.

reaction.) The labeling of the untailed plasmid is somewhat expensive, and it is not absolutely necessary if a method is available to determine the size of the bands on the gel. Since the *Pst*I site is 34 bases from the first added thymidine, the tail lengths can be determined. I have successfully used preparations with average tail lengths as short as 45 T's and as long as 85 T's, with no obvious differences in efficiency.

After two passes on an oligo(dA)–cellulose column, 91 μg of T-tailed vector primer remained. Yields usually range from 40–60% of the linear plasmid starting material. If the tailed vector is not affinity purified on oligo(dA)–cellulose, the background in the transformations is unacceptably high (50% or greater), probably due to the presence of undigested or untailed plasmid. One pass lowers the background to only around 20%, while two passes results in backgrounds of less than 10% (usually less than 5%).

The homopolymer cytidine tails in this batch of pARC5 averaged about 14 residues (Fig. 4B), with the bulk of the sample in the 10–20 residue range. I have noticed somewhat better cloning efficiencies with linker having shorter C-tails (8–10 residues), although well-controlled comparisons have not been done. Again the absolute length, when digested with *Pst*I, is 34 bases plus the tails. If the sample is digested with *Bam*HI before the gel analysis, instead of *Pst*I, the length will be 53 plus the tails.

Following *Bam*HI digestion and preparative gel electrophoresis, a smear was evident in the area above the 50-bp region (Fig. 4C). This area was excised and recovered by electroelution. An additional purification was done to remove residual acrylamide. NACS tips or NENSORB 20 columns have been used for this and similar purifications with good results.

In some batches of pARC5 we have noticed difficulty in obtaining complete *Sst*I digests. It is not the result of a heterogeneous population of plasmids, since linearization of the plasmid with another enzyme renders the site susceptible to *Sst*I. We assume it is a secondary structure effect, possibly caused by residual ethidium bromide. This has not been further investigated at this time.

The preparation of the vector–primer and linker pieces is the most laborious part of this method. However, successful preparations yield enough material for constructing 40–50 cDNA libraries.

cDNA Synthesis

Once the vector–primer and linker pieces have been prepared and checked, cDNA cloning can be a routine and simple procedure. Several mRNA samples may be cloned at once, and a library can be made and amplified for storage by a single worker in 6–7 working days.

The synthesis is done with a relatively small amount of radioactive nucleotide. The only purpose is to provide a tracer that may be followed with a Geiger counter during subsequent steps, since the ethanol pellets are often too small to see. It is also a helpful indicator of success in the first strand synthesis; a poor reaction will yield almost no counts after two ethanol precipitations in the DNA workup.

The mRNA used in this experiment was from tomato leaves and was purified as described previously.[2] It is important that the mRNA be free of impurities that might interfere with the reverse transcription reaction. The mRNA preparations have been by far the greatest source of variability in cDNA cloning with this system. Pilot reactions using oligo(dT) primer may be performed to test the mRNA with reverse transcriptase. (Alternatively, if one wishes to risk 2 μg of the vector–primer, the first strand reaction may be performed as described here, analyzed by the simple observation on an agarose gel, and either used or discarded.) Most animal tissues do not present problems with such impurities, but plant tissues often have polysaccharides which copurify with the mRNA on oligo(dT)–cellulose. Activity in an *in vitro* translation system is not a sure indication that an mRNA preparation will reverse transcribe, although a lack of *in vitro* translational activity probably will result in no reverse transcription. Oligosaccharides may be removed by centrifugation in a CsCl step gradient[8]; it is possible to perform this operation using the poly(A) fraction, but it is usually done with the total RNA fraction.

The products of the first strand synthesis are shown in Fig. 5. Ten percent of the products was digested with *Bam*HI and electrophoresed. Lanes 1 and 2 were from reactions with 2 μg of vector–primer and 5 μg of mRNA, except that reverse transcriptase was omitted from the reaction in lane 2. The smear in lane 1, beginning at the monomer vector size (3.1 kb) is the result of cDNA synthesis, making the annealed message double stranded. The presence of annealed message alone does not affect the mobility of the vector plasmid (lane 2). Smearing is clearly visible up to the 4–5 kb range or higher, indicating large numbers of cDNAs of 1–2 kb or larger. Some very active mRNA preparations leave very little material at the 3.1 kb vector size. A poor mRNA preparation often results in most of the vector remaining at 3.1 kb, possibly with a minimal smear above.

[8] T. Maniatis, E. F. Fritsch, and J. Sambrook, "Molecular Cloning: A Laboratory Manual." Cold Spring Harbor Lab., Cold Spring Harbor, New York, 1982.

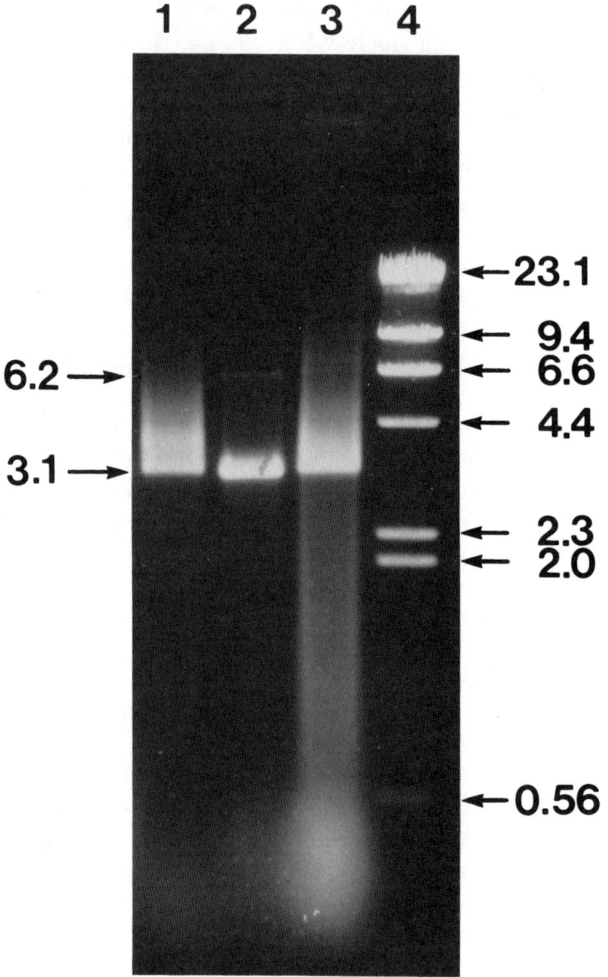

FIG. 5. Agarose gel of cDNA–mRNA–vector complexes. Lane 1 contains 0.2 μg of vector–primer and 0.5 μg of mRNA following first strand synthesis and *Bam*HI digestion. Lane 2 is the same as lane 1 except that the reverse transcriptase is eliminated. Lane 3 is the same as lane 1 except that there is 2.5 μg of mRNA rather than 0.5 μg. Lane 4 contains size standards (kilobases). The 1% agarose minigel (TAE) was stained with ethidium bromide.

Lane 3 (Fig. 5) was from a reaction with a large excess of mRNA (25 μg). The cDNA synthesis on the vector is not greatly different in this reaction, but note the large smear below 3.1 kb, representing aberrant initiation of the nonannealed RNA. Thus no advantage is gained by using a very large excess of mRNA, and problems result later when the linker is added; the unattached cDNAs become G-tailed and can anneal to the C-tailed linker in the cyclization step, lowering the cloning efficiency greatly. For this reason it is necessary to use similar amounts of mRNA when determining the best linker dilution (see below) and when doing the final cloning. In the absence of reverse transcriptase this large amount of RNA is seen only at the bottom of the gel and not in a smear (not shown).

More detailed analysis of cDNAs is certainly possible.[2] However, the simple agarose gel analysis gives a very good and reliable estimation of the average insert length for a library; the attached vector allows a staining of the cDNAs that approaches a number average.

The G-tailing reaction is usually done "blind" because it is quite reliable if the terminal transferase enzyme has been used recently and is known to be active. It is possible to include radioactive dGTP in the reaction to monitor success; the counts in the pellet will increase greatly, but few quantitative conclusions may be drawn, since some incorporation occurs as tailing on excess RNA. The time of reaction does not appear to be critical; all of our cDNAs sequenced to present have G–C tracts in the 13–18 bp range, even when some of the G-tailing reaction is extended up to 30 min.

The *Bam*HI digestion may be monitored from the large reaction rather than an analytical digestion after the first strand reaction, but in the case of a poor reaction an extra day has been spent. The preparative reaction is often not monitored (if the analytical reaction is successful); the enzyme is in fair excess, and it is unlikely that inhibitors of the digestion will be introduced in the G-tailing reaction. It is not known if it is possible to digest RNA–DNA hybrids with *Bam*HI; several clones have been sequenced which contain an intact *Bam*HI site within the cDNA insert.

Cyclization, Repair, and Plating

It is important to note that the annealing of the G–C tracts at 42°, followed by dilution for ligation at 12–15°, results in almost all of the cDNAs being single inserts into single plasmids. This strategy is also responsible for the directionality of the insert, which would be especially important in expression plasmids of this type. Strategies which use four-base sticky ends on both ends, even if they are different, require other approaches to cut down on concatenation and/or self-closure.

Serial dilutions of the linker preparation were ligated with aliquots of the cDNA–vector molecules to determine the optimal amount of linker for the library construction. Following the standard repair and transformation, 100-μl portions of the 4-ml outgrowths were plated on L plates containing ampicillin. The results are shown in Table I. The number of

TABLE I
EMPIRICAL DETERMINATION OF OPTIMAL LINKER CONCENTRATION FOR cDNA CYCLIZATION[a]

Volume of linker preparation (μl)[b]	Colonies per plate (av.)[c]	Transformants/pmol of vector ends
0	3	13,000
0.06	50	220,000
0.125	108	480,000
0.25	138	610,000
0.5	210	930,000
1.0	190	840,000

[a] Each reaction was performed with approximately 0.1 pmol of cDNA–vector complexes (10% of standard reaction) and increasing amounts of linker preparation. Following ligation and repair, 10% of this mix (1% of standard reaction) was used to transform *E. coli*.

[b] The preparation consists of that C-tailed linker remaining, after purification, of 100 pmol starting material, dissolved in 100 μl.

[c] Average of triplicate plates, each receiving 100 μl of the 4-ml outgrowth.

transformants increases with increased linker up to about 0.5 μl of linker preparation per reaction. Most linker preparations give an optimum in the 0.25–0.5 μl range, indicating either a fairly low recovery of tailed linker or the presence of G-tailed contaminants in the cDNA reaction (or both).

The result with the optimum linker concentration, compared to the reaction without linker, shows a 66-fold increase over background (i.e., background is 1.5% of signal). This represents a "library size" of 8.4×10^5, or an efficiency of 9.3×10^5 transformants per picomole of vector ends (note that 10% of the sample was sacrificed for the evaluation of first strand synthesis). Many mRNA preparations, using the methods described here, have often yielded efficiencies of greater than 4×10^6 transformants/pmol; occasionally preparations have yielded up to 2×10^7 transformants/pmol. Our initial description of the dimer–primer approach[2] reported efficiencies on the order of 2×10^5 transformants per pmol of vector. Three changes have been made since that publication which may be responsible for the increased efficiency, although controlled tests have not yet been done to determine which change, or combination of changes, is responsible for the effect. They are (1) use of M-MLV reverse transcriptase instead of the AMV enzyme, (2) use of T4 DNA ligase instead of *E. coli* DNA ligase, and (3) additional purification of the linker piece (using NACS or NENSORB 20). We also previously reported using MM294 as the recipient strain, but we have used TB-1 and DH-5α in the protocol described here with similar or slightly improved results.

The transformation of the cells used in the example described in this chapter with 1 ng of supercoiled pBR322 yielded 1.2×10^7 transformants/μg. This is somewhat below optimum, and one would normally perform preparative plating with freshly made cells of high efficiency. The three cell lines mentioned above, made competent by the protocol described in General Methods, normally yield 5×10^7 to 2×10^8 transformants/μg when fresh. They keep their top efficiency for a few weeks and then tend to decline slowly. They are quite useful, for routine transformations that do not require the highest efficiency, for at least a year.

A library size of 5×10^5 transformants was chosen for amplification and storage. This represents 60% of the total cDNAs. The cyclization and repair reaction was scaled up 6-fold and the resultant 6 ml of diluted repaired cDNA mix was transformed into 12 ml of competent cells (in 12 tubes). Following the heat shock the cells were pooled into 225 ml of 2×L broth in a 2-liter Erlenmeyer flask for the outgrowth. The cells were centrifuged and resuspended in 4 ml of L broth, then 200-μl aliquots were spread on each of 20 large plates (aiming for 25,000 colonies/plate). Analytical dilutions were also done to confirm the plating density. The following morning the cells were scraped from the plates with L broth. The final volume was approximately 100 ml, and the A_{600} was 39. Multiple freezer stocks were prepared, and another portion was used to inoculate 2 liters of supplemented M9 medium for plasmid preparation. The remaining volume was made 10% in DMSO and frozen in bulk. Plating of serial dilutions revealed a titer of 5×10^{11} ampicillin-resistant transformants per milliliter of the fresh cells. This titer decreased only 10–20% after thawing of frozen stocks.

Comments

Vector-primed cDNA cloning has several advantages over other cDNA cloning methods. These include (1) much higher efficiency than the older methods of cloning double-stranded cDNA into a plasmid; (2) high recovery of full-length cDNAs[1]; (3) it is simple, fast, and reliable once a supply of vector and linker have been prepared and tested; and (4) analysis and recovery of the inserts is much easier than in phage cDNA cloning methods, which it now rivals in efficiency. One apparent disadvantage is that there is no easy size selection of cDNAs prior to cloning; sizing can be done, but the length of the attached vector makes it more difficult to resolve the small differences in cDNA lengths. However, the high efficiency and the large percentage of clones with relatively long inserts compensates somewhat for this. The plasmids isolated from the amplified library can be used in several types of size-selection strategies. The selected fraction is then transferred back into *E. coli* for screening.

There is nothing special about the particular plasmid vector described here, compared to analogous constructs with pUC plasmids or other similar plasmids. Dimer–primer vectors can easily be constructed from any pair of plasmids which contain polylinkers in opposite orientations. Vector-primed plasmid systems for the expression of cDNAs in specific hosts have been described.[9–11] Some things to consider when making analogous vectors are listed below.

1. The cut restriction site used for homopolymer addition must have a protruding 3′ end.
2. Cutting of the restriction site at the "hinge" (*Bam*HI in this case) should be tested following the appropriate steps in the protocol. The Okayama–Berg method uses *Hin*dIII at this site. *Eco*RI has also been used at a similar point in the cloning, using a slightly modified approach with RI-methylase and linkers,[12] although we have experienced difficulty obtaining complete digestion of a dimer–primer with *Eco*RI at the hinge position.
3. If expression involving the vector body is desired, orientation of the constructs must be taken into account. In the method described here the cDNAs are always in the same orientation with respect to the linker, but not the vector body.
4. It is convenient to have common restriction sites at both ends of the cDNAs, as well as some unique sites at either end. This allows for easy insert analysis and subcloning. The pARC5–pARC7 system has four sites common to both ends of the insert (*Sma*I, *Pst*I, *Xba*I, and *Sal*I). The 5′ end of the cDNAs (i.e., 5′ sequence of mRNA) also contains *Bam*HI, *Hin*dIII, and *Sst*I sites, but there are no unique sites at the 3′ end. *Pst*I and *Hin*dIII sites also occur in the vector, as does *Eco*RI.

A recent innovation has taken advantage of the unique *Sst*I site directly adjacent to the G–C tract in cDNAs made by this method; immunological screening is performed by expressing these cDNAs as fusion pro-

[9] H. Okayama and P. Berg, *Mol. Cell. Biol.* **3,** 280 (1983).

[10] A. Miyajima, N. Nakayama, I. Miyajima, N. Arai, H. Okayama, and K. Arai, *Nucleic Acids Res.* **12,** 6397 (1984).

[11] G. L. McKnight and B. L. McConaughy, *Proc. Natl. Acad. Sci. U.S.A.* **80,** 4412 (1983).

[12] J. Hunsperger, personal communication.

teins in a λ phage.[13] The plasmid library may be linearized by *Sst*I digestion and subcloned directly into the unique *Sst*I site of the β-galactosidase gene in λCharon 16. This allows antibody screening directly analogous to the method of Young and Davis,[14] and, importantly, the cloned cDNA is easily and quickly recovered from the phage as an intact plasmid, by virtue of its ampicillin-resistance marker.

Addendum

Since the first preparation of this chapter, several improvements have been made in our system which may be of general interest to anyone wishing to clone with vector-primer cDNA systems. It has become clear that high efficiency cloning may be done with a variety of vectors, and making a vector to meet one's specific needs is relatively simple. Some general "rules" that apply include the following:

1. Monomer vectors work just as well as dimer vectors *if they are treated like dimer vectors*. The advantages of eliminating the gel-purification step and its preceding digestion (i.e., the *Hpa*I digestion and agarose gel-purification of the Okayama and Berg method[1]) are preserved if both T-tails are left on the prepared vector; the many-fold increase in vector activity more than offsets the 2-fold decrease in efficiency per microgram of mRNA. In our new vectors the *Bam*HI "hinge" site is near or adjacent to the "extra" T-tail; cDNAs are made on both ends, but after the *Bam*HI digestion the extra cDNA merely cyclizes with the linker to form a minicircle that does not transform the host. The advantage to monomer vectors over the dimer is that functionalities of the vector may be used, such as eucaryotic promotors, T7 or SP6 promoters to make RNA probes, or M13 intergenic regions to allow expression of the cDNA as single stranded DNA; also, pUC-based parents give the advantage of high copy number and a minimum of restriction sites in the vector body. The creative use of synthetic polylinkers, the synthesis of which is now available to most labs, makes possible a variety of designs to suit specific needs.

2. It is desirable to remove the *lac* promoter from these vectors. Expression of the cDNA inserts in host bacteria proved to be somewhat toxic to the host, resulting in highly variable colony morphologies and growth rates; removal to the *lac* promoter completely solved this problem. The problem was especially acute when the promoter caused transcription into the poly(A) : poly(T) tract at the 3′ end of the cDNA insert, and plating the cDNAs on a *lacI*^{q} host did not significantly help.

[13] A. L. Genez, D. C. Alexander, J. M. Rejda, V. M. Williamson, and B. G. Williams, submitted.

[14] R. A. Young and R. W. Davis, *Proc. Natl. Acad. Sci. U.S.A.* **80,** 1194 (1983).

3. The cyclizing linker used here, or analogous linkers, work very well when made synthetically. A single synthesis yields on the order of 0.2–1 μmol of product, a huge amount considering that only picomole amounts are used to plate a complete cDNA library. The following points should be considered for this approach:

a. *N,N*-diisopropylcyanoethylphosphoramidite chemistry must be used to synthesize the linker strands. This chemistry yielded highly active linker, whereas synthesis with the *O*-methylphosphoramidite chemistry yielded totally inactive linker, a result reported previously by Ureda *et al.*[15]

b. Using 10 C-residues works as well as using 20, so we chose 10 residues to minimize the length of the G : C tract in the cDNA.

c. Neither strand of the linker is phosphorylated; this resulted in much better efficiencies, presumably by preventing linker–linker ligation.

d. Only the strand bearing the C-tail need be purified (usually by polyacrylamide gel electrophoresis). Variability in length due to synthesis failure occurs at the 5′ end, the end forming the sticky end used for ligation to the vector; the other strand may vary at the 5′ end since the repair reaction will fill in any gaps.

e. The linker need be only long enough to withstand the 65° step in the cyclization reaction, but may be much longer. It is possible to incorporate promoters or primer sequences into this segment.

Details of the modifications described here will be published elsewhere.

[15] M. S. Urdea, L. Ku, T. Horn, Y. G. Lee, and B. D. Warner, *Nucleic Acids Res. Symp.* **16,** 257 (1985).

[4] Use of Primer–Restriction End Adapters in cDNA Cloning

By CHRISTOPHER COLECLOUGH

Extensive fractionation of cDNA is often desirable prior to cloning so that copies of mRNA species of particular interest will be more frequently represented in resultant clone banks. Efficient fractionation schemes typically yield of the order of nanograms or tens of nanograms of enriched cDNA. Most existing methods for insertion of cDNA into cloning vectors demand that it be first rendered double-stranded and then be provided with some means of attachment to vector molecules, e.g., restriction enzyme ends or single-stranded terminal homopolymers. Such manipulation of minute quantities of cDNA may result in its total loss, as carrier free single-stranded DNA at low concentration tends to bind tightly to surfaces suffering further loss during partition at solvent interfaces. With a view to alleviating this difficulty, F. Erlitz and I developed a scheme whereby single-stranded cDNA can be directly inserted into double-stranded cloning vehicles, without the necessity of additional manipulation.[1] The method depends on the use of synthetic oligonucleotides called primer–restriction end adapters, or PRE-adapters, which can serve both as primers for polymerase reactions and as ligation substrates.

Figure 1 shows a schematic diagram of the use of a set of PRE-adapters. Note that the only biochemical reaction performed with the cDNA, before its ligation to vector molecules and after its synthesis, which is primed by a PRE-adapter, is its extension by a short homopolymer tract using terminal deoxynucleotidyltransferase. This is done shortly after synthesis, prior to any fractionation, so that significant quantities of cDNA (usually micrograms or hundreds of nanograms) are used and fractional losses are minimal. Vector molecules are so prepared as to have a short terminal homopolymer tail, complementary to that on the 3′ end of the cDNA, and a staggered restriction enzyme end capable of ligation to the PRE-adapter used to prime cDNA synthesis. Because the termini are different, the insertion of cDNA molecules into vectors always takes place in a predictable orientation. The method of insertion involves a strong selection for both termini of the original cDNA molecules, so that the size of inserts is limited chiefly by the length of the primary reverse transcripts, favoring the production of "full-length" clones. We have described the development of the technique elsewhere[1] and demonstrated its efficiency in producing full-length cDNA clones. Here I will discuss the detail of the technical steps and describe some variations.

Methods

Sources

Reverse transcriptase was from Life Sciences Inc.; terminal deoxynucleotidyltransferase from Pharmacia-P.L.; human placental RNase inhibitor (RNasin) from Promega-Biotec; calf intestinal phosphatase from Boehringer Mannheim. T4 DNA ligase used in the experiments shown

[1] C. Coleclough and F. Erlitz, *Gene* **34**, 305 (1985).

A

5' pGATCCACGCGTTTTTTTTTTTTT 3'

3' GTGCGCAAAAAAAA 5'

B

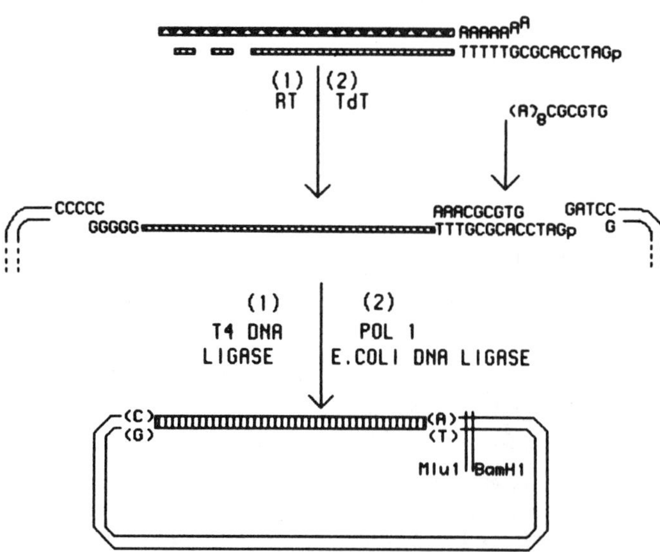

FIG. 1. Use of PRE-adapters in cDNA cloning. (A) Pair of PRE-adapters designed for attaching cDNA copied from poly(A)-containing mRNA, to *Bam*HI-digested vector molecules. The adapters incorporate a *Mlu*I site (ACGCGT). (B) The 3′-oligo(dT)-containing adapter hybridizes to the poly(A) tail of mRNA and can prime cDNA synthesis by reverse transcriptase (RT). cDNA molecules are lengthened at the 3′ end by a short homopolymer tail [shown here as oligo(dG)] using terminal transferase (TdT). cDNA can now be fractionated as desired and can be inserted directly into cloning vectors. Recipient vector molecules have a short 3′ single-stranded homopolymer [here, oligo(dC)] complementary to the cDNA tail and a *Bam*HI end. cDNA is annealed with prepared vector and the complementary adapter, and chimeric molecules are formed by ligating the *Bam*HI terminus of the vector to the *Bam*HI end formed at the 5′ terminus of the cDNA using T4 DNA ligase. Finally the second strand of the cDNA is synthesized by *E. coli* polymerase I (POL 1) as a gap-filling reaction, and recombinant molecules sealed with *E. coli* DNA ligase.

here was either from New England Biolabs or Anglian Biotechnology. DNA polymerase I, T4 polynucleotide kinase, *Escherichia coli* DNA ligase, λ exonuclease, and restriction enzymes were obtained from New England Biolabs.

Actinomycin D was from Calbiochem; methylmercury hydroxide from Alfa Products; unlabeled nucleotides from Pharmacia-P.L.; ^{32}P-labeled nucleotides from Amersham. Ampicillin and normal (grade III) or low melting temperature (grade VII) agarose were obtained from Sigma; hydroxylapatite (HAP, DNA grade) was from Bio-Rad. The plastic resin base used for preparation of a chromatography matrix similar to RPC-5 (Voltalef 300 micro P.-L.) was obtained from Ugine Kuhlmann. Adogen 464 was a gift from Ashland.

Design of PRE-Adapters

We use the PRE-adapters shown in Fig. 1 for most cDNA cloning in plasmids. They are designed for ligation to a *Bam*HI end and incorporate a *Mlu*I restriction site, but clearly a great variety of sequences could be used. In any case, the sequence of the adapters should provide for the exact construction of the desired restriction end on annealing the two partners. As this is important in ensuring efficient ligation, highly redundant sequences which offer the possibility of misalignment should therefore be avoided. Short sequence motifs which would impart some phenotype to recombinants could be incorporated. PRE-adapters can be designed for more specialized purposes. Figure 2 shows a set which we have used to clone cDNA copies of mouse T-cell receptor β chain[2] mRNA molecules. Note that in this case a restriction enzyme end, *Kpn*I, with a 3′ overhang was used: this allowed the primer used to initiate reverse transcription to be 100% complementary to mRNA. These oligonucleotides were synthesized by a solid phase phosphoamidite method.

The PRE-adapter molecules which should provide 5′ phosphate groups for ligation to restriction ends on vectors must be, as far as possible, fully 5′ phosphorylated by T4 polynucleotide kinase. Incubate adapters at 1–5 μM with 50–100 U/ml of T4 polynucleotide kinase in 1 mM ATP, 50 mM Tris–HCl (pH 7.6), 10 mM $MgCl_2$, 5 mM DTT, 1 mM spermidine, and 1 mM EDTA for 3 hr at 37°. Efficient phosphorylation can be checked by incubating a small amount of oligonucleotide with a known molar excess (say 5-fold) of ATP to which some fresh [γ-^{32}P]ATP has been added, in the above conditions, then sieving through a small Sephadex G-50 superfine column to separate oligonucleotides from free

[2] S. M. Hedrick, E. A. Nielsen, J. Kavaler, D. I. Cohen, and M. M. Davis, *Nature (London)* **308,** 153 (1984).

```
Tβ mRNA      5' ...TGTACTCCACCCAAGGTCTCCTTGTTTGAGCCATCA... 3'

PRE-adapters { (i)  3' GAGGTGGGTTCCAGAGGAACAAACTCp 5'
               (ii)      5' CCCAAGGTCTCCTTGTTTGAGGTAC 3'

                                  CCCAAGGTCTCCTTGTTTGAGGTAC       C=========
cDNA         3' ...ACATGAGGTGGGTTCCAGAGGAACAAACTCp→ CATGG========= Vector
```

FIG. 2. PRE-adapters which can be used to clone only mouse T-cell receptor β cDNA molecules. A section of the mRNA sequence of the mouse T_β constant region[2] is shown above. Adapter (i) is complementary to part of this mRNA and thus can hybridize to it and prime the specific reverse transcription of this mRNA. For cloning the T_β-specific cDNA produced in this way, the cDNA is first tailed with a homopolymer stretch at its 3′ end, as shown in Fig. 1. Tailed cDNA can then be annealed with adapter (ii) together with a plasmid vector prepared so as to have a 3′ homopolymer terminus complementary to that of the cDNA and a *Kpn*I terminus. Adapter (ii) is complementary to adapter (i) except for four nucleotides at the 3′ terminus, which, when the two adapters are paired, form a *Kpn*I staggered end. T4 DNA ligase can be used to join vector and cDNA molecules at their *Kpn*I ends and to produce recombinant plasmids in a manner similar to that shown in Fig. 1.

ATP. The fraction of ATP incorporated into polymer should approach the stoichiometric ratio. This small amount of labeled oligonucleotide can be concentrated by butanol extraction and added as a tracer to the preparative kinase mix before phenol–chloroform extraction and similar recovery through G-50 superfine in 1 m*M* EDTA. Phosphorylated primer can be stored at −20°. It is not necessary to phosphorylate that member of the PRE-adapter set whose 3′ terminus forms the restriction end.

cDNA Synthesis

Titrating the 3′-oligo(dT)-containing PRE-adapter shown in Fig. 1 against poly(A)-containing mRNA in reverse transcription assays shows saturation at a mass ratio of about 1 : 40 primer : mRNA, roughly a 1.5-fold molar excess of primer. Addition of much more primer does not increase the cDNA yield, wastes primer, and might cause problems in subsequent steps. A typical cDNA synthesis reaction was set up as follows: mix 25 μl of 1 mg/ml mRNA in water with 25 μl of freshly diluted 20 m*M* methylmercury hydroxide in water; hold at room temperature for about 3 min. Add 40 μl of 15 μg/ml phosphorylated oligo(dT)-containing PRE-adapter, followed by 25 μl 40 m*M* 2-mercaptoethanol. Chill to 0°. Add 25 μl 10× RT buffer (0.7 *M* KCl, 0.5 *M* Tris–HCl, pH 8.78, at 25°, 0.1 *M* $MgCl_2$, 0.3 *M* 2-mercaptoethanol); 35 μl of a solution 10 m*M* each in dATP, dCTP, dGTP, and dTTP; 10 μl [α-^{32}P]dCTP and 10 μl [α-^{32}P]dATP (both 410 Ci/mmol, 10 mCi/ml in aqueous solution); 15 μl 1 mg/ml actinomycin D; 10 μl RNase inhibitor (40,000 U/ml); 1 μl T4 polynucleotide kinase (200 U/ml); 15 μl of water; and 13 μl reverse transcriptase (25,000 U/ml). Incubate at 16° for 2 min, 43° for 20 min, 48° for 5 min; then terminate the reaction by adding 10 μl 20% SDS, 10 μl 0.5 *M* EDTA, and 1 μl 10 mg/ml proteinase K. Incubate at 43° for 20 min, then extract twice with phenol–chloroform. Recover polynucleotides by sieving through a small Sephadex G-50 medium column in 50 m*M* NaCl. Concentrate the excluded material to 100 μl by extraction with butanol or isobutanol and precipitate nucleic acids by addition of 200 μl ethanol and chilling. Collect the precipitate by centrifugation and resuspend it in 50 μl 0.2 *M* NaOH, 1 m*M* EDTA; hold at 60° for 20 min then cool and neutralize by adding 10 μl of 1 *M* Tris (pH 8) and 2 μl of 5 *M* HCl.

Methyl mercury is a strong, reversible denaturant of nucleic acids, and preincubation of mRNA in its presence helps increase the proportion of complete reverse transcripts.[3] It probably also inactivates trace amounts of nuclease bound to the RNA. Methyl mercury must then be sequestered in a mercaptoethanol complex to prevent its inactivating reverse transcriptase. High concentrations of deoxynucleoside triphosphates promote complete reverse transcription,[4] and radioactive tracers help follow the cDNA through subsequent steps and allow easy estimation of the cDNA yield. Actinomycin D prevents extensive hairpin formation at cDNA termini. RNase inhibitor provides bulk protein and may help reduce nuclease action. T4 polynucleotide kinase, which can use deoxynucleoside triphosphates as well as ATP as substrates, promotes maintenance of the crucial 5′ phosphate group on the PRE-adapter. The brief incubation at 16° is meant to allow transcriptase to initiate and copy mRNA through the 3′-poly(A) stretch and into heteropolymeric sequence, avoiding slippage and recopying of poly(A), which can produce excessively long poly(dT) stretches at higher temperatures. Proteinase treatment and phenol–chloroform extraction remove protein which might enhance the stickiness of cDNA.

cDNA yield is easily estimated from the molar incorportion of deoxynucleotide, but for fairly large-scale reactions like the one above it is useful to check it spectrophotometrically as well, because the fractional incorporation of ^{32}P-labeled deoxynucleotides may underestimate the yield, especially if the label is not fresh. The mass yield of cDNA is frequently as high as 65% of the input mRNA, and should always be at

[3] F. Payvar and R. T. Schimke, *J. Biol. Chem.* **254,** 7637 (1979).

[4] A. Efstratiadis, T. Maniatis, F. Kafatos, A. Jeffrey, and J. N. Vournakis, *Cell* **4,** 367 (1975).

least 20%. Variations in yield are more likely to be due to overestimates of mRNA purity than to inefficient copying.

cDNA could now be tailed with terminal transferase directly, but it is preferable to remove any unincorporated primer before tailing. If tailed primer is carried through to the final ligation to vector, pseudo-recombinants lacking cDNA inserts will be common. Low molecular weight primer could be removed from cDNA by gel filtration, but we have found it more convenient to do this by brief agarose gel electrophoresis. cDNA in quantities greater than 1 μg can be loaded immediately after alkali treatment, without neutralization, onto a 0.7% gel formed from low melting temperature agarose in 25 m*M* NaOH, 2 m*M* EDTA,[5] briefly separated by electrophoresis, and molecules longer than about 300 nucleotides (nt) recovered in a small cube of agarose. Sufficient water and NaCl solution is added so as to produce a mixture 0.5 *M* in NaCl, 12.5 m*M* NaOH, 1 m*M* EDTA, and the agarose is melted at 65°. The resultant solution is forced under positive pressure through a small (20–50 μl) bed of an RPC-5 analog which was previously washed with 0.5 *M* NaCl, 12.5 m*M* NaOH, 1 m*M* EDTA. The matrix is then washed with about 1 ml of 0.5 *M* NaCl; cDNA is eluted with a minimal volume (about 30 μl) of 2 *M* NaCl and desalted through Sephadex G-50 medium or by ethanol precipitation. We have prepared a matrix similar to RPC-5 by coating plastic particles (Voltalef 300 micro P.-L., from Ugine Kuhlmann) with Adogen 464 according to method C of Pearson *et al.*[6] Probably commercially available analogs of RPC-5 could be used in the same way (see Ref. 7).

Loading at salt concentrations less than 0.5 *M* can result in increased losses of cDNA due to irreversible binding to the matrix. Such losses are also a problem with smaller quantities of cDNA. In this case, we recover cDNA from a neutral 1% agarose gel formed in 40 m*M* Tris–acetate buffer lacking EDTA, by electrophoresis into a narrow trough containing a minimal volume of HAP, as described by Tabak and Flavell.[8] HAP should be boiled in 0.5 *M* pH 7 sodium phosphate buffer (PB) and stored in the same buffer after decanting fines. Wash the HAP several times in water, followed by equilibration with electrophoresis buffer, before use. After the cDNA has been run into the HAP remove it and wash it in 40 m*M* Tris–acetate buffer several times by centrifugation and resuspension. Elute the cDNA with two or three small (about 20 μl) aliquots of 0.15 *M* PB at 60°

[5] M. W. McDonnell, M. N. Simon, and F. W. Studier, *J. Mol. Biol.* **110,** 119 (1977).

[6] R. L. Pearson, J. F. Weiss, and A. D. Kelmers, *Biochim. Biophys. Acta* **228,** 770 (1971).

[7] J. A. Thompson, R. W. Blakesley, K. Doran, C. J. Hough, and R. D. Wells, this series, Vol. 100, p. 368.

[8] H. F. Tabak and R. A. Flavell, *Nucleic Acids Res.* **5,** 2321 (1978).

by centrifugation and collect the supernatant. The pooled material is then desalted using a Sephadex G-50 column. Recovery of cDNA is very efficient, especially if the HAP has been stored in suspension for some time. This method is also useful if a specific narrow size range of cDNA (e.g., that containing complete copies of some mRNA of interest) is desired. We prefer chromatographic methods of recovering cDNA, rather than removing molten agarose by phenol extraction, because cDNA losses into the phenol phase, though variable, may approach 100%.

When a sequence-specific PRE-adapter like that shown in Fig. 2 is used, one expects a very small mass yield of cDNA. In this case electrophoretic separation and problems due to unincorporated primer can be avoided by hybridizing mRNA with an estimated 5- to 10-fold molar excess of specific primer over the particular mRNA species of interest, in 0.5 *M* NaCl, 10 m*M* Tris–HCl (pH 8), 1 m*M* EDTA, 0.1% SDS at an appropriate temperature (50° for the primer shown in Fig. 2; T_m of this hybrid in this buffer is about 62°), to a C_0t of about 10^{-2} with respect to the primer; i.e., if the concentration of a 20-mer primer is 5 μg/ml, incubate for 10 min. Then pass the mixture through a small oligo(dT)–cellulose column, wash the column with 0.5 *M* NaCl, 10 m*M* Tris, 1 m*M* EDTA to remove excess primer, and elute mRNA in 10 m*M* Tris, 1 m*M* EDTA. Most primer–mRNA hybrids of about 15 base pairs or more should be stable in this buffer at room temperature. Add NaCl to 0.2 *M* and ethanol precipitate. Dissolve the dried ethanol pellet in 10 m*M* NaCl and set up the reverse transcription reaction as above, omitting the methyl mercury treatment and the 16° incubation. Tail the cDNA product directly with terminal transferase, without electrophoretic separation.

cDNA at 60 μg/ml or less is extended with short tails of oligo(dC) or oligo(dG) by incubation in 100 m*M* potassium cacodylate, 25 m*M* Tris base, 1 m*M* $CoCl_2$, 0.2 m*M* DTT (pH ~6.9) (see Ref. 9), containing 50 μ*M* dCTP or dGTP and 600 U/ml terminal transferase, for 3 min at 16°. To cDNA dissolved in water, add dCTP or dGTP, then 10× reaction buffer and hold at 37° for 3 min, 0° for 3 min, and 16° for 3 min before adding the enzyme. Terminate the reaction after exactly 3 min by adding EDTA to 5 m*M* and SDS to 0.5%; extract with phenol–chloroform and remove free nucleotides by filtering through Sephadex G-50 medium.

Preparation of Vector

The only structural prerequisite of vectors is that they should have two different restriction enzyme sites, at least one of which should pro-

[9] R. Roychoudhury and R. Wu, this series, Vol. 65, p. 43.

duce staggered ends, in a nonessential region of the genome. We have used small plasmids of the pUC series[10] and the related pSP series,[11] or derivatives, as cDNA cloning vectors. Recipient vector molecules must have a 3′ terminal single-stranded homopolymer tract and a staggered restriction end, both compatible with the cDNA termini. The preparation of vectors to receive cDNA inserts then has two steps: the first to produce the terminal homopolymer; the second, a restriction enzyme digest. We have investigated three techniques of producing single-stranded homopolymer termini: terminal transferase extension, λ exonuclease treatment, and the use of PRE-adapters with T4 DNA ligase.

Extension of Vector Molecules by Terminal Transferase Treatment. Use of calf thymus terminal deoxynucleotidyltransferase has recently been reviewed in detail by Deng and Wu.[12] In using the enzyme to add dC or dG tracts to cDNA and vector molecules, one is chiefly concerned with optimizing three variables: the efficiency of utilization of primer molecules, which should be close to 100%; the average length of the added homopolymer; and the dispersion of tail lengths around that mean, which should be minimal. The 3′-dC and -dG tails on the vector and cDNA molecules must be long enough to produce a specific and stable interaction. However, dC–dG tracts longer than about 25 nt can cause problems in the utilization of cDNA clones (see below). Although tail lengths as short as 4–5 nt could be used for cDNA insertion, this would involve the loss of the selection for maintenance of the original 3′ ends of cDNA molecules in recombinant clones. The interaction of the 3′ terminal dC- and dG-tails of vector and cDNA molecules in this case would be no more probable than that of the vector tail with an internal oligo(dC) or oligo(dG) stretch in the cDNA. The optimal tail length is probably 8–10 nt.

The three relevant variables can be most conveniently monitored for model reactions by using synthetic oligonucleotides, labeled at the 5′ end with ^{32}P-phosphate, as described recently.[12] Using such assays, we settled on the conditions described here, which are similar to those used by Land *et al.*[13] We have extended *Pst*I-cut linear DNA of the plasmid pUC9 or pSP64 with dG residues using conditions similar to those described above for cDNA tailing: linear plasmid at 200 μg/ml (about 200 n*M* 3′ termini, for these plasmids) is incubated with 600 U/ml terminal transferase (about 200 n*M* enzyme) in 100 m*M* potassium cacodylate, 25 m*M* Tris base, 1

[10] J. Messing, this series, Vol. 101, p. 20.

[11] D. A. Melton, P. A. Krieg, M. R. Rebagliati, T. Maniatis, K. Zinn, and M. R. Green, *Nucleic Acids Res.* **12,** 7035 (1984).

[12] G.-R. Deng and R. Wu, this series, Vol. 100, p. 96.

[13] H. Land, M. Grez, H. Hauser, W. Lindenmaier, and G. Schutz, *Nucleic Acids Res.* **9,** 2251 (1981).

mM $CoCl_2$, 0.2 mM DTT (pH ~6.9), and 50 μM dGTP for 3 min at 16°, and the reaction is terminated by addition of EDTA and phenol–chloroform extraction, as above.

After recovery by ethanol precipitation, tailed plasmid is digested with *Bam*HI under standard conditions to generate the *Bam*HI termini necessary for ligation to PRE-adapters. *Bam*HI termini can be dephosphorylated at the same time simply by adding calf intestinal phosphatase[14] to the reaction mix. Recipient vector molecules can finally be separated from the short DNA fragments released by *Bam*HI digestion by electrophoresis through low melting temperature agarose. For oligo(dC)-tailed plasmids this preparative electrophoresis is performed in a neutral pH buffer. dG-tailed plasmid molecules, however, may associate with each other at neutral pH to produce dimers and trimers.[15] Preparative electrophoresis of oligo(dG)-tailed plasmids is therefore performed in alkali, as above. DNA is recovered from low melting temperature agarose by the addition of NaCl to 0.5 M, neutralizing with HCl and Tris buffer when necessary, and holding at 65° for 20 min to melt the agarose and allow the DNA to reanneal. Molten agarose is removed by extracting 4 times with an equal volume of phenol (90%, saturated with 0.5 M NaCl, 1 mM EDTA), and duplex DNA recovered from the aqueous phase by ethanol precipitation.

An aliquot of the reaction mixture was treated with [γ-^{32}P]ATP and polynucleotide kinase. This labeled material was then subjected to electrophoresis on a 12% polyacrylamide–urea sequencing gel in order to monitor the addition of unlabeled dG residues. Figure 3 shows such assays and also demonstrates some of the vagaries of terminal transferase tailing. The two pSP64 samples were prepared, as far as possible, identically. One (lanes 1 and 2), has an average dG tail of about 10 nt with most molecules having tails within the range of 8–13 nt, very few longer: this preparation proved excellent for cloning purposes. The second sample (lanes 3 and 4) shows a major population of extended molecules with tails varying from about 11 to greater than 26 nt long, on average around 15. Two less abundant discrete populations can be seen, one with tails apparently about 70 nt long and the other apparently with even longer tails, of over 100 nt. As these other two sets are roughly twice and three times the apparent molecular weight of the major fraction, it is conceivable that they are dimers and timers of the latter, incompletely denatured by urea, held together by dG–dG interactions, which are known to be strong.

We had switched from plasmids of the pUC series to those of the pSP series for use as routine cDNA cloning vehicles because the latter, having

[14] G. Chaconas and J. Van de Sande, this series, Vol. 65, p. 75.

[15] A. Dugaiczyk, D. L. Robberson, and A. Ullrich, *Biochemistry* **19,** 5869 (1980).

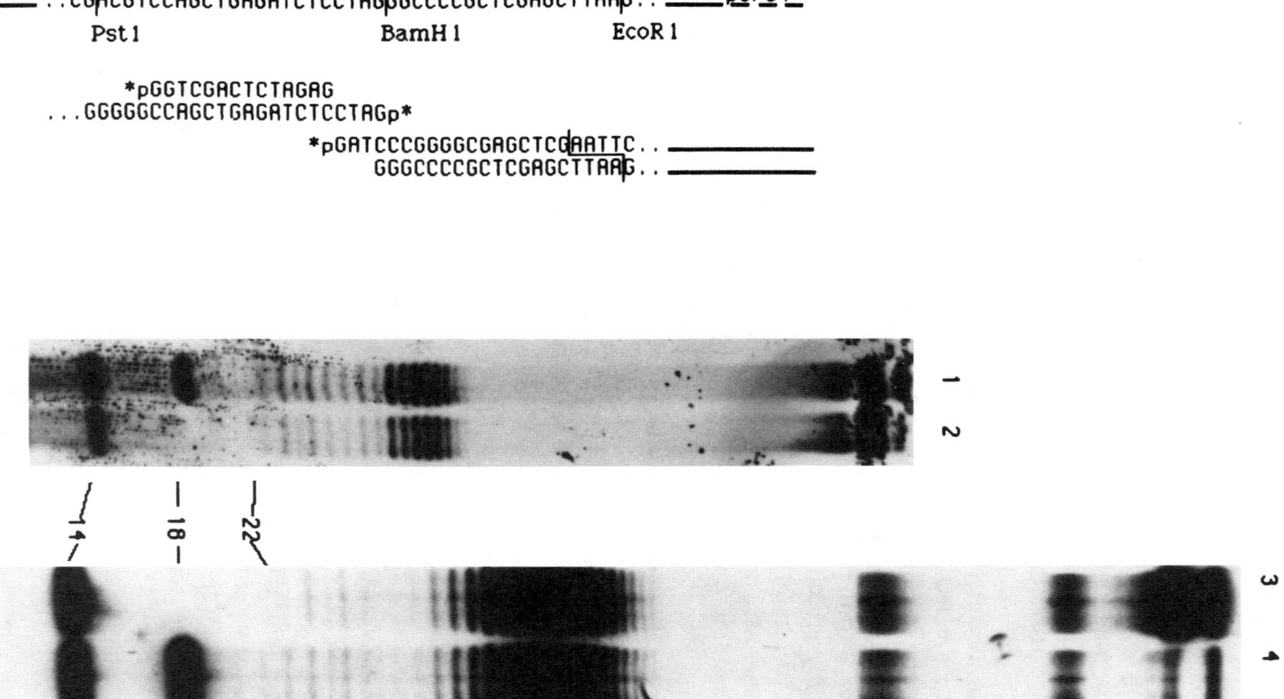

FIG. 3. 3′-dG tailing plasmids using terminal transferase. Plasmids like pSP64, shown here, can be prepared to receive cDNA by linearizing with *Pst*I, adding a short homopolymer tail with terminal transferase, and then digesting with *Bam*HI. The length of the added tail can be monitored by 5′ end labeling of the *Bam*HI digestion products. As shown, this would produce a 22-nt fragment from nontailed molecules, so that the tail length of terminal transferase-treated molecules can be determined simply by counting up from 22. A 14-nt discrete fragment is released from the complementary DNA stand. Digestion with *Eco*RI additionally releases a discrete labeled 18-nt fragment from the other terminus of the plasmid and can be used to provide an internal size standard. Lanes 1 and 2, and 3 and 4 show such an analysis of two separate oligo(dG)-tailed preparations of pSP64. Fragments were displayed on 12% polyacrylamide–urea sequencing gels. DNA in lanes 1 and 4 was additionally digested with *Eco*RI. Fragments of 14, 18 and 22 residues are indicated.

promoters for *Salmonella* phage SP6 RNA polymerase immediately adjacent to the cDNA insertion site, allow easy *in vitro* production and translation of model mRNA molecules.[11,16] Although the dG-tailed pSP64 vector preparation analyzed in Fig. 3, lanes 3 and 4, was effective in producing cDNA clones, these proved very difficult to exploit in this way. On analysis of individual clones, it was found that many had dG–dC homopolymer stretches in excess of 30 nt long. We found that such clones are poor templates for SP6 polymerase, at least when the dG stretch is on the sense strand. Moreover, these long dC–dG stretches were difficult to sequence through, because of hybridization during sequencing gel electrophoresis, and when they are close to the ends of restriction fragments they can cause aberrant mobility on native agarose gels. A tailing method which would reliably produce only clones with short dC–dG tracts is clearly desirable.

It is difficult to control the terminal transferase reaction precisely. There is no obvious alternative which could efficiently produce short, precise 3′ homopolymer tails on single-stranded cDNA molecules (though some scheme involving T4 RNA ligase might conceivably be possible; see Ref. 17). However, we considered it should be possible at least to produce vector molecules with precise homopolymer tails. Two alternative tailing schemes are described below.

Use of λ Exonuclease to Produce 3′ Tails. λ exonuclease is a double-strand specific 5′–3′ processive exonuclease (reviewed in Ref. 18). Treatment of a linear duplex which has a dC–dG homopolymer block at one of its termini should produce a molecule with a 3′ single-stranded homopolymer tail. Because the homopolymer block preexists in the plasmid, it is possible to specify its exact length. We constructed a plasmid, pSPCG, which can be prepared to receive cDNA inserts in this fashion (see Fig. 4). It contains the SP6 promoter adjacent to which is an *Eco*RI site followed by a block of 9 dC–dG base pairs terminating in an *Sma*I site (CCCGGG), the dC tract being on the sense strand with respect to the SP6 promoter. An irrelevent "stuffer" fragment from phage λ follows, separating the homopolymer block from a polylinker region. Preparation of this vector to receive cDNA molecules primed with the *Bam*-oligo(dT) PRE-adapter of Fig. 1 and elongated with dG residues consists of *Sma*I digestion, λ exonuclease treatment, and *Bam*HI digestion. We argued that, even if the exonuclease treatment is difficult to control precisely, and produces a population of molecules having 3′ single-stranded tails

[16] P. A. Krieg and D. A. Melton, *Nucleic Acids Res.* **12,** 7057 (1984).
[17] C. A. Brennan, A. E. Manthey, and R. I. Gumport, this series, Vol. 100, p. 38.
[18] A. Kornberg, "DNA Replication." Freeman, San Francisco, California, 1980.

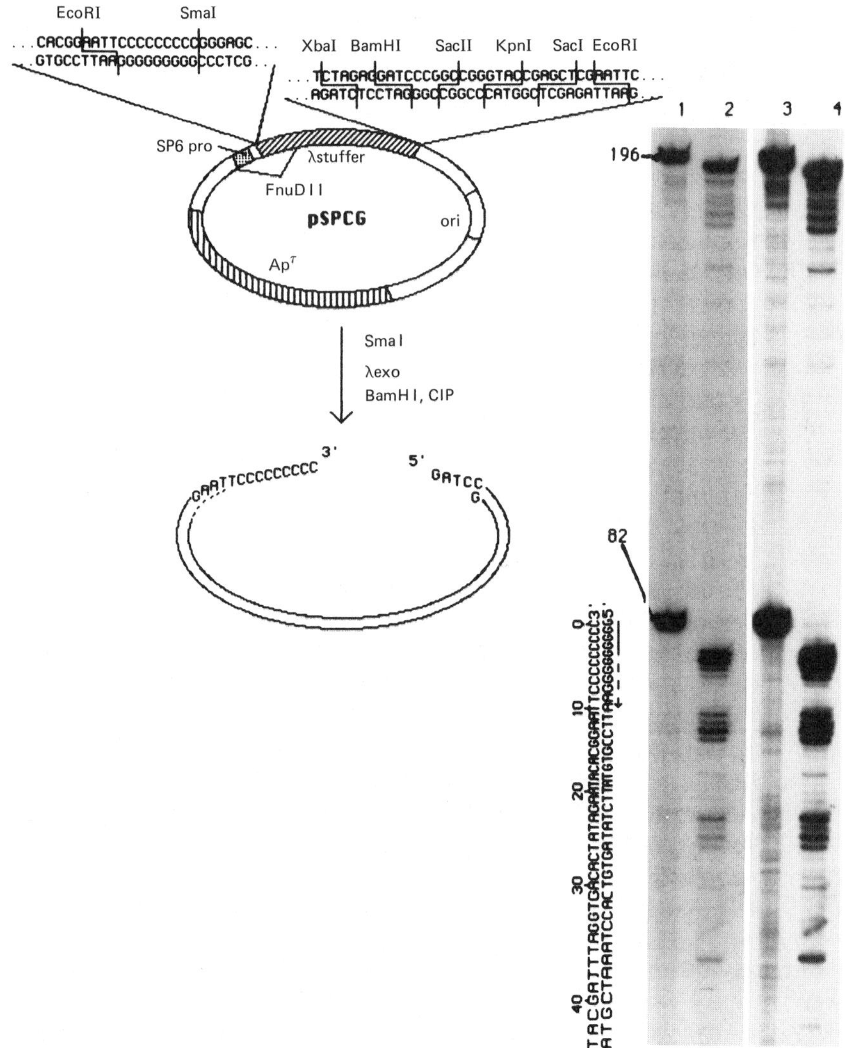

FIG. 4. Tailing by λ exonuclease digestion. pSPCG is a derivative of pSP65.[11] It contains a promotor for *Salmonella* phage SP6 RNA polymerase, adjacent to which is a dC–dG homopolymer box flanked by *Eco*RI and *Sma*I sites. A 654-bp disposable "stuffer" fragment of phage λ DNA (nt 25162–nt 24508) separates the *Sma*I site from a polylinker sequence. Only the *Fnu*DII sites closest to the homopolymer box are indicated. Preparation of pSPCG to receive cDNA inserts consists of *Sma*I digestion, λ exonuclease treatment, then *Bam*HI digestion. The extent of exonuclease digestion can be determined on an aliquot of DNA by 5′ labeling with ^{32}P and digesting with *Fnu*DII, which releases the homopolymer terminus from untreated plasmid as an 82-nt fragment and from exonuclease-digested DNA as fragments shorter than 82 nt. Lanes 1–4 show pSPCG DNA thus treated then electrophoresed through an 8% polyacrylamide–urea gel; lanes 3 and 4 are a longer exposed version of 1 and 2. Lanes 1 and 3 show the 82- and 196-nt terminal fragments from control DNA untreated with exonuclease; lanes 2 and 4 show the effect of digestion with a molar excess of λ exonuclease for 6 min at 0°. The sequence of the homopolymer terminus is written alongside lane 1, and the 5′–3′ exonuclease action indicated.

varying in length, the maximum number of dC residues exposed in the homopolymer tract will be 9, and any gaps will be filled in by DNA polymerase in the second phase of the insertion scheme (see Fig. 1).

*Sma*I-cut linear pSPCG DNA at 50 μg/ml was digested with 250 U/ml λ exonuclease in 67 m*M* glycine–KOH (pH 9.4), 2 m*M* $MgCl_2$, 3 m*M* 2-mercaptoethanol for 6 min on ice. The reaction was quenched by the addition of EDTA to 5 m*M* and Tris–HCl (pH 7.0) to 100 m*M*. DNA was recovered by phenol–chloroform extraction and ethanol precipitation. It was digested with *Bam*HI, dephosphorylated, and the 3-kb fragment purified by agarose gel electrophoresis and recovered, as described above.

λ exonuclease is strictly processive, so it is crucial to provide a molar excess of enzyme over DNA ends. These incubation conditions were selected after pilot experiments to determine the rate of digestion. The length of the single-stranded tails could be determined by taking an aliquot of the exonuclease-treated DNA prior to *Bam*HI digestion, denaturing it at 100°, treating with calf intestinal phosphatase, heating to 100° again, then labeling with polynucleotide kinase and [γ-^{32}P]ATP, and finally digesting with *Fnu*DII after allowing strand reassociation at 60° (Fig. 4). Such treatment of non-exonuclease-treated *Sma*I-cut linear molecules releases the two termini as 82- and 196-bp fragments, the smaller containing the homopolymer block. Exonuclease-digested DNA should release fragments shorter than 82 bp as compared to those released from untreated DNA. The degree of reduction in chain length is determined by the extent of λ exonuclease digestion and gives, by implication, length of the 3′ single-stranded tails.

Such an analysis is shown in Fig. 4. Clearly a molar excess of enzyme was achieved, as all molecules are resected to some degree; however, the distribution of digestion products is very wide and not continuous. A fraction, about one-third, of molecules has lost only 4–7 nt. A similar fraction has lost 12–15 nt, and there is another cluster of molecules which have lost 24–27 nt. Presumably these clusters indicate "pause" sites through which digestion proceeds relatively slowly. These sites have no obvious correlation to primary sequence. Evidently, the extent of digestion and thus the length of the 3′ single-stranded tails is rather hard to control. Nevertheless, the primary objective, of producing a vector sample in which most of the molecules display short 3′-oligo(dC) tails, was achieved. This vector preparation has proved highly efficient in cDNA cloning experiments.

Tailing Vectors with PRE-Adapters. We have recently begun to use a third tailing method which provides a flexible and precise scheme of vector preparation. T4 DNA ligase can be used to attach PRE-adapters containing 3′ homopolymers to restriction enzyme-digested vector mole-

```
                    pAGCTTGCCGGCGTTCCCCCCCCC3'
--  ─────A ↙          ACGGCCGCA
-- vector─TTCGAp
```

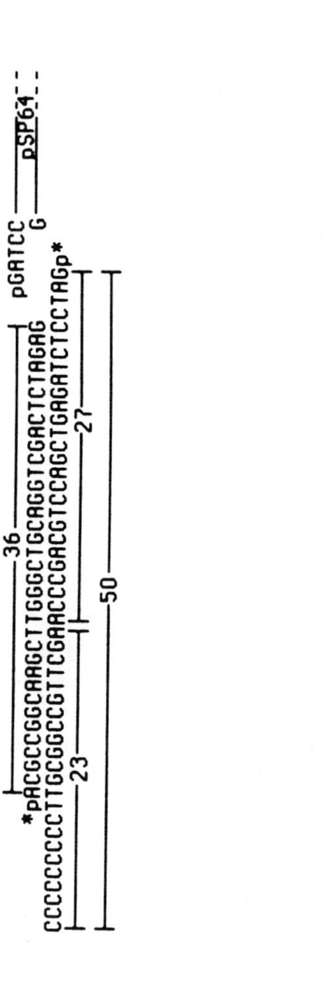

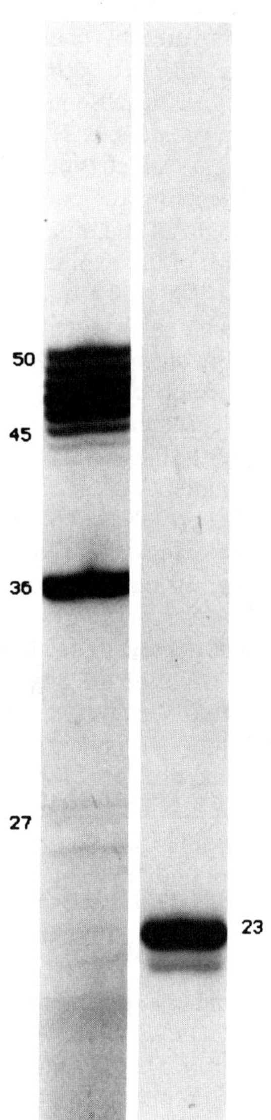

cules. Here again it should be possible to specify an exact homopolymer length, and indeed, to specify the exact structure of the termini. Figure 5 shows a pair of PRE-adapters designed for ligation to *Hin*dIII-cut DNA to produce a single-stranded 3′-dC stretch exactly 9 nt long.

The 23-mer, 3′-oligo(dC) PRE-adapter was preparatively phosphorylated as described above. About 300 pmol was annealed with the 9-mer adapter at an approximately 2 : 1 ratio in 200 μl 50 m*M* NaCl, 100 m*M* Tris–HCl (pH 7.8), 1 m*M* EDTA by heating to 50° and cooling slowly to 4°. Then 20 μg (about 20 pmol termini) of *Hin*dIII-digested pSP64 DNA was added, the solution brought to 12.5 m*M* NaCl, 100 m*M* Tris–HCl (pH 7.8), 10 m*M* $MgCl_2$, 10 m*M* DTT, 1m*M* ATP, 1200 U of T4 DNA ligase, and 20 μg of BSA in a total volume of 800 μl, and held at 16° for 16 hr. Tailed DNA was digested with *Bam*HI, dephosphorylated, and purified as before. Again, the extent of reaction could be assessed by polyacrylamide gel electrophoresis of an aliquot of the reaction mixture after *Bam*HI digestion followed by labeling of the 5′ ends with ^{32}P-phosphates.

In this case, termini which have not received an adapter release a 27-nt labeled *Hin*dIII–*Bam*HI fragment (see Fig. 5), while those which have been ligated to adapters are expected to produce a longer 50-nt fragment and a shorter 36-nt fragment. Figure 5, lane 1, shows this analysis: evidently the ligation was highly efficient, as very few unutilized 23-nt molecules are evident. At the gel position for the expected 50-nt fragment there appears a ladder of fragments sized from 50 nt down to 45 nt. Electrophoresis of an aliquot of the ^{32}P-labeled original PRE-adapter showed that this heterogeneity was not a consequence of heterogeneity preexisting in the ligase substrate (lane 2). This was probably due to a 3′ exonuclease activity contaminating the ligase preparation. The net effect was to produce a population of vector molecules having 3′-dC tails on average about 6–7 nt long, none longer than 9. This preparation has been very effective in cDNA cloning.

This approach may prove a useful alternative in cloning schemes to the use of terminal transferase for dC or dG tailing. It has the advantages

FIG. 5. Tailing by ligation of PRE-adapters. The sequence of a pair of adapters which can convert a *Hin*dIII end to a 3′-oligo(dC)-tailed end is shown above. Ligation of this adapter pair to *Hin*dIII-linearized pSP64, followed by *Bam*HI digestion and 5′ ^{32}P-labeling should release a moiety which on denaturation will yield a 50-nt 3′-oligo(dC)-containing fragment and a 36-nt fragment from the recessed 5′ end. Lane 1 shows pSP64 DNA thus treated and separated on a 10% polyacrylamide–urea gel. The mobility of fragments of 27, 36, and 45–50 nt is indicated. Lane 2 shows a sample of the 23-mer oligo(dC)-containing adapter, 5′ ^{32}P-labeled.

of precision, flexibility in that with appropriate adapters any staggered restriction end can be tailed, and it avoids the possibility of tailing at internal nicks which can cause problems, especially with large vectors.

Discussion

Following any fractionation scheme, or, when nonselected libraries are desired, after terminal transferase treatment, cDNA should be desalted and freed from contaminants. For example, if cDNA has been selected by hybridization and HAP chromatography, residual RNA should be removed by alkali digestion. Neutralized cDNA can be desalted by Sephadex G-50 and concentrated by freeze-drying or, at concentrations of greater than 10 μg/ml, by ethanol precipitation.

In a typical ligation, 200 ng of cDNA was annealed with 60 ng of the partner PRE-adapter [5′-oligo(dA)-containing; see Fig. 1] and 1 μg of prepared plasmid vector, in this case pSPCG, in 50 μl of 30 m*M* NaCl, 10 m*M* Tris–HCl (pH 7.8), 1 m*M* EDTA, by heating to 60° and cooling slowly to less than 10°. This is conveniently done by placing the reaction tube in a 250 ml beaker of water at 60° and allowing it to stand at room temperature for 1 hr, then 4° for 1 hr. Six microliters of 1 *M* Tris–HCl (pH 7.8), 0.1 *M* $MgCl_2$ was added, followed by 1.5 μl of 1 *M* DTT, 1.5 μl of 10 m*M* ATP, 1 μl of 1 mg/ml BSA, and 0.5 μl of T4 DNA ligase (4×10^5 U/ml; see below), and incubation continued at 16° for 16 hr. Four microliters of 1 *M* KCl, 0.5 *M* Tris–HCl (pH 7.6), 40 m*M* $MgCl_2$, 1 *M* $(NH_4)_2SO_4$ was added, followed by 7.5 μl of a solution 10 m*M* in each of the four deoxynucleotide triphosphates, 1.5 μl of 10 m*M* NAD, sufficient water to bring the final volume to 100 μl, 2 μl of *E. coli* DNA ligase (5000 U/ml), and 0.5 μl of DNA polymerase I (5000 U/ml), and the reaction held at room temperature for 6 hr.

If desired, the vector DNA can be added to the cDNA to act as carrier during the final round of desalting and concentration. Two hybridization reactions occur during the annealing step: the dC–dG homopolymer tail interaction and the pairing of the PRE-adapter partners. Slow annealing promotes correct alignment in both reactions. A considerable molar excess of the partner adapter is used, but a rough molar equivalence of vector, to minimize the background of nonrecombinant colonies. DNA concentration during the ligation reaction must be low enough to ensure that cyclization of cDNA–plasmid hybrids is much more likely than dimerization (see Ref. 19): this is easy to achieve with the small plasmids used here.

[19] A. Dugaiczyk, H. W. Boyer, and H. M. Goodman, *J. Mol. Biol.* **96,** 171 (1975).

DNA is now ready to be used to transform *E. coli* cells. We use strain DH1, rendered competent for transformation as described by Hanahan.[20] We find that the competent bacteria tolerate only up to about 25 μM nucleotide in the transformation mix. Higher concentrations profoundly inhibit transformation. The nucleotide concentration in the final polymerase–ligase reaction buffer is slightly over 3 mM, so free nucleotides must be removed or greatly diluted for efficient transformation. For fairly large quantities of DNA, like those used in this example, we use gel filtration and ethanol precipitation, or precipitation from 40% isopropanol in 0.2 M KCl, to remove nucleotides. For small amounts of DNA, simple dilution in the transformation cocktail is easier: add a maximum of 8 μl to 1 ml of bacteria.

One nanogram of cDNA routinely produces between 100 and 500 (genuine) recombinants. The efficiency of clone formation and, importantly, the background of non- or pseudo-recombinant clones depends on the quality, amounts and ratios of enzymes used, as Table I demonstrates. Here we have assayed the efficiency of clone formation and the frequency of nonrecombinant background clones by performing the biochemical operations with various amounts of the enzymes and scoring the number of clones produced from duplicate reactions either containing or lacking only cDNA. Such assays suggest worse cases for background clone frequency, as the presence of cDNA in molar equivalence with vector itself decreases the probability that the vector will cyclize without an insert. However, it is an easy and conservative way of assessing the nonrecombinant background. Most phenotypic screens for nonrecombinant parental types are sensitive to the deletion of a single nucleotide and would suggest a misleadingly low background in reactions where exonucleases might be active.

Titration of T4 DNA ligase is crucial in maintaining an acceptably low background of nonrecombinant clones. We have repeatedly noticed with many lots of this enzyme, used at high concentration, that control reactions which lack cDNA actually produce more colonies than reactions containing cDNA (Table I, lines 1 and 2). Such background colonies contain plasmids which always lack the restriction sites used for cDNA insertion and often lack flanking sites in the polylinker region as distant from the original termini as 10–15 nt. Again it seems likely that contaminating exonuclease chews away the single-stranded plasmid termini, ultimately allowing cyclization by ligase, which may be relatively efficient at high enzyme concentrations. The exonuclease is unlikely to be a phage T4

[20] D. Hanahan, *J. Mol. Biol.* **166,** 557 (1983).

TABLE I
EFFICIENCY OF CLONE FORMATION[a]

	Recipient psPCG vector (ng)	5′-Oligo(dA) partner adapter (ng)	cDNA (ng)	T4 DNA ligase (units)	*E. coli* DNA polymerase (units)	*E. coli* DNA ligase (units)	Number of colonies
1	30	2	6	200	0.8	3	4,500
2	30	2	—	200	0.8	3	6,050
3	30	2	6	14	0.8	3	1,364
4	30	2	—	14	0.8	3	760
5	30	2	6	2	0.8	3	1,078
6	30	2	—	2	0.8	3	194
7	30	2	6	14	5.0	3	402
8	30	2	—	14	5.0	3	190
9	30	2	6	14	2.5 (Klenow)	3	172
10	30	2	—	14	2.5 (Klenow)	3	40

[a] Two annealing mixtures were set up, both containing 150 ng of dC-tailed, *Bam*HI-digested pSPCG vector and 10 ng of the 5′-oligo(dA)-containing partner PRE-adapter (see Fig. 1); one mix additionally contained 30 ng of PRE-adapter primed, dG-tailed cDNA (copied from hybridoma mRNA). After annealing, each mix was divided into five equal aliquots, ligase buffer components and various amounts of T4 DNA ligase added as shown, and held at 16° overnight. Deoxynucleotides and NAD were added, and the reaction buffer adjusted as described in the text before the addition of *E. coli* DNA polymerase I and *E. coli* DNA ligase as indicated; the incubation was continued at room temperature for a further 6 hr. Reactions 9 and 10 received the Klenow fragment of DNA polymerase I, rather than the holoenzyme. Reactions, volume now 12.5 μl, were each added to 1 ml of competent *E. coli* strain DH1 in transformation buffer, prepared as described by Hanahan.[20] After holding on ice, heat shock, and dilution as described,[20] bacteria were spread on L broth agar plates containing 35 μg/ml ampicillin, and colonies counted after overnight incubation at 37°.

gene product, as the ligase lots we have used were purified from bacteria lysogenic for a λ phage carrying the T4 DNA ligase gene.

Some suppliers define ligase activity in units according to Weiss *et al.*,[21] others in units based on the rate of intermolecular ligation of restriction fragments; we have used the latter in Table I. It is reported that 1 Weiss unit is equivalent to about 60 "intermolecular ligation" units. We recommend that the enzyme be stored at the high activity at which it is supplied, then freshly diluted with carrier protein just before use. Though individual lots vary, we have yet to find a supplier whose product is reliably free from exonuclease as assayed in this sensitive fashion.

[21] B. Weiss, A. Jacquemin-Sablon, R. Live, G. C. Fareed, and C. C. Richardson, *J. Biol. Chem.* **243,** 4543 (1968).

More interesting is the dependence of cloning efficiency on the ratio of polymerase I to *E. coli* DNA ligase in the second phase of the insertion reaction. Inspection of Fig. 1 will show that polymerase I will copy parts of both strands of the cDNA–plasmid hybrid: on one strand it will use oligo(dC) as a primer initially to fill the gap opposite the cDNA, then continue in nick translation "rightward"; on the other strand it will use oligo(dG) as a primer to initiate nick translation "leftward." In the absence of ligation these nick translations may continue toward one another around the circular plasmid to produce a linear molecule, ineffective in transformation, when the nicks are opposite one another. Ligase activity must be high enough to ensure that this does not happen. The processivity of polymerase I under these conditions is probably of the order of 10 nucleotides (see Ref. 18), so the rate of nick translation can be increased by increasing the polymerase concentration. At a given ligase concentration one might then expect an increase in polymerase concentration to cause a decrease in the cloning efficiency, and this, indeed, is what happens (Table I, lines 3 and 7). Klenow fragments cannot effectively substitute for polymerase I in this reaction (lines 9 and 10).

Acknowledgments

Much of the work described here was done at the Roche Institute of Molecular Biology, Nutley, New Jersey. Both the Basel Institute for Immunology and the Roche Institute of Molecular Biology are entirely supported by Hoffmann–La Roche. I wish to thank F. Erlitz, K. Collier, R. Schulze, and H. Kiefer for synthesizing the oligonucleotides.

Section II

Identification of Cloned Genes and Mapping of Genes

[5] Mapping of RNA Using S_1 Nuclease and Synthetic Oligonucleotides

By A. A. REYES and R. BRUCE WALLACE

The S_1 nuclease mapping method developed by Berk and Sharp[1] has become a standard technique in the analysis of RNA structure. The two most common probes used for this method are restriction fragments and single-stranded DNA fragments derived from M13 recombinant clones. In the case of double-stranded restriction fragment probes, annealing of probe to RNA must be done under conditions where formation of DNA : RNA rather than DNA : DNA duplexes is preferred. Strand separation of the probe prior to annealing with RNA is one way to eliminate this problem. On the other hand, construction of single-stranded probes from M13 recombinant clones requires an additional cloning step wherein the gene of interest is first inserted into an M13 vector.

The use of synthetic oligodeoxyribonucleotides as probes for S_1 nuclease mapping offers several advantages. First, because of recent advances in synthetic DNA chemistry, large amounts of probes are readily available and can be targeted to any region of a gene of known sequence. The choice of probe is not dependent on the presence of convenient restriction sites. Second, probes which are otherwise impossible to obtain from cloned sequences can be designed. (This might include, for example, sequences which are hypothesized to exist, but have not yet been cloned.) Such probes could be particularly useful in the analysis of alternative splice patterns, as will be described below. Third, a large excess of oligonucleotide probe can be added during the annealing reaction. This results in higher hybridization rates and shorter hybridization times.

Probe Design

The length of the oligonucleotide probe depends on the complexity of the RNA population to which it is annealed. In order for an oligonucleotide probe to be complementary to a unique RNA species in a given population it should be longer than a minimum calculated length.[2] However, if the target RNA is transcribed from a gene that is a member of a homologous multigene family, it may be necessary to use a longer probe.

[1] A. J. Berk and P. A. Sharp, *Proc. Natl. Acad. Sci. U.S.A.* **75,** 1274 (1978).
[2] R. Lathe, *J. Mol. Biol.* **183,** 1 (1985).

Thus, such a probe should encompass nucleotide positions that are polymorphic among the different members of the gene family. This ensures specificity of hybridization between the probe and the specific target RNA.

The synthesis of a long probe from two shorter, complementary (but only partially overlapping) oligonucleotides in an enzymatic polymerization reaction eliminates the need to chemically synthesize long oligonucleotides. Furthermore, such a scheme allows the production of novel probes by simply mixing and matching different pairs of short oligonucleotides (see below).

Furthermore, the oligonucleotide can be labeled at its 5′ end with T4 polynucleotide kinase and [γ-^{32}P]ATP, or, if a higher specific activity is desired, internally labeled with DNA polymerase I and [α-^{32}P]dNTPs.[3] The latter is possible if the probe is enzymatically synthesized from two shorter, complementary oligonucleotides. A uniformly labeled probe also allows detection of all S_1 nuclease-protected fragments by autoradiography.

Oligonucleotide : RNA Annealing and S_1 Nuclease Digestion

To a first approximation, the dissociation temperature, T_d (in degrees Celsius), of a short (less than 20 base pairs) oligonucleotide : RNA duplex can be calculated using the empirical formula[4]

$$T_d = 4(G + C) + 2(A + T)$$

(where the bases in parentheses refer to the composition of the oligonucleotide). This formula was originally derived for hybridization of an oligonucleotide to target DNA immobilized on a solid matrix such as nitrocellulose. In S_1 analysis, the T_d for an oligonucleotide : RNA duplex may be different, since hybridization occurs in solution and DNA : RNA duplexes may be more stable than DNA : DNA duplexes. Nevertheless, the annealing temperature can be set initially at 2–5° below the calculated T_d. For longer oligonucleotides, T_d's are harder to estimate and appropriate annealing temperatures have to be determined empirically. In our experience, probes 47 to 201 nucleotides long can be hybridized to RNA at 65°.

The oligonucleotide probe can be added in large excess such that the hybridization reaction obeys pseudo-first-order kinetics with respect to probe concentration. For example, when a 47-mer probe is used at a

[3] A. Studencki and R. B. Wallace, *DNA* **3,** 7 (1984).

[4] S. V. Suggs, T. Hirose, T. Miyake, E. H. Kawashima, M. J. Johnson, K. Itakura, and R. B. Wallace, *in* "Developmental Biology Using Purified Genes" (D. Brown and C. F. Fox, eds.), pp. 683–693. Academic Press, New York, 1981.

concentration of 10 ng/ml, the $t_{1/2}$ is 11 min.[5] Because there is no competing DNA : DNA hybridization, it is not necessary to add formamide in the annealing buffer. This is important, since the short hybridization times will avoid RNA degradation.

S_1 nuclease digestion is usually done at a temperature between 20 and 45°. When using short oligonucleotide probes, the S_1 digestion temperature should be below the T_d of the probe : RNA duplex. The optimum S_1 nuclease concentration and incubation temperature should be determined empirically for each hybrid.

Analysis of *H-2K* Splice Patterns

To illustrate the technique, the use of synthetic oligonucleotides to map the 3′-end splice patterns in *H-2K* mRNAs will be discussed. There are two possible splice acceptor sites located 27 bases apart in exon 8 of the *H-2K* gene. In the typical *H-2K*^{b} transcript (*H-2K*^{b} is the allele of the *K* gene in a b strain mouse), exon 7 is spliced to exon 8a. However, a small fraction (5–10%) of *H-2K*^{b} transcripts displays the alternative pattern where exon 7 is joined to exon 8b. S_1 nuclease mapping using oligonucleotide probes was used to determine whether the same splicing patterns are used by a second allele, *H-2K*^{k}.

Probe K^{b}47B bridges the exon 7–exon 8 junction and is complementary to the *H-2K*^{b} mRNA in which exon 7 is spliced to exon 8b. Therefore, K^{b}47B would be either fully or partially protected from S_1 nuclease digestion, depending on whether it hybridizes to the atypical or to the typical *H-2K*^{b} mRNA, respectively. A second probe, K^{b}47A, is complementary to the mRNA in which exon 7 is spliced to exon 8a. When annealed to either type of *H-2K*^{b} mRNA, the S_1 protection pattern of K^{b}47A is the reverse of that of K^{b}47B. The two probes used in conjunction give an accurate determination of the splicing patterns used in *H-2K*^{b}. Since *H-2K*^{b} and *H-2K*^{k} are homologous in this region of the gene, S_1 protection patterns using K^{b}47A and K^{b}47B probes should indicate the splice patterns used in *H-2K*^{k} transcripts.

Probe K^{b}47B is prepared from oligonucleotides K^{b}27 and K^{b}29B by a polymerization reaction (Fig. 1). First, K^{b}27 is 5′ end labeled with T4 polynucleotide kinase and [γ-^{32}P]ATP. It is then annealed to K^{b}29, which contains a dimethoxytrityl (DMT) group at its 5′ end. Both strands of the resulting duplex are extended in the 3′ direction by adding DNA polymerase I and all four unlabeled dNTPs. The desired probe strand, K^{b}47B, is separated from the complementary strand (47U) by electrophoresis in a denaturing gel (Fig. 2). The greatest separation between the two strands is

[5] J. Meinkoth and G. Wahl, *Anal. Biochem.* **138,** 267 (1984).

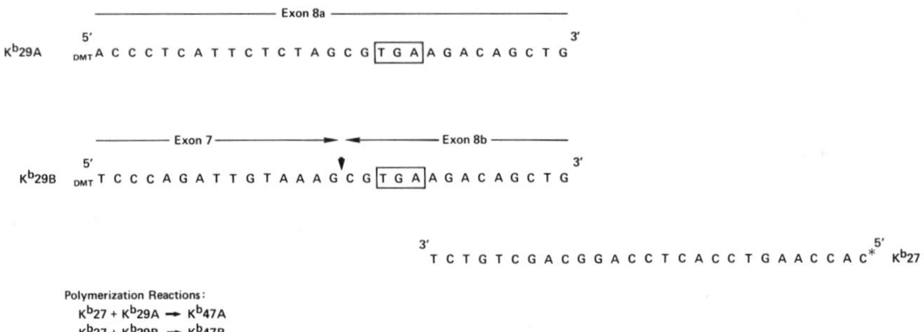

FIG. 1. Synthesis of 5′ ^{32}P-labeled oligonucleotide probes from two shorter sequences. Two oligonucleotides complementary over the terminal nine 3′ nucleotides are annealed and extend in a DNA polymerase reaction. Two different 47-base-long probes (K^b47A + K^b47B) can be synthesized using the same 27-base-long lower strand and two different 29-base-long upper strands (K^b29A + K^b29B).

obtained when 47U contains a 5′-DMT group. Some of the DMT groups are probably hydrolyzed in the course of electrophoresis; however, 5′-OH-47U and K^b47 are also adequately resolved in this gel system.

To synthesize K^b47A, K^b27 is again used as the lower strand primer. However, the upper strand oligonucleotide is different (Fig. 1). This is an example of how one can use one short sequence (K^b27) as starting material to synthesize two different longer probes (K^b47A and K^b47B).

The result of an S_1 nuclease mapping experiment using probes K^b47A and K^b47B is shown in Fig. 3. *H-2K*^{b} and *H-2K*^{k} transcripts use both splice patterns. However, the predominant pattern for *H-2K*^{b} is the rare pattern for *H-2K*^{k}, and vice versa.

Experimental Procedures

5′ End Labeling of K^b27

Set up the following reaction mixture in a 0.5-ml Eppendorf tube:
- 5′-OH-K^b27, 50 ng (5.6 pmol)
- [γ-^{32}P]ATP, 50 pmol (300 μCi, crude preparation, New England Nuclear)
- 70 m*M* Tris–HCl, pH 8
- 10 m*M* $MgCl_2$
- 10 m*M* dithiothreitol (DTT)
- 3 units T4 polynucleotide kinase
- H_2O to 5 μl.

Incubate at 37°, 30 min.
Heat at 90°, 5 min.

FIG. 2. Strand separation of 5′ ^{32}P-labeled oligonucleotide probes. The products of the DNA polymerase reaction are subjected to electrophoresis on a denaturing polyacrylamide gel. The two strands of the 47-base-long duplex have a different mobility. When K^b27 (Fig. 1) is 5′ ^{32}P-labeled and used in the polymerase reaction the desired product, K^b47B (used in the S_1 nuclease experiments), has the mobility indicated (^{32}P47L). When the upper strand is 5′ ^{32}P-labeled (K^b29B) the 47-base-long product has the mobility indicated (^{32}P47U), demonstrating the difference in mobility of the two strands.

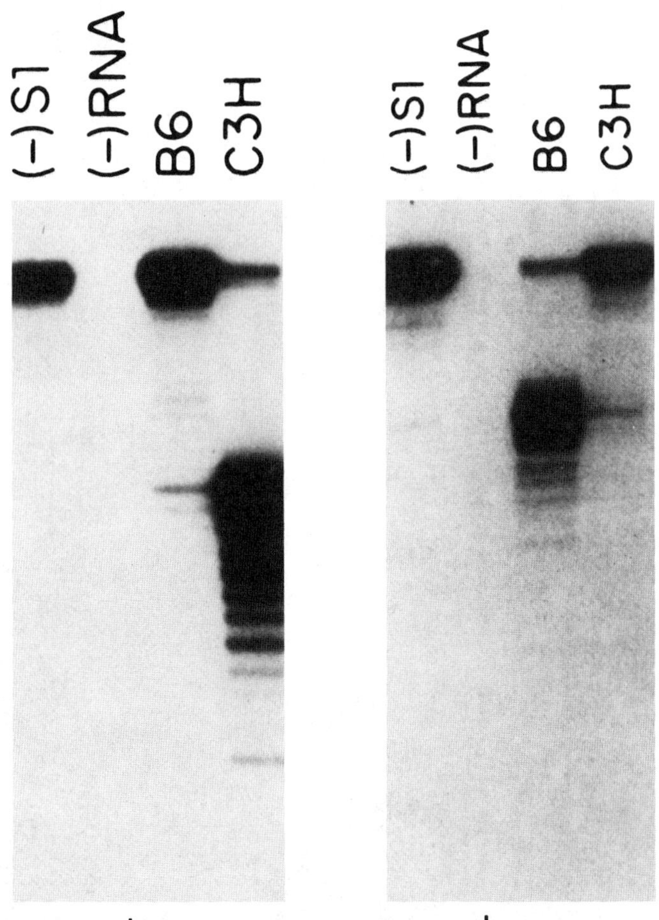

FIG. 3. S_1 nuclease mapping of poly(A)$^+$ RNA from the b strain (B6) and k strain (C3H) mouse using 5′ ^{32}P-labeled K^b47A and K^b47B. S_1 nuclease mapping was done as described in the text. Controls include reactions done without S_1 nuclease [(−)S1] and without poly(A)$^+$ RNA [(−)RNA]. The reverse patterns of S_1 nuclease protection are seen in the two strains of mice indicating that the product of the *H-2K*^{k} allele is spliced differently than the product of the *H-2K*^{b} allele.

Synthesis of K^{b}47B

To the tube containing the K^b27 5′ end labeling reaction mixture, add
5′-DMT-K^b29B, 250 ng (26 pmol)
NaCl to 7 m*M*
H_2O to 10 μl.

Incubate at 65°, 10 min.
Cool in ice water.
Add
1 μl 2.5 m*M* (dATP, dCTP, dGTP, dTTP)
1 unit DNA polymerase I (Klenow fragment).
Incubate at room temperature, 15 min.
Heat at 65°, 5 min.
Add 10 μl formamide loading buffer [80% formamide, 1/20× TBE, 0.1% xylene cyanol, 0.1% bromphenol blue (1× TBE is 90 m*M* Tris, 90 m*M* boric acid, 2 m*M* EDTA)].

Probe Purification

Load the sample onto a 40-cm-long 20% acrylamide [20 : 1 acrylamide : bis(acrylamide)]/8 *M* urea/1× TBE gel; run at constant power (35 watts) until the bromphenol blue dye is at the bottom of the gel.
Expose the gel briefly to X-ray film.
Excise probe band from gel.
Crush the gel slice and soak in 0.5 ml 10 m*M* Tris–HCl (pH 8), 0.1 m*M* EDTA at 37° for 12–16 hr.
Recover the supernatant; rinse the gel pieces twice with soaking buffer; combine rinses with supernatant.
Dialyze against water at room temperature for 1 hr.
Filter through an Acrodisc membrane (Gelman).
Concentrate to 2.5–5 × 10^5 cpm/μl by evaporation. Typical yield is 5 × 10^6 cpm.

Annealing and S_1 Nuclease Digestion

The probe concentration during the annealing reaction is 1–2 pmol/ml (5–10 ng/ml). The K^b47B : *H-2K* mRNA molar ratio is approximately 100 : 1.
Ethanol precipitate the following:
1 μg poly(A)$^+$ RNA
20 μg yeast tRNA.
Resuspend in 10 μl of
40 m*M* PIPES, pH 6.4
400 m*M* NaCl
1 m*M* EDTA.
Add
2.5–5 × 10^5 cpm K^b47B.
Incubate at 65° for 2 hr.

Add to 100 μl of S_1 buffer the following:
- 280 m*M* NaCl
- 50 m*M* NaOAc, pH 4.5
- 4.5 m*M* $ZnSO_4$
- 2 μg single-stranded *Escherichia coli* DNA
- 50 units S_1 nuclease.

Incubate at 37° for 30 min.
Stop the reaction by adding 17 μl 4 *M* NH_4OAc, 100 m*M* EDTA.
Ethanol precipitate S_1-resistant nucleic acids.
Dissolve pellet in formamide loading buffer.
Heat at 90°, 2 min.
Electrophorese on a 12% acrylamide/8 *M* urea/1× TBE sequencing type gel until the bromphenol blue dye is at the bottom of the gel.

Comments

The gel conditions described here are for oligonucleotide probes 47 bases long. For other probes, appropriate electrophoretic conditions may be different. Similarly, the S_1 nuclease digestion conditions should be determined empirically for each probe : RNA hybrid.

Acknowledgments

The work described was supported by a grant from the National Institutes of Health (GM31261) to RBW. RBW is a member of the Cancer Center at the City of Hope (NIH CA33572).

[6] Oligonucleotide Hybridization Techniques

By C. GARRETT MIYADA and R. BRUCE WALLACE

Synthetic oligodeoxyribonucleotides (henceforth referred to as oligonucleotides) have been incorporated into a number of molecular biology protocols in recent years. Reductions in synthesis time and cost have made oligonucleotides available to most any laboratory. Oligonucleotides have been used in the construction of synthetic gene sequences, for the addition of new restriction enzyme sites, for sequence-directed mutagenesis, and as primers for either primary nucleotide sequence determination or complementary DNA (cDNA) synthesis. Oligonucleotides have also been extensively developed as probes in various hybridization techniques.

Synthetic oligonucleotides may be used in most hybridization procedures that require nick-translated DNA probes; however, oligonucleotide hybridization probes require a stringent sequence specificity that provides them with one clear advantage over nick-translated probes. Oligonucleotide hybridization probes, under the appropriate experimental conditions, enable the discrimination of nucleic acid sequences that differ by as little as a single nucleotide.[1] This sequence specificity has led to the development of oligonucleotides as probes for cloned genes based solely on protein sequence information.[2,3] Oligonucleotides have also been used to detect single-base pair differences in sequences as complex as a human genomic restriction digest.[4] This technique has proven extremely powerful in monitoring certain human genetic diseases.

This chapter will provide protocols for the use of synthetic oligonucleotides as hybridization probes to nucleic acid sequences. The labeling of oligonucleotides with [γ-^{32}P]ATP and their subsequent use in Southern,[5] Northern,[6] plaque,[7] and colony-screening[8] hybridization procedures will be described. This chapter will also provide procedures that improve the sensitivity of detecting a single-copy gene sequence within a mammalian chromosomal restriction digest, which had been size fractionated by electrophoresis and hybridized directly in the agarose gel matrix. The protocol describes the detection of a single-copy gene sequence with a kinased oligonucleotide in as little as 1.2 μg of a genomic restriction digest.

Materials

The 23-base K^b oligonucleotide[9] (5′-CTGAGTCTCTCTGCTTCACCAGC-3′) was synthesized by the solid-phase phosphotriester method[10]

[1] R. B. Wallace, J. Shaffer, R. F. Murphy, J. Bonner, T. Hirose, and K. Itakura, *Nucleic Acids Res.* **6,** 3543 (1979).

[2] K. L. Agarwal, J. Brunstedt, and B. E. Noyes, *J. Biol. Chem.* **256,** 1023 (1981).

[3] R. B. Wallace, M. J. Johnson, T. Hirose, T. Miyake, E. H. Kawashima, and K. Itakura, *Nucleic Acids Res.* **9,** 879 (1981).

[4] B. J. Conner, A. A. Reyes, C. Morin, K. Itakura, R. L. Teplitz, and R. B. Wallace, *Proc. Natl. Acad. Sci. U.S.A.* **80,** 278 (1983).

[5] E. M. Southern, *J. Mol. Biol.* **98,** 503 (1975).

[6] J. C. Alwine, D. J. Kemp, and G. R. Stark, *Proc. Natl. Acad. Sci. U.S.A.* **74,** 5350 (1977).

[7] W. D. Benton and R. W. Davis, *Science* **196,** 180 (1977).

[8] M. Grunstein and D. S. Hogness, *Proc. Natl. Acad. Sci. U.S.A.* **72,** 3961 (1975).

[9] C. G. Miyada, C. Klofelt, A. A. Reyes, E. McLaughlin-Taylor, and R. B. Wallace, *Proc. Natl. Acad. Sci. U.S.A.* **82,** 2890 (1985).

[10] Z.-K. Tan, S. Ikuta, T. Huang, A. Dugaiczyk, and K. Itakura, *Cold Spring Harbor Symp. Quant. Biol.* **47,** 383 (1983).

on a Systec Microsyn 1450 automated DNA synthesizer. The oligonucleotide was purified by HPLC by using a PRP-1 reverse-phase column (Hamilton) and a linear gradient of 5–35% acetonitrile.[10] SeaKem ME agarose (FMC) was used for gel electrophoresis. Polynucleotide kinase was from either New England Nuclear, Boehringer Mannheim Biochemicals, or Bethesda Research Laboratories. Restriction endonuclease *Bam*HI was from New England Biolabs or Boehringer Mannheim Biochemicals. [γ-^{32}P]ATP (7000 Ci/mmol) was obtained from ICN.

Methods

DNA Isolation and Digestion with Restriction Endonuclease BamHI. DNA was isolated from mouse strain C57BL/6 livers by freezing the tissue in liquid nitrogen, pulverizing the frozen tissue into small fragments, and suspending the tissue from one liver in 10 ml of 100 m*M* Na_2EDTA, 50 m*M* Tris–Cl (pH 8), 0.5% SDS, and 500 μg/ml Proteinase K (Boehringer Mannheim Biochemicals). The mixture was incubated overnight at 55°. Proteins were removed with sequential extractions of phenol, phenol : chloroform (50 : 50), and chloroform. Glycogen was removed from the DNA preparation by banding the DNA in an isopycnic CsCl–ethidium bromide gradient.[11] The DNA was digested at a concentration of 0.1 μg/μl in 150 m*M* NaCl, 10 m*M* Tris–Cl (pH 7.5), 10 m*M* $MgCl_2$, 1 m*M* dithiothreitol, 100 μg/ml bovine serum albumin, and 0.4 units/μl *Bam*HI at 37°. The digests were determined to be complete, and then the DNA was extracted once with phenol : chloroform (50 : 50), twice with chloroform, and ethanol precipitated. The DNA was resuspended in TE (10 m*M* Tris–Cl, pH 8, 1 m*M* EDTA) at a DNA concentration of 0.5 μg/μl and stored at 4° until subjected to agarose gel electrophoresis.

Agarose Gel Electrophoresis. Agarose gel electrophoresis was performed in TBE buffer (89 m*M* Tris–borate, pH 8.3, 2.5 m*M* EDTA) in a commercially available horizontal gel apparatus. The gel size was 10 × 14 cm, and the gel volume was 60 ml. Genomic DNA restriction digests were subjected to electrophoresis on the above system at 25 volts for 16 hr at room temperature.

Preparation of DNA Samples for Dried-Gel Hybridization.[12] After electrophoresis, gels were stained with ethidium bromide and then were photographed. The DNA samples were denatured by soaking the gel in 500 m*M* NaOH, 150 m*M* NaCl for 30 min at room temperature with gentle shaking. The samples were then neutralized by soaking the gel in 500 m*M*

[11] E. Lacy, S. Roberts, E. P. Evans, M. D. Burtenshaw, and F. D. Costantini, *Cell* **34,** 343 (1983).

[12] S. G. S. Tsao, C. F. Brunk, and R. E. Perlman, *Anal. Biochem.* **131,** 365 (1983).

Tris–Cl (pH 8), 150 mM NaCl for 30 min at 4° with occasional shaking. The gel was then placed on two sheets of Whatman 3M paper. The gel was trimmed at this point to remove any unused lanes. The gel (still on the Whatman paper) was covered with SaranWrap, placed in a gel dryer (Bio-Rad), covered with only the neoprene rubber sheet, and dried with only the vacuum until the gel was nearly flat (approximately 30 min). Then the gel dryer heater was turned on to 60°, and the gel was dried an additional 0.5–1 hr. The vacuum was then released, and the gel, which had been reduced to a thin membrane on the Whatman paper, was now ready for hybridization. Dried gels could be stored at room temperature indefinitely on the paper after wrapping it in SaranWrap.

Oligonucleotide Labeling and Purification. The K^b oligonucleotide was labeled with ^{32}P for use as a hybridization probe. The reaction contained in 10 μl the following: 18.5 pmol DNA, 30 pmol [γ-^{32}P]ATP, 50 mM Tris–Cl (pH 9), 10 mM $MgCl_2$, 10 mM dithiothreitol, 50 μg/ml bovine serum albumin, and 5–6 units of polynucleotide kinase. The reaction was incubated at 37° for 40 min. An equal volume of deionized 98% formamide containing 0.15% bromophenol blue and 0.15% xylene cyanole was added, the sample was heated 5 min at 95°, and then was subjected to electrophoresis on a 14.5% acrylamide, 0.5% bis(acrylamide) gel (15 cm × 30 cm × 1 mm) containing 7 M urea in TBE buffer. Electrophoresis was done at 27.5 mA until the bromophenol blue reached the gel bottom. The labeled oligonucleotide was located through autoradiography, the radioactive band was excised, and the DNA was eluted by soaking the gel slice in two 300 μl changes of TE over a period of at least 12 hr at 37°.

If the kinased oligonucleotide had been previously gel-purified, the unincorporated [γ-^{32}P]ATP may be removed by chromatography over DE-52 cellulose (Whatman).[13] After the kinase reaction, the labeled oligonucleotide is diluted 10-fold in TE. The sample is then applied to a small column of DE-52 (bed volume, approximately 0.2 ml), previously equilibrated with TE. The column is then washed with 5 bed volumes of TE followed by 5 bed volumes of 0.2 M NaCl in TE, which removes the unincorporated label. The labeled oligonucleotide is then eluted with 0.5 ml of 1 M NaCl in TE and used directly in hybridization experiments. Kinased oligonucleotides prepared by either method may be used for at least 1 week when stored at −20°.

Dried-Gel Hybridizations. The dried gel was removed from its paper backing by soaking it in a shallow dish of distilled water. Although the gel was no longer stuck to the paper, the gel was supported by the paper to facilitate its handling. After blotting the gel to remove excess water, the

[13] R. B. Wallace, M. Schold, M. J. Johnson, P. Dembek, and K. Itakura, *Nucleic Acids Res.* **9,** 3647 (1981).

gel and supporting paper were placed in a Seal-a-Meal (Dazey) plastic bag. The gel was pressed against the Seal-a-Meal bag, which led to the preferential adherence of the gel to the plastic and allowed the paper backing to be removed. The dried gel was hybridized in a solution that contained the following: 5× SSPE [1× SSPE = 180 m*M* NaCl, 10 m*M* $(Na_{1.5})PO_4$, 1 m*M* Na_2EDTA, pH 8], 0.1% SDS, 10 μg/ml sonicated, denatured *Escherichia coli* or salmon sperm DNA, and 2×10^6 cpm/ml of labeled oligonucleotide. The hybridization volume for a 10 × 14 cm dried gel was 6 ml. After the hybridization period, the gel was washed with 6× SSC (1× SSC = 150 m*M* NaCl, 15 m*M* Na citrate) first at room temperature then at a stringent temperature. Usually two 15-min washes at room temperature were done followed by a 4-hr wash at room temperature. For the K^b oligonucleotide, the stringent wash was performed at 67° for 2.5 min. Gels were reused in hybridizations by merely subjecting them to the alkali denaturation and neutralization as described in the section on drying down the gels.

Autoradiography. After the appropriate washes, the hybridized gel was wrapped with SaranWrap and used to expose Kodak XAR-5 X-ray film between two Cronex Lightning Plus intensifying screens at −70°.

Densitometric Analysis. Autoradiographs were analyzed with a Perkin–Elmer LCI-100 Laboratory Computing Integrator that was connected to Hoefer Scientific Instruments GS-300 Transmittance/Reflectance Scanning Densitometer. The area under each densitometric peak was used to quantitate the amount of radioactivity in a hybridizing restriction fragment.

Experimental Section

Several parameters of oligonucleotide hybridization to genomic restriction digests were investigated. The length of hybridization, the wash conditions, and the sensitivity of the assay were studied. The K^b oligonucleotide and *Bam*HI-digested C57BL/6 DNA were used in all of the following experiments. The K^b oligonucleotide hybridized to a 4.4-kb restriction fragment, which contained a portion of the *H-2K*^{b} transplantation antigen gene, in these digests.

Length of Hybridization

Previous experiments published from this laboratory had used hybridization times of 2–3 hr in oligonucleotide hybridizations to genomic restriction digests in dried gels.[4,9,14] Although these experiments were sensi-

[14] C. G. Miyada and R. B. Wallace, *Mol. Cell. Biol.* **6**, 315 (1986).

tive enough to detect a single-copy gene in 5–10 μg of a mammalian genomic restriction digest, it was not known if the hybridization times used provided for the maximal amount of binding of the labeled oligonucleotide. Different hybridization times (1, 2, 4, and 16 hr) were tested with 5 and 10 μg amounts of DNA to determine which time gave the optimal hybridization signal. As seen in Fig. 1, the 16-hr hybridization time resulted in the strongest radioactive signal in the oligonucleotide hybridiza-

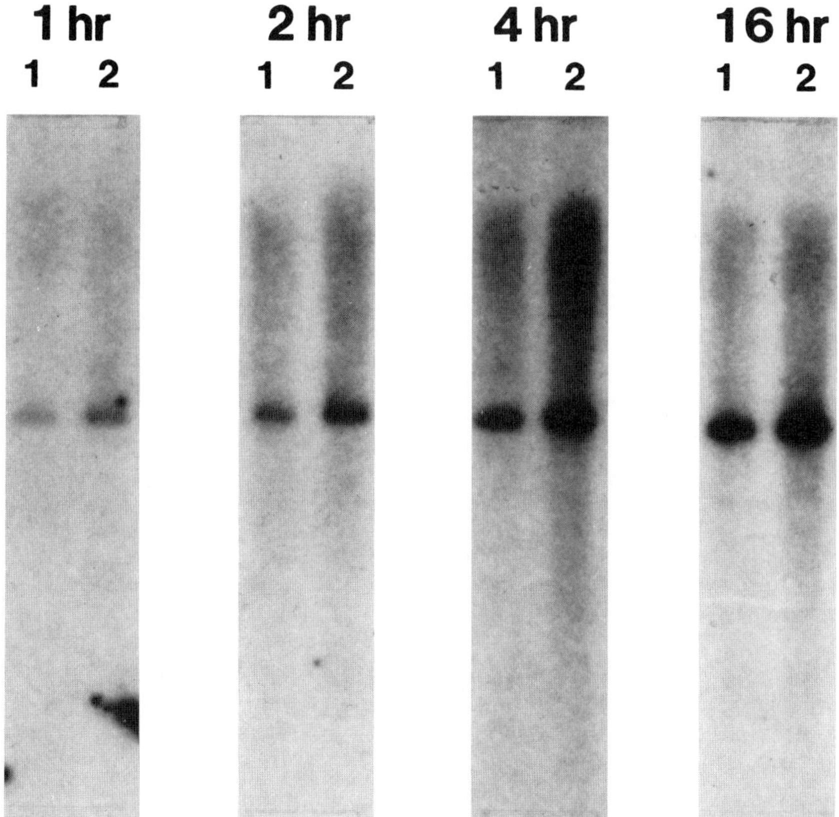

FIG. 1. Different hybridization times. Five (lanes 1) and 10 μg (lanes 2) amounts of *Bam*HI-digested C57BL/6 liver DNA were subjected to electrophoresis on 0.7% agarose gels in quadruplicate. Samples were hybridized directly within the gel matrix with the K^b oligonucleotide as described in the text. The time of hybridization was either 1, 2, 4, or 16 hr. After hybridization the dried gels were washed twice for 15 min each followed by a 4-hr wash (with all washes done in 6× SSC at room temperature). The samples were then given a 2.5-min stringent wash at 67° in 6× SSC. The samples were then exposed for 4 days as described in the text to produce the autoradiographic exposures shown above.

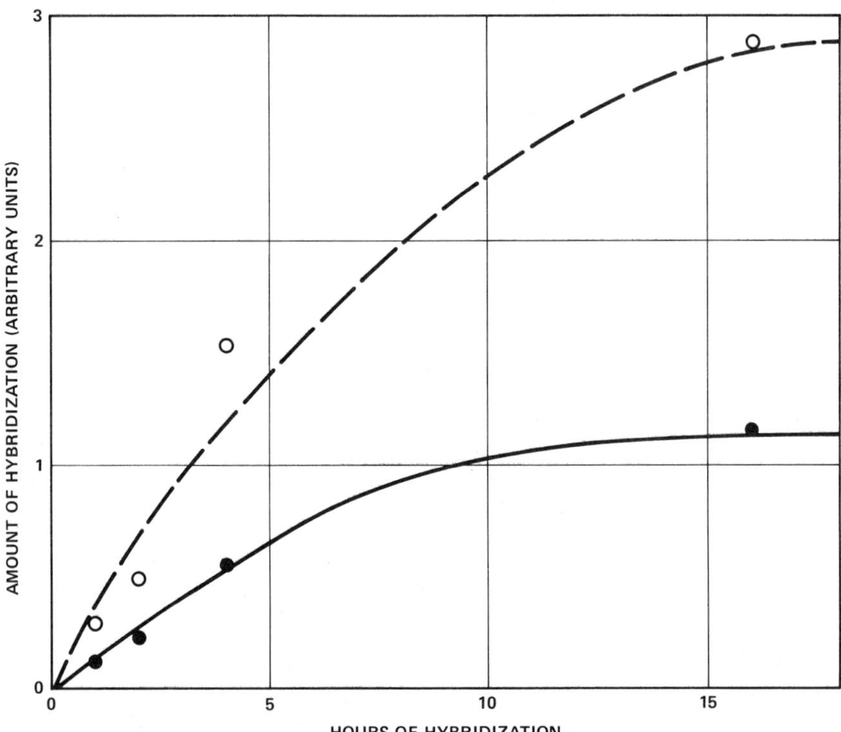

FIG. 2. Kinetics of hybridization. The autoradiographic exposures generated from the data shown in Fig. 1 were scanned with a densitometer, and the area under each peak was used to quantitate the amount of hybridization in each lane. The amount of hybridization was plotted against the hybridization time for 5 μg (●) and 10 μg (○) amounts of genomic restriction digest. Other data described in the text were used to generate the two curves shown above.

tions. The amount of hybridization for each time was quantitated by using a densitometric scan of autoradiogram. The results are shown in Fig. 2. Other hybridization times were tested in a separated experiment (data not shown). Eight hours of hybridization provided for a less than optimal hybridization signal, while a 36-hr hybridization time gave a similar autoradiographic signal as the 16-hr hybridization. These results demonstrated that hybridization times of 16 hr or greater resulted in the strongest autoradiographic signal in oligonucleotide hybridizations to genomic restriction digests in dried gels.

Wash Conditions

In earlier experiments, it was found that a long wash at room temperature prior to the high temperature stringent wash alleviated some of the nonspecific binding of the oligonucleotide to high molecular weight DNA. The benefit of a long room temperature wash was reexamined while using 16-hr hybridization times. In this experiment, one dried gel sample was given two 15-min washes at room temperature in 6× SSC after the hybridization. An identical sample was given the same treatment and an additional 4-hr wash at room temperature in 6× SSC. Both samples were then exposed to X-ray film; the resulting autoradiogram is shown in Fig. 3A. For both samples the majority of the oligonucleotide hybridization signal was due to the nonspecific binding of the oligonucleotide to large (>4 kb) molecular weight DNA. The additional 4-hr wash did little to remove the nonspecific oligonucleotide binding.

Both samples were then given a stringent wash and reexposed. The resulting autoradiogram (Fig. 3B) clearly shows that the stringent wash

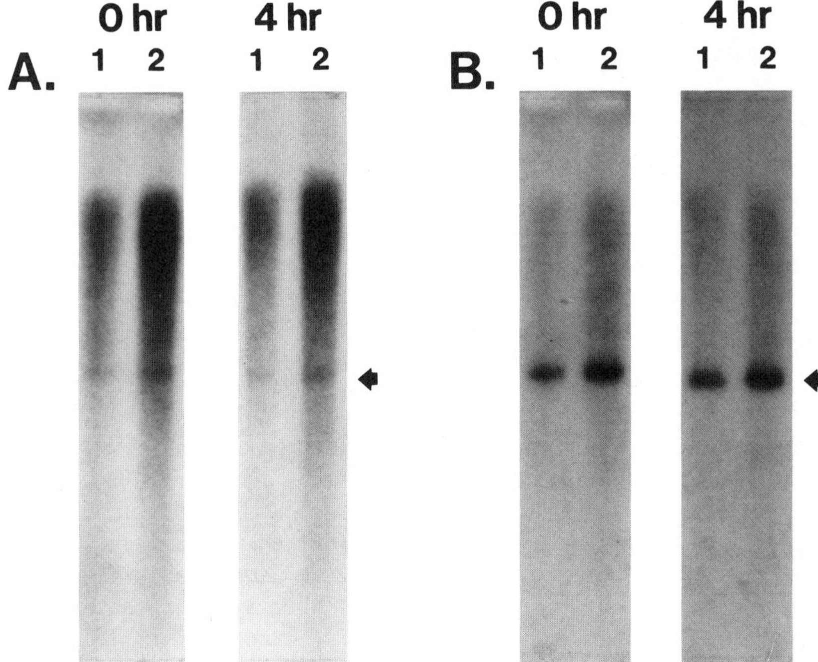

FIG. 3. Effect of different wash conditions. Five (lanes 1) and 10 μg (lanes 2) amounts of *Bam*HI-digested C57BL/6 liver DNA were prepared and hybridized as described in Fig. 1. After the hybridization the dried gels were washed twice for 15 min each in 6× SSC. One gel was then given an additional 4-hr wash in 6× SSC at room temperature. Both gels were then exposed to X-ray film for 4 hr as described in the text. The resulting autoradiogram is shown in (A). Both gels were then given a 2.5-min stringent wash at 67° in 6× SSC, and the gels were reexposed to X-ray film for 1.5 days. The resulting autoradiogram is shown in (B). The arrow denotes the position of the 4.4-kb hybridizing restriction fragment in both panels.

removed most of the nonspecific hybridization as seen by the clear signal from the 4.4-kb hybridizing restriction fragment (denoted with an arrow). The additional 4-hr room temperature wash did not appreciably lessen or enhance the radioactive signal produced with a 16-hr hybridization.

Sensitivity of Oligonucleotide Hybridization to Genomic Restriction Digests in Dried Gels

Since the sensitivity of oligonucleotide hybridization is increased with a 16-hr hybridization time, it should be possible to detect a radioactive signal in smaller amounts of a genomic restriction digest than used previously. An experiment was done in which 10, 5, 2.5, and 1.2 μg amounts of the genomic restriction digest were hybridized with the kinased oligonucleotide. The resulting autoradiogram (Fig. 4) shows that a hybridization signal was easily detected in the 1.2-μg sample after a 3.5-day exposure. The result demonstrated that the method described here has the sensitivity to detect as little as 1×10^{-18} mol of a sequence in a genomic restriction digest.

Discussion

Suggs *et al.* have suggested an empirical formula for the calculation of the T_d (the temperature at which one-half of an oligonucleotide duplex becomes associated in 1 *M* NaCl) for a given oligonucleotide.[15] The formula is as follows:

$$T_d = 4°(G + C) + 2°(A + T) \quad (1)$$

where (G + C) is the number of G and C bases in the oligonucleotide and (A + T) the number of A and T bases in the oligonucleotide. This formula was used to determine hybridization and wash temperatures for the K^b oligonucleotide (calculated $T_d = 72°$) in the described experiments. In this case the hybridizations were done at 12° below the T_d and the stringent wash at 5° below the T_d. Similar conditions should be employed when using an oligonucleotide in hybridizations for the first time. The stringent wash time should be reduced to 1.5 min initially. A second stringent wash of 1 min may be added after the first autoradiographic exposure. In the experiments presented in this chapter, a 2.5-min wash time was used since it gave roughly the same result as the two shorter washes. It must be emphasized, however, that finding optimal hybridization and wash conditions is empirical in nature. These suggestions are merely guidelines, and

[15] S. V. Suggs, T. Hirose, T. Miyake, E. H. Kawashima, M. J. Johnson, K. Itakura, and R. B. Wallace, *in* "Developmental Biology Using Purified Genes" (D. Brown and C. F. Fox, eds.), p. 683. Academic Press, New York, 1981.

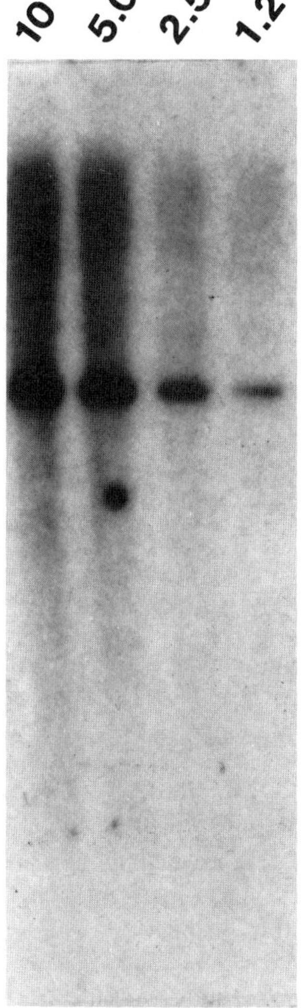

FIG. 4. Sensitivity of oligonucleotide hybridizations to genomic restriction digests in dried gels. *Bam*HI-digested C57BL/6 liver DNA samples (in 10, 5, 2.5, and 1.2 μg amounts) were subjected to electrophoresis in a 0.7% agarose gel and prepared for hybridization as described in the text. The gel was hybridized with the K^b oligonucleotide for 16 hr at 60°. The gel was then given two 15-min washes followed by one 4-hr wash; all washes were done in 6× SSC at room temperature. The gel was then given a 2.5-min stringent wash at 67° and subjected to autoradiography as described in the text. The X-ray film was exposed for 3.5 days, and the resulting autoradiogram is shown above.

one should be willing to experiment in determining the optimal hybridization conditions for any new oligonucleotide.

To improve hybridization signals and reduce nonspecific oligonucleotide binding, different gel concentrations and buffers have been used in the electrophoresis of genomic restriction digests.[14] To analyze hybridization to restriction fragments greater than 10 kb, the agarose concentration is reduced to 0.4–0.5%, and Tris–acetate buffer (40 m*M* Tris base, 20 m*M* Na acetate, 2 m*M* EDTA, adjusted to pH 7.8 with acetic acid) is used. This gel system, when run at low voltages (1 volt/cm) and with buffer recirculation, produces better separation of restriction fragments greater than 10 kb when compared to the TBE buffer system. Agarose gels of 0.4% are poured with 1.5 times the normal gel volume to strengthen the resulting dried gel and make it easier to handle in subsequent manipulations.

The 16-hr hybridization time that was used in the described experiments with the K^b oligonucleotide produced the strongest radioactive signals in hybridizations to genomic restriction digests. The data in Fig. 2 indicate that a 16-hr hybridization should result in a 2.5-fold stronger signal when compared to the previously used 3-hr hybridization time.[9,14] The 16-hr hybridization time also produced other benefits in the given hybridization protocol although the reason for these benefits is not known. The longer hybridization time reduced the amount of spottiness, which is quite common in this type of hybridization, in the resulting autoradiographic exposure. The longer hybridization time also reduced nonspecific binding of the oligonucleotide as seen in the lack of hybridization to phage lambda DNA size markers present in the gel.

The probe concentration used in these experiments is approximately 1 ng/ml. According to formulations of Wetmur and Davidson,[16] the hybridization of the K^b oligonucleotide should have been half-complete after 0.92 hr if this were a liquid-phase hybridization. The data in Fig. 2 show that in these experiments the oligonucleotide hybridization was half-complete after 4–5 hr. Wallace *et al.*[1] had also shown that a 14-base oligonucleotide hybridized to DNA bound to nitrocellulose at a rate 3–4 times slower than that predicted by the Wetmur and Davidson calculations.

Other Oligonucleotide Hybridization Techniques

Genomic Restriction Digests Bound to Hybridization Membranes

In our hands, oligonucleotide probes produce a hybridization signal approximately 5 times stronger with a genomic restriction digest hybrid-

[16] J. G. Wetmur and N. Davidson, *J. Mol. Biol.* **31,** 349 (1968).

ized directly in the agarose gel matrix than the same digest that has been transferred to nitrocellulose or similar hybridization membrane. There are times, however, when it is absolutely necessary to transfer the DNA to a hybridization membrane prior to its hybridization with an oligonucleotide. Oligonucleotide hybridizations to restriction fragments less than 1.5 kb in size require the initial transfer of the DNA to a hybridization membrane since restriction fragments of this size tend to leach out of the agarose gel matrix during the hybridization and subsequent washes. Acid depurination (nicking)[17] of the DNA, which facilitates its transfer out of the gel, should be avoided since oligonucleotides do not hybridize efficiently to DNA treated in this manner. The following protocol was used to detect a 600-bp restriction fragment with the K^b oligonucleotide in 10 μg of a genomic restriction digest that had been transferred to GeneScreen *Plus* (New England Nuclear) hybridization membrane.[14]

Following electrophoresis in a 2% agarose gel buffered in TBE, the DNA samples are stained in ethidium bromide and photographed. The gel is then soaked in 0.4 *M* NaOH, 0.6 *M* NaCl for 30 min at room temperature followed by an incubation in 1.5 *M* NaCl, 0.5 *M* Tris–Cl (pH 7.5) for 30 min at room temperature. The DNA is then transferred to GeneScreen *Plus* by electrophoretic blotting in Tris–acetate buffer. The transfer is done in a commercially available apparatus at 4° for 4 hr at 10 volts followed by 2 hr at 50 volts. The membrane is prehybridized in 10× Denhardt's solution (Denhardt's solution = 0.02% bovine serum albumin, 0.02% polyvinylpyrrolidone, and 0.02% Ficoll) containing 0.5% SDS for 1 hr at 60°. The membrane is then washed in 2× SSC and blotted dry. The membrane is hybridized in a solution that contains 5× Denhardt's solution, 5× SSPE, 0.5% SDS, and 2×10^6 cpm/ml of labeled oligonucleotide at 60°. Although a 3-hr hybridization was originally used, recent experiments with genomic restriction digests bound to GeneScreen *Plus* hybridization membrane have shown that a 16-hr hybridization will provide a stronger signal. After hybridization the membrane is washed 3 times for 15 min each in 6× SSC containing 0.5% SDS at room temperature. The membrane was then given a 1.5-min stringent wash in 6× SSC containing 0.5% SDS at 67° prior to autoradiography.

Cloned Restriction Fragments in Southern or Dried-Gel Hybridizations

The same techniques used to detect restriction fragments in genomic restriction digests may be used for cloned DNAs. The hybridization time may be reduced to a couple of hours, however, since the amount of the hybridizing restriction fragment is much greater and results in a much stronger radioactive signal.

[17] G. M. Wahl, M. Stern, and G. R. Stark, *Proc. Natl. Acad. Sci. U.S.A.* **76,** 3683 (1979).

RNA in Northern Hybridizations[9]

RNA or control DNA samples are denatured in 50% formamide, 2.2 *M* formaldehyde and MOPS running buffer (20 m*M* MOPS, pH 7, 5 m*M* sodium acetate, and 1 m*M* EDTA) for 15 min at 55°. Samples are then subjected to electrophoresis in a 2.2 *M* formaldehyde, 1% horizontal agarose gel in MOPS running buffer for 3 hr at 3.5 volts/cm. Following electrophoresis, the gel is rinsed twice with distilled water, and the samples are transferred to GeneScreen (New England Nuclear) hybridization membrane overnight by capillary action in 1× SSC. The next day the filter is baked under reduced pressure for 2 hr at 80°. The RNA blots are prehybridized in 10× Denhardt's solution containing 0.1% SDS for 1 hr at 60°. The prehybridized filters are washed in 2× SSC and blotted dry. The hybridization is done in a solution containing 5× Denhardt's solution, 5× SSPE, 0.1% SDS, and 1×10^6 cpm/ml of labeled oligonucleotide for 3 hr at 60°. The blots are then washed 3 times for 15 min each in 6× SSC at room temperature. The blot is then given a 1.5-min stringent wash followed by a second 1-min stringent wash. Probes are removed by giving the blot two 15-min washes in 0.1× SSC at 60° prior to hybridization with a different probe.

In the above protocol, the hybridization and wash temperatures are determined as described for genomic restriction digests. The position of the 28 and 18 S rRNAs (when total RNA is being hybridized) may be determined if the blot is exposed to X-ray film after a single stringent wash, because of nonspecific binding of the oligonucleotide. Transcripts from the *Q10* gene, which represent 0.015% of the clones in a liver cDNA library,[18] were detected in 5 μg of LiCl-precipitable liver RNA after a 3-day autoradiographic exposure with the above protocol.[9]

Colony Screening

Colonies containing recombinant plasmids are transferred to Whatman 540 filter circles and prepared for hybridization as described by Gergen *et al.*[19] The filters are then rinsed in a solution containing 6× NET (1× NET = 150 m*M* NaCl, 15 m*M* Tris–Cl, pH 8.3, 1 m*M* EDTA), 10× Denhardt's solution, and 0.1% SDS. The filters are hybridized in the above solution containing 1×10^6 cpm/ml of labeled oligonucleotide at the appropriate temperature ($T_d - 5°$) for 2 hr. The filters are then washed 3 times for 5 min each in 6× SSC at 30–40° below the hybridization temper-

[18] J.-L. Lalanne, C. Transy, S. Guerin, S. Darche, P. Meulien, and P. Kourilsky, *Cell* **41,** 469 (1985).

[19] J. P. Gergen, R. H. Stern, and P. C. Wensink, *Nucleic Acids Res.* **7,** 2115 (1979).

ature. The filters are then exposed to X-ray film 1–4 hr at −70° with two intensifying screens. The filters can then be washed in 6× SSC for 1 min at the hybridization temperature and reexposed to X-ray film. Comparison of the two autoradiograms will show the effect of the higher temperature wash and help identify positive colonies.

Plaque Screening

A prehybridization step is beneficial in the screening of recombinant phage by the method of Benton and Davis[7] with oligonucleotide probes. The filters are prehybridized in a solution containing 6× NET, 10× Denhardt's solution, 0.1% SDS, and 50 μg/ml of sonicated, denatured *E. coli* DNA at 65° for 4 hr. The prehybridization solution is then removed, and the filters are hybridized as described in the colony-screening section. It is necessary to expose the filters to X-ray film about 10 times longer than for the colony screening owing to the lower signals obtained with this procedure.

Acknowledgments

The work described was supported by a grant from the National Institutes of Health (GM31261) to RBW. RBW is a member of the Cancer Center at the City of Hope (NIH CA33572).

[7] λgt 11: Gene Isolation with Antibody Probes and Other Applications

By MICHAEL SNYDER, STEPHEN ELLEDGE, DOUGLAS SWEETSER, RICHARD A. YOUNG, and RONALD W. DAVIS

Genes can be isolated with either nucleic acid probes[1] or antibody probes. For cloning genes that are expressed at a low level or not at all in the organism of study, oligonucleotide probes can be prepared if protein sequence data are available. Alternatively, antibodies can be used as probes to isolate directly recombinant clones producing proteins of interest regardless of whether protein sequence is available. This chapter describes methods for the isolation of eukaryotic and prokaryotic genes by

[1] T. Maniatis, E. F. Fritsch, and J. Sambrook, "Molecular Cloning: A Laboratory Manual." Cold Spring Harbor Lab., Cold Spring Harbor, New York, 1982.

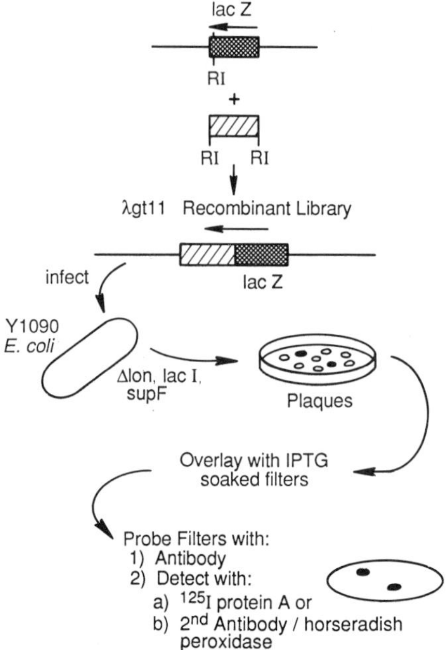

FIG. 1. Screening λgt 11 expression libraries with antibody probes. A recombinant DNA library is first constructed in λgt 11. The library is plated on bacterial cells to form plaques. After initial propagation of the phage, antigen production is induced by overlaying the plates with IPTG-treated filters, and phage growth is continued. The antigen-coated filters are then removed and probed with antibody. The bound antibody is detected with ^{125}I-labeled protein A or second antibody techniques to find the desired recombinant.

screening *Escherichia coli* expression libraries with antibody probes using the bacteriophage expression vector λgt 11.[2,3] Other uses of λgt 11, including preparation of foreign proteins expressed in *E. coli,* mapping epitope coding regions, and transplason mutagenesis are also described.

Gene Isolation by Immunoscreening

The general scheme for immunoscreening is shown in Fig. 1.[3-5] A recombinant DNA library is first constructed in λgt 11, an *E. coli* expres-

[2] R. A. Young and R. W. Davis, *Proc. Natl. Acad. Sci. U.S.A.* **80,** 1194 (1983).

[3] R. A. Young and R. W. Davis, *Science* **222,** 778 (1984).

[4] R. A. Young and R. W. Davis, *Genet. Eng.* **7,** 29 (1985).

[5] M. Snyder and R. W. Davis, *in* "Hybridomas in the Biosciences and Medicine" (T. Springer, ed.), p. 397. Plenum, New York, 1985.

sion vector, and foreign antigens are expressed from the DNA inserts in particular *E. coli* host cells. The antigens are transferred onto nitrocellulose filters and then probed with antibodies to detect the desired recombinant.

λgt 11

The features of λgt 11 are shown in Fig. 2.[3] λgt 11 contains the *lacZ* gene of *E. coli* which can express foreign DNA inserts as β-galactosidase fusion proteins. cDNA or genomic DNA fragments are inserted into the unique *Eco*RI site located in the 3′ end of the *lacZ* gene, 53 bp upstream of the translation termination codon. When the DNA fragments are introduced in the proper orientation and reading frame, the foreign DNA is expressed as a β-galactosidase fusion protein. Fusion of a foreign protein to a stable *E. coli* protein enhances the stability of the foreign protein in at least several cases.[6,7] However, antigens can also be expressed and not fused to *lacZ* (see below).

Since the *lacZ* gene has a strong *E. coli* promoter, and since λgt 11 propagates to high copy number during lytic infection, a significant percentage of total *E. coli* cellular protein is produced from the *lacZ* gene. For *lacZ* alone (unfused to any other sequences), approximately 5% of the total cellular protein is β-galactosidase.[3,5] For the λgt 11 fusions examined thus far, the intact fusion proteins comprise from less than 0.1% up to 4% of the total *E. coli* protein.[3,5]

In addition to lytic functions, λgt 11 contains the genes necessary for lysogeny. Conversion of growth from the lysogenic state to lytic is readily controlled by the presence of the *cI857* gene, which encodes a temperature-sensitive *cI* repressor. λgt 11 also contains *S100*, an amber mutation in the *S* lysis gene. As described further below, this mutation is useful for producing proteins from isolated clones, because large amounts of protein can be accumulated in the bacterial cells without extensive cell lysis.

Bacterial Host

The *E. coli* host, Y1090, has three features that are useful for screening λgt 11 expression libraries with antibody probes. The bacterial strain used is deficient in the *lon* protease. In lon^- cells β-galactosidase fusion proteins often accumulate to much higher levels relative to wild-type cells.[3]

[6] H. Kupper, W. Keller, C. Kurtz, S. Forss, H. Schaller, R. Franze, K. Strommaier, O. Marquardt, V. Zaslavsky, and P. H. Hofschneider, *Nature (London)* **289,** 555 (1981).

[7] K. Stanley, *Nucleic Acids Res.* **11,** 4077 (1983).

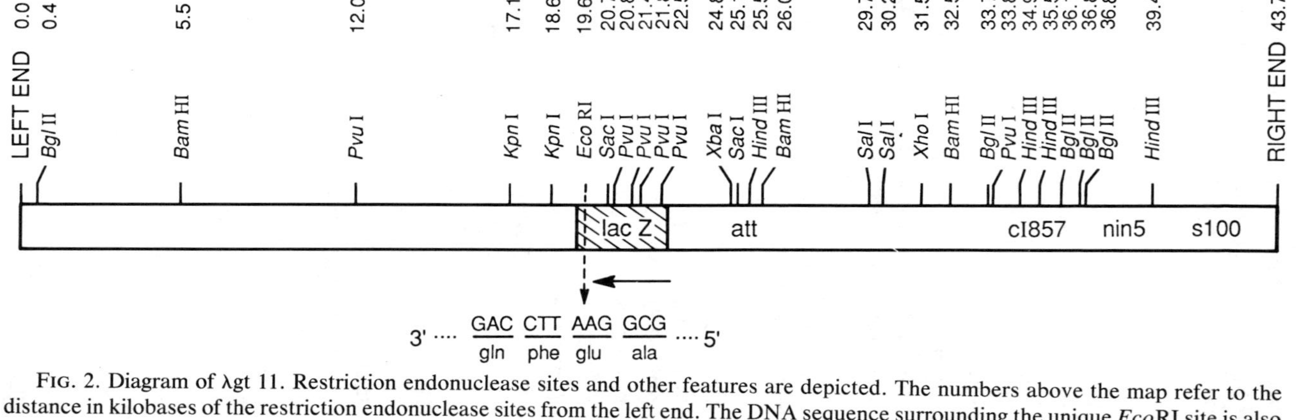

FIG. 2. Diagram of λgt 11. Restriction endonuclease sites and other features are depicted. The numbers above the map refer to the distance in kilobases of the restriction endonuclease sites from the left end. The DNA sequence surrounding the unique *Eco*RI site is also shown.

The bacterial strain contains pMC9, a pBR322 plasmid harboring the *lacI* gene which encodes the *lac* repressor.[8] This allows the regulated expression of β-galactosidase fusion proteins. Expression of foreign proteins, some of which may be harmful to growth of the bacterial host, is repressed during initial growth of the phage, but is expressed later after induction with isopropyl-β-D-thiogalactoside (IPTG). In the presence of the *lon* mutation and absence of *lac* repressor plasmids, λgt 11 recombinants typically form smaller plaques.

The third feature of the *E. coli* host is the presence of the *supF* suppressor tRNA. *supF* suppresses the *S100* mutation of λgt 11 and allows plaque formation. Plaques are preferable for immunoscreening because they provide a much better signal/noise ratio than colony screening methods.

λgt 11 Recombinant DNA Libraries

The most important components of successful immunoscreening are the recombinant DNA library and the antibody probe. Two types of recombinant DNA libraries can be used: cDNA libraries and genomic DNA libraries. For organisms which have a large genome size, such as mammals (3×10^9 bp), cDNA libraries are screened because currently it is not technically feasible to screen a sufficient number of genomic recombinants to find the desired gene (see below). cDNA libraries may also be preferable for screening when a gene is abundantly expressed. The construction of cDNA libraries is described elsewhere.[9] Near full-length cDNAs provide the most antigenic determinants and are preferable for immunoscreening.

However, λgt 11 cDNA expression libraries suffer from the fact that nonabundant mRNAs are represented very rarely, if at all, and often encode only a 3′-terminal portion of the gene. It is useful to be able to clone any gene, independent of its level of expression. This can be achieved by using a genomic DNA library, constructed by randomly shearing and inserting genomic DNA into the λgt 11 *lacZ* gene. For organisms with a genome size equal to *Drosophila* (1.8×10^8 bp) or smaller, the construction and screening of genomic libraries is possible.

In genomic DNA libraries, the probability of having a particular gene fused to *lacZ* depends on the length of the gene and the size of the organism's genome. For instance, fusion of any portion of a 1.5-kb yeast

[8] M. P. Calos, T. S. Lebkowski, and M. R. Botchan, *Proc. Natl. Acad. Sci. U.S.A.* **80,** 3015 (1983).

[9] T. V. Huynh, R. A. Young, and R. W. Davis, *in* "DNA Cloning Techniques: A Practical Approach" (D. Glover, ed.), p. 49. IRL Press, Oxford, England, 1984.

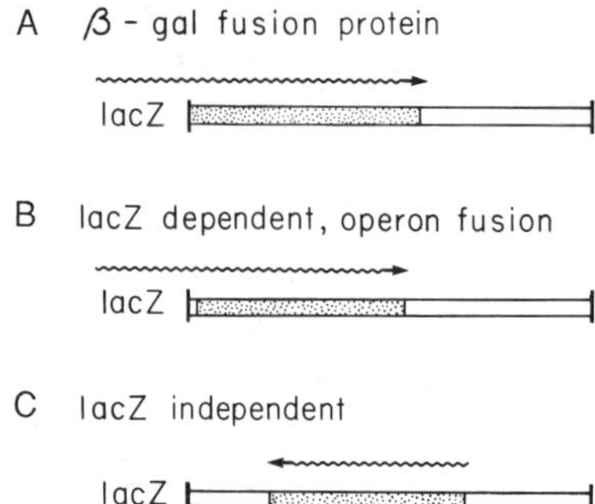

FIG. 3. Modes of expression from λgt 11 genomic clones. Expression can be (A) as β-galactosidase fusion proteins or (B and C) not fused to β-galactosidase. In Case B antigen production is still *lacZ* dependent and expression occurs via an operon-type structure. In Case C expression is *lacZ* independent and insert sequences serve for transcription and translation in *E. coli*.

gene (yeast genome = 14,000 kb) to *lacZ* will occur once in every 9,333 times, and one out of six fusions will reside in the proper orientation and reading frame. Therefore, an inframe fusion will occur an average of once in every 5.6×10^4 recombinants. Fusions nearest the amino-terminal protein coding sequences will encode more antigenic determinants than carboxy-terminal protein fusions. In addition, the frequency of detecting a gene within a genomic library may be affected by intervening sequences. Organisms of small genome size, such as *Drosophila* and yeast, usually contain few if any introns within their coding sequences.

One interesting feature of λgt 11 libraries is that not all of the recombinants isolated encode β-galactosidase fusion proteins, a feature often encountered when using genomic libraries (Fig. 3).[10–13] For yeast, greater than 50% of the genes isolated from a genomic library by immunoscreening are expressed but not fused to β-galactosidase. In some instances, the expression of these genes may still be *lacZ* dependent and presumably use the *lacZ* promoter for expression (Fig. 3, Case B). Alternatively, expression can be *lacZ* independent, and insert sequences serve for transcription and translation in *E. coli* (Fig. 3, Case C). The mode of expression will also affect the relative frequency with which a gene is isolated from a genomic DNA library.

A general method for constructing a random shear genomic DNA

[10] J. L. Kelly, A. L. Greenleaf, and I. R. Lehman, *J. Biol. Chem.* **261,** 10348 (1986).

[11] T. Goto and J. C. Wang, *Cell* **36,** 1073 (1984).

[12] E. Ozkaynak, D. Finley, and A. Varshavsky, *Nature* (*London*) **312,** 663 (1984).

[13] M. Snyder, S. Elledge, and R. W. Davis, *Proc. Natl. Acad. Sci. U.S.A.* **83,** 730 (1986).

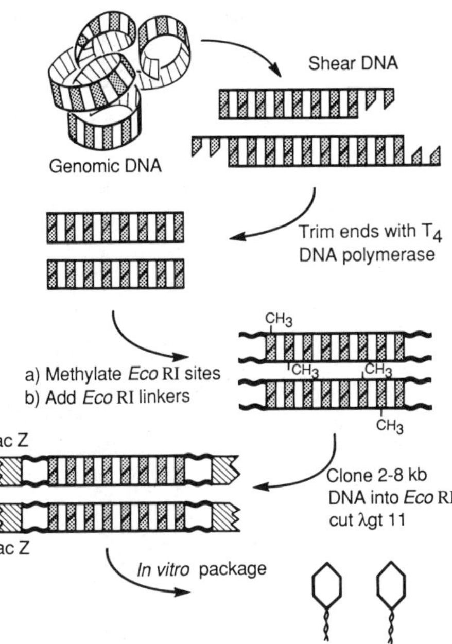

FIG. 4. Construction of a genomic DNA library. Genomic DNA is randomly sheared through a syringe needle to an average size of 5 kb. The ends are trimmed with T4 DNA polymerase and the DNA methylated with *Eco*RI methylase. *Eco*RI linkers are ligated on the ends, cleaved with *Eco*RI, and the DNA size fractionated on a 0.7% agarose gel. The 2- to 8-kb DNA is collected and cloned into *Eco*RI-cleaved λgt 11 that has been treated with calf alkaline phosphatase.

library is diagramed in Fig. 4. DNA is randomly sheared to an average size of 5 kb, and the ends made flush with T4 DNA polymerase I. The DNA is subsequently methylated, *Eco*RI linkers ligated and cleaved, and the DNA is size fractionated on an agarose gel. The 2- to 8-kb fraction is then cloned into the *lacZ* gene of λgt 11. A protocol is as follows:

1. Genomic DNA preparation: Shear the genomic DNA (15–20 μg at 100–200 μg/ml) to an average size of 5 kb by passing DNA vigorously through a 27-gauge syringe needle with a 1-ml syringe. The sheared DNA size is monitored by gel electrophoresis using a 0.7% agarose gel.[1]
2. Trim the DNA ends in 100 μl of 50 m*M* Tris–HCl (pH 7.8), 5 m*M* $MgCl_2$, 50 μg/ml BSA, 10 m*M* 2-mercaptoethanol, 20 μ*M* of each dNTP, and 5 units of T4 DNA polymerase. Incubate for 30 min at room temperature.
3. Methylate the DNA in 50 μl of 50 m*M* Tris–HCl (pH 7.5), 1 m*M*

EDTA, 5 m*M* DTT, 10 μM *S*-adenosyl-L-methionine plus 2 μg *Eco*RI methylase. Incubate at 37° for 15 min, then inactivate the enzyme by heating for 10 min at 70°.

4. (Optional step) To ensure blunt DNA ends, add $MgCl_2$ to 10 m*M* and dNTP to a final concentration of 20 μM. Add 2 units of *E. coli* DNA polymerase I large (Klenow) fragment. Incubate for 30 min at room temperature. Extract once with phenol : $CHCl_3$ (50 : 50), then once with $CHCl_3$ alone. Ethanol precipitate DNA.
5. Add phosphorylated *Eco*RI linkers on DNA using 50 μl of 70 m*M* Tris–HCl (pH 7.5), 10 m*M* $MgCl_2$, 200 μg/ml gelatin, 1 m*M* DTT, 5% polyethylene glycol, 1 m*M* rATP, 100 μg/ml phosphorylated *Eco*RI linkers plus 0.5 μl of 1.5 mg/ml T4 DNA ligase. Incubate at room temperature for 6–12 hr. Ethanol precipitate DNA.
6. Dissolve pellet in 100 μl *Eco*RI restriction digest buffer [50 m*M* NaCl, 100 m*M* Tris–HCl (pH 7.5), 7 m*M* $MgCl_2$, 1 m*M* DTT]. Add 50 units *Eco*RI and incubate 1.5 hr at 37° then 10 min at 70°. Ethanol precipitate the DNA by adding 2.5 volumes of ethanol and chilling at −20° for 1 hr. Repeat ethanol precipitation.
7. Purification of DNA inserts from *Eco*RI linkers and size fractionation: The DNA is completely dissolved in electrophoresis buffer and size-fractionated on a 0.8% agarose gel. The 2- to 8-kb DNA is collected from the gel. Two convenient methods are (1) Cut an empty well in front of the DNA to be recovered. As the gel is run at high current, remove aliquots from the well every 45–60 sec. Recover the DNA by adding sodium acetate (pH 6) to 0.3 *M* and precipitating with ethanol. (2) Place Schleicher and Schuell NA45 membrane in a gel slot cut just in front of the DNA to be collected. Run the DNA into the membrane at high current. Rinse the NA45 membrane in electrophoresis buffer and elute DNA in 1 *M* NaCl, 50 m*M* arginine (free base) at 70° for 2–3 hr. Extract the DNA with phenol : $CHCl_3$ as in step 4 and recover DNA by ethanol precipitation. *Note:* Removing the linkers is a crucial step. If linker contamination presents a problem, repeat the gel isolation step or fractionate the DNA on a P60 column prior to the gel isolation step.
8. λgt 11 vector preparation: Ligate the λgt 11 DNA to form concatamers according to the buffer conditions in step 5 (without linkers) for 2 hr at room temperature. Incubate at 70° for 15 min and ethanol precipitate DNA. Cleave DNA with *Eco*RI. Treat with calf intestine alkaline phosphatase (CAP) in 50 m*M* Tris–HCl (pH 9.0) using high quality enzyme that has been titered to yield less than 10% vector in the presence of insert DNA. Do not overtreat.

Inactivate the CAP by heating at 70° for 10 min. Extract DNA with phenol : $CHCl_3$ as in step 4 and ethanol precipitate DNA.

9. Library construction: Ligate the genomic insert DNA into λgt 11 using a ratio of insert DNA : vector DNA of 0.5–1 : 1 by mass using buffer conditions in step 5. A final DNA concentration of 4–800 μg/ml is used. Incubate at room temperature for 8–20 hr.
10. *In vitro* package DNA and titer library according to Huynh *et al.*[9] Yield is approximately 5×10^5 phage/μg λ DNA.
11. Amplify library on *E. coli* strain Y1088 on LB plates containing 50 μg/ml ampicillin [Y1088 = Δ*lacU169 supE supF hsdR*$^-$ *hsdM*$^+$ *metB trpR tonA21 proC*::Tn*5* + pMC9]. This bacterial strain is *hsdR*$^-$ and *hsdM*$^+$ and contains the *lacI* plasmid in order to prevent the expression of proteins that are harmful to *E. coli*.

Using these methods λgt 11 genomic DNA libraries containing 10^6–10^7 recombinants from yeast[5] and mycobacteria[14] have been constructed.

Antibody Probes

Successful immunoscreening depends on the quality of the antibody. Antibodies that produce good signals on immunoblots or "Westerns" usually work well for antibody screening. Both monoclonal and polyclonal antibodies have been used successfully. Polyclonal antisera have the advantage that they can recognize multiple epitopes on any given protein. The ability to detect a variety of epitopes on a protein is important, since genomic and cDNA inserts often will not be full length and hence only a portion of the polypeptide will be expressed. Since a single epitope can be shared by multiple proteins, clones other than the desired recombinants can also be detected. Therefore, immunoscreening with polyclonal antisera or with several monoclonal antibodies recognizing different determinants on the same protein is recommended.

For a polyclonal serum it is usually not necessary to use antibodies that have been affinity purified for the protein of interest. However, this can lead to the isolation of clones other than the recombinant of interest, particularly if the original serum is made against an impure protein.

Most rabbit and human sera contain significant amounts of anti-*E. coli* antibodies. These are easily removed by "pseudoscreening" *E. coli* protein lysates. λgt 11 phage (without inserts) are plated and protein-coated filters prepared according to the screening conditions described in the

[14] R. A. Young, B. R. Bloom, C. M. Grosskinsky, J. Ivanyi, D. Thomas, and R. W. Davis, *Proc. Natl. Acad. Sci. U.S.A.* **82,** 2583 (1985).

next section. The filters are treated and incubated with antibodies to absorb the anti-*E. coli* antibodies from the serum. It is usually necessary to repeat the pseudoscreening 2 or more times to remove *all* anti-*E. coli* antibodies.[15] The same λgt 11 plates can be reused: after the first filter is removed the plates are overlayed with a second filter and incubated for an additional 2 hr at 37°.

Screening λgt 11 Libraries with Antibody Probes

1. Plate out library: Grow a culture of Y1090 cells at 37° to saturation in LB media[16] containing 50 μg/ml ampicillin and 0.2% maltose. Per 150-mm LB plate infect 0.1 ml of the saturated culture with 1×10^5 phage. Allow phage to absorb for 25 min at room temperature or 15 min at 37°. Add 6.5 ml LB top agar and plate immediately. Neither top agar nor plates contain ampicillin. Plates work best 3–5 days after preparation. Incubate plates for 3.0 hr at 42° until lawn just becomes visible.
2. Overlay plates with IPTG-treated nitrocellulose filters prepared as described below. Incubate the plates at 37° for 8–10 hr (overnight). At the end of the incubation period, remove the lid from the plates and incubate at 37° for an additional 10–15 min. This step helps prevent top agar from sticking to the filter.
3. Mark the position of the filters with a syringe needle containing ink. Carefully remove filters from plates. For duplicate screening, overlay the plates with a second IPTG-treated filter and return to 37° for an additional 2–4 hr.

   For subsequent steps, filters are treated in such a manner that they are well exposed to the washing and probing solutions. One convenient method is to use two filters per Petri dish.
4. Rinse filters 1–2 times in TBS (15 ml/filter) (TBS = 150 m*M* NaCl, 50 m*M* Tris–HCl, pH 8.1).
5. Incubate filters for 30 min or longer in TBS plus 0.05% Tween 20 plus 0.5% BSA or TBS plus 20% fetal calf serum (FCS) (15 ml/filter). (These solutions may be saved and reused several times.)
6. Probe filters in the antibody solution diluted in TBS plus 0.05% Tween 20 plus 0.1% BSA or TBS plus 20% FCS (10 ml/filter) for 1–8 hr.
7. Wash filters 10 min each with (1) TBS plus 0.1% BSA, (2) TBS plus 0.1% BSA plus 0.1% NP40, (3) TBS plus 0.1% BSA (Tween

[15] T. St. John, personal communication.

[16] R. W. Davis, D. Botstein, and J. R. Roth, "Advanced Bacterial Genetics." Cold Spring Harbor Lab., Cold Spring Harbor, New York, 1980.

20 optional). The antibody probes can then be detected by either methods 8A or 8B.

8A. ^{125}I-Labeled protein A probe:
   a. Treat filters with ^{125}I-labeled protein A in TBS plus 0.1% BSA for 1.5–2.5 hr. Use 1 μCi of >30 mCi/mg specific activity ^{125}I-labeled protein A (ICN or Amersham) per 132 mm filter.
   b. Wash filters 10 min each with (1) TBS plus 0.1% BSA, (2) TBS plus 0.1% BSA plus 0.1% NP40, repeat once, (3) TBS plus 0.1% BSA.

8B. Horseradish peroxidase/alkaline phosphatase conjugated probes:
   a. Incubate filters with second antibody according to the manufacturer's recommendations (see below).
   b. Wash filters and incubate with substrate solution according to manufacturer's instructions.

Sample positives of these screens using ^{125}I-labeled protein A or horseradish peroxidase probes are shown in Fig. 5.

Comments

1. General methods for plating and handling bacteriophage λ can be found in Maniatis *et al.*[1] and Davis *et al.*[16]
2. IPTG-treated filters are prepared by wetting nitrocellulose filters in solution of 10 m*M* IPTG dissolved in distilled water and allowed to air dry on plastic wrap.
3. Incubation times for most of the steps can be varied greatly. Incubation of IPTG-treated nitrocellulose filters on the plates for 8–10 hr produces signals 5–10 times stronger than for a 2-hr incubation. Overnight incubation with antibody yields a 3- to 5-fold stronger signal relative to a 2-hr incubation.
4. Most antibodies produce good signals at room temperature. Signals generally increase 3- to 5-fold if the antibody treatments and subsequent incubations are performed at 4°. This is the recommended temperature for low-affinity antibodies.
5. The antibody solution can be reused many times and is stored with 0.01% sodium azide after each use. For anti-DNA polymerase I antibodies, the serum was reused 12 times with little reduction in intensity.[17] Every time the serum is used, the background is reduced, presumably because of the further removal of anti-*E. coli* antibodies from the serum.

[17] L. M. Johnson, M. Snyder, L. M. S. Chang, R. W. Davis, and J. L. Campbell, *Cell* **43,** 369 (1985).

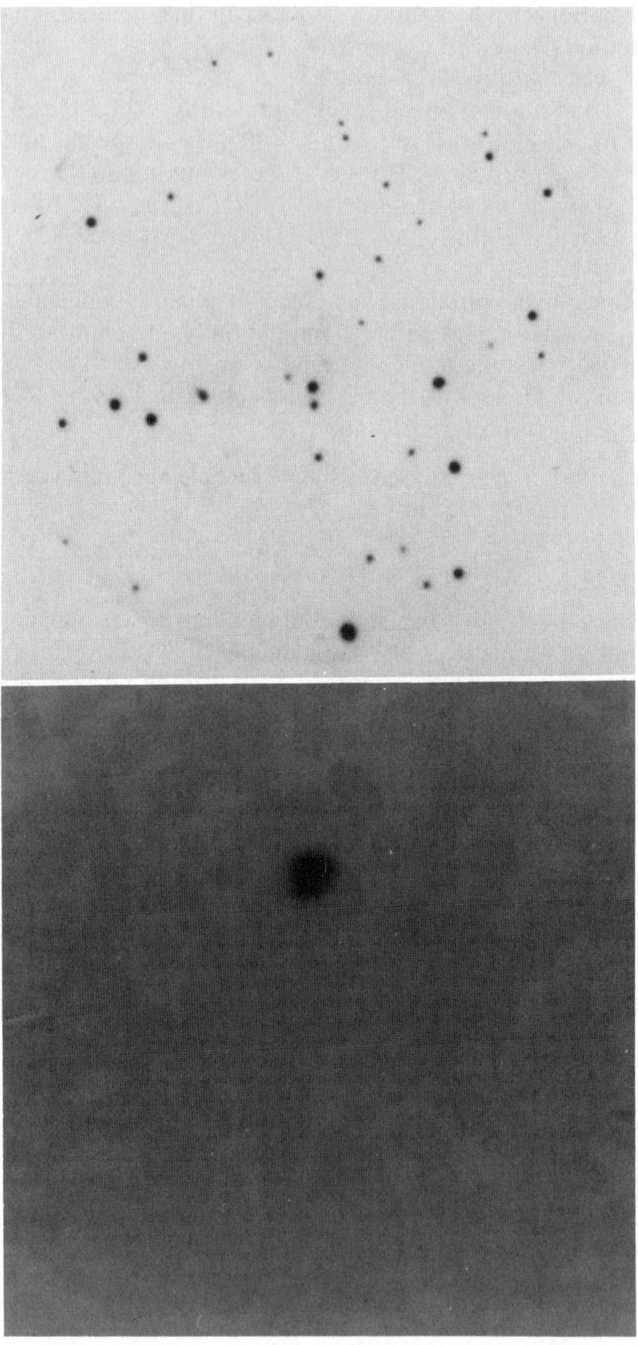

6. For primary screens it is recommended that plates be screened in duplicate to avoid false positives. For ^{125}I-labeled protein A these usually occur at a frequency of one or two per 132-mm filter. The number of phage to be used should be low enough such that the lawn is not completely lysed prior to the application of the second filter.
7. Filters can be stored for several weeks in TBS plus 20% FCS prior to probing with antibody.
8. Control experiments where known amounts of protein antigen have been spotted onto filters and screened with antibody as above indicate that the limit of detection for one high-titer serum is approximately 100 pg of protein.[18] Actual amounts of antigen transferred from a single plaque vary up to approximately 100 times higher than this for the positives thus far isolated.
9. Ampicillin, which is used to select for cells containing the *lacI* plasmid, is omitted in the screening plates because it slows cell growth.
10. High specific activity ^{125}I-labeled protein A (>30 mCi/mg) is often necessary for detecting positive clones.
11. Numerous methods exist for enzymatic detection of the antibody using second antibodies. These include horseradish peroxidase probes (Bio-Rad, Vector Laboratories) or alkaline phosphatase probes (Promega Biotec, Vector Laboratories).
12. Common substitutes for 0.05% Tween 20/0.5% BSA or 20% FCS protein blocks (step 5) are 3% BSA or 5% solution of powdered milk. Substrates that work well for immunoblots also work well for immunoscreening.

Verification of Gene Identity

The most difficult part of isolating genes by antibody screening is to determine whether the immunoreactive clones encode the gene of inter-

[18] C. Glover and M. Snyder, unpublished results.

FIG. 5. Exemplary positives from immunoscreening. Top: A positive yeast DNA polymerase–λgt 11 clone (approximately 30/plate) was plated with 10^4 nonreactive phage on a 90-mm plate. The antigens were immunoscreened with an anti-DNA polymerase I polyclonal serum and detected with ^{125}I-labeled protein A. Bottom: High magnification of a positive signal from a *Mycobacterium leprae* clone immunoscreened with a monoclonal antibody and detected with horseradish peroxidase conjugated probes (Vectastain probes from Vector Laboratories).

est. Several useful methods for determining the identity of clones are as follows.

A. Affinity Purification of Antibodies Using λgt 11 Clones. Proteins produced from the λgt 11 clones can be used to affinity purify antibody from the screening serum. The affinity-purified antibody is then used to probe immunoblots of a purified or crude preparation of the protein of interest. This will determine if the isolated clone is antigenically related to the protein of interest, and is particularly useful if the original antiserum was not made against a homogeneous protein. A protocol is as follows[5]:

1. Plate the positive recombinants at 2–5 × 10^4 phage/90-mm LB plate. Use 2.5 ml of LB top agar and 50 μl of a saturated culture of Y1090 cells. Follow steps 1–7 of the immunoscreening protocol for preparing filters, and probing them with antibody and washing. One plate per clone produces sufficient antigen for most sera.
2. Place the antibody-treated filter in a capped tube. Add 3.5 ml of 0.2 *M* glycine–HCl (pH 2.5). Mix the tube for 2 min and no longer. Remove filter.
3. *Immediately* add 1.75 ml of 1.0 *M* KPO_4 (pH 9.0) plus 5% FCS and mix.
4. Dilute to lower ionic strength by adding 6 ml of distilled water plus 2 ml of TBS. Add FCS to 20% and use directly for immunoblots.

Immunoblot protocols are described by Burnette.[19] Alternative antibody elution protocols are detailed by Earnshaw and Rothfield[20] and Smith and Fisher.[21]

B. Gene Expression. The insert from the λgt 11 recombinant can be used to recover an intact gene by hybridization techniques. The intact gene can be overexpressed in *E. coli* or other organisms and assayed for gene activity. Overexpression in *E. coli* has been used for yeast topoisomerase II[11] and yeast DNA polymerase I.[17] Overexpression in yeast has also been performed for yeast DNA polymerase I.[17]

C. DNA Sequencing. The DNA sequence of the λgt 11 recombinant insert can be determined, and the deduced amino acid sequence compared with the known protein sequence. Primers are available (New England Biolabs) that are complementary to the *lacZ* portion of λgt 11 adjacent to the *Eco*RI cloning site. These primers can be used to direct dideoxy sequencing by the method of Sanger[22] (see below).

[19] W. N. Burnette, *Anal. Biochem.* **112,** 195 (1981).

[20] W. C. Earnshaw and N. Rothfield, *Chromosoma* **91,** 313 (1985).

[21] D. E. Smith and P. A. Fisher, *J. Cell Biol.* **99,** 20 (1984).

[22] F. Sanger, A. R. Coulson, B. G. Barrell, A. J. H. Smith, and B. Rose, *J. Mol. Biol.* **143,** 161 (1980).

Protein Preparation from λgt 11 Clones

Proteins can be prepared from Y1089 cells [Δ*lacU169 proA*$^+$ Δ*lon araD139 strA hflA* [chr::Tn*10*] (pMC9)] or CAG456 cells [*lac*(*am*) *trp*(*am*) *pho*(*am*) *supC*^{ts} *rpsL mol*(*am*) *htpR165*]. The relevant features of Y1089 are that it contains the *lacI* plasmid, Δ*lon* mutations, and the *hflA* mutations for forming lysogens at a high frequency, and the strain contains no suppressor—this latter feature allows λgt 11 protein products to accumulate to high levels without cell lysis. CAG456 cells also will not be lysed by λgt 11 and contains the *htpR* mutation which is a mutation in the heat shock sigma factor. The *htpR* mutation causes temperature-sensitive growth of the cells, and prevents the synthesis of heat shock proteins, many of which are proteases. This protease deficiency is observed at both 30 and 37°.[23] The level of protein production from two distinct yeast λgt 11 recombinant clones were compared in the two strains. Both fusion proteins were found to accumulate 10- to 100-fold higher in CAG456 cells relative to Y1089 cells.[24] However, the stability of different fusion proteins varies greatly among different strains of *E. coli* so it is useful to test several bacterial strains.

A. Preparation of Lysogens in Y1089

1. Infect a saturated culture of Y1089 cells grown in LB plus 0.2% maltose plus 50 μg/ml ampicillin with λgt 11 clones using an m.o.i. (multiplicity of infection) of 10. Absorb 20 min at room temperature.
2. Plate 200 cells per LB plate and incubate at 30°.
3. Pick individual colonies with a toothpick and test for growth at 42 and 30°. λgt 11 lysogens will not grow at 42°. Usually 30–50% of the colonies are lysogens.

B. Small-Scale Preparation of Protein Lysates from λgt 11 Lysogens

1. Four milliliters of lysogens are grown with vigorous shaking at 30° to OD_{600} = 0.4.
2. Lysogens are induced by shifting the temperature to 44° for 15 min with vigorous shaking. A rapid temperature shift is important for maximal induction.
3. IPTG is added, and the culture is incubated for 1 hr at 37°. The time of incubation varies with the particular recombinant; some clones promote rapid cell lysis and so must be harvested earlier.

[23] T. A. Baker, A. D. Grossman, and C. A. Gross, *Proc. Natl. Acad. Sci. U.S.A.* **81,** 6779 (1984).

[24] M. Snyder, M. Cai, and R. W. Davis, unpublished results.

4. Cells are harvested as quickly as possible by sedimentation at 10,000 *g* for 30 sec and freezing immediately at −70°. For SDS gel analysis, SDS sample loading buffer is added just prior to freezing. Four milliliters of cells yields sufficient protein for 8–10 gel lanes.

C. Preparation of Protein Lysates by Infection. This procedure is similar to that above. Cells (Y1089 or CAG456) are grown to $OD_{600} = 0.40$ and infected at a multiplicity of infection of 5 instead of inducing the culture. The culture is then incubated at 37° for the appropriate length of time: Y1089 cells, about 1 hr as above; CAG456 cells, 1–4 hr. (Determine the appropriate length of time.) Cells are then harvested as above.

The lysogen induction and the infection procedure produce comparable protein yields. However, the infection procedure requires much more phage, which can present a problem if the procedure is performed on a larger scale.

Mapping Epitopes on λgt 11 Clones

λgt 11 can be used to locate the boundaries of the antigenic determinants of a protein.[25] The strategy involves (1) isolating a DNA clone that encodes the entire antigen of interest and determining its nucleotide sequence, and consequently, the encoded amino acid sequence; (2) constructing a λgt 11 sublibrary containing fragments of the gene; (3) detecting the expression of epitope-coding sequences with monoclonal antibody probes; and (4) isolating and determining the precise nucleotide sequences of the cloned DNA fragments via primer-directed DNA sequence analysis. A comparison of shared sequences among the antibody-positive clones locates the epitope-coding region.

A λgt 11 clone that is capable of expressing the antigen of interest is isolated by immunoscreening or specifically constructed. It is important to confirm that the antigenic determinants of interest can be expressed. The foreign DNA is then sequenced, and the protein amino acid sequence is deduced.

The next step is to construct a λgt 11 gene sublibrary that contains small random DNA fragments from the gene of interest. The aim is to produce recombinant phage in sufficient numbers to obtain DNA insert end points at each base pair in the gene, such that all possible overlapping segments of the coding sequence are expressed. A protocol is as follows:

1. DNA fragments with random end points are generated by digesting with DNase I (1 ng DNase I/10 μg DNA/ml) in a buffer containing

[25] V. Mehra, D. Sweetser, and R. A. Young, *Proc. Natl. Acad. Sci. U.S.A.* **83,** 7013 (1986).

20 m*M* Tris–HCl (pH 7.5), 1.5 m*M* $MgCl_2$, and 100 μg/ml BSA at 24° for 10–30 min to produce short random fragments.

2. The DNA is fractionated on a 1% agarose gel, and fragments of 200–1000 bp are isolated and purified.
3. These DNA fragments are end repaired by treatment with T4 polymerase in the presence of dNTPs and then ligated to phosphorylated *Eco*RI linkers (Collaborative Research).
4. This material is then digested with *Eco*RI, heat inactivated at 70° for 5 min, and fractionated on a P60 column (Bio-Rad) to remove unligated linkers.
5. The linkered DNA fragments are further purified on an agarose gel, from which they are eluted, phenol extracted, and ethanol precipitated.
6. The *Eco*RI-linkered DNA fragments are ligated onto phosphatase-treated λgt 11 arms (Promega Biotec), and the ligated DNA is packaged *in vitro*.
7. The resultant recombinant phage are amplified on *E. coli* Y1088.

These steps are described in detail above.

The gene sublibrary is screened with monoclonal antibodies to isolate DNA clones that express the epitope-coding sequences. The limits of the sequences that encode an epitope are located by subjecting the recombinant clones to two types of analysis. First, the sequences of the DNA insert end points are determined for each clone. Second, the lengths of the DNA insert fragments are ascertained by restriction analysis.

The sequence of DNA insert end points in λgt 11 is determined as follows:

1. The recombinant DNA is first isolated from phage purified by CsCl block gradient centrifugation.[16]
2. DNA (1–5 μg) is digested with the restriction endonuclease *Kpn*I and *Sac*I, then phenol extracted, ethanol precipitated, and resuspended in 20 μl H_2O.
3. The DNA is denatured by adding 2 μl of 2 *M* NaOH, 2 m*M* EDTA, and the solution is incubated for 10 min at 37°. The solution is neutralized with 6.5 μl of 3 *M* sodium acetate (pH 5.2), 6.5 μl H_2O is added, and the DNA is ethanol precipitated, washed twice with 70% ethanol, and resuspended in 10 μl H_2O.
4. To the DNA (optimum amount should be determined empirically), 1 μl of 10 μg/ml DNA primer and 1.5 μl sequencing buffer (75 m*M* Tris–HCl, pH 7.5, 75 m*M* DTT, 50 m*M* $MgCl_2$) are added and incubated at 55° for 15 min. The two primers used (New England Biolabs) are complementary to *lacZ* sequences adjacent the *Eco*RI

site in λgt 11; the sequence of the "forward primer" is GGTG-GCGACGACTCCTGGAGCCCG, that of the "reverse" primer is TTGACACCAGACCAACTGGTAATG.

5. Primer extension and dideoxy termination reactions are performed as described by Sanger *et al.*,[22] and the products are subjected to electrophoresis on an 8% polyacrylamide–8 *M* urea gel.

The recombinant subclones are characterized further to ascertain whether they contain multiple inserts or rearranged insert DNA, either of which could complicate the interpretation of the data. DNA from each of the clones is digested with the restriction endonuclease *Eco*RI and is subjected to agarose gel electrophoresis to determine the number and sizes of inserted DNA fragments. Subclones that contain insert DNAs whose sequenced end points predict a DNA fragment length that agrees with the size determined by agarose gel electrophoresis of the *Eco*RI DNA fragments can be used for subsequent analysis.

The amino acid sequences that contain an epitope are those that are shared by all of the subclones that produce positive signals with a particular antibody (Fig. 6). Only clones that produce positive signals with an antibody can be used to deduce the position of the determinants. Clones that produce no signal could contain and even express the appropriate amino acids; however, the antigenic determinant might not be detectable because it is susceptible to proteolysis or it lacks the ability to form the correct antigenic structure. Since the epitope lies within the amino acid sequences shared by signal-producing clones, the resolution of the boundaries of an epitope should improve as larger numbers of recombinant clones are analyzed.

An example of this technique for mapping mycobacterial antigenic determinants is presented in Fig. 6. It is striking that all of the recombinant clones studied thus far that contain coding sequences for the mycobacterial antigenic determinants express detectable levels of that determinant (Fig. 6). As discussed above, clones containing DNA encoding an epitope might not express that epitope at detectable levels, reflecting different stability or structural constraints. The λgt 11 system, coupling the expression of fusion proteins with the use of *lon* protease-deficient host cells, may help express all encoded epitopes at detectable levels. The particular monoclonal antibodies that are used and whether they recognize segmental or assembled topographic determinants may also influence this result.

One application of this method is that recombinant clones from these sublibraries can also be used to elucidate determinants to which T cells respond. *Escherichia coli* lysates containing antigen expressed by λgt 11

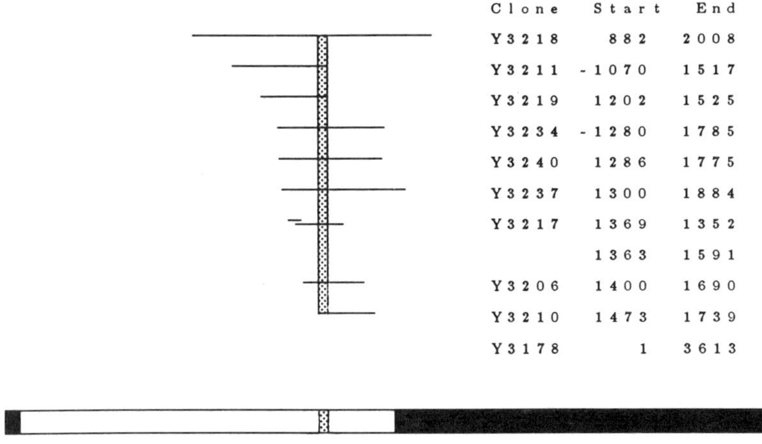

FIG. 6. Deducing an epitope in the *M. leprae* 65-kDa antigen. The heavy horizontal line at bottom depicts the Y3178 insert DNA in which the open box represents the 65-kDa antigen open reading frame. The thin horizontal lines illustrate the extent of the insert DNA fragments from Y3178 subclones. The vertical stippled region indicates the extent of the epitope-coding sequence (15 amino acids) as defined by the minimum overlap among clones that produce a positive signal with the anti-*M. leprae* monoclonal antibody MLIIIE9. To the right is tabulated the precise insert end points for each DNA clone.

recombinants can be used to assay antigen-specific T-cell stimulation *in vitro*.[26]

Transplason Mutagenesis of λgt 11 Clones

Another method for mapping antigenic determinants uses transposon mutagenesis, which was developed in collaboration with Elledge.[13] Mini-Tn*10* transposons were constructed containing *E. coli* selectable markers, *tet*^{R}, *kan*^{R}, and *supF*. They are present on a high copy number pBR322-type plasmid and in a strain, BNN114, that overproduces Tn*10* transposase.[13,27] A library of transposon insertions is easily constructed for a λgt 11 clone by growing the phage on the mutagenesis strain and selecting for the transposition events. After mutagenesis the clones are screened with polyclonal or monoclonal antibody probes to determine the effect of the

[26] A. S. Mustafa, H. K. Gill, A. Nerland, W. J. Britton, V. Mehra, B. R. Bloom, R. A. Young, and T. Godal, *Nature* (*London*) **319,** 63 (1986).

[27] D. J. Foster, M. A. Davis, D. E. Roberts, K. Takeshita, and N. Kleckner, *Cell* **23,** 201 (1981).

insertion on the level of antigen production. Tn*10* transposons when inserted in the middle of a coding segment obliterate downstream antigen expression. Correlation of the position of the insertion and the effect on the immunoscreening signal can be used to locate the antigen-coding region (Fig. 7). This procedure is useful for mapping antigenic coding segments, regardless of whether the recombinant clone is synthesizing a β-galactosidase fusion protein or not.

Transposons containing yeast selectable markers (*URA3*, *TRP1*) have been constructed.[13] These elements can be used to rapidly mutagenize yeast clones to map coding segments. The mutated genes can be directly

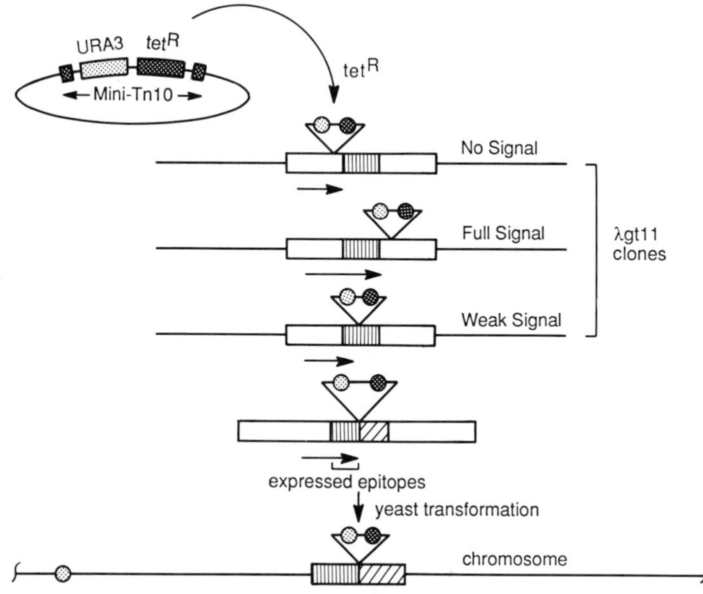

FIG. 7. Transplason mutagenesis of λgt 11 clones. Mini-Tn*10* transposon containing 70 bp of Tn*10* ends, *E. coli* selectable markers, and yeast selectable markers are carried on high copy number plasmids and in an *E. coli* strain that overproduces Tn*10* transposase. λgt 11 clones are grown on these strains, and transposition events into the phage are selected. A correlation of the position of the insertion with the effect on the immunoscreening signal can be used to locate the antigenic coding region. For a polyclonal serum, insertions between the promoter and antigenic coding region result in no immunoscreening signal, insertions outside the promoter and antigen coding region yield full level signals, and insertions within a coding region produce reduced signal. Reduced signals are probably the result of failure to express downstream epitopes, but other explanations are also possible.[13] For yeast *all* relevant insertions can be substituted for the chromosomal sequences by one-step gene transplacement[26] to inactivate the gene.

substituted for genomic copies by one-step gene transplacement[28] (Fig. 7). Because of their use for both transpo*son* mutagenesis in *E. coli* and for *transpla*cement of genomic sequences in yeast, these elements are called transplasons.

A protocol for mTn*10*/*URA3*/*tet*^{R} mutagenesis—a mini-Tn*10* transplason containing the *E. coli tet*^{R} selectable marker and the *URA3* gene of yeast—is as follows:

1. Grow 0.5 ml of the Tn*10* mutagenesis strain in LB medium containing 50 μg/ml ampicillin, 0.2% maltose plus 10 m*M* $MgSO_4$ to OD_{600} = 0.4.
2. Infect cells at m.o.i. 2–4 with λgt 11 recombinant phage. Incubate 2 hr at 37°.
3. Lyse the cells by adding 150 μl of chloroform. Incubate 15 min at 37° with shaking.
4. Clear the lysate of unlysed bacteria and debris by centrifugation for 5 min in an Eppendorf centrifuge. Repeat once. Store the lysate over 100 μl of chloroform.
5. Mix 100 μl of lysate (2–10 × 10^9 phage) with 0.4 ml of a saturated culture of BNN91 grown in LB medium plus 0.2% maltose plus 10 m*M* $MgSO_4$. Absorb 20 min at room temperature [BNN91 = Δ*lacZ hflA150 strA*].
6. Plate a serial dilution of cells on LB plates containing 15 μg/ml tetracycline to select for lysogens. For insert sizes of 3–4 kb, 25–70% of the transposition events occur in the insert DNA.
7. Prepare phage by growing a 5 ml culture of lysogens in LB plus 10 m*M* $MgSO_4$ (without drug) to OD_{600} = 0.50. Induce phage by heating culture at 44° for 15 min with vigorous shaking. Incubate at 37° for 1.5–2 hr. Add 150 μl of chloroform to lyse cells and pellet bacterial debris as above.
8. Minipreparations of DNA are prepared using polyethylene glycol precipitation according to Snyder *et al.*[13] From a 5 ml culture approximately 5 μg of DNA is obtained.
9. Yeast transformation: Cut 2 μg of DNA with a restriction enzyme that cuts on one or both sides of the yeast insert flanking the transplason. Ethanol precipitate the DNA and transform yeast colonies using the lithium acetate procedure.[29] Approximately 10–60 transformants/2 μg DNA are obtained.

[28] R. J. Rothstein, this series, Vol. 101, p. 202.

[29] H. Ito, Y. Fukuda, K. Murata, and A. Kimura, *J. Bacteriol.* **153,** 163 (1983).

Examples

Using these procedures, antibodies directed against pure proteins have been used to isolate a large number of genes. These include the yeast genes for RNA polymerases I, II, and III,[3,30] clatherin,[31] topoisomerase II,[10] and ubiquitin.[12] Examples from other organisms include *Plasmodium* antigens,[32] *Mycobacterium tuberculosis* antigens,[14] human terminal deoxynucleotidyltransferase,[33,34] rat fibronectin,[35] and human factor X.[36]

Knowledge of the function a protein is not obligatory for gene isolation. Use of a combination of the methods described above may help determine the function of that gene. For instance, Kelly, Greenleaf, and Lehman purified a yeast protein whose function was unknown *in vivo*.[37,38] The gene was cloned using the above techniques. Inactivation of that gene by transplason mutagenesis led to the discovery that the protein was a mitochondrial RNA polymerase subunit.

Antibodies made against impure proteins can also be used to isolate genes. For example, the gene for yeast DNA polymerase I was isolated using an antiserum directed against an impure preparation of that enzyme (estimated to be 50% homogeneous).[17] This was made possible by (1) using a genomic DNA library and isolating *all* sequences of the yeast genome that would react with the antibody and (2) identifying the correct DNA polymerase clones among a collection of other reactive phage using the independent assays described above.

Acknowledgments

We thank Tom St. John, Andrew Buchman, Stewart Scherer, and Carl Mann for their contributions to the development of these methods. We are grateful to C. Glover, J. Reichardt, S. Cotterill, and others for their contributions. This work was supported by grants from the National Institutes of Health (GM34365, AI23545, and GM21891) and the World Health Organization/World Bank/UNDP Special Program for Research and Training in Tropical Diseases.

[30] J. M. Buhler and A. Sentenac, unpublished results.

[31] G. S. Payne and R. Schekman, *Science* **230,** 1009 (1985).

[32] D. J. Kemp, R. L. Coppel, A. F. Cowman, R. B. Saint, G. V. Brown, and R. F. Anders, *Proc. Natl. Acad. Sci. U.S.A.* **80,** 3787 (1983).

[33] R. C. Peterson, L. C. Cheung, R. V. Mattaliano, L. M. S. Chang, and F. J. Bollum, *Proc. Natl. Acad. Sci. U.S.A.* **81,** 4363 (1984).

[34] N. R. Landau, T. P. St. John, I. L. Weissman, S. C. Wolf, A. E. Silverstone, and D. Baltimore, *Proc. Natl. Acad. Sci. U.S.A.* **81,** 4363 (1984).

[35] J. E. Schwartzbauer, J. W. Tamkun, I. R. Lemischka, and R. O. Hynes, *Cell* **35,** 421 (1983).

[36] S. P. Leytus, D. W. Chung, W. Kisiel, K. Kurachi, and E. W. Davie, *Proc. Natl. Acad. Sci. U.S.A.* **81,** 3699 (1984).

[37] J. Kelly, A. Greenleaf, and I. R. Lehman, submitted for publication.

[38] A. Greenleaf, J. Kelly, and I. R. Lehman, submitted for publication.

[8] Searching for Clones with Open Reading Frames

By MARK R. GRAY, GAIL P. MAZZARA, PRANHITHA REDDY, and MICHAEL ROSBASH

We have devised a simple strategy to quickly locate and express open reading frames. Using this approach, it is possible to identify protein-coding regions without any extensive information about the products. Typically, in order to identify and characterize a region of DNA, its sequence, RNA, or protein products are used to construct a molecular map.

Frequently, it is not easy to generate such a map. The approach described in this chapter can be used to locate coding regions whose products are either unknown or undetectable, starting with cloned or viral DNA. The same approach can be used to express any portion of either a poorly characterized or a well-characterized DNA or cDNA. Antibodies can be made against polypeptide determinants encoded by the open reading frame sequences of the starting DNA.

Principle

One of the distinguishing features of DNA in protein-coding regions is the presence of at least one open reading frame (ORF). DNA sequences outside of protein-coding regions, such as intergenic regions and introns, rarely have long open reading frames. In some organisms, such as *Drosophila,* the DNA outside of protein-coding units is enriched for A and T bases, increasing the chance of having one of the three stop codons TGA, TAG, and TAA.[1]

The open reading frame cloning strategy discussed here utilizes the pMR series of plasmid vectors; in these plasmids, a *lac*Z fusion gene is used to select small (100–1000 bp) DNA fragments that have continuous open reading frames.[2] The enzyme β-galactosidase, encoded by the *lac*Z gene of *Escherichia coli,* is often biologically active when an additional polypeptide sequence is attached to its amino terminus.[3] The pMR plasmids carry a strong bacterial promoter driving the expression of a

[1] P. O'Connell and M. Rosbach, *Nucleic Acids Res.* **12,** 5495 (1984).

[2] M. Gray, H. Colot, L. Guarente, and M. Rosbash, *Proc. Natl. Acad. Sci. U.S.A.* **79,** 6598 (1982).

[3] J. Beckwith, *in* "The Operon" (J. Miller and W. Reznikoff, eds.), Cold Spring Harbor Lab., Cold Spring Harbor, New York, 1978.

cI::*lacIZ* fusion gene; a cloning site is located between the *cI* and the *lacIZ* portions of the gene. When there is a frameshift mutation downstream from the cloning site, the production of a high level of β-galactosidase activity by bacterial transformants is eliminated. This frameshift can be corrected by the insertion (into the cloning site) of a DNA fragment that restores a continuous open reading frame through the *lacIZ* portion of the gene. This results in the production of a large amount of β-galactosidase activity by the transformants. When the frame-shifted vector is ligated with appropriate DNA fragments, and the ligated plasmids used to transform lac^- bacteria, transformants that produce high levels of β-galactosidase activity often contain plasmids with inserts having continuous open reading frames. The recombinant plasmids can be used as probes to map the genomic location of open reading frames. These transformants express the inserted DNA sequence as part of a β-galactosidase fusion protein. Thus, with this procedure, the protein-coding portion of a gene can be identified, mapped, and expressed.

Materials

All enzymes, chemicals, and apparatus needed to locate and express open reading frames are available from commercial sources as indicated below. The sources of the bacterial strains and DNA fragments used to construct the plasmid vectors have been previously reported.[2] The *Ubx* DNA was a gift from Welcome Bender; the CPV DNA was a gift from Solon Rhode.

Methods

A diagram of the structure of all of the pMR vectors is shown in Fig. 1A. Each of the plasmids pMR1, pMR2, pMR100, and pMR200 differ from each other only in the region of the cloning site between the *cI* and *lacIZ* portions of the fusion gene; the different sequences for each of the plasmids are shown in Fig. 1B. The plasmids pMR1 and pMR200 confer on host bacteria a strong lac^+ phenotype; the plasmids pMR2 and pMR100 give a lac^- phenotype.

The plasmid pMR2 has a frameshift in the *cI*::*lacIZ* fusion gene upstream from the *lacIZ* part of the gene; it is designed to select open reading frame DNA fragments with *Hin*dIII termini. The plasmid pMR1 has no frameshift and causes the production of a high level of full-length and enzymatically active *cI*::*lacIZ* fusion protein when used to transform the *lac* operon deletion strain LG90.[4] The plasmid pMR1 is used to esti-

[4] L. Guarente, G. Lauer, T. Roberts, and M. Ptashne, *Cell* **20,** 543 (1980).

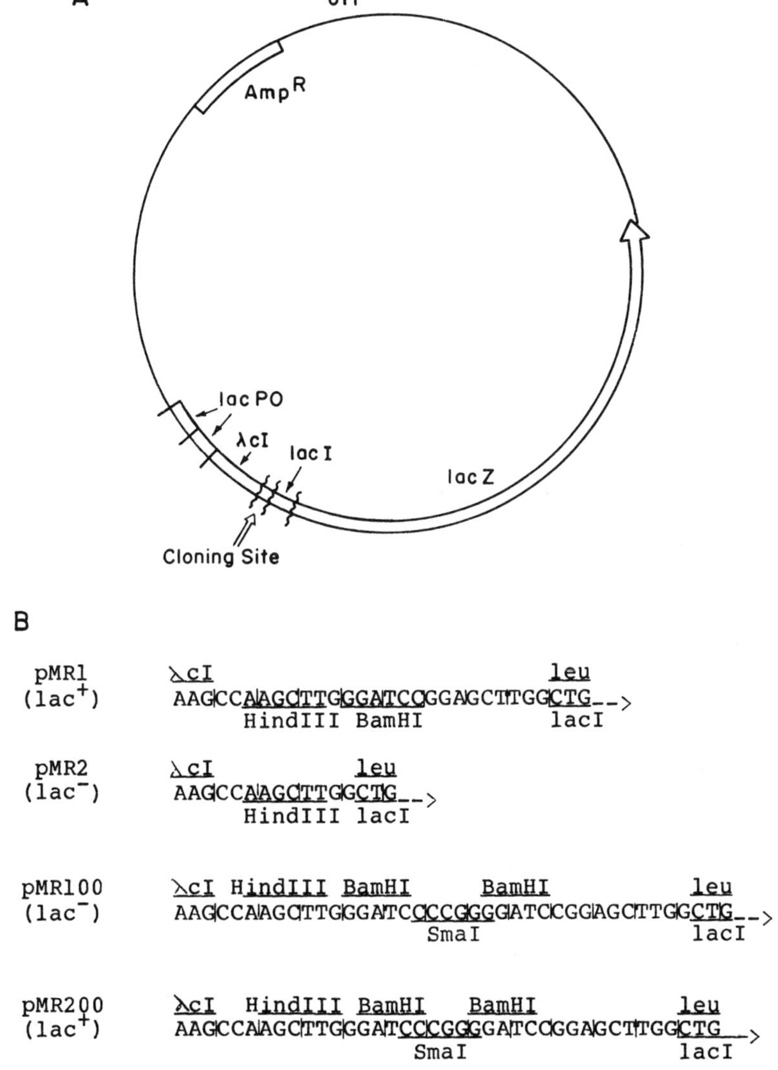

FIG. 1. (A) Diagram of the structure of pMR vectors; all four versions are identical except for the sequence at the cloning site between the *cI* and *lacIZ* portions of the fusion gene. (B) DNA sequences at the cloning sites of pMR1, pMR2, pMR100, and pMR200.

mate the insert frequency in parallel ligations of the same fragments with pMR2; the fraction of the total number of pMR1 transformants that become lac$^-$ after ligation with insert fragments can be used to estimate the number of transformants that have inserts in an identical ligation using pMR2.

The plasmid pMR100 was derived from pMR1 by the insertion of a 10-bp *Sma*I/*Bam*HI adapter at the *Bam*HI site (Fig. 1B). This insertion introduces a frameshift in the fusion gene of pMR1, creating a plasmid that can be used for the selection of any blunt-ended fragment that can correct the frameshift by insertion into the *Sma*I site. The plasmid pMR200 is a single base-pair deletion derivative of pMR100 that retains the *Sma*I site and removes the frameshift (Fig. 1B). As with pMR1, LG90/pMR200 transformants produce large amounts of full-length *cI*::*lacIZ* fusion protein (Table I; see Fig. 2). The amounts of β-galactosidase activity produced by transformants of each of the pMR plasmids, and their transformant *lacZ*

TABLE I
β-Galactosidase Expression in LG90 Host Bacteria Transformed with pMR Plasmids[a]

Clone	MacConkey phenotype	XGal phenotype	Enzyme activity	Insert size
LG90	White	White	0	—
pMR1	Red	Blue	4471	—
pMR2	White	Blue	121	—
pMR100	White	Blue	60	—
pMR200	Red	Blue	9800	—
UBX7	Red	Blue	2083	~305
UBX17	Red	Blue	508	~375
UBX35	Red	Blue	687	~335
UBX27	Red	Blue	401	317
UBX20	Red	Blue	342	305
UBX10	Red	Blue	1173	~315
UBX12	Red	Blue	2316	~525
UBX13	Red	Blue	2091	569
CPV3	Red	Blue	6126	295
CPV39	Red	Blue	666	439
CPV170	Red	Blue	1724	431
UBX102	Weakly red	Blue	278	503
UBX103	Red	Blue	461	429
CPV6	Red	Blue	8301	312
CPV27	Red	Blue	844	315

[a] The UBX and CPV plasmids are derived from pMR100 as described in the text. The MacConkey phenotype refers to the color of the colony on MacConkey agar plates (red = lac^+); the XGal phenotype refers to the color on XGal/minimal plates (blue = lac^+). The enzyme activity is the number of β-galactosidase activity units (Miller units) produced by each cell line. The insert size is the length in base pairs, determined by DNA sequencing or comparison to DNA fragment size standards by analytical gel electrophoresis.

phenotypes on 5-bromo-4-chloro-3-indolyl-β-D-galactoside (XGal) and MacConkey agar plates are listed in Table I.

The methods described below refer to the ligation of blunt-ended DNA fragments into pMR100; the same procedures are applicable to ligations using pMR1, pMR2, and pMR200. In order to map open reading frames in uncharacterized cloned DNA or express portions of a well-characterized gene, the simplest approach is to randomly fragment the DNA, repair the fragment termini to blunt ends, ligate size-fractionated fragments into the *Sma*I site of pMR100, transform, and analyze lac$^+$ LG90 transformants. Since all restriction and DNA modification enzymes have some substrate specificity (and therefore possible sequence specificity), sonication is the fragmentation method discussed below. Sonication does not cut DNA entirely at random; AT-rich DNA fragments are disrupted by sonication faster than GC-rich fragments.[5] Many eukaryotic sequences within protein-coding exons may be GC-rich, such as in those of *Drosophila*[1]; in these cases, sonication should disrupt exon sequences less quickly than AT-rich intron and spacer sequences.

The random fragmentation approach has several advantages over using restriction enzymes to make the insert fragments. (1) There is no need to have the appropriate restriction sites at both ends of the fragment in order to maintain the correct open reading frame. (2) Even the most carefully selected restriction fragment may not necessarily be the best one to use to isolate a stable fusion protein in bacteria. (3) The use of randomly cut DNA fragments bypasses the confusion and frustration caused by frameshift errors in the DNA sequence determination of the starting DNA. (4) If a variety of fusion proteins are desired from a single gene, randomly cut fragments from all regions of the substrate DNA will be ligated to pMR100 and the resultant lac$^+$ clones selected by their production of easily detectable amounts of stable fusion protein.

Preparation of DNA Fragments

DNA used to prepare small fragments for the detection of open reading frames should be free of contaminating DNA, such as that from the bacterial chromosome or lambda phage and plasmid vectors; these contaminating DNAs are rich in open reading frame sequences. Since most cloned DNAs are prepared by gel electrophoresis, it is important to digest the DNA under conditions that minimize degradation, to avoid vector fragments.

To follow each of the subsequent preparative steps for the insert fragments, start with a large excess (e.g., 20 μg) of the source DNA. If the

[5] P. Deininger, *Anal. Biochem.* **129,** 216 (1983).

DNA is in a lambda phage vector, or a relatively low copy number plasmid, it is worth the small investment of time to reclone it in a high copy number plasmid such as those of the pUC and pEMBL series.[6,7] In this way, 0.5–1.5 mg of plasmid DNA containing the target DNA can be easily prepared from a single 500 ml culture of transformed cells, providing more than enough DNA to prepare small fragments for ligation into pMR vectors.

Purification of Starting DNA Fragments. Any reliable method can be used to purify the starting DNA; we have found that electrophoresis on small agarose gels to be the simplest and fastest. A DNA fragment can be easily removed from the gel by first trapping the stained band on a small piece of DEAE membrane (NA-45, Schleicher and Schuell) by electrophoresis, and then eluting the DNA fragment from the membrane in 400 μl of 50 mM arginine plus 1.0 M NaCl (two incubations at 65°, 200 μl each). We have observed that DNA fragments purified in this way are not resistant to further enzyme treatments, unlike those purified by electroelution from agarose gel slices in dialysis bags; the DEAE method is much faster and less cumbersome as well. The purified fragments can be stored in any volume until sonication, or precipitated in ethanol; in either case, it is useful to check for recovery by electrophoresis of a small fraction of the purified target DNA on minigels. Throughout all of the fragment preparation steps, it is important to avoid irradiation and damage of DNA by short-wave ultraviolet light; photograph the ethidium bromide-stained preparative gels using only long-wave (greater than 300 nm) ultraviolet light sources.

Sonication. The purified DNA fragments are sonicated to a size range of 100–1000 bp, with most of the fragments 400–500 bp, as follows. Suspend the DNA in TE (10 mM Tris–HCl, 1 mM EDTA, pH 8.0) plus 0.2 M NaCl, in a tube suitable for sonication; plastic tubes are less likely than glass tubes to shatter accidentally. Carefully clamp the tube so that the sonicator tip does not touch the wall of the tube. Surround the tube with ice water so that the DNA solution does not overheat during sonication. Set the power output to a level just below the maximum recommended for the tip. Position the sonicator tip so that when the power is on, the DNA solution does not splash and foam; when the tip is not fully submerged in the liquid during sonication, the DNA is not fragmented well and the tip may become damaged. Sonicate for 30 sec at a time, with at least 30 sec in between bursts to allow time for cooling. To reduce most of the fragments to the length range 100–1000 bp, with most of the DNA around 450 bp,

[6] J. Vieira and J. Messing, *Gene* **19,** 259 (1982).

[7] L. Dente, G. Cesareni, and R. Cortese, *Nucleic Acids Res.* **11,** 1645 (1983).

sonicate for a total of 3 min. Small (<1.5 kb in length) target DNA fragments occasionally require longer times of sonication than larger (~4 kb) molecules.

After sonication, concentrated (at least 20 μg/ml) DNA fragments may be recovered by ethanol precipitation. More dilute DNA solutions should be passed over a DEAE–cellulose column (DE-52; Whatman); alternatively, disposable columns that attach to syringes (Elutips; Schleicher and Schuell) can be used. After binding on DEAE–cellulose in low salt buffer (TE plus 0.2 *M* NaCl or less), the DNA fragments are eluted in a small (300–400 μl) volume of high salt buffer (TE plus 1 *M* NaCl). Precipitate the DNA fragments in a microfuge tube by adding 2 volumes of 95% ethanol, leaving the tube on dry ice for 10 min, then centrifuging for 5 min; wash the pellet in 70% ethanol, recentrifuge, and dry the pellet. Dissolve the DNA fragments in a small (10–50 μl) volume of the buffer used for the end repair steps to follow.

Repair of Termini to Blunt Ends. After sonication, the ends of the DNA fragments are virtually unligatable since they are most probably a random mixture of 5′ and 3′ overhangs of variable lengths. There are many enzymes that can be used to repair the ends of the fragments to blunt ends. Nucleases that remove single-stranded DNA ends or polymerases that fill in by using the single-strand overhangs as templates work well. Preparations of each enzyme vary over a wide range in their efficiency of making blunt ends; it is best to evaluate several different approaches with some abundant source of sonicated DNA before selecting a repair strategy for the purified target DNA fragments of interest.

We have achieved good results in making blunt ends by a short digestion of the fragments with S_1 nuclease (to remove single-stranded termini), followed by a fill-in treatment using T4 polymerase; treatment of the sonicated fragments with either of these enzymes alone has been less successful. In this approach, the precipitated sonicated DNA (2–10 μg) is resuspended in 50 μl of S_1 buffer, placed at 37°, and digested with S_1 nuclease (final concentration 500 units/ml) for 15 min. Stop the reaction by adding EDTA to 25 m*M*; dilute in 200 μl of TE plus 0.2 *M* NaCl, extract with 250 μl of buffered phenol/chloroform (1 : 1), and precipitate the S_1-digested fragments with ethanol. Dissolve the lyophilized pellet in 20 μl of TA buffer (33 m*M* Tris–acetate, 66 m*M* potassium acetate, 10 m*M* magnesium acetate, 0.5 m*M* DTT, pH 7.9). Place the DNA at 37° and add one unit of T4 polymerase for each microgram of DNA fragments. In the absence of added nucleotides, the 3′ to 5′ exonuclease activity of T4 polymerase converts all single-stranded termini to 5′ overhangs. After 3 min, add all four nucleotides to 0.1 m*M* for the fill-in reaction (at least 30 min at 37°). The reaction is stopped by phenol/chloroform extraction and reprecipitation in ethanol of the repaired fragments. The success of the end repair reactions should be tested by self-ligating a small aliquot of the repaired and unrepaired fragments; compare ligated and unligated samples by electrophoresis on minigels.

Size Fractionation of Sonicated Fragments. To avoid cloning very small random ORFs from genomic sequences that do not encode proteins, it is necessary to select by size the DNA fragments desired for ligation into pMR100. The probability that a fragment of random sequence of length x codons is the correct size ($3n + 2$) and entirely open reading frame is $1/3 \times (61/64)^x$ (without correcting for GC composition differences). In this calculation, there is a 1/3 probability that the fragment is the proper length, and a 61/64 chance that any codon will not be a stop codon. If the fragments (of random sequence) are in the range of 200 bp (or 66 codons), then the probability of the fragment being suitable for correcting the frameshift in pMR100 is 4.6×10^{-3}. If the fragments are 300 bp long, then the probability is 9.1×10^{-4}, and for 500 bp, 3.5×10^{-5}. The probability that a fragment of DNA from a large *open reading frame* sequence will give rise to a pMR100 lac^+ clone is considerably higher. It is the product of the probability of having the correct length (1/3) times that of starting in the correct frame (1/3) times the probability of being cloned in the proper orientation (1/2). For completely open reading frame DNA, the probability is 5.5×10^{-2} and is independent of the length of the fragment. For a 500-bp fragment, the difference in the probabilities for the detection of an ORF fragment in random sequence versus open reading frame is large enough to strongly suggest that it is derived from a bona fide open reading frame from the protein coding region of a gene. Alternatively, for a 150-bp fragment, the difference is very small.

The very small sonicated fragments can be removed by size selection after electrophoresis. Electrophorese the DNA on a 10% acrylamide/TBE gel (0.6 mm thick), along with appropriate size markers, at 200 V until the bromphenol blue has traveled 10–15 cm (about 2 hr).[8] Up to 10 μg of DNA can be loaded into a well 1.6 cm wide. Stain the gel in ethidium bromide, destain in distilled water, and photograph using a long-wave ultraviolet light source. Electrotransfer the DNA fragments from the gel to DEAE–cellulose paper (Whatman DE-81) as follows. Assemble a sandwich consisting of the following: a Scotchbrite pad (3M Corporation), two layers of filter paper (Whatman 3MM), the gel, a piece of DEAE–cellulose paper, two more layers of filter paper, and a second Scotchbrite pad. Submerge the sandwich between the two rigid supports of the electrotransfer apparatus after it has been filled with enough transfer buffer (20 mM Tris–HCl,

[8] T. Maniatis, A. Jeffrey, and H. Van de Sand, *Biochemistry* **14,** 3787 (1975).

1 m*M* EDTA, pH 8.0) to cover the gel. The sandwich must be oriented so that the DEAE–cellulose paper is between the gel and the positive electrode. Electrophoresis at 150 mA (about 8–15 V) for 2 hr is sufficient to transfer completely all fragments under 800 bp in length. For larger fragments, increase the current (up to 600 mA at room temperature) or increase the time of electrophoresis.

Remove the very fragile wet DEAE paper by covering the gel and DEAE paper with a piece of plastic wrap, inverting the gel so that the plastic wrap is on the bottom, and then lifting the gel away from the DEAE paper. Cover the DEAE paper with another layer of plastic wrap if it is not possible to elute the DNA immediately; if the DEAE paper is allowed to dry, the DNA fragments do not elute easily. Photograph the DEAE paper using a long-wave UV light source; compare the photograph of the gel to that of the DEAE paper in order to check for recovery. Mark on the plastic wrap the region that contains the DNA fragments of the desired size, using the size standards as a guide. We usually discard any fragments under 200 bp, and save size fractions of 200–400 bp and 400–700 bp. Cut out the DEAE paper that contains the fragments and pack it into a 0.5 ml microfuge tube. Cover the paper with 100 μl of TE plus 0.2 *M* NaCl to prevent drying. Make a small hole in the tip of the tube using a 26-gauge needle and place this tube into a 1.5-ml microfuge tube. This arrangement is used to wash the DEAE paper and elute the DNA by centrifugation in a microfuge. Wash the paper 3 times with 100 μl of TE plus 0.2 *M* NaCl. After all of the wash buffer is spun out and discarded, elute the DNA with three 100 μl washes of TE plus 1 *M* NaCl. Pool the 300 μl of DNA solution and precipitate in ethanol as described above. Resuspend each size fraction of DNA in 10–20 μl of TE; electrophorese 1 μl of each DNA sample on a minigel along with appropriate size and concentration standards in order to estimate recovery. It is sufficient to have 100 ng of each size fraction for ligation with pMR100.

Ligation of Sonicated Fragments to pMR100

For each sample of sonicated DNA to be ligated to pMR100, 0.5 μg of *Sma*I-digested pMR100 is used; it is convenient to prepare 10–20 μg at one time for many ligations. In order to prevent recircularization of the linearized vector molecules, digest the *Sma*I-cut DNA with calf intestinal phosphatase (CIP; Boehringer Mannheim) in order to remove the terminal 5′ phosphate groups. This treatment reduces transformation (by greater than 99%) by plasmid molecules which lack inserts and makes it possible to transform with much more ligated DNA on each plate. In our experience, phosphatasing the vector does not affect the efficiency of ligation of

DNA fragments to vector molecules, nor does it increase the frequency of false positives due to frameshifts at the *Sma*I site.

After digestion of pMR100 with *Sma*I, check for complete digestion by electrophoresis of a small portion of the DNA. If there are no supercoiled or open circular bands present, increase the volume of the DNA to 200 μl in 10 m*M* Tris–HCl (pH 8.0) and add approximately one unit of CIP for each microgram of vector DNA. Digest at 37° for 30 min, and then phenol/chloroform extract and precipitate the phosphatased vector in ethanol; resuspend the DNA to 0.5 μg/μl in TE.

Mix 0.5 μg pMR100 vector DNA, 50–100 ng of insert DNA fragments, and no more than 1 unit of T4 DNA ligase (Boehringer Mannheim) in a 10 μl volume of ligase buffer. Addition of more than 1 unit of ligase results in fewer transformants and a large increase in false positives (discussed below). When ligating pMR100 with sonicated fragments, always set up an identical ligation that is missing the insert fragments. Incubate for 12–16 hr at 15° before transformation. It is informative to compare the ligated DNA samples by electrophoresis of very small aliquots (0.5 μl of the 10 μl total) on minigels. The intensity of the ligated open-circle vector band above the unligated linear band provides a reliable prediction of the expected number of transformants.

Transformation of Host Bacteria

Any transformation method that will consistently yield 10^6 or more transformants from 1 μg of supercoiled plasmid DNA is sufficient. We have used the method of Dagert and Ehrlich,[9] with some minor modifications, for transforming LG90 cells with ligated pMR100 DNA.

Prepare competent LG90 cells by diluting a saturated culture (less than 7 days old) 1 : 100 in 100 ml of L broth in a 500-ml culture flask. Grow the cells to early log phase (about 90 min at 37°) with vigorous shaking for maximal aeration; if the cells are grown too long, the transformation efficiency is decreased by 2- to 10-fold. Put the cells on ice for 10 min, and then centrifuge at 7,000 rpm (4°) for 5 min. Discard all of the L broth, and resuspend the bacteria in one-half of the original culture volume of ice-cold 100 m*M* $CaCl_2$. Leave the cells on ice for 20 min, and then centrifuge again; resuspend the pellet in 1% of the original culture volume of ice-cold 100 m*M* $CaCl_2$. For optimal efficiency, store the cells at 0° for 12–20 hr before transformation.

In our hands, phenol extraction and ethanol precipitation of ligated DNA increases the transformation efficiency 5- to 10-fold. Add 200 μl of

[9] M. Dagert and S. Ehrlich, *Gene* **6,** 23 (1979).

TE plus 0.2 *M* NaCl and 10 μg of tRNA carrier to each ligated DNA sample before phenol/chloroform extraction and precipitation. Resuspend the dry pellet in TE at 10 ng of vector DNA/μl.

Pipet the DNA into sterile tubes on ice. For each plate, use 50 ng of ligated vector DNA (or 10 ng if the vector was not phosphatased); it is best to have about 500 colonies on each plate. After 5 min, add 50 μl of competent LG90 cells to each tube; mix and leave the tubes on ice for another 5 min. Temperature shock the cells by putting the tubes in a 37° water bath for 5 min. Add 1 ml of L broth (with no ampicillin) to each tube, and shake at 37° for 1 hr. Transfer each sample of transformed cells to 1.5-ml microfuge tubes, and centrifuge for 10 sec. Discard all of the L broth and resuspend the pellet of transformed cells in one drop of sterile distilled water. Spread the cells on a MacConkey agar (Difco Labs) plate (50 μg ampicillin/ml) until the surface of the plate appears dry. When the plates have lower (25–40 μg/ml) concentrations of ampicillin, often more transformants are recovered, but satellite nontransformed colonies frequently appear after overnight incubation; if a larger number of transformants is desired, a lower concentration of ampicillin should be tested.

Incubate the plates at 37° upside down for at least 24 hr. The transformed colonies will be visible at 12 hr, but the red color of the lac$^+$ transformants may take as long as 40 hr to develop. There is a wide range of expression of β-galactosidase in open reading frame clones, depending on the DNA sequence of the insert; this results in lac phenotypes ranging from a slowly developing (40–48 hr) slightly red to a very rapidly developing (less than 12 hr) dark red. After incubation at 37°, mark the positions of the lac$^+$ colonies on the back of the plate with a felt-tip pen. The plates can be stored at room temperature for up to 5 days without any change in the red color phenotype; when the plates are stored at 4°, all colonies, including lac$^-$ ones, will turn red.

The number of transformants resulting from each ligated DNA sample may vary over a wide range and is dependent on the efficiency of each of the preparative steps preceding transformaton. Typical results for different types of experiments are shown in Table II. For ligation of sonicated DNA inserts, the efficiency of the insert fragment end-repair steps is the most important variable.

Selection of lac$^+$ Transformants

Some of the red lac$^+$ colonies are not always the result of the insertion of an open reading frame fragment in the frame-shifted pMR100 fusion gene. The proportion of artifact red colonies (discussed below) is much higher if the number of transformants is low (when there is little or no

TABLE II
TYPICAL TRANSFORMATION RESULTS USING pMR100 AND pMR200 VECTOR DNA

Vector	Inserts	# Transformants/ μg vector	% lac⁺
pMR100 uncut	None	$1.0–1.5 \times 10^6$	0
pMR200 uncut	None	$1.0–1.5 \times 10^6$	100
pMR100 *Sma*I-cut	None	$0.5–1.0 \times 10^6$	0.5–1.0
pMR200 *Sma*I-cut	None	$0.5–1.0 \times 10^6$	99.0–99.5
pMR100 *Sma*I-cut	Fragments with *Hae*III ends	0.5×10^6	0.5–5.0[a]
pMR200 *Sma*I-cut	Fragments with *Hae*III ends	0.5×10^6	Up to 75
pMR100 *Sma*I/CIP	None	$5–10 \times 10^3$	0.5–1.0
pMR200 *Sma*I/CIP	None	$5–10 \times 10^3$	99.0–99.5
pMR100 *Sma*I/CIP	Fragments with *Hae*III ends	$0.25–0.5 \times 10^6$	0.5–5.0[a]
pMR200 *Sma*I/CIP	Fragments with *Hae*III ends	$0.25–0.5 \times 10^6$	0.5–5.0[a]
pMR100 *Sma*I/CIP	Sonicated with repaired ends	$0.5–2.5 \times 10^4$	0.5–5.0[a]
pMR200 *Sma*I/CIP	Sonicated with repaired ends	$0.5–2.5 \times 10^4$	1.0–75[a]

[a] Yield depends on the insert frequency, the length of the insert fragments, the presence of open reading frame sequence, and the presence of sequences that can cause initiation of translation in bacteria.

increase in the number of transformants obtained with the ligation of phosphatased pMR100 and insert fragments over that of vector alone). For this reason, it is best to check each lac⁺ transformant for the presence of the correct inserts by Grunstein–Hogness colony screening, using a probe made from the target DNA.[10] Since it is about as much work to screen 50 transformants as it is to check 500, it is useful to transform LG90 cells with all of the ligated pMR100 DNA on many plates (e.g., 10 plates from one ligation of 0.5 μg of vector DNA) before the colony hybridization experiments.

Inoculate a fresh MacConkey agar plate containing ampicillin (50 μg/ml) with small (1–2 mm in diameter) patches of cells from each lac⁺ transformant in a numbered grid pattern; use a paper template taped to

[10] M. Grunstein and D. Hogness, *Proc. Natl. Acad. Sci. U.S.A.* **72,** 3961 (1975).

the underside of a plate. At the same time, inoculate 2–3 known positive and 2–3 negative control transformants in some asymmetric pattern among the 80–100 unknown transformants on each plate. Grow the cells at 37° for at least 16 hr, and record the intensity of the red color phenotypes of the transformants. The range of the intensity of the red color is often a reflection of the total amount of β-galactosidase proteins produced by the transformant line; this information is helpful in the electrophoretic analysis of the fusion proteins (discussed below). Replica plate the transformants to several nitrocellulose filters and one new plate. Grow the transformants on the nitrocellulose for at least 12 hr and then lyse the cells with alkali and fix the DNA.[11] Hybridize the filters with probes made by nick translation of the purified fragments used to make the open reading frame clones. Probes made from whole recombinant lambda phage vectors and most plasmids will hybridize with pMR100, because of shared *lac* operon, lambda *cI*, or pBR322 DNA sequences. Any purified fragment greater than 400 bp in length can be labeled by nick translation; small fragments can be nick translated easily by making concatemers of them by self-ligation. Expose the hybridized and washed filters to X-ray film with an intensifying screen at −70° for 3–4 hr; positive signals should be clearly distinguishable from the negative controls.

The proportion of the lac$^+$ transformants that have inserts derived from the correct DNA source is highest when the total insert frequency is high and there is some open reading frame sequence present in the target DNA. The lac$^+$ clones that do not hybridize to the target DNA probe are caused by the following artifacts: (1) A single base deletion at the *Sma*I cloning site of pMR100 that results in correction of the frameshift in the *cI*::*lacIZ* fusion gene. Red transformants of the control plasmid pMR200 are an example of these plasmids; these transformants are distinctive in that they always produce high levels (higher than most open reading frame clones) of β-galactosidase and have a very dark red MacConkey agar phenotype. These colonies are almost certainly caused by exonuclease contaminants either in the *Sma*I restriction enzyme or the T4 DNA ligase. The proportion of these clones is usually less than 1% in transformations with the ligated vector without inserts. (2) Rare plasmid deletions that put the fusion gene in frame. (3) Correction of the frameshift by the insertion of an open reading frame fragment derived from the wrong DNA. Common sources of contaminating DNA are the largely open reading frame vector sequences and bacterial chromosomal DNA.

An alternative approach which is occasionally available is screening

[11] T. Maniatis, E. F. Fritsch, and J. Sambrook, "Molecular Cloning: A Laboratory Manual." Cold Spring Harbor Lab., Cold Spring Harbor, New York, 1982.

the lac$^+$ transformants by using polyclonal antibody preparations directed against the protein sequences that are expected to be expressed as fusion proteins in the open reading frame clones.[12]

Analysis of Fusion Proteins

Some of the insert-containing lac$^+$ pMR100 transformants do not produce full-length (*cI*::inserted polypeptide sequence::*lacIZ*) fusion proteins, because of translation initiation within the inserted RNA sequence (discussed below). For this reason, it is important to examine the β-galactosidase polypeptides by electrophoresis on Laemmli SDS gels.[13]

Prepare protein samples from 2 ml overnight cultures (L broth plus 50 μg/ml ampicillin) of each transformant. Transfer the cells to 10 × 75 mm Pyrex glass tubes and centrifuge at 7,000 rpm for 5 min. Discard all of the L broth and completely resuspend the cells in 100 μl of 1.2× Laemmli sample buffer. Lyse the cells by incubation at 100° for 3 min. If the lysate is not clear, the cells are not completely lysed; large amounts of cell debris interfere with resolution of large proteins on the SDS gels. Often, the cause of incomplete lysis is decomposition of the 5× Laemmli sample buffer. The clear protein lysates are extremely viscous and hard to pipette because of bacterial chromosomal DNA. The high viscosity can be reduced either by aspiration of the lysate through a long 22 gauge needle several times, or by brief sonication. The protein lysates can be stored for several months at −20°.

Prepare a 0.6-mm-thick 7.5% acrylamide SDS gel with a 4.5% acrylamide stacking gel, using a comb with wells 4 mm wide. Electrophorese 4 μl of each protein sample (along with samples prepared from both pMR100 and pMR200 transformants) as follows: (1) Run the tracking dye all the way through the stacking gel at 100 V (about 30 min). (2) Increase the voltage to 200 V for 90 min. (3) Increase again to 250 V; let the dye front run off and electrophorese for 30 min more (if the running gel is about 11 cm long). Stain the gel for at least 1 hr in 0.25% Coomassie blue in 50% methanol plus 10% acetic acid. Destain in 50% methanol plus 10% acetic acid until the background is clear (several hours). Fix the gel in 10% acetic acid for long-term storage or drying.

An example of a protein gel prepared as described above is shown in Fig. 2. The β-galactosidase band is visible as the only high molecular weight protein that varies both in size and abundance among the samples.

[12] U. Ruther, M. Koenen, A. Sippel, and B. Muller-Hill, *Proc. Natl. Acad. Sci. U.S.A.* **79,** 6852 (1982).

[13] U. Laemmli, *Nature* (*London*) **227,** 680 (1970).

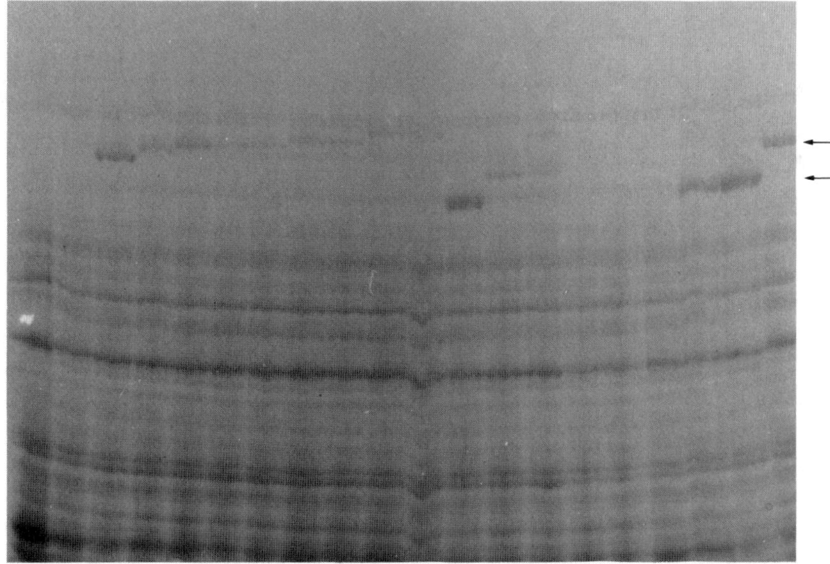

FIG. 2. Gel electrophoresis of fusion proteins. Samples (4 μl) of protein minilysates prepared from various transformants were electrophoresed on a 7.5% acrylamide/SDS gel and stained with Coomassie blue as described. Lanes: 1, LG90; 2, 17, pMR100; 3, 20, pMR200; 4–11, *Ubx* ORF clones UBX7, UBX17, UBX35, UBX27, UBX20, UBX10, UBX12, and UBX13, respectively (Table I; Fig. 3); 12–14, *CPV* ORF clones CPV3, CPV39, and CPV170, respectively; 15, UBX102; 16, UBX103; 18, CPV6; 19, CPV27. Arrows: lower, the position of wild-type β-galactosidase; upper, the pMR200 *cI*::*lacIZ* fusion protein (lanes 3, 20).

The staining intensity of the β-galactosidase proteins often correlates with the degree of red color of the transformant colony on MacConkey plates.

The β-galactosidase proteins produced by transformants with weak lac$^+$ phenotypes are occasionally difficult to see by Coomassie blue staining of the gel. Protein blots (also termed Western blots) are useful for the analysis of clones of this type. In Western blot analysis there is no confusion caused by co-migrating proteins; long exposure times allow visualization of very small amounts of β-galactosidase fusion proteins. We have used anti-β-galactosidase and anti-λ cI as probes to visualize proteins encoded by the pMR100 fusion gene.

Any protein blot procedure is sufficient; we have used the procedure of Towbin with consistent success.[14] SDS gels are prepared as described

[14] H. Towbin, T. Staehelin, and J. Gordon, *Proc. Natl. Acad. Sci. U.S.A.* **76,** 4350 (1979).

above, and then electroblotted to nitrocellulose after electrophoresis. The protein blots are incubated with anti-β-galactosidase, anti-cI, or antibodies against the polypeptide sequence encoded by the inserted DNA sequence. Finally, the blots are incubated with ^{125}I-labeled protein A, washed, and exposed to X-ray film.

Analysis of the Open Reading Frame DNA Insert Fragments

For several reasons, it is useful to make plasmid DNA from 1-ml cultures of each open reading frame clone. This DNA can be used to regenerate the clone even after years of storage. The length of the ORF insert fragment can be determined easily by digestion of the plasmid with *Bam*HI. The *Sma*I cloning site is between the only two *Bam*HI sites in pMR100; the small *Bam*HI fragment from each recombinant plasmid contains only 10 bp of vector sequence. In this way, the blunt-ended, randomly cut fragments now have *Bam*HI ends. (If the inserted ORF fragment has a *Bam*HI site, then two small fragments will be found after digestion with *Bam*HI.) The insert fragment can be easily sequenced by recloning the *Bam*HI fragments into sequencing vectors such as those of the M13 or pEMBL series.[6,7] Alternatively, the inserts can be sequenced in pMR100, without recloning, by using double-stranded DNA sequencing methods.[15] Primers homologous to sequences in pMR100 on either side of the *Sma*I site can be used. The miniprep DNA can also be used to make nick-translated probes for Southern or Northern blots.

A reliable plasmid DNA miniprep protocol is the alkaline lysis method.[16] Resuspend the DNA pellet (about 1 μg of DNA) from a 1 ml saturated culture of transformed cells in 25 μl of TE. In order to visualize the DNA insert easily, digest 5 μl of the plasmid DNA with 2 units of *Bam*HI in a total volume of 10 μl for 60 min at 37° in the presence of a small amount (0.5 μl of a 0.1 mg/ml stock) of RNAse A. Electrophorese the digested DNA samples along with the appropriate size markers on a 10% acrylamide/TBE gel.[8]

Preparation of Antibodies Specific for the Inserted Polypeptide Sequence of pMR100 Fusion Proteins

The β-galactosidase fusion proteins produced by pMR100 ORF clones can be purified easily and used as immunogens for the production of antibody probes. Many different protein purification strategies can be

[15] R. Wallace, M. Johnson, S. Suggs, K. Miyoshi, R. Bhatt, and K. Itakura, *Gene* **16,** 21 (1981).

[16] D. Ish-Horowicz and J. Burke, *Nucleic Acids Res.* **9,** 2989 (1981).

Precipitate the fusion protein by adding 4 volumes of cold (−20°) acetone and leaving the mixture at −20° overnight. Centrifuge at 18,000 rpm for 30 min; discard the supernatant and resuspend the dry protein pellet in a volume of PBS suitable for immunization.

Immunization of Rabbits. We have immunized rabbits with pMR100 fusion proteins, using the protocol below. Monoclonal antibodies specific for determinants of the inserted polypeptide sequence have also been successfully isolated using pMR100 fusion proteins as antigens.[19]

Before the initial injection, collect a pre-immune blood sample. For the initial injection, mix 200 μg (in 1–2 ml PBS) of the fusion protein with an equal volume of complete Freund's adjuvant (Difco Labs, 0639-60-6); inject the protein subcutaneously into several regions of the back and neck of the rabbit. Six weeks after the initial immunization, inject 50–70 μg of fusion protein in 1 ml PBS mixed with an equal volume of incomplete Freund's adjuvant (Difco Labs, 0638-60-7). Later booster injections can be given 3–4 weeks apart.

Rabbits are bled (40 ml) 10 days after each booster immunization; immunoreactivity of the serum is estimated by ELISA assay using purified β-galactosidase (Sigma) and/or purified fusion protein as antigen.[20] Immunoreactivity is usually observed after the first booster injection; maximal immune response levels are usually achieved after the third boost.

Purification and Characterization of the Antibodies. Allow the blood samples to clot at room temperature for 2 hr and then at 4° for 12–16 hr. Spin down the clot at 10,000 rpm for 10 min, and remove the serum. Precipitate the proteins from 1 ml of serum by mixing in 1 ml of saturated ammonum sulfate; incubate at 4° for 1–2 hr with constant stirring. Centrifuge at 10,000 rpm for 20 min; resuspend the pellet in 20 ml of phosphate buffer (17.5 m*M* sodium phosphate, pH 6.5) and dialyze in phosphate buffer as described above.

The serum proteins can be further enriched for IgG molecules by passing the dialyzed sample over a DEAE–cellulose (Whatman DE-52) column; many non-IgG proteins bind to DEAE. Prepare a 1-ml column; equilibrate with phosphate buffer. Check the pH and conductivity of the phosphate buffer before and after the column in order make sure that the column is properly equilibrated. Add the dialyzed serum proteins to the column and collect 0.5-ml fractions; assay the fractions for proteins by optical absorbance (A_{280}). Pool the protein fractions and dialyze overnight in BSB (150 m*M* NaCl, 17.5 m*M* boric acid, pH 7.5) as described above.

[19] R. White and W. Wilcox, *Cell* **39,** 163 (1984).

[20] A. Voller, D. Bidwell, and A. Bartlett, "The Enzyme Linked Immunosorbent Assay (ELISA)." Dynatech Laboratories, Alexandria, Virginia, 1979.

Antibody molecules specific for the pMR100 recombinant fusion protein are purified by their ability to bind to the fusion protein used as the immunogen. Prepare an Affigel-10 (Bio-Rad Chemicals, #153-6046) column with the fusion protein coupled to the gel. Couple 500 μg of substrate analog-purified fusion protein to 0.5 ml of Affigel-10 following the manufacturer's directions. Wash the column with 200 m*M* glycine–HCl plus 500 m*M* NaCl, pH 2.5, and then neutralize with BSB. Treat the column with 1 ml of BSB plus 0.1% SDS (SDS treatment is discussed below) and equilibrate the column with BSB. Pass the rabbit serum proteins (in BSB) over the column 3 times at 4°; only antibody molecules specific for any portion of the fusion protein will bind to the column. Wash the column with 10 ml of BSB; wash the column again with 2 ml of BSB plus 500 m*M* NaCl in order to remove nonspecifically bound or low-avidity antibodies. Elute the bound antibody molecules with 5 ml of 200 m*M* glycine–HCl, pH 2.5 plus 500 m*M* NaCl; collect 12 0.5-ml fractions; neutralize (pH 7.0) each of the antibody fractions immediately by the addition of 200 m*M* Tris base. Add bovine serum albumin (BSA) to a final concentration of 0.5 mg/ml to each of the fractions; dialyze the first 6 fractions separately in BSB as described above. Assay all fractions and the flowthrough sample using the ELISA assay with both β-galactosidase and recombinant fusion proteins as antigens. Usually, fractions 2–5 have most of the anti-fusion protein antibody; pool all the antibody-containing fractions after dialysis.

To remove antibody molecules specific for the λ *cI* repressor and *E. coli* β-galactosidase antigens and recover only those specific for the inserted polypeptide sequence, the dialyzed antibody preparation above is allowed to adsorb to Affigel-10 coupled with the pMR200 fusion protein. The pMR200 fusion protein contains only the *cI* and *lacIZ* portions of pMR100 recombinant fusion proteins (Fig. 1). Prepare the pMR200 Affigel-10 according to the manufacturer's directions; equilibrate with BSB. Mix 250 μl of pMR200–Affigel-10 with the antibody preparation; shake gently for 1 hr at 4°. Centrifuge at 2,000 rpm for 5 min; repeat the adsorption of the supernatant with a second aliquot of pMR200–Affigel-10. Antibodies specific for *cI* and *lacIZ* antigens will bind to the pMR200–Affigel-10 and those specific for the inserted polypeptide will be left in the supernatant. Assay the final antibody preparation using the ELISA method with the recombinant fusion protein, pMR200 fusion protein, and β-galactosidase as antigens. The specificity of the antibody can be tested using Western blots, as discussed above,[14] or Western dot blots[21] (Fig.

[21] H. Stahl, R. Coppel, G. Brown, R. Saint, K. Lingelbach, A. Cowman, R. Anders, and D. Kemp, *Proc. Natl. Acad. Sci. U.S.A.* **81,** 2456 (1984).

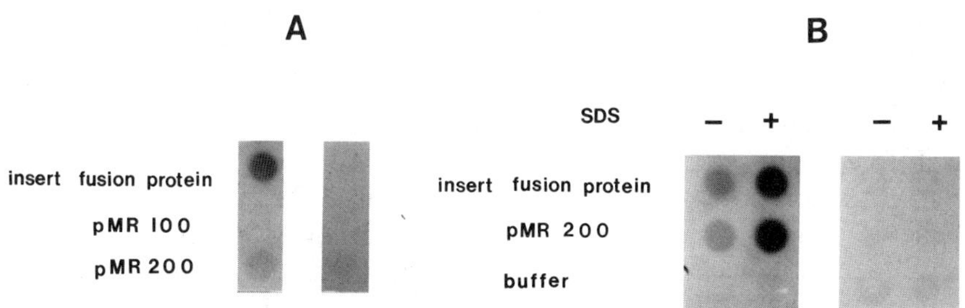

FIG. 3. (A) Preparation of antibodies specific for the inserted polypeptide sequence of pMR100 fusion proteins. Duplicate bacterial cell lysate samples (each containing approximately 1 ng of fusion protein) treated with 0.1% SDS were blotted onto nitrocellulose. The blot on the left was incubated with affinity-purified antibodies to the inserted polypeptide sequence. The blot on the right was incubated with pre-immune serum subjected to the same purification procedure as that of the immune serum. Both blots were then incubated with ^{125}I-labeled protein A, washed, and exposed to X-ray film with an intensifying screen at −70°. (B) Effect of SDS on the antigenicity of pMR200 fusion proteins. Duplicate fusion protein samples (as above) in either PBS or PBS plus 0.1% SDS were blotted onto nitrocellulose; samples of PBS and PBS plus 0.1% SDS with no proteins were also blotted. The blot on the left was incubated with a 1 : 500 dilution of rabbit immune serum; the blot on the right was incubated with pre-immune serum diluted 1 : 500. Both blots were then incubated with ^{125}I-labeled protein A, washed, and exposed to X-ray film.

3A). The dot blots are prepared using a minifold apparatus (Millipore). The final volume of purified antibody is usually 1–2 ml; dilute to 1 : 300 for Western blots.

It has been reported that proteins purified by SDS–gel electrophoresis are strong immunogens and can be used to generate high-titer antisera.[22] In one well-studied case in our experiments, when the immunogen used was pMR100 fusion protein purified on SDS gels, the resultant polyclonal rabbit antiserum recogized SDS-treated antigens much better than untreated antigens. When the antigen is mixed with 0.1% SDS, the antibody–antigen reaction is 5-fold higher using either ELISA assays or Western type dot blots (Fig. 3B).

Examples

Location of Open Reading Frames in Uncharacterized DNA. Selection of open reading frame fragments using pMR100 has been used to characterize DNA from the *bithorax* locus in *Drosophila melanogaster*.[23]

[22] K. Knudson, *Anal. Biochem.* **147,** 285 (1985).

[23] W. Bender, M. Akam, F. Karch, P. Beachy, M. Peifer, P. Spierer, E. Lewis, and D. Hogness, *Science* **221,** 23 (1983).

Over 95 kb of cloned DNA from the *Ubx* portion of the locus was sonicated and fragments longer than 200 bp ligated to pMR100 as described above. Lac$^+$ transformants were screened for the correct DNA inserts by colony hybridization using the appropriate *Ubx* probes. From 60,000 transformants, 1200 were lac$^+$ (2% of the total). Approximately one-half of these had *Ubx* DNA inserts, and the proportion of these that made full-length fusion proteins varied from 0 to 65%, depending on the region of the *Ubx* DNA from which the insert originated. All clones with inserts longer than 300 bp, as well as some smaller ones, were selected for fusion protein analysis; 61 of these produced full-length fusion proteins that include the inserted *bithorax* polypeptide sequence. Each of these ORF DNA inserts was mapped to restriction fragments in *Ubx* DNA by Southern blot analysis; the results are summarized in Fig. 4.

The only known protein-coding regions of *Ubx* are found within 4 exons spread over 75 kb (Fig. 4).[24,25] The 5′ exon contains the first 840 bp of the long ORF found in *Ubx* cDNA clones. The two central miniexons are each 51 bp in length; the 3′ exon contains only 200 bp of open reading frame sequence. There are some sequence data from other regions within the *Ubx* transcription unit; no additional open reading frames greater than 350 bp in length have been found.[26]

There are two substantial clusters of ORF fragments found within *Ubx*. One cluster is in the restriction fragment that includes the 840-bp ORF at the 5′ exon; the longest ORF fragments (including one of 569 bp) map at this location. The other cluster of fragments maps in a region 15 kb downstream from the 5′ exon within the second intron; the longest of these clones is 335 bp in length. There is no other molecular evidence of which we are aware that suggests that a protein-coding exon is present at this position. Most of the remaining fragments map outside of the DNA known to be part of the *Ubx* transcription unit; those within *Ubx* are all less than 300 bp in length and are scattered throughout the *Ubx* region. These results predicted that a long open reading frame exists at the genomic position shown by cDNA mapping to contain the main part of the protein-coding region of the gene. Few or no significant ORF sequences exist at other positions in *Ubx*, consistent with current molecular data indicating a lack of other long protein-coding regions in *Ubx*.[25] These results suggest that the approach of "shotgun" cloning into pMR100 to locate ORF sequences in a large region of uncharacterized DNA can provide a reasonably accurate picture of the open reading frame sequence

[24] P. Beachy, S. Helfand, and D. Hogness, *Nature* (*London*) **313,** 545 (1985).

[25] M. O'Connor and W. Bender, personal communication.

[26] M. Gray, Doctoral thesis. Brandeis University, Waltham, Massachusetts, 1985.

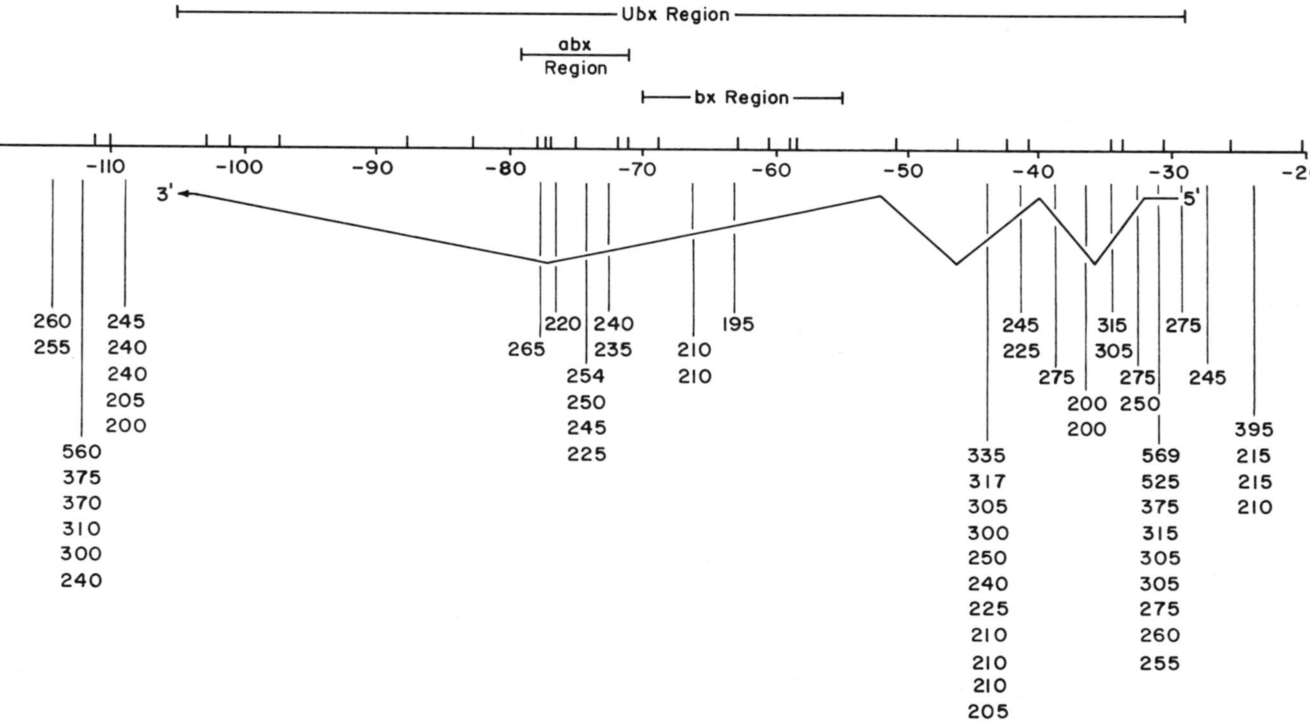

FIG. 4. Position of ORF fragments on the molecular map of the *Ubx* region of the *bithorax* complex in *Drosophila melanogaster*. The map coordinates are in kilobases; the short vertical lines on the map denote *Eco*RI sites. The positions of the four exons found in *Ubx* cDNA clones are shown below the restriction map; the diagonal lines represent the intron regions. The positions of the *Ubx*, *abx*, and *bx* regions are summarized above the map. The numbers indicate the positions and lengths of the ORF fragments.

distribution. This approach for mapping of ORF regions has also been useful in the analysis of the *serendipity* locus of *Drosophila*.[27,28]

Expression of Portions of a Well-Characterized Open Reading Frame Sequence. In order to see if many (or all) regions of an entirely open reading frame DNA fragment can be expressed in pMR100 as described above, a 1423-bp fragment that encodes part of the virus coat protein of canine parvovirus (CPV) was used to prepare insert fragments.[29] All lac$^+$ transformants with CPV inserts (300 bp or longer) were analyzed further. Of 192 lac$^+$ transformants with CPV inserts, 44 made full-length fusion proteins. These were mapped within the 1423-bp CPV fragment by restriction site mapping or by DNA sequencing. The open reading frame fragments mapped to all regions of the sequence, suggesting that no part of CPV sequence necessarily gives rise to an unstable fusion protein (Fig. 5).

Analysis of lac$^+$ Transformants That Produce Small Fusion Proteins. In both the *Ubx* and CPV experiments, most of the lac$^+$ clones that were shown to have the appropriate inserts by colony hybridization produced β-galactosidase proteins that were shorter than expected—frequently the size of β-galactosidase without any additional fused polypeptide sequences, including the λ *cI* portion of the pMR200 fusion protein (Fig. 2). Three of the *Ubx* and 7 of the CPV lac$^+$ clones that made only small proteins were studied further, by recloning the inserts in pEMBL9 and sequencing the inserts. All 10 fragments contained the wrong number of bases needed to correct the pMR100 frameshift and lacked a continuous open reading frame in either orientation. All 10 fragments carried an internal ATG codon followed by an open reading frame in phase with that of the downstream *lacIZ* portion of the pMR100 fusion gene. Upstream from these ATG codons were short sequences that matched or closely resembled bacterial Shine–Delgarno sequences.[30] This suggested that β-galactosidase proteins lacking the *cI* part of the protein sequence and much of the insert sequence might be the result of the initiation of translation within the insert. Indeed, the sizes of the small proteins suggested that this was the case; the distance between the ATG codon and the 3′ end of the insert correlated well with the increase in the size (relative to that of the wild-type protein) of the β-galactosidase protein (Fig. 6). The data suggest that if the DNA fragment has a sequence that can cause the initiation of translation in bacteria at a nearby downstream ATG codon, it

[27] A. Vincent, P. O'Connell, M. Gray, and M. Rosbash, *EMBO J.* **3,** 1003 (1984).
[28] A. Vincent, H. Colot, and M. Rosbash, *J. Mol. Biol.* **186,** 149 (1985).
[29] S. Rhode, *J. Virol.* **54,** 630 (1985).
[30] J. Steitz and K. Jakes, *Proc. Natl. Acad. Sci. U.S.A.* **72,** 4734 (1975).

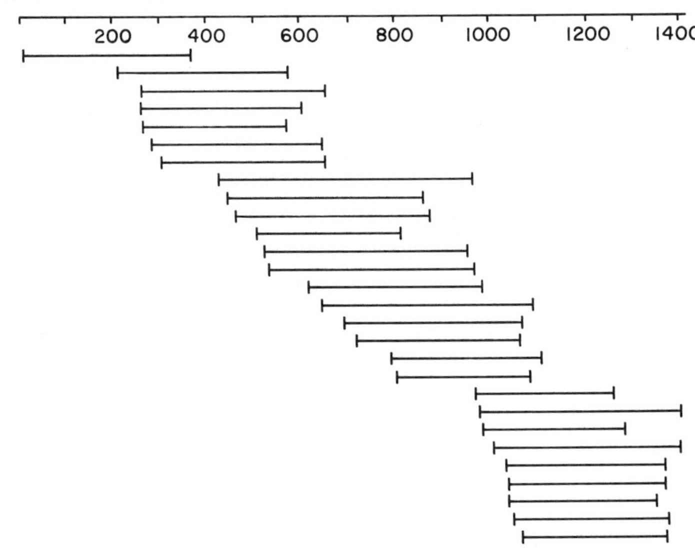

FIG. 5. Molecular map positions in the 1423-bp CPV coat protein gene fragment of ORF fragments expressed in pMR100.

will give rise to a lac$^+$ pMR100 transformant that makes a small β-galactosidase protein. If an open reading frame fragment of the correct size (with $3n + 2$ bp) and in the correct frame at both ends has a Shine–Delgarno sequence upstream from a nearby ATG codon which is in the correct reading frame, then two sizes of β-galactosidase proteins might be present—a large full-length protein and a small protein with an amino terminus corresponding to the ATG codon within the insert. The open reading frame insert of one of the CPV clones (CPV170) that produced both large and small proteins was sequenced and shown to fulfill these sequence requirements (Figs. 2 and 6).

The only requirements for fragments to cause the initiation of translation in the insert sequence of the pMR100 fusion gene mRNA are the insertion in the right orientation and in the proper reading frame at the junction of the 3′ end with the *lacIZ* portion of the fusion gene. The variability in the fraction of the total lac$^+$ clones that make small proteins with the DNA source is most likely a reflection of the relative abundance of the Shine–Delgarno-like sequences and open reading frames in the source DNA.

The clones that produce only small β-galactosidase proteins often are weakly lac$^+$ on MacConkey plates and produce small amounts of β-galactosidase; the CPV example discussed above is exceptional (Table I; Fig.

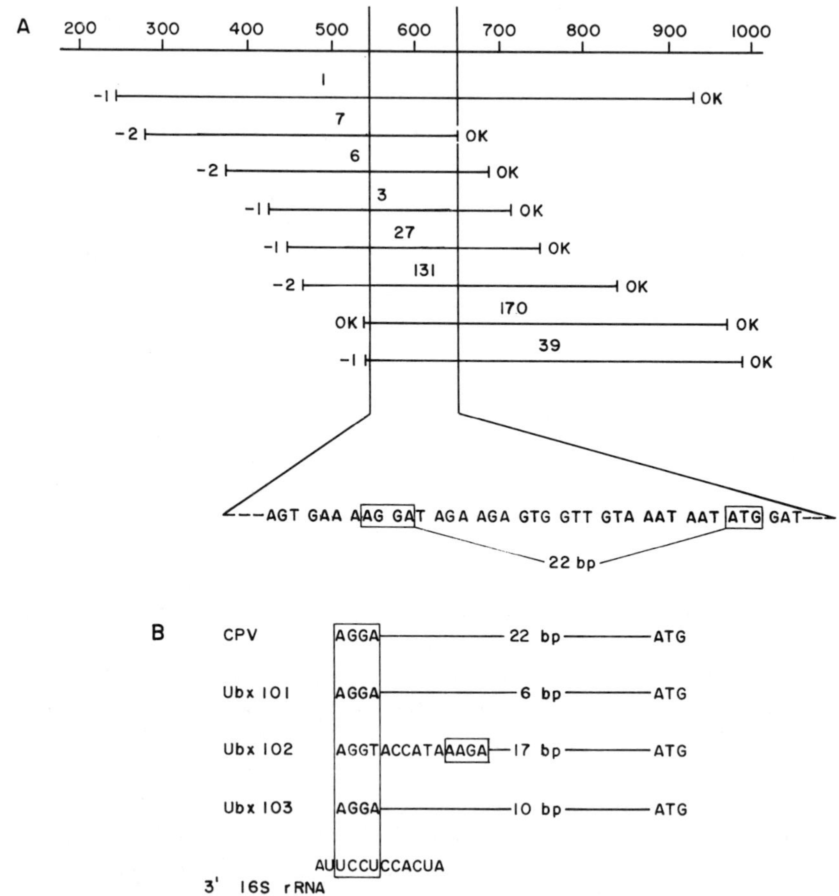

FIG. 6. (A) Molecular map positions of CPV fragments that cause initiation of translation within the inserted sequence of the pMR100 fusion gene mRNA. The numbers above each fragment identify the clone; the numbers at the left end of each fragment represent the number of bases needed to join correctly the CPV ORF with that of the *cI* portion of the pMR100 fusion gene. The designation "OK" indicates that the CPV and pMR100 fusion gene sequences are joined in the correct frame. (B) DNA sequences that might act as translation start sites in pMR100 recombinant plasmids. The CPV sequence is the same as that in (A); the *Ubx* sequences are from three different cloned fragments, as indicated. It has been proposed that the translational start sequences, termed Shine–Delgarno sequences, form hybrids with the 3′ end of 16 S rRNA during translation; the sequence of the *E. coli* 16 S rRNA thought to participate in mRNA binding is shown.

2). The ratio of the intensity of the β-galactosidase bands in the CPV170 sample suggests that the CPV translation start sequence is as efficient as that of the pMR100 fusion gene. The variability in the amount of the small β-galactosidase protein made by a lac$^+$ transformant might be a reflection of how efficiently the sequence in the insert can initiate translation in bacteria. The exact sequence of the Shine–Delgarno sequence and the neighboring DNA, as well as the distance from a downstream ATG codon, affect the rate of translation to a large extent.[30] Alternatively, the *in vivo* stability of the small and large proteins may vary significantly.

To date, all of the clones we have analyzed that make small proteins have inserts with sequences that match those found associated with the initiation of translation. Thus, it is likely that this is the most prevalent cause for the small proteins in pMR100 lac$^+$ transformants, rather than protein degradation *in vivo*, as suggested previously.[2] The incidence of an ATG codon with a nearby upstream Shine–Delgarno sequence should vary with each DNA source used; if these sequences are present in a larger region of DNA that includes an open reading frame, then the frequency of small protein clones should be at least 3 times higher than that of the large protein clones, because the latter is dependent on the proper reading frame at both the 5′ and 3′ ends of the insert fragment. The translation-initiating fragments need to have the correct configuration only at the 3′ end. Because of the large variety of very short sequences that have been identified as Shine–Delgarno sequences in bacteria, it is not surprising that eukaryotic DNA has many sequences that can act as translation initiators when placed in the appropriate context in bacterial genes.[30]

Comments

As discussed above, "shotgun" cloning of randomly cut fragments from uncharacterized DNA into pMR100 can be used for the mapping and expression of many and in some cases all of the open reading frames in the initial DNA. The only limitation of which we are aware is that with some DNA sources many of the lac$^+$ pMR100 clones synthesize β-galactosidase without expressing an open reading frame DNA insert, due to internal translation starts. In most of our experiments, only about 25% of the insert-containing lac$^+$ pMR100 clones make full-length fusion proteins (because of the presence of translation-initiating sequences); this proportion may vary from 0 to 100% depending on the relative abundance of open reading frame and translation-initiating sequences in the substrate DNA. In some cases, the transformants that synthesize the shortest of the small β-galactosidase proteins can be avoided by screening the lac$^+$ colonies with antibodies directed against the polypeptide sequence expected from the insert.[12]

The β-galactosidase fusion proteins produced by recombinant pMR100 transformants can be used as antigens for the production of antibodies; this has been successful for *Drosophila* proteins encoded by the *Ubx* and *per* loci,[19,31] trypanosome surface glycoproteins,[32] and canine parvovirus coat protein.[33] The strategy of shotgun cloning in pMR100 has also been used to quickly express parts of genes from uncharacterized viral genomes, such as that of HTLV III.[34]

Acknowledgments

We would like to thank our colleagues Hildur Colot, Linda Gritz, Julia Sue, and Tobie Tishman for their advice and assistance, and Gabrielle Peterson for performing ELISA assays. This work was supported by grants from the National Institutes of Health (GM33205 and HD08887).

[31] P. Reddy and M. Rosbash, unpublished experiments.

[32] M. Lenardo, S. Brentano, and J. Donelson, *Nucleic Acids Res.* **12,** 4637 (1984).

[33] G. Mazzara, unpublished experiments.

[34] N. Chang, P. Chanda, A. Barone, S. McKinney, D. Rhodes, S. Tam, C. Shearman, J. Huang, T. Chang, R. Gallo, and F. Wong-Staal, *Science* **228,** 93 (1985).

[9] Use of Open Reading Frame Expression Vectors

By GEORGE M. WEINSTOCK

Introduction

Producing foreign polypeptides in *Escherichia coli* poses a variety of problems. These include the proper joining of bacterial transcription and translation initiation signals to the foreign coding sequence and assaying for expression. Open reading frame (ORF) expression vectors provide a general solution to these problems. ORF vectors do not express an intact gene; rather, only a part of the coding sequence, lacking translation termination codons, is expressed to produce a polypeptide representing a part of the complete foreign protein. Such partial proteins can be used to produce antibodies which can then be used to detect the complete protein in its natural host. It is not necessary to know the DNA sequence of the foreign coding region in order to use ORF vectors to express it in *E. coli.*

Moreover, DNA from cloned genes, cDNA, or genomic DNA can be used. Thus, the ORF vectors provide a very general method for heterologous gene expression. Strategies for using ORF vectors as well as some of the available vectors have been reviewed previously.[1] In this chapter the procedure for using ORF vectors containing the initiator region from the *E. coli ompF* gene[2] will be reviewed. Some examples of the use of these vectors are the production of antibodies against Herpes virus thymidine kinase,[2] *Agrobacterium* indoleacetamide hydrolase,[3] human T-cell leukemia virus envelope proteins,[4] a trypanosome kinetoplast protein,[5] the *Drosophila melanogaster engrailed* protein,[6] and the hepatitis B virus X product.[7]

Principle

The plasmid vectors pORF1 and pORF2 have two important parts: the 5′ end of the *E. coli ompF* gene and an *E. coli lacZ* gene that has been deleted for its promoter and translation start site. The *ompF* gene fragment includes the promoter, translation start site, and beginning of the structural gene encoding the N-terminal signal sequence and first 12 amino acids of the mature OmpF protein. Immediately downstream from this is the *lacZ* sequence, encoding an active β-galactosidase but dependent on the *ompF* transcription and translation initiation signals for its expression. The *lacZ* sequence is out of frame with respect to the *ompF* coding sequence. Hence, translation initiating in *ompF* does not read the *lacZ* sequence in the correct frame to produce β-galactosidase. As a result, the pORF vectors confer a $LacZ^-$ phenotype. Between the *ompF* and *lacZ* sequences are restriction enzyme recognition sites. When an open reading frame DNA fragment of the appropriate length is inserted in one of these sites, such that the ORF is in frame with both the *ompF* and *lacZ* sequences, a tripartite gene is created in which the *lacZ* translational reading frame is aligned with that of the *ompF* sequence. Such plasmids

[1] G. M. Weinstock, *in* "Genetic Engineering" (J. K. Setlow and A. Hollaender, eds.), Vol. 6, p. 31. Plenum, New York, 1984.

[2] G. M. Weinstock, C. Aprhys, M. L. Berman, B. Hampar, D. Jackson, T. J. Silhavy, J. Weisemann, and M. Zweig, *Proc. Natl. Acad. Sci. U.S.A.* **80,** 4432 (1983).

[3] L. S. Thomashow, S. Reeves, and M. F. Thomashow, *Proc. Natl. Acad. Sci. U.S.A.* **81,** 5071 (1984).

[4] T. Kiyokawa, H. Yoshikura, S. Hattori, M. Seiki, and M. Yoshida, *Proc. Natl. Acad. Sci. U.S.A.* **81,** 6202 (1984).

[5] J. Shlomai and A. Zadot, *Nucleic Acids Res.* **12,** 8017 (1984).

[6] S. DiNardo, J. M. Kuner, J. Theis, and P. H. O'Farrell, *Cell* **43,** 59 (1985).

[7] E. Elfassi, W. A. Haseltine, and J. L. Dienstag, *Proc. Natl. Acad. Sci. U.S.A.* **83,** 2219 (1986).

confer a LacZ$^+$ phenotype on the host and produce a tribrid protein containing the polypeptide encoded by the inserted ORF sandwiched between the N-terminal OmpF and C-terminal β-galactosidase sequences. With this system, then, the LacZ$^+$ phenotype serves as a simple and sensitive assay for proper joining of expression signals and production of a foreign polypeptide. Because of this, the expression of a foreign sequence in ORF vectors does not require knowing the DNA sequence or an assay for its product.

The *ompF* initiator region in these vectors is derived from a gene encoding one of the major outer membrane proteins in *E. coli*.[8] There are about 100,000 molecules of OmpF in the cell, thus the *ompF* initiator region provides strong expression signals. The expression of *ompF* is also regulated by environmental factors such as media composition and osmotic strength and requires the product of the *ompR* gene as a positive regulator, making it possible to control the level of expression. This is important since high level production of OmpF–β-galactosidase hybrid proteins is lethal to the cell, presumably as a consequence of the blocking of the export apparatus by β-galactosidase. This lethality can be prevented by using an *ompR*$^-$ mutant host to reduce expression. During insertion of ORF DNA into these vectors, an *ompR* mutant is used that allows enough expression to observe the LacZ phenotype while preventing lethality. However the overproduction lethality is also exploited to determine whether translation initiates in *ompF* or within the inserted ORF sequence in LacZ$^-$ clones. Only when translation starts at *ompF* and proceeds through the insert will lethality be observed in an *ompR*$^+$ host. This genetic test serves to eliminate clones that are LacZ$^+$ due to fortuitous translation starts in the insert. Such clones are unwanted since their inserts may not be ORFs.

Materials

The *E. coli* strains used are as follows:

MH3000 = *araD139* Δ(*ara, leu*)*7697* Δ(*lac*)*X74* *galU*$^-$ *galK*$^-$ *rpsL*(Str^r) *ompR101*
MH3000/pORF1
MH3000/pORF2
TK1046 = *araD139* Δ(*argF–lac*)*U169* *rpsL150*(Str^r) *relA1* *flbB5301* *deoC1* *ptsF25* *malPQ*::Tn*5*(Kan^r, λ^r) *ompR1*(Cs)

MH3000 is a highly transformable strain with a deletion of the *lac* operon (to allow the LacZ phenotype to be tested) and a null mutation in *ompR*

[8] M. N. Hall and T. J. Silhavy, *Annu. Rev. Genet.* **15,** 91 (1981).

(to prevent lethality). This strain is used to construct clones. TK1046 has a cold-sensitive mutation in *ompR* that is used to test for lethality and produce large amounts of tribrid protein.

The LacZ phenotype is tested using XG (5-bromo-4-chloro-3-indolyl-β-D-galactoside). A stock solution of XG is prepared at 20 mg/ml in dimethylformamide, and 0.1 ml is spread on an agar plate and allowed to dry before spreading cells. L medium (contains 10 g tryptone, 5 g yeast extract, 5 g NaCl per liter of water) is used for liquid cultures and agar (1.5%) plates, and ampicillin is added to 150 μg/ml for growth of strains containing plasmids. Use of these media is described in Silhavy *et al.*[9]

Methods

Insertion of DNA into ORF Vectors. The structure of pORF1 is *ompF*-*Sal*I-*Bam*HI-*Sma*I-*Bam*HI-*lacZ* while pORF2 is *ompF*-*Bgl*II-*Bam*HI-*Sma*I-*Bam*HI-*lacZ*. Thus both vectors have a blunt-ended cloning site (*Sma*I) flanked by restriction sites for excising the insert. In general the DNA sequence of the foreign fragment will not be known. In order to insure that an ORF of the correct length and reading frame is present, it is best to generate a population of blunt-ended fragments with random end points and insert these into the *Sma*I site. A variety of methods have been used to accomplish this including digestion with DNase I[10] or nuclease *Bal*31[3] or sonication.[6,11] If it is possible to isolate a restriction fragment that is known to be an ORF, alternative methods for randomizing the length such as tailing with terminal deoxynucleotidyltransferase or adding a mixture of 8-mer, 10-mer, and 12-mer *Bam*HI linkers can be used. Finally, if the DNA sequence is known, it is usually possible to generate a restriction fragment, and specifically modify its ends if necessary, so that it will create the desired tripartite gene upon insertion into one of the sites in the vector.[2,4]

Identification of Clones. Following ligation of fragments and the plasmid vector, strain MH3000 is used as the host for transformation and colonies are selected on L agar containing ampicillin and XG. The indicator XG is used to detect the LacZ phenotype rather than minimal-lactose or MacConkey-lactose agar for several reasons.The strains are *lacY*$^-$ (no lactose permease), and XG does not require permease to enter the cell. In addition, XG is a more sensitive indicator, and, since MH3000 allows only a low level of expression to prevent lethality (it is *ompR*$^-$) maximal sensitivity is important.

[9] T. J. Silhavy, M. L. Berman, and L. W. Enquist, "Experiments with Gene Fusions." Cold Spring Harbor Lab., Cold Spring Harbor, New York, 1984.

[10] S. Anderson, *Nucleic Acids Res.* **9,** 3015 (1981).

[11] P. L. Deininger, *Anal. Biochem.* **129,** 216 (1983).

Following transformation of MH3000 with a ligation, one usually observes a range of colony colors from white to dark blue. White colonies do not contain *lacZ* fusions. Light blue colonies contain weakly expressed fusions and thus most have inserts. However, a light blue colony could carry a strongly expressed fusion that contains an insert that reduces the specific activity of β-galactosidase in the tribrid protein. Dark blue colonies contain strongly expressed fusions and some of these are frameshifts creating *ompF–lacZ* fusions without inserts.

There are a number of factors that affect expression to keep in mind. A weakly expressed fusion (light blue) may be a true tribrid gene, with translation initiating in *ompF*, but the insert may decrease transcription, translation, stability, etc. Alternatively, the 5′ end of the insert may not be in frame with *ompF* and translation of *lacZ* may be initiating within the insert. In this case translation initiation is limiting but such an insert may not be an ORF, only a fortuitous ATG codon in frame with *lacZ*. The ideal is a dark blue clone with an insert that does not reduce expression. In general, however, it is best not to limit the colonies chosen for subsequent analysis to a particular type.

Controls. There are two important controls that should be done at the same time as the cloning and transformation. The first is to digest the ORF vector with *Sma*I (if this is the enzyme being used) and ligate and transform without adding insert. The frequency of blue colonies observed represents the background due to frameshifts generated by the *in vitro* reactions. The second control is to isolate an *ompF–lacZ* fusion that does not contain an insert. This can be from the control experiment above. Alternatively a *Bam*HI digest of pORF1 or a *Bam*HI/*Bgl*II double-digest of pORF2, followed by ligation, creates a small deletion that puts *ompF* in frame with *lacZ*. These LacZ$^+$ (blue) pORF plasmids are used to test lethality and to provide size markers on SDS–polyacrylamide gels (below).

Test for Tribrid Genes. The three possible ways for a LacZ$^+$ (blue) phenotype to occur are (1) frameshift the vector without an insert, (2) insert a DNA fragment but translation initiates from within the inserted sequence, and (3) create a tribrid gene fusion. To distinguish these, the first step is to prepare plasmid DNA from LacZ$^+$ colonies. Then, restriction enzyme digestion followed by polyacrylamide or agarose gel electrophoresis can determine whether an insert is present. When the *Sma*I site is used for cloning, digestion with *Bam*HI will excise the insert, allowing its size to be determined.

The next step is to transform plasmids containing an insert into strain TK1046 in order to test for lethal overproduction and distinguish between possibilities (2) and (3) above. For unknown reasons, TK1046 transforms at about 1% of the frequency of MH3000. Hence it is necessary to construct clones with MH3000 and subsequently transform the candidates into TK1046. Since TK1046 contains a cold-sensitive *ompR* mutation, it is necessary to incubate plasmid-containing strains at low temperature at all times in order to maintain the $OmpR^-$ phenotype and prevent lethality. Generally the heat pulse is omitted during the transformation and the selective L, amp, XG plates are incubated at room temperature. Colonies should appear after 2–3 days.

To test for temperature sensitivity, streak for single colonies on an L, amp, XG plate at room temperature (control) and 42°. Simply making a line of cells on the plate is not sensitive enough since a few temperature-resistant mutants can give a positive response. Different inserts have given different responses. The ideal response is no growth at all at 42° (although some blue color, "ghosts," may be seen). In other cases fewer or smaller colonies are seen at 42° than at low temperature, and some of these may be white (deletions or other rearrangements). There have been cases where a tribrid was formed that produced a lot of protein but was not lethal. The sandwiched polypeptide in such a clone may prevent lethality by preventing export of β-galactosidase, but this appears to be rare. In general, when there is no lethality there is either a low level of expression and/or translation initiates within the insert and not from *ompF*.

Production of Tribrid Protein. As described above, the degree of protein production is usually, but not always, correlated with color and lethality, depending on effects of the insert on β-galactosidase specific activity or export lethality. Using β-galactosidase activity as a measure of synthesis after inducing a TK1046 culture has not been very reliable—the activity can decrease even though protein is accumulating. This may be related to export from the cytoplasm or aggregation when there is a lot of protein made. The most reliable measure is SDS–polyacrylamide gel electrophoresis. Be sure to include a control without a fusion (e.g., TK1046 with a parental $LacZ^-$ pORF plasmid). There are some endogenous proteins which migrate in the tribrid protein region that are turned on when cultures get too dense.

To induce expression from *ompF* in TK1046, first grow the strain with the plasmid overnight in L medium with ampicillin at room temperature. Make a 1/100 dilution the next day into fresh medium and grow at room temperature until the OD_{600} reaches 0.2. Shift the culture to 42° to induce and incubate with aeration for 1–2 hr. Measure the final OD_{600} of the culture, then centrifuge a total of 9–10 OD_{600} units of cells; i.e., volume of cells centrifuged × final OD_{600} = 9 to 10. Resuspend the pellet in 0.7 ml of loading buffer, then place in a boiling water bath for 5–10 min. After boiling, spin out debris in a microfuge for 2–5 min. Load 30 to 35 μl of sample on an SDS–polyacrylamide gel[12] for an analytical gel or load 700 μl for a preparative gel. Polyacrylamide concentrations from 4.5 to 9% have been used.

Several variations on this procedure have been successful. For induction, temperatures from 37 to 42° are effective. It is possible that some proteins may be more susceptible to degradation at 42°. If difficulties are encountered, it is best to do a preliminary time course for induction at 37 and 42°.

The recovery of tribrid protein after cell lysis can also be variable for some proteins. In some cases,[4,6] the cells were sonicated before extraction with the gel loading buffer, a treatment that increased recovery. Presumably these proteins aggregate and/or are tightly bound to cell debris. In general, the tribrid protein is observed as one of the major cellular proteins, and yields of tribrid protein as high as 15–20% of total cellular protein have been reported.

Large-scale production of tribrid protein from TK1046 follows the same procedure as described above. For a tribrid protein that is produced at the usual level, only 20–50 ml of culture is required.

Purification of Tribrid Protein and Production of Antibodies. Using a preparative polyacrylamide gel, protein bands are visualized by staining with Coomassie blue, and the tribrid protein band is cut out and broken up. Electroelute the protein from the gel slice as follows:

1. Place a piece of latex tubing on the bottom of a 10-ml Bio-Rad Econocolumn. The tubing should fit tightly and extend just past the bottom of the stem of the column.
2. Take approximately 5 inches of Spectropor No. 6 (10 mm × 10 m) dialysis tubing and push one end over the rubber tubing.
3. Seal dialysis tubing and rubber tubing to Econocolumn with a piece of parafilm.
4. Fasten bottom of dialysis bag with a clip.
5. Fill dialysis bag and column with running buffer.
6. Break up gel slice into a few pieces. Place in column and allow them to float to the bottom. Knock out any bubbles, particularly those at the interface between the dialysis tubing and the column.
7. Insert a single pole male plug through the lid of the column and attach approximately 5 inches of platinum wire to it on the inside. Twist the wire into a spiral of two or three turns. Put the lid onto the column, extending the wire well into the running buffer.
8. Support the column vertically. Place the bottom of the dialysis bag

[12] U. K. Laemmli, *Nature (London)* **227,** 680 (1970).

in an electrophoresis chamber filled with running buffer. The dialysis bag should extend straight down into the running buffer, then bend upward with the clip hanging over the side of the chamber.

9. Attach a positive lead to this chamber and a negative lead to the lid. Electrophorese at 1–2 mA overnight.
10. Remove the buffer in the Econocolumn and discard.
11. Save the buffer from the dialysis bag. There should be at least 2 ml.

The protein solution is emulsified with an equal volume of Freund's adjuvant and used to immunize a rabbit. Generally three inoculations are given, at 2-week intervals, each containing 100–150 μg of protein. Antisera are collected starting 1–2 weeks after the last inoculation. A somewhat simpler approach[3] is to visualize protein in the preparative gel by staining in ice-cold 0.2 *M* KCl,[13] then excise and mince the tribrid band and emulsify with an equal volume of Freund's incomplete adjuvant. This sample can be used directly for immunization.

Comments

This chapter has focused on the use of ORF vectors that contain the *ompF* initiator region. Several other vectors have been developed[1] that use other initiator regions for expression but are otherwise analogous in the use of β-galactosidase to monitor expression. The *ompF* vectors provide a convenient test for proper inserts (overproduction lethality) and an as yet untested possibility for increased tribrid protein stability due to export from the cytoplasm. However, the behavior of foreign nucleic acid and protein sequences in *E. coli* is far from predictable, and thus not all open reading frame sequences may behave in the desired fashion in a particular vector. For this reason it is sometimes necessary to try several expression systems before one is found that is useable for a particular insert.

ORF vectors have also found other applications beside the production of antibodies. When the initiator region in the tribrid gene is replaced with the 5′ end of the foreign sequence, the β-galactosidase activity of the resulting *lacZ* gene fusion can be used to optimize expression in *E. coli* from the foreign gene's initiation sequences. Then, replacement of the *lacZ* region with the 3′ end of the foreign gene results in a vector capable of producing the intact foreign polypeptide in *E. coli*.[3] In addition, the tribrid proteins have been used as a diagnostic tool to detect antibodies present in serum samples.[4,7] These examples demonstrate the versatility of this simple gene expression system.

[13] D. A. Hager and R. R. Burgess, *Anal. Biochem.* **109,** 76 (1980).

[10] 5-Fluoroorotic Acid as a Selective Agent in Yeast Molecular Genetics

By JEF D. BOEKE, JOSHUA TRUEHEART, GEORGES NATSOULIS, and GERALD R. FINK

Genetic techniques that permit the selection of mutant cells in the presence of large numbers of wild-type cells are extremely useful. These selective techniques usually depend on the conversion of a nontoxic compound to one that is toxic to wild-type cells. Mutant cells lacking the ability to form the toxic compound grow in the presence of the inert precursor. In yeast only a few selective compounds of this type have been discovered: α-aminoadipate prevents the growth of *LYS2*$^+$ and *LYS5*$^+$ cells[1]; methyl mercury prevents the growth of *MET2*$^+$ and *MET5*$^+$ cells[2]; and both ureidosuccinate[3] and 5-fluoroorotic acid[4] prevent the growth of *URA3*$^+$ cells. Spontaneous or induced mutations can be selected in each of these genes by plating large numbers of wild-type cells on the appropriate inhibitor.

The 5-fluoroorotic acid (5-FOA) selection is extremely useful for a number of genetic and molecular biological manipulations requiring the detection of rare *ura3*$^-$ cells. The utility of the 5-FOA selection is due to several factors: the availability of a large collection of *URA3*-based cloning vectors of various types for yeast; the small size and known sequence of the *URA3* gene[5]; the availability of numerous well-studied mutations in the *URA3* gene; the specificity, ease, and efficacy of the selection.

The 5-Fluoroorotic Acid Selection

Yeast cells are generally pregrown in liquid YPD medium or on YPD plates[6] prior to selection on 5-FOA. As many as 10^7 cells may be plated on a single, 50 mm Petri dish containing 5-FOA medium (Table I). Resistant colonies will grow up within 4–7 days at 30°. The most useful concentra-

[1] B. B. Chattoo, F. Sherman, D. A. Azubalis, T. J. Fjellstedt, D. Mehvert, and M. Ogur, *Genetics* **93,** 51 (1979).

[2] A. Singh and F. Sherman, *J. Bacteriol.* **118,** 911 (1974).

[3] M. Bach and F. LaCroute, *Mol. Gen. Genet.* **115,** 126 (1972).

[4] J. D. Boeke, F. LaCroute, and G. R. Fink, *Mol. Gen. Genet.* **197,** 345 (1984).

[5] M. Rose, P. Grisafi, and D. Botstein, *Gene* **29,** 113 (1984).

[6] F. Sherman, G. R. Fink, and C. Lawrence, "Methods in Yeast Genetics." Cold Spring Harbor Lab., Cold Spring Harbor, New York, 1977.

TABLE I
5-FLUOROOROTIC ACID MEDIUM[a]/2× CONCENTRATE[b]

Ingredient	Amount
Yeast nitrogen base	7 g
5-Fluoroorotic acid[7]	1 g
Uracil	50 mg
Glucose	20 g
H_2O	to 500 ml

[a] Add ingredients to 500 ml 4% agar (autoclaved). Mix and pour into Petri dishes.
[b] Heat to 65° to dissolve and filter sterilize.

tion of the drug is generally 1 mg/ml 5-fluoroorotic acid,[7] but in the interest of economy it is possible to use as little as 500 μg/ml 5-FOA. When lower concentrations of 5-FOA are used, however, background growth of the Ura$^+$ cells is considerably greater. This background growth is less of a problem when Ura$^-$ derivatives occur at a high frequency (e.g., when a *URA3*$^+$ marker is located between the elements of a direct duplication). In some instances it may be desirable to use a rich medium such as YPD as opposed to the minimal medium normally used with 5-FOA selection. Unfortunately, addition of 5-FOA to YPD medium at 1 mg/ml prevents the growth of some Ura$^+$ strains but not others, whereas the 5-FOA minimal medium in Table I inhibits all Ura$^+$ yeast strains we have tested. The reason for the variability on YPD is not clear. On the other hand, the addition of 5-FOA at 1 mg/ml to SC medium,[6] which contains all 20 amino acids and other supplements added to minimal medium, gives satisfactory results with all of the strains we have tested.

Transplacement of Mutant Alleles: Use of 5-Fluoroorotic Acid

Selection of Ura$^-$ cells with 5-FOA is used frequently in the replacement of resident chromosomal genes by transformation.[8] Generally the goal is to replace the gene of interest with a mutant allele of that gene generated *in vitro* (Fig. 1). Replacement of the wild-type gene with a deletion mutation permits the assessment of the effect of a null mutation on the phenotype of the cells. If there is a possibility that the gene is

[7] Available from Pharmacia/P-L Biochemicals, Piscataway, New Jersey, and SCM Specialty Chemicals, Gainsville, Florida.
[8] F. Winston, F. Chumley, and G. R. Fink, this series, Vol. 101, p. 211.

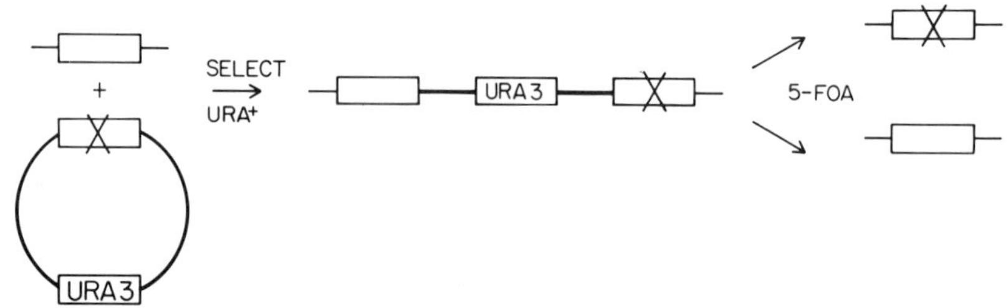

FIG. 1. Transplacement of a yeast gene with a mutant allele. The heavy lines indicate pBR322 sequences. The X symbolizes a mutation introduced into the cloned DNA. The gene of interest (open box) is cloned into Yip5, a yeast integrating vector carrying the *URA3* gene as the selectable marker. The plasmid is used to transform a yeast strain containing a nonrevertable *URA3* allele (e.g., *ura3-52*) to Ura$^+$. The structure of the chromosome in the Ura$^+$ transformant is shown in the center: the *URA3* gene and pBR322 sequences are flanked by the wild-type (left) and mutated (right) versions of the gene. When the cells containing the integrated plasmid are plated onto medium containing 5-fluoroorotic acid (5-FOA), derivatives which lose the *URA3* gene and the flanking plasmid DNA (Ura$^-$) via a homologous recombination event between the two homologous segments flanking *URA3*$^+$ survive, whereas the parent cells (Ura$^+$) do not.

essential, these experiments are normally performed in a diploid homozygous for a nonreverting *ura3* mutation such as *ura3-52*. If the gene in question is known to be nonessential, the experiments can be performed in the corresponding haploid strain. The deletion mutation is introduced in a derivative of Yip5[9] or a similar yeast integrating plasmid bearing both the *URA3* gene and the gene of interest. Transformation of the target strain with the deleted Yip5 derivative by either the spheroplast[10] or lithium acetate[11] procedures, followed by selection for the Ura$^+$ phenotype, results in integration of the mutant form of the gene in a tandem arrangement with the wild-type allele (Fig. 1). The *URA3* gene and plasmid sequences lie between the duplicated sequences in the transformants, which are referred to as integrants.

The next step in the allele replacement process is a selection for cells that have undergone a reversal of the integration process (this reversal may be viewed as a single-crossover recombination event). This step is greatly facilitated by the use of 5-FOA. Cells which undergo the desired recombination event will lose the plasmid and *URA3* sequences, thereby

[9] D. Botstein, S. C. Falco, S. E. Stewart, M. Brennan, S. Scherer, D. T. Stinchcomb, K. Struhl, and R. W. Davis, *Gene* **8,** 17 (1979).

[10] A. Hinnen, J. B. Hicks, and G. R. Fink, *Proc. Natl. Acad. Sci. U.S.A.* **75,** 1929 (1978).

[11] H. Ito, Y. Fukuda, K. Murata, and A. Kimura, *J. Bacteriol.* **153,** 163 (1983).

becoming resistant to 5-FOA (a procedure variously referred to as "popping out," "looping out," or excision of the plasmid). The fraction of popouts which leave behind the mutant allele will depend on the relative recombination frequencies of the two segments of DNA flanking the mutation. (This frequency is usually related to the size of the homology regions flanking the mutation.) If a particular phenotype is expected or predicted, the presence of the mutant allele can be ascertained by phenotypic screening. Where one does not know what phenotype to expect, the mutant allele may be identified by Southern hybridization to the chromosomal DNA from several 5-FOA^r derivatives in order to identify the mutant allele. This analysis assumes that the allele has some characteristic restriction site or pattern on a gel.

Plasmid Shuffling: A General Technique for Isolating Conditional Mutations in an Essential Gene

Essential genes present special problems in mutational analysis because their deletion is lethal. To obtain viable cells with nonfunctional alleles of an essential function, investigators have turned to temperature-sensitive alleles. A typical mutant hunt for such alleles involves the identification of strains that fail to form colonies at the restrictive temperature, but allow normal growth at the permissive temperature. If the gene has been cloned, then one can mutagenize the cloned gene *in vitro* and transform it back into yeast. The best recipient would be a strain containing a deletion of the gene on the chromosome so that the alleles produced on the *in vitro* mutagenized copy will be revealed after transformation. The recipient strain must possess a functional copy of the gene until it obtains a second copy introduced on the mutagenized plasmid. However, a functional copy of the gene will mask the phenotype of any temperature-sensitive allele borne by a plasmid. The resident gene must then be lost or destroyed so that the function of the mutagenized gene may be assayed. This Byzantine conundrum has several solutions.

One solution involves transformation of a wild-type strain with a mutagenized integrating plasmid that carries a truncated version of the cloned gene.[12] The appropriate integration event leads to disruption of the wild-type chromosomal copy and restoration of the missing sequences to the mutagenized copy (Fig. 2A). The only intact copy of the gene contains mutagenized sequences, so if a mutation has been induced, the phenotype corresponding to that mutation will be expressed even if it is recessive. This technique suffers from the constraints imposed on the integration

[12] D. Shortle, P. Novick, and D. Botstein, *Proc. Natl. Acad. Sci. U.S.A.* **81,** 4889 (1984).

reaction. First, integrative transformation is a low frequency event. Second, if the crossover occurs on the incorrect side of the mutation, or if gene conversion by the wild-type allele accompanies the integration, the mutation may not be recovered (Fig. 2B, C). Moreover, in at least one study[12] the majority of temperature-sensitive mutations obtained by this procedure did not map to the gene in question. Experiments using this procedure in our laboratory also gave a high proportion of unlinked temperature-sensitive mutations. Apparently, the transformation process itself has a general mutagenic effect on the recipient cell.

We devised a plasmid shuffling technique to obviate the problems in obtaining temperature-sensitive mutations in the *DLF* (desired lethal function) gene. Mark Rose was extremely helpful in the development of these ideas. The procedure requires construction of a *ura3*$^-$ *leu2*$^-$ recipient containing a lethal deletion of *DLF* (Δ*dlf*) on the chromosome and a wild-type copy (*DLF*$^+$) on a *URA3*$^+$ plasmid. This strain is then transformed by a second plasmid containing the selectable *LEU2*$^+$ marker and a copy of the *DLF*$^+$ gene. The *LEU2* plasmid is mutagenized *in vitro* prior to transformation. The mutations induced in the *DLF*$^+$ gene on the *LEU2* plasmid are revealed when the *DLF*$^+$ gene on the *URA3*$^+$ plasmid is lost upon 5-FOA selection, because the cells then have only a deleted Δ*dlf* gene on the chromosome. The actual manipulations will be described in subsequent sections.

The plasmid shuffling strategy has been used successfully to obtain temperature-sensitive mutations in the *CDC27* gene. Previously isolated

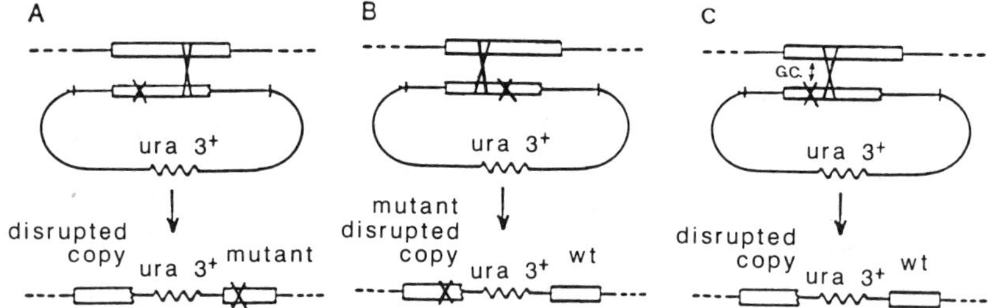

FIG. 2. Integration of a mutagenized, truncated gene. An integrating plasmid is constructed bearing a truncated version of the gene of interest. The gene can be deleted at either the 5′ or 3′ end. If a mutation is induced in the truncated segment, and integration occurs as in (A), the complete copy of the gene will harbor the mutation. However, if the integration occurs as in (B), the complete copy of the gene will not contain the mutation. If the integration is accompanied by gene conversion (GC) of the mutated segment back to its wild-type state, the mutation will not be recovered (C).

temperature-sensitive mutations in *cdc27* cause the cell to arrest late in nuclear division upon growth at the elevated temperature.[13] We have mapped this gene to the left arm of chromosome II, 13 centimorgans distal to *ils1*. Sporulation and dissection of a diploid containing a wild-type *CDC27* gene and a disrupted *cdc27Δ1* gene yielded two viable and two inviable spores per tetrad. When the same diploid was transformed with pSB17, a 2μ-based plasmid carrying both the *URA3* and the *CDC27* gene, seven out of seven tetrads tested gave four viable spores. Moreover, in each of the seven tetrads two of the spores failed to segregate 5-FOA-resistant derivatives, an indication that strains containing the deletion cannot survive without the *URA3*$^+$ *CDC27*$^+$ plasmid. Southern analysis confirmed the presence of the *cdc27Δ1* allele in the chromosomes of these strains. This experiment shows that at least one copy of *cdc27* is required for growth.[14]

The plasmid shuffle required a second plasmid carrying *CDC27* on a centromere vector which contained the *LEU2* gene as the selectable marker. The *CDC27*$^+$ gene on this vector is mutagenized by hydroxylamine and introduced by transformation into a strain containing a disrupted chromosomal *cdc27Δ1* gene and an intact episomal *CDC27* gene on a *URA3*–2μ vector. Leu$^+$ transformants were selected on plates containing uracil but lacking leucine. This combination of nutrients selects for the presence of the mutagenized *LEU2–CDC27* plasmid, and simultaneously releases any selection for the *URA3–CDC27* plasmid. These transformants are grown at the permissive temperature, during which time a proportion of the transformed cells lose the unmutagenized *CDC27*$^+$ gene carried on the *URA3* plasmid. These Ura$^-$ cells are now capable of expressing only the phenotype of the *CDC27* gene on the *LEU2* plasmid. After the transformed colonies are replica-plated onto 5-FOA medium, one can assay the properties of these Ura$^-$ cells. If the *LEU2* plasmid carries a nonmutant *CDC27*$^+$ gene, the Ura$^-$ cells will be able to grow at both permissive and restrictive temperatures. Since the original colony is a mixture of Ura$^+$ and Ura$^-$ cells, the replica on 5-FOA will give rise to Ura$^-$ papillae that grow out of the background. If, on the other hand, the *LEU2* plasmid bears a temperature-sensitive *cdc27* gene, the Ura$^-$ cells within the colony will be able to grow at the permissive but not the restrictive temperature. A null mutation in the *LEU2* plasmid-borne gene will render any Ura$^-$ segregant inviable: no papillation of 5-FOA^r colonies will be seen at any temperature (Fig. 3).

[13] L. H. Hartwell, R. K. Mortimer, J. Culotti, and M. Culotti, *Genetics* **74,** 267 (1973).

[14] J. Trueheart, unpublished results.

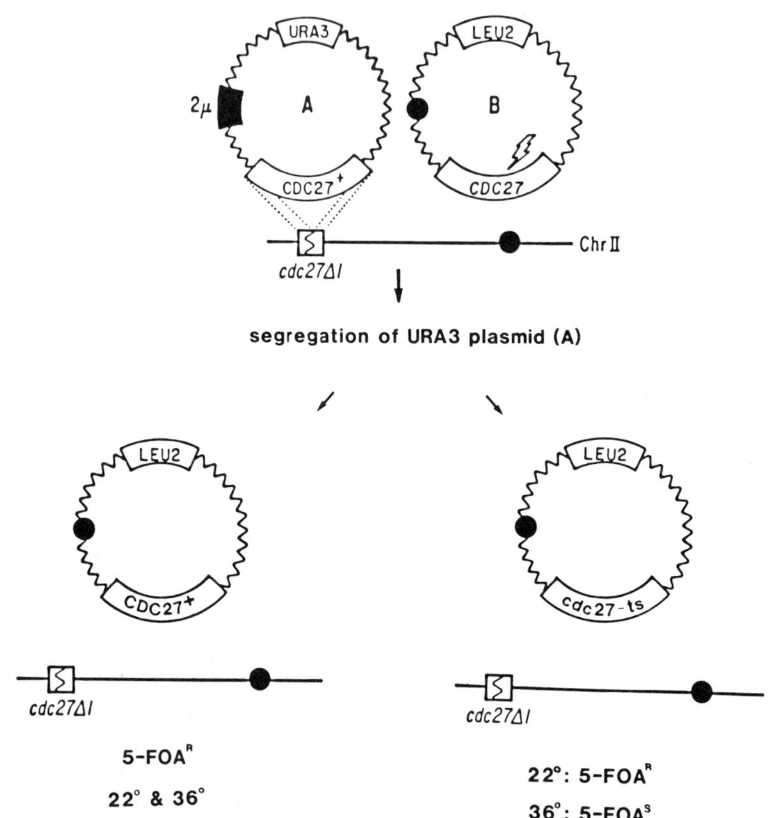

FIG. 3. Plasmid shuffling: a general method for introducing mutations into a cloned essential gene. A strain containing the *cdc27Δ1* allele integrated into chromosome II (solid circles represent centromeres) can be propagated if Yep24 (which carries the *URA3* gene as its selectable marker) bearing the *CDC27* gene (plasmid A) is present in the cell. Because of the *cdc27Δ1* deletion these cells depend on the Yep24 derivative for the essential *CDC27* function. A second, mutagenized plasmid containing the *CDC27* gene and the *LEU2* selectable marker (plasmid B) is subsequently introduced into the same cell. If the *LEU2* plasmid (B) carries a functional *CDC27* gene, the Yep24 derivative (A) becomes dispensable to the cell; 5-FOA-resistant derivatives arise frequently which lack the Yep24 derivative. If the *LEU2* plasmid carries a mutation which inactivates the *CDC27* gene, the cell remains dependent on plasmid A and cannot papillate 5-FOA-resistant derivatives. Any conditional defect in the *CDC27* gene on plasmid B, such as temperature sensitivity, will be reflected in the 5-FOA phenotype, i.e., cells with a temperature-sensitive *cdc27* allele will form 5-FOA^R papillae only at the permissive temperature.

TABLE II
PLASMID SHUFFLING: MUTAGENESIS OF THE *CDC27* GENE

Mutagenized plasmids tested	*cdc27* nonconditional	*cdc27* temperature sensitive
800	11	2

Strain JY72 [*MAT***a**, *ura3-52*, *leu2-3,112*, *trp1Δ1*, *cdc27Δ1* (pSB17 = 2μ–*URA3–CDC27*)] was transformed with 9 μg of hydroxylamine-mutagenized[15] pSB31 (*LEU2–CEN4–CDC27*) DNA, giving approximately 800 Leu$^+$ colonies after incubation of the transformed cells at 22° for 4 days. These colonies were replica-plated onto 5-FOA plates with added uracil and tryptophan, and incubated at 36° for 1 day. The phenotypes of the colonies obtained are summarized in Table II. Of 35 colonies that initially failed to papillate 5-FOA-resistant derivatives at high temperature, eleven failed to grow on 5-FOA at any temperature; these were presumed to carry completely nonfunctional copies of the *CDC27* gene. Two isolates displayed the desired phenotype: ability to grow on 5-FOA at 22° but not at 36°, and no growth defect on rich or leucine-deficient media at either temperature. The remaining colonies either failed to retest or displayed an intermediate phenotype at the high temperature upon retesting.

Any temperature-sensitive mutation that arises in an unlinked chromosomal gene can be identified easily and discarded at the beginning of the experiment. Such mutants would fail to grow at the high temperature regardless of the media employed. In the *CDC27* experiment, none of the 35 isolates showed any growth defect on rich or leucine-deficient media at either temperature. Therefore, this strategy seems not to suffer the high rate of unlinked mutation which plagues the integrative technique.

The two putative temperature-sensitive mutants were analyzed further with respect to their cell division cycle (*cdc*) phenotype. Both were cured of pSB17 by growth on YPD medium at 22°, and Ura$^-$ colonies were identified by replica plating to SC–Ura medium. When a segregant from each isolate was grown at 36° in YPD medium, both temperature-sensitive mutants arrested as a uniform population of large-budded cells, as do other previously identified alleles of the *CDC27* gene.[13] This analysis shows that the temperature sensitivity of papillation to 5-FOA resistance observed initially was not merely an artifact stemming from the presence of 5-FOA, but that the drug merely prevented the growth of those cells that still contained the original wild-type gene.

[15] M. Rose and G. R. Fink, *Cell* **48,** 1047 (1987).

Introduction of ura3 Deletions into the Chromosome

Because of the widespread use of *URA3*$^+$ as a selectable marker, the availability of *ura3*$^-$ deletion mutations of defined structure would be extremely useful. Large deletions of this region do not exist in the current mutant collection. Apparently, deletion of the entire 1.1 kb *Hin*dIII fragment on which the *URA3* gene resides is lethal to the cell.[16] There is circumstantial evidence that an essential gene lies 5′ to the *URA3* gene.[17] The *ura3-52* mutation is nonreverting, but its utility is compromised. The mutation is caused by the insertion of a Ty element and the complete wild-type *URA3* sequences are all still present.[18]

As a first step to constructing a series of defined mutations in the *URA3* gene, we have made a 220-bp deletion extending from the *Pst*I site just 5′ of the *URA3* ATG initiation codon to the *Nco*I site at position 432.[5] The mutant allele was constructed in an integrating plasmid containing the *TRP1* gene as a selectable marker (the structure of the resultant plasmid outlined in Fig. 4 was confirmed by restriction enzyme digestion). The 5-FOA selection was invaluable in replacing *URA3*$^+$ with the deletion of *ura3* that we made *in vitro*.

We tested two methods of introducing the *ura3Δ1* allele into yeast cells, both of which make use of the 5-FOA selection. One involved an attempt to introduce the deletion mutation directly by transforming with a linear piece of DNA carrying the deletion and homologous to *URA3* at both ends, followed by direct selection on 5-FOA. The second was to integrate the *ura3Δ1* mutation (using the *TRP1* gene as a selectable marker) at the *URA3* locus and subsequently to select for popouts of the wild-type allele by growth on 5-FOA medium. Only the second method gave the desired insertion of the *ura3Δ1* gene.

Yeast cells (strain JBX169-10B; *MAT***a**, *trp1Δ1*, *lys2*, *GAL*$^+$) were prepared for transformation by the lithium acetate[11] procedure. The cells were transformed with plasmid pJEF1332 (Fig. 4) which had been cleaved with one of two restriction enzymes: *Hin*dIII, which releases the *URA3* fragment completely from the plasmid, and *Stu*I, which cuts within the *URA3* segment but does not cut the plasmid elsewhere. When the *Hin*dIII-cut plasmid is used in transformation, one expects direct gene replacement[19] of the *URA3* gene on the chromosome by the deletion allele. Presumably the *ura3Δ1* transformants could then be selected on 5-FOA. However, it seemed possible that the cells would require a period

[16] F. Winston and T. Donahue, personal communication.

[17] M. Rose and J. Cappello, personal communication.

[18] M. Rose and F. Winston, *Mol. Gen. Genet.* **193,** 557 (1984).

[19] R. J. Rothstein, this series, Vol. 101, p. 202.

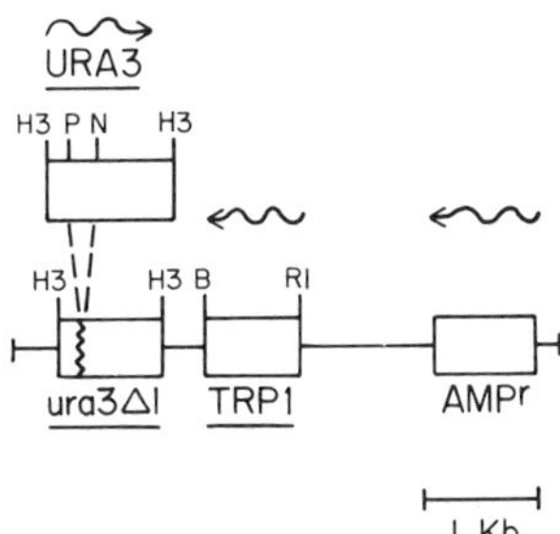

FIG. 4. An integrating plasmid bearing the *ura3Δ1* mutation, pJEF1332. The *ura3Δ1* mutation and its wild-type counterpart are diagramed. Boxes represent genes; wavy lines are transcripts. Restriction sites are abbreviated as follows: H3, *Hin*dIII; P, *Pst*I; N, *Nco*I; B, *Bam*HI; RI, *Eco*RI. The plasmid backbone (all sequences to the left of *URA3* and to the right of *TRP1*) is from pUC18, drawn as though linearized at position 1 (as defined in the New England BioLabs 1985/86 catalog). The polylinker sequences between *Hin*dIII and *Eco*RI have been substituted with the yeast genes; the short *Hin*dIII–*Bam*HI fragment between the *TRP1* and *URA3* genes is derived from pBR322.

of nonselective growth in order for the Ura$^-$ (5-FOA^r) phenotype to be expressed. Therefore, the cells were cultured in liquid YPD medium overnight prior to plating them onto 5-FOA medium. The results of this experiment were rather disappointing. Although there were about 3 times as many 5-FOA^r colonies on plates where *ura3Δ1* DNA was added as on control plates to which no DNA was added, the presence of colonies on control plates suggested that spontaneous *ura3*$^-$ mutants had arisen in the culture without added DNA. Nevertheless, 12 5-FOA^r colonies from the plate to which DNA had been added were analyzed by Southern hybridization in order to examine the structure of the *URA3* locus. All 12 strains contained the wild-type 1.1-kb *Hin*dIII fragment rather than the expected 0.9-kb mutant *Hin*dIII fragment corresponding to the deleted *URA3* gene. The presence of only the wild-type *URA3* fragment suggested that none of the 12 5-FOA^r colonies resulted from direct replacement by the deletion allele. Probably, each carried some spontaneously arising point mutation in the *URA3* gene. Similar results were obtained in another laboratory in experiments designed to replace the *URA3* locus with cloned DNA derived from the Ty element insertion allele *ura3-52*.[20]

The alternative approach was to introduce the deletion mutation by the integration–excision or popout technique described earlier. Linearized pJEF1332 plasmid was integrated by selection for the Trp$^+$ phenotype and then segregants that had lost the wild-type *URA3* gene, the

[20] R. J. Rothstein, personal communication.

plasmid sequences, and the *TRP1* gene were obtained on 5-FOA medium. Yeast cells (JBX169-10B) were grown in YPD and transformed with 1 μg of pJEF1332 DNA previously linearized with the restriction enzyme *Stu*I, which cuts only once in the plasmid, within the remaining coding sequences of the *URA3* gene. The transformed cells were plated on SC − Trp medium in order to select for integration. Five transformants were chosen for further study. Several 5-FOA^r segregants were isolated from each transformant by simply streaking the transformant colony onto 5-FOA medium (containing tryptophan and other nutritional requirements of the strain). Of 30 5-FOA^r derivatives, 27 were Trp$^+$ and 3 were Trp$^-$. All 3 Trp$^-$ derivatives proved to be simple popouts of the pJEF1332 plasmid as judged by Southern analysis and by reversion analysis (as expected, no reversion of the deletion mutation was observed, even after extensive UV irradiation). Presumably, the Trp$^+$ derivatives resulted from gene conversion of the wild-type copy of the *URA3* gene by the deleted copy. Why this event should be so much more frequent than the popping out of the plasmid, which requires only a single crossover event, is unclear.

The direct selection of Ura$^-$ using 5-FOA in transformation needs further study, in view of a recent report of the successful use of this procedure. A plasmid containing the *Bam*HI fragment of *URA3*, which is 5 kb in length, was used as the starting material for the construction of a 446 bp deletion which removes the 3′ portion of the *URA3* coding sequences (extending from the *Stu*I site to the *Sma*I site).[21] Apparently, transformation of Ura$^+$ yeast with *Bam*HI-linearized plasmid DNA results in 10 to 50 5-FOA^r transformants per microgram, all of which carry the deletion.[21] Perhaps the increased homology to chromosomal DNA in such an experiment allows for a higher frequency of transformants. No outgrowth in YPD was used in these experiments.[21]

Additional Applications

The 5-FOA selection for Ura$^-$ cells is very useful in the study of repeated sequences and their relative rates of recombination. It is a simple matter to measure the mitotic recombination frequency of any yeast duplication using 5-FOA. The sequence is simply cloned into Yip5 and then integrated by cutting within the region of homology to the sequence of interest, with selection for the Ura$^+$ phenotype. The frequency of Ura$^-$ segregants (detected as 5-FOA^r colonies) can then be easily measured. Winston *et al.*[22] used the selection to obtain accurate estimates of the rate

[21] K. Frohlich, personal communication.

[22] F. Winston, D. T. Chaleff, B. Valent, and G. R. Fink, *Genetics* **107,** 179 (1984).

of δ–δ recombination in Ty elements by inserting the *URA3*$^+$ marker between the δ elements.

Summary

5-FOA is an extremely useful reagent for the selection of Ura$^-$ cells amid a population of Ura$^+$ cells. The selection is effective in transformation and recombination studies where loss of *URA3*$^+$ is desired. A new plasmid shuffling procedure based on the 5-FOA^R selection permits the recovery of conditional lethal mutations in cloned genes that encode vital functions.

Acknowledgments

J.D.B. was supported by a Helen Hay Whitney Fellowship, G.N. is a fellow of the Belgian Fonds National de la Recherche Scientifique, G.R.F. is an American Cancer Society Professor.

[11] Transposon Tn5 Mutagenesis to Map Genes

By FRANS J. DE BRUIJN

Introduction

Transposable (Tn) elements (or transposons) are extremely useful tools in both classical as well as molecular bacterial genetics. Their utility is based on a number of specific characteristics:

1. Transposition (or insertion) of a Tn element into a structural gene or its regulatory region usually leads to insertional inactivation of the gene, and these mutations are stable.
2. Tn elements carry genetic markers, such as antibiotic resistance genes, which greatly facilitate genetic mapping of Tn element-induced mutations, as well as the transduction and cloning of mutated regions.
3. Tn element-induced insertion mutations are generally nonleaky, highly polar on genes located downstream in operons, and revert (by precise excision) at very low frequencies.
4. Tn elements usually have a characteristic physical structure, including inverted repeats of varying lengths as well as unique restriction endonuclease cleavage sites, which facilitate mapping of their insertion

site by restriction analysis or DNA heteroduplex analysis using the electron microscope.

5. Tn elements can induce a variety of genomic rearrangements at their insertion site, such as deletions and inversions, which can lead to the fusion of genes to other nearby genes or promoters. They can also create novel Tn elements consisting of genomic segments flanked by two copies of the original transposon.

These general properties of Tn elements and their utility in "classical" bacterial genetics have been the subject of several reviews and monographs.[1-3]

Tn*5*[4] has been particularly useful for the mutagenesis and mapping of genes. Tn*5* is 5818 bp in length and consists of two inverted repeats of 1535 bp, which are mobile elements in their own right (IS*50*L and IS*50*R[5]), flanking a unique region which carries three genes organized in one operon that confer upon certain hosts' resistance to the antibiotics kanamycin/neomycin (Km, Neo[4]), bleomycin (Bleo[6]), and/or streptomycin (Str[7,8]). The entire nucleotide sequence of Tn*5* has been determined, and the locations of the coding regions of the antibiotic resistance genes, transposition functions, as well as the sites involved in transposition have been mapped[9-13] (see Fig. 1).

Tn*5* transposes at a high frequency in a variety of gram-negative bacteria with a low insertional specificity and has a low (precise) excision frequency. A variety of genetic and physical markers make it useful for mutagenesis and combined genetic/physical mapping of genes.[13,14]

Tn*5* mutagenesis experiments can generally be divided in two major

[1] N. Kleckner, J. Roth, and D. Botstein, *J. Mol. Biol.* **116,** 125 (1977).

[2] M. P. Calos and J. H. Miller, *Cell* **20,** 579 (1980).

[3] J. A. Shapiro, "Mobile Genetic Elements." Academic Press, New York, 1983.

[4] D. E. Berg, J. Davies, B. Allet, and J. D. Rochaix, *Proc. Natl. Acad. Sci. U.S.A.* **72,** 3628 (1975).

[5] D. E. Berg, L. Johnsrud, L. McDivitt, R. Ramabhadran, and B. J. Hirschel, *Proc. Natl. Acad. Sci. U.S.A.* **79,** 2632 (1982).

[6] O. Genilloud, M. C. Garrido, and F. Moreno, *Gene* **32,** 225 (1984).

[7] P. Mazodier, E. Giraud, and E. Gassner, *FEMS Microbiol. Lett.* **13,** 27 (1982).

[8] P. Putnoky, G. B. Kiss, I. Ott, and A. Kondorosi, *Mol. Gen. Genet.* **191,** 288 (1983).

[9] E. A. Auerswald, G. Ludwig, and H. Schaller, *Cold Spring Harbor Symp. Quant. Biol.* **45,** 107 (1981).

[10] E. Beck, G. Ludwig, E. A. Auerswald, B. Reiss, and H. Schaller, *Gene* **19,** 327 (1982).

[11] P. Mazodier, P. Cossart, E. Giraud, and F. Gassner, *Nucleic Acids Res.* **13,** 195 (1985).

[12] W. S. Reznikoff, *Cell* **34,** 362 (1982).

[13] D. E. Berg and C. M. Berg, *Bio/Technology* **1,** 417 (1983).

[14] F. J. de Bruijn and J. R. Lupski, *Gene* **27,** 131 (1984).

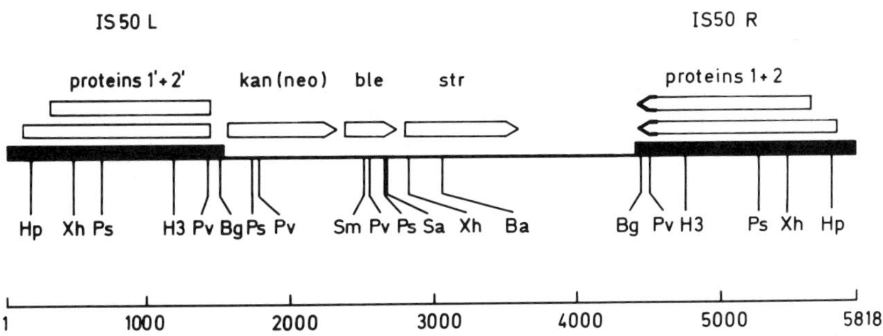

FIG. 1. Physical map of Tn*5*. The abbreviations used for restriction enzymes are as follows: Hp, *Hpa*I; Xh, *Xho*I; Ps, *Pst*I; H3, *Hin*dIII; Pv, *Pvu*II; Bg, *Bgl*II; Sm, *Sma*I; Sa, *Sal*I; Ba, *Bam*HI. The position of the coding regions for the Tn*5*-encoded proteins, and their direction of transcription is indicated by open arrows. Proteins 1 + 2 are involved in Tn*5* transposition, and proteins 1′ + 2′ represent truncated versions of 1 + 2. The data presented are taken from Refs. 9–13. For further details see text. The scale provided is in base pairs.

categories: "random Tn*5* mutagenesis" and "site-directed Tn*5* mutagenesis." The first category involves the introduction of Tn*5* into the bacterial species of interest via transformation, transduction, or conjugation with plasmid or phage vectors carrying Tn*5*. These Tn*5*-containing vectors are called "suicide vectors," because of their instability (or inability to be stably maintained) in the recipient bacteria. Selection for one or more of the antibiotic resistance markers carried on Tn*5* allows the detection of transposition of Tn*5* to the recipient bacterial genome as well as the loss of the vector by segregation. Screening of the Tn*5*-harboring bacterial colonies for Tn*5*-induced insertion mutants with the desired auxotrophic or other genetic phenotype can be used to tag, target, map, and clone the corresponding gene(s). The vectors used for such experiments, protocols/parameters, and tables of some bacterial species mutagenized with Tn*5* are presented in recent papers.[13–18] These techniques will not be the subject of this chapter.

The second category involves the Tn*5* mutagenesis of DNA segments cloned into (multicopy) plasmids in *Escherichia coli*. This is followed by physical mapping of the Tn*5* insertion and whenever possible determination of the "phenotype" of the Tn*5*-mutated recombinant plasmid to con-

[15] R. Simon, M. O'Connell, M. Labes, and A. Puehler, Vol. 118, p. 641.

[16] R. Simon, U. Priefer, and A. Puehler, *Bio/Technology* **1,** 784 (1983).

[17] G. Selvaray and V. N. Iyer, *J. Bacteriol.* **156,** 1292 (1983).

[18] J. E. Beringer, J. L. Beynon, A. V. Buchanan-Wollaston, and A. W. B. Johnston, *Nature* (*London*) **276,** 633 (1978).

struct a correlated physical and genetic map (see Fig. 2). The method is often coupled to the reintroduction of specific Tn*5*-mutated segments into their original bacterial background followed by replacement of the wild-type gene or region with its Tn*5*-mutated homolog via forced double reciprocal recombination (gene replacement or homogenotization; see Fig. 5).

This chapter will focus on the latter category of Tn*5* mutagenesis experiments. Application to the study of the structure of cloned genes with a readily identifiable phenotype in *E. coli* (or related bacteria) and of genetic organization in agrobacteria and (fast growing) rhizobia will be emphasized. Two protocols for Tn*5* mutagenesis of cloned DNA segments in *E. coli* and two protocols for gene replacement with Tn*5*-mutated segments will be outlined and discussed (Figs. 2 and 5). Examples of the experimental procedures will be given. Further examples and an in-depth discussion of various experimental parameters can be found in previous reviews[13–15] and sources quoted therein. Moreover, the analysis presented here will be expanded on, with a special emphasis to the study of bradyrhizobia and related gram-negative bacteria, in Chap. [12].

Transposon Tn*5* Mutagenesis of DNA Segments Cloned into (Multicopy) Plasmids

The outline for Tn*5* mutagenesis of DNA segments cloned into multicopy plasmids is shown in Fig. 2. Starting material can consist of any segment of prokaryotic (or eukaryotic) DNA (region X) cloned in any plasmid or cosmid vector (pMX, with antibiotic resistance marker gene *M*) which has a measurable "phenotype" in *E. coli* (*Ec*) or related bacterial species. The host must be amenable to genetic manipulation (strain Q, any bacterial strain allowing detection of the pMX phenotype) and/or originate from an organism amenable to gene replacement experiments. The measurable "phenotype" of pMX, in strain Q, can be

1. Genetic complementation of a mutation (e.g., auxotrophy) in the bacterial species of choice (Q) by region X cloned in pMX.
2. Expression of a certain enzymatic activity or synthesis of a protein of distinct molecular weight by region X on pMX in strain Q.
3. Expression of a certain antigenic determinant by region X on pMX which can be identified in lysates of strain Q harboring pMX (/pMX) with appropriate antibodies.
4. Hybridization of a complementary region X on plasmid pMX with a heterologous DNA (or RNA) probe carrying an analogous gene region from another organism.

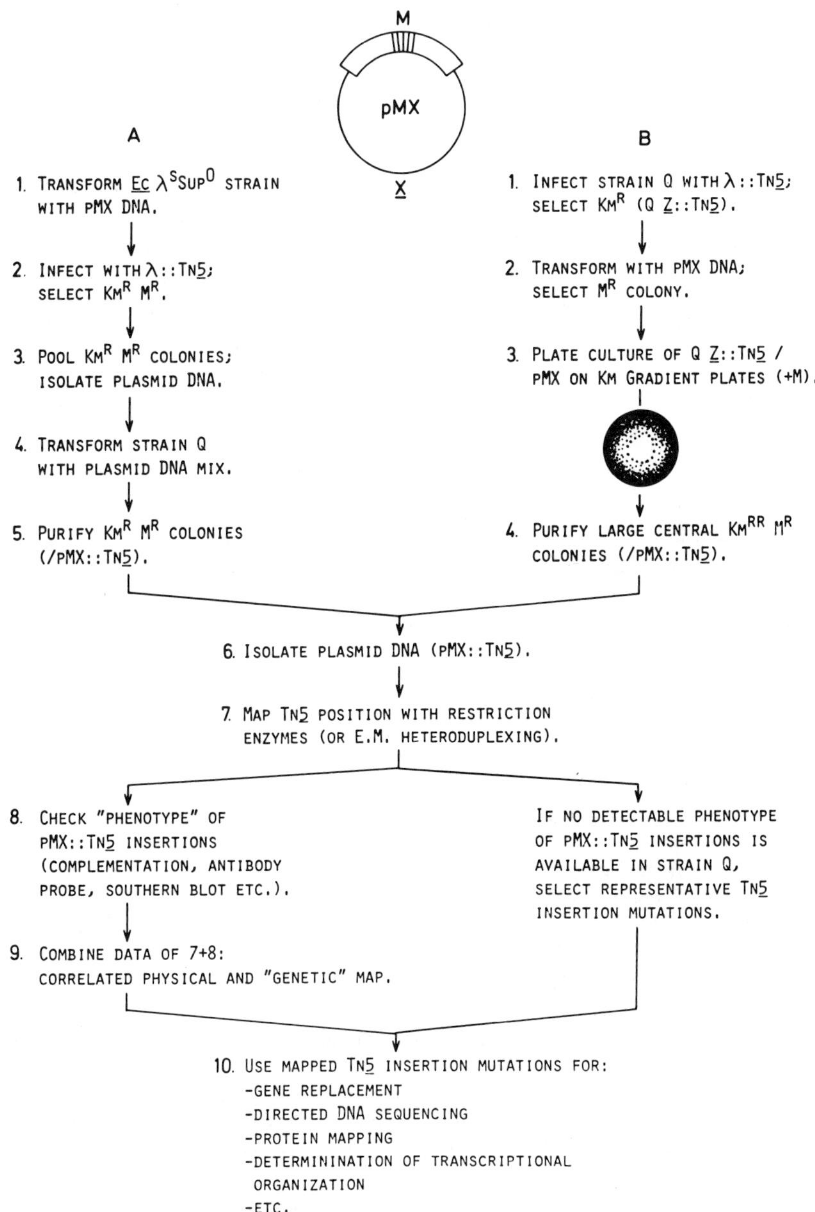

FIG. 2. Tn*5* mutagenesis scheme. For details see text.

Experimental Protocol A

Strains and media used are shown in Table I. In protocol A the Tn*5* donor used for the mutagenesis of region X carried by pMX is a "suicide vector," λ467. It was obtained from Dr. N. Kleckner and is analogous to the λ::Tn*5* phage originally described by Berg *et al.*[4] The genotype of λ467 is *b221 rex*::Tn*5 cI857*, *Oam29*, *Pam80*. The *b221* mutation is a deletion in the λ genome that removes *att*, the phage attachment site. This prevents the bacteriophage from undergoing lysogeny. Gene *rex* is nonessential for λ growth and in this case contains the Tn*5* donor copy. The *cI857* mutation results in a temperature-sensitive repressor. Both the *O* and *P* genes are involved in phage λ DNA replication. Amber mutations in these genes will prevent phage replication in a Su^0 (suppressor zero) background. Any Su^0, λ-sensitive bacterial strain can be used as host for plasmids (pMX) to

TABLE I
STRAINS AND MEDIA

***Escherichia coli* strains, phages, and plasmids**	
LE392	F^-, *hsdR514* (r_k^-, m_k^-) *supE44 supF58 lacY1 galK2 galT22 metB1 trpR55* λ^- [14]
GJ23	AB1157 (*arg his leu pro thr lac gal ara* [Str^R]) + R64*drd*11 (Str^R Tc^R) + pGJ28 (Km^R)[36]
λ467	*b221 rex*::Tn*5 cI857 Oam29 Pam80*
pRK290	Tc^R, RK2 *bom*[34]
pRK2013	Km^R, RK2 *mob*, *tra*[34]
pPH1JI	Gm^R, incompatible with RK2[35]
Media	
λ broth	10 g Bactotryptone, 2.5 g NaCl per liter
λ plates	Solidified with 1.1% agar
λ top agar	Solidified with 0.6% agar
YM broth	λ Broth with 0.2% maltose and 0.01% yeast extract
LB broth	10 g Bactotryptone, 5 g yeast extract, 5 g NaCl per liter, pH 7.5
LB plates	Solidified with 1.5% agar
TY broth	1.3 g $CaCl_2$, 10 g Bactotryptone, 5 g yeast extract per liter
TY plates	Solidified with 1.5% agar
Antibiotics[a] are used at the following concentrations (in μg/ml) (may be dependent on bacterial species):	
For *E. coli*	Tc10, Km20, Amp40, Gm20, Cm60, Neo20, Str20
For rhizobia	Tc5–15, Km200, Amp200, Gm100, Cm10–30, Neo200, Str100–250
For agrobacteria	Tc2–4, Km25, Gm25, Cm25–40, Neo25–40, Str200–300
SM buffer	
100 m*M*	NaCl, 20 m*M* Tris (pH 7.5), 10 m*M* $MgSO_4$, 0.01% gelatin

[a] Tc, Tetracycline; Km, kanamycin; Amp, ampicillin; Gm, gentamycin; Cm, chloramphenicol; Neo, neomycin; Str, streptomycin.

be mutagenized with λ467. The λ::Tn*5* lysate is prepared as described below.[14]

Cells of *E. coli* strain LE392 are infected with λ::Tn*5* phage and incubated overnight at 37° to obtain fresh plaques. Simultaneously, a fresh culture of LE392 is grown in 5 ml YM broth to saturation. The next day the LE392 culture is diluted in LB and regrown to an A_{600} (Bausch and Lomb Spectronic 20) of approximately 0.8. One fresh plaque from the above plate is picked with a sterile pasteur pipette and placed into 1 ml of SM buffer. Then 0.25 ml of SM buffer containing phage is added to 0.5 ml of the regrown LE392 cells. After mixing, the cells are incubated at 37° for 20 min to allow phage absorption. To the cell and phage mixture 7.5 ml of melted λ top agar equilibrated at 45° is added. Aliquots of 2.5 ml are then spread per λ plate. An equivalent amount of LE392 cells with no phage added is plated as a control. The plates are incubated at 37° for 6–8 hr and inspected for lysis by comparing with the LE392 control. When confluent lysis is observed on the plates containing the λ::Tn*5*-infected cells, and the control plates show a visible bacterial lawn, the phage is harvested by scraping the top agar into a sterile screw-top Corex centrifuge tube and adding 2.5 ml SM buffer plus 0.5 ml chloroform. The mixture is then vortexed thoroughly, incubated on ice for 5 min, and centrifuged in a Beckman JA-17 rotor at 10,000 rpm, 4°, for 5 min. After decanting the supernatant into a sterile screw-top tube, a few drops of chloroform are added and the lysate is stored at 4°. The phage are then titered on LE392.

Step A1. A λ-sensitive Su^0 strain is transformed with plasmid pMX DNA[19] selecting for the antibiotic resistance marker gene on the vector (*M*). This marker cannot be Km^R or Neo^R. An M^R transformant is single-colony purified once, and a small-scale plasmid DNA preparation is made (see below, Step AB6) to verify the monomeric structure of pMX.

Step A2. Plasmid pMX-carrying cells are then grown to saturation in 5 ml YM broth (with antibiotic M) at 37°. The culture is diluted 100-fold and grown at 37° in 5 ml YM broth plus antibiotic M to an A_{600} of 0.8 [10^9 colony-forming units (cfu)/ml]. Then 1 ml of cells is mixed with λ::Tn*5* phage at an multiplicity of infection (m.o.i.) of 1–10 and placed at 37° for 2 hr. Infected cells are then plated in 0.2-ml aliquots on LB agar plates containing 20 μg/ml kanamycin and antibiotic M. The plates are incubated at 40° for 48 hr.

Step A3. Km^R M^R colonies are washed off the plates by adding 5 ml 20% sucrose resuspension buffer (20% sucrose, 10 m*M* Tris, pH 7.8, 1 m*M* EDTA) to each plate using a sterile glass spreading rod. The cell

[19] D. A. Morrison, this series, Vol. 68, p. 326.

suspension is decanted into a sterile 30-ml Corex centrifuge tube, and the cells are collected by centrifugation at 7000 rpm for 10 min in a Beckman JA-17 rotor. The cells are resuspended in 5 ml sucrose resuspension buffer or LB broth and plasmid DNA is isolated using a scaled-up (10×) version of the alkaline lysis protocol described below (Step AB6).

Step A4. Plasmid DNA prepared above is used to transform[19] competent cells of the desired bacterial strain (Q). Km^R M^R transformants are selected on LB Km M plates.

Step A5. The Km^R M^R transformants are purified by restreaking for single colonies on LB Km M plates. These represent strain Q/pMX::Tn*5* colonies to be subjected to further analysis (Steps AB6–10).

Experimental Protocol B

Strains and media are listed in Table I. In protocol B the Tn*5* donor, used for mutagenesis of pMX directly in strain Q, is a chromosomal copy of Tn*5* (Q *Z*::Tn*5*, where *Z* is an unknown nonessential gene). This form of Tn*5* mutagenesis is primarily useful for a multicopy target (pMX) plasmid, since it is based on the observation that Tn*5* carried by a multicopy plasmid confers a much higher level of kanamycin and neomycin resistance on its host strain than a single chromosomal copy. Thus by selecting for very high levels of Km^R/Neo^R, derivatives of Q *Z*::Tn*5*/pMX are selected in which Tn*5* has transposed from the chromosomal locus *Z*::Tn*5* into the pMX multicopy plasmids. This selection is done on kanamycin (neomycin) gradient plates, namely, LB plates, supplemented with M to select for the M^R of pMX, in the center of which a drop of highly concentrated (20 mg/ml) Km (Neo) solution has been placed and allowed to diffuse overnight. Bacteria which harbor pMX::Tn*5* plasmids will appear as discrete (large) colonies near the total killing zone in the center of the plate and can be immediately used for further analysis.[20,21]

Step B1. The Tn*5* donor strain Q *Z*::Tn*5* is constructed by infecting strain Q with λ::Tn*5* (see Step A2) and selecting Km^R colonies. Several Km^R colonies are purified once and screened to verify that Tn*5* has inserted in a nonessential (unknown) locus *Z* and that all other markers of strain Q needed to screen for the pMX::Tn*5* phenotype (see below, Step AB8) remain intact.

Step B2. Strain Q *Z*::Tn*5* is transformed with plasmid pMX DNA,[19] an M^R transformant is selected and purified once on LB M medium. The

[20] W. Klipp and A. Puehler, *in* "Advanced Molecular Genetics" (A. Puehler and K. N. Timmis, eds.), p. 224. Springer-Verlag, Berlin, 1984.

[21] R. Simon, *Mol. Gen. Genet.* **196,** 413 (1984).

Km^R marker of Tn*5* in the Q *Z*::Tn*5*/pMX transformant is checked, and a small-scale plasmid DNA preparation is made (see below, Step AB6) to verify the (monomeric) structure of pMX.

Step B3. A culture of Q *Z*::Tn*5*/pMX is grown in LB medium plus M and Km, and the cells are plated on M-containing Km/Neo gradient plates. The plates are incubated for approximately 2 days.

Step B4. Larger, isolated colonies appearing near the middle of the gradient plates are picked and single colony purified once on plates containing M and a high concentration of Km/Neo (depending on species; for *E. coli,* 200–400 μg/ml). These represent Q isolates carrying pMX::Tn*5* derivatives, to be used for further analysis in an identical fashion as those Q/pMX::Tn*5* produced by Steps A1–5.

DNA Isolation and Mapping Protocol

To map the physical location of the Tn*5* insertions on pMX::Tn*5*, a small-scale plasmid DNA preparation is made from the individual Q/pMX::Tn*5* colonies and digested with a restriction enzyme that cuts within Tn*5* and within the pMX plasmid DNA.

Step AB6. The DNA isolation protocol is adapted from Refs. 14 and 22. Plasmid pMX::Tn*5*-bearing strains are inoculatd from a single colony into 3 ml of LB plus antibiotics M and Km. The cells are grown to A_{600} (Spectronic 20) of approximately 0.4–0.5, and 300 μg chloramphenicol is added to amplify the plasmid when possible. After allowing approximately 5 hr for amplification, the cells are collected in a 1.5-ml Eppendorf tube by performing two 30-sec centrifugations in an Eppendorf tabletop centrifuge. The pellet is resuspended in 200 μl resuspension buffer (25 m*M* Tris, pH 8.0, 50 m*M* glucose, 10 m*M* EDTA, 2 mg/ml lysozyme, 50 μg/ml RNase) and incubated on ice for 5 min or until the viscosity changes. It is then mixed with 200 μl of freshly prepared alkali lysing buffer (0.5% SDS, 0.2 *N* NaOH) until the solution becomes clear. Then 200 μl 3 *M* potassium acetate (pH 4.8) is added, and a white precipitate appears. This is mixed well, but gently, and incubated for 5 min at 0°. The solution is centrifuged for 5 min at 4°, and the supernatant is transferred to another Eppendorf tube. The DNA is precipitated by adding 500 μl isopropanol and placing it at −20°.

After centrifuging for 5 min in the cold room, the pellet is resuspended in 50 μl H_2O and passed over a G-50 Sephadex minicolumn equilibrated with 10 m*M* Tris (pH 7.5), 1 m*M* EDTA. The Sephadex minicolumns are made by puncturing the bottom of a 400-μl Eppendorf tube, using a 20-

[22] H. C. Birnboim and J. Doly, *Nucleic Acids Res.* **7,** 1513 (1979).

gauge needle heated over a bunsen burner, covering this hole with a small glass wool plug, followed by sterilization and placing the tube inside a larger 1.5-ml Eppendorf tube which acts as a support. The 400-μl Eppendorf is then filled to overflowing with equilibrated G-50 Sephadex and centrifuged at top speed for 75 sec in an IEC tabletop centrifuge to pack the column. Then 120 μl of resuspended plasmid DNA is loaded on top of the packed column and the tube spun again in a 1.5-ml Eppendorf tube at top speed for 45 sec. In this manner 24 minipreps can be done simultaneously. Enough purified plasmid DNA is usually recovered from 5 ml of liquid culture to perform 10–12 restriction enzyme digests.

Step AB7. A number of convenient restriction sites are available for the physical mapping of Tn*5* insertions in pMX::Tn*5* plasmids (Fig. 1). Depending on the restriction sites already mapped on the pMX plasmid and the size of the cloned DNA segment mutagenized, a combination of usually two enzymes are selected for the mapping. Enzyme 1 is usually a representative of the group which does not cleave Tn*5* (Fig. 2) and is used to determine the approximate position of Tn*5*. In the example shown in Fig. 4, *Eco*RI was used since the map position of the *Eco*RI sites had been previously determined on the cloned DNA segment. Insertion of Tn*5* into one of the *Eco*RI fragments (F) of pFB6162 leads to the disappearance of F and generation of a fragment of size F + 5818 bp (see also Fig. 6). Enzyme 2 is a representative of the group which cleaves within the inverted repeats (IS*50*L and IS*50*R; Fig. 1) of Tn*5* or in (approximately) the middle of the transposon (*Bam*HI; Fig. 1). These enzymes allow orientation-independent mapping of Tn*5* insertions.

In the example shown (Fig. 4) *Bam*HI was used to map Tn*5* insertions within the cloned DNA segment. This enzyme cleaves the target plasmid (pMX = pFB6162) twice, generating the vector and the 11.5-kb insert. Since it cleaves Tn*5* once in the middle, three fragments are generated from pFB6162::Tn*5* DNA consisting of the vector fragment (III, Fig. 3) and two "junction fragments" (I, II; Fig. 3). By measuring the sizes of the fragments I and II and subtracting approximately 2900 bp, the distance of Tn*5* from the *Bam*HI sites flanking the cloned DNA segment can be determined with an accuracy of ±100–150 bp. With the combination of mapping data obtained from enzymes 1 and 2 (*Eco*RI and *Bam*HI) the physical location of Tn*5* can be unambiguously determined (see Fig. 4). To determine the relative orientation of Tn*5* on plasmid pMX::Tn*5*, a third class of enzymes which cleave Tn*5* asymmetrically (such as *Sma*I and *Sal*I, see Fig. 1) can be used.

Step AB8. Depending on the method used to determine the phenotype of the gene product encoded on plasmid pMX in strain Q (see above), a number of different experiments are carried out to determine the "pheno-

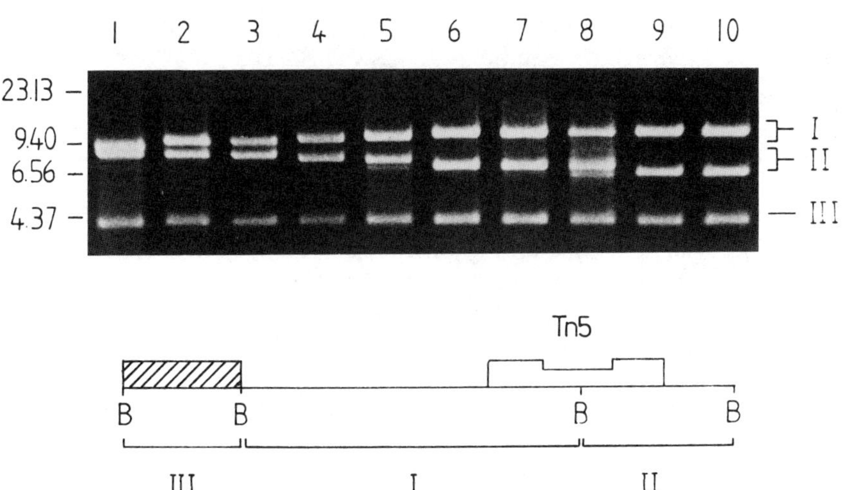

FIG. 3. Mapping of Tn*5* insertions in pFB6162::Tn*5* (pMX::Tn*5*) using *Bam*HI. A 1% agarose gel, stained with ethidium bromide to visualize the DNA, of *Bam*HI digestions of 10 different pMX::Tn*5* plasmids is shown. Molecular weight standards are indicated on the left in kilobases. The structure of pMX::Tn*5* plasmids is schematically diagramed below the gel. The hatched region represents the vector DNA, the open box Tn*5*, and B *Bam*HI restriction sites. For details see text.

type" of the pMX::Tn*5* insertions. Genetic complementation tests,[22a] hybridization experiments using DNA or RNA probes, analysis of pMX(::Tn*5*)-encoded polypeptides, enzyme activity tests, or antibody assays are some of the ways which have been used to measure phenotypic expression. In the example shown in Fig. 4, genetic complementation was employed. Plasmid pMX (=pFB6162) carries a segment of *Rhizobium meliloti* DNA, capable of complementing a glutamine synthetase-deficient mutant strain of *E. coli* (strain Q = GlnA$^-$; glutamine auxotroph).[23] Thus Q/pMX::Tn*5* colonies derived from Steps A1–5 and B1–4 were screened for their GlnA phenotype directly, and a number of GlnA$^-$ Tn*5* insertions in a 4-kb region of pMX were identified (region X). These were flanked on both sides by GlnA$^+$ Tn*5* insertions (see Fig. 4 and Ref. 24). This region was shown to contain an operon, encoding several polypeptides involved

[22a] L. Clarke and J. Carbon, this series, Vol. 68, p. 396.

[23] F. J. de Bruijn, V. Sundaresan, W. W. Szeto, D. W. Ow, and F. M. Ausubel, *in* "Advances in Nitrogen Fixation Research" (C. Veeger and W. E. Newton, eds.), p. 627. Nijhoff/Junk, The Hague, and Pudoc, Wageningen, 1984.

[24] F. J. de Bruijn, S. Rossbach, M. Schneider, F. M. Ausubel, and J. Schell, manuscript in preparation.

in glutamine biosynthesis. The X::Tn*5* insertions were used for gene replacement experiments (see below).

If direct genetic complementation cannot be utilized to determine the "phenotype" of pMX::Tn*5* insertions, the other measurable "phenotypes" of pMX in strain Q listed above must be used. For example when pMX encodes a polypeptide of distinct molecular weight, strain Q could be the "Maxicell" strain CSR603[25] or the "Minicell" strain DS410[25a] and the pMX(::Tn*5*)-encoded polypeptides analyzed by SDS–PAGE[26] to determine which Tn*5* insertions abolished synthesis of the polypeptide of interest.[27] Alternatively, if a certain readily assayable enzymatic activity is encoded by pMX, this assay can be carried out on Q/pMX(::Tn*5*) derivatives to screen for those Tn*5* insertions abolishing the enzymatic activity.

Yet another method of screening for a pMX(::Tn*5*) "phenotype" involves the use of monoclonal antibodies against the protein product of the cloned cDNA copy of a eukaryotic gene. In this case lysates of different Q/pMX(::Tn*5*) derivatives are screened for their ability to bind antibodies. Thus the approximate region encoding the epitope can be mapped (see Ref. 28, for example). Although this method only works efficiently for insertion mutations coming in from the 3′ end of the gene, it will probably not work on genes whose product contains a nonlinear epitope. Lastly an indirect method involves the use of Southern (Northern) blotting.[29] It is usually coupled to the gene replacement protocol discussed below to determine actual "phenotype" of X::Tn*5* insertions.

Step AB9. Regardless of the method used to determine the "phenotype" of pMX::Tn*5* insertions in strain Q, the data obtained in Step AB8 can now be combined with the physical mapping data obtained in Step AB7 to construct a correlated physical and genetic map (see Fig. 4). The example shown represents a preliminary analysis of an 11.5-kb DNA segment cloned in a multicopy plasmid. Only Tn*5* insertions in the insert are shown. Their position is mapped ±100 bp (Step AB7) and their phenotype determined by direct genetic complementation (Step AB8). The results of a single Tn*5* mutagenesis experiment thus facilitated the delimitation of the region responsible for complementation (region X). As can be readily observed, the distribution of Tn*5* insertions in this example is not entirely random, but in most cases characterization of a larger number of Tn*5* insertion leads to Tn*5* saturation mapping of the cloned DNA

[25] A. Sancar, A. M. Hack, and W. D. Rupp, *J. Bacteriol.* **137,** 692 (1979).

[25a] J. Reeve, this series, Vol. 68, p. 493.

[26] U. K. Laemmli, *Nature* (*London*) **227,** 680 (1970).

[27] F. J. de Bruijn and F. M. Ausubel, *Mol. Gen. Genet.* **192,** 342 (1983).

[28] J. R. Lupski, L. S. Ozaki, J. Ellis, and G. N. Godson, *Science* **220,** 1285 (1983).

[29] J. Meinkoth and G. Wahl, *Anal. Biochem.* **138,** 267 (1984).

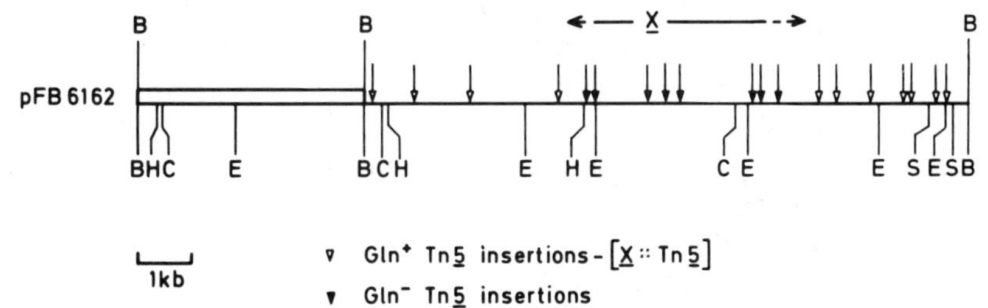

FIG. 4. Correlated physical and genetic map of plasmid pFB6162 (pMX). The open box represents vector DNA sequences, the line represents the cloned DNA segment. The abbreviations used for restriction enzymes are as follows: B, *Bam*HI; H, *Hin*dIII; C, *Cla*I; E, *Eco*RI; S, *Sal*I. Region X responsible for complementation is delimited by a broken, double headed, horizontal arrow. For details see text.

segments.[14] The preliminary mapping as shown in Fig. 4 constitutes a rapid means of delimiting a region of interest. This is especially true with large segments of DNA, i.e., cosmid clones. This facilitates subcloning and provides an important link between the initial cloning/identification of a genetic region and a detailed molecular analysis by other means.

Step AB10. The Tn*5* insertions mapped and characterized in Steps AB6–9, as well as random pMX::Tn*5* insertions in a region for which no direct observable phenotype in *E. coli* exists, can now be used for a variety of secondary analyses, as listed in Step 10 (Fig. 2). Gene replacement protocols will be discussed in detail below. Of the other types of analyses, a brief discussion with examples will be given.

Directed DNA Sequencing

Tn*5* insertion mutations obtained can be readily used for determining the nucleotide sequence of a DNA segment X. Having mapped an individual Tn*5* insertion, one can take a directed as opposed to a "shotgun" approach to nucleotide sequencing. The presence of a unique *Hpa*I site 185 bp from the end of Tn*5*[9] (Fig. 1) can be very helpful for this purpose. A plasmid (pMX) containing Tn*5* in a known position is digested with *Hpa*I and another enzyme (*Bam*HI, *Sal*I, *Pst*I, *Hin*dIII or any enzyme that leaves blunt ends), which cuts the cloned DNA. This fragment mixture is ligated into M13mp9[29a] and digested with *Sma*I plus the appropriate second enzyme; competent cells are transformed and transformants screened by hybridization with a plasmid probe that contains only the terminal 185 bp of Tn*5* to identify subclones of interest. DNA sequence analysis using

[29a] J. Messing and J. Vieira, *Gene* **19**, 269 (1982).

the dideoxy technique[30] is then carried out by priming of the M13mp9 recombinants with a synthetic M13 primer. Extension of the primer with DNA polymerase I (Klenow fragment) yields DNA sequences that extend through known Tn*5* sequence into unknown sequence. This "unknown sequence" (X) corresponds to the DNA into which the Tn*5* originally inserted. The method can be further generalized with the use of a synthetic primer consisting of the terminal 15–20 nucleotides of Tn*5* (B. Horvath and A. Kondorosi, personal communication).

Protein Mapping and Determination of the Direction of Transcription

Tn*5* contains translational stop codons in all three reading frames within the terminal 30 bp of its inverted repeats.[9] Therefore, Tn*5* insertions can be used to identify gene products of gene *X* contained within region X and to determine the direction of transcription. The polypeptides synthesized by plasmids can be examined in maxicells[25] or minicells.[25a] Since Tn*5*-mutated genes will direct the synthesis of truncated polypeptides, a correlation between the physical location of Tn*5* insertions in the *X* gene on pMX and the observed M_r of the truncated polypeptides yields information which can be used to determine the direction of transcription of the cloned gene *X* and the approximate translational start site.[20,27,31] In the example shown in Fig. 4 this analysis was used to show that the 4.5-kb region X comprises an operon transcribed from right to left encoding four different polypeptides involved in complementing the GlnA phenotype of a Gln$^-$ *E. coli* strain ("strain Q").[32]

Gene Replacement Experiments Using Tn5-Mutated DNA Segments

A powerful extension of the Tn*5* mutagenesis techniques is the use of "gene replacement" protocols to replace the wild-type gene *X* in the original organism with its *X*::Tn*5* analog generated and characterized in *E. coli*. This procedure allows the determination of the genetic phenotype of *X*::Tn*5* mutations in the original bacterial species. The general principle is to introduce the *X*::Tn*5* fragment back into its organism of origin and force a double reciprocal recombination event between the wild-type *X* sequences in the bacterial genome and those flanking the Tn*5* insertions

[30] F. Sanger, S. Nicklen, and A. R. Coulson, *Proc. Natl. Acad. Sci. U.S.A.* **74,** 5643 (1977).

[31] M. Giphart-Gassler, R. H. A. Plaskerk, and P. van de Putte, *Nature (London)* **297,** 339 (1982).

[32] F. J. de Bruijn, S. Rossbach, and J. Schell, *in* "Nitrogen Fixation Research Progress" (H. J. Evans, P. J. Bottomley, and W. E. Newton, eds.), p. 218. Nijhoff Dordrecht, 1985.

on the *X*::Tn*5* fragment (Fig. 5), followed by loss of the wild-type *X* segment.

Two methods generally applicable to gene replacement in a variety of gram-negative bacteria have been developed. Ruvkun and Ausubel[33] described one such method. The wide host range replicon pRK290[34] was used to introduce *X*::Tn*5* segments back into rhizobia via conjugation. This was followed by the introduction into the *Rhizobium* sp./pRK290–*X*::Tn*5* of a plasmid incompatible with pRK290 (pPH1JI[35]). Selection for the antibiotic resistance markers of Tn*5* and pPH1JI (Gm^R) simultaneously forces the double reciprocal recombination event(s) described above and results in the loss of the pRK290 replicon.

The second method, described by Van Haute *et al.*,[36] was first used in *Agrobacterium* sp. Plasmids carrying the *Col*EI-type *bom* sequence (site essential for mobilization), such as pBR322[37] or pACYC184,[38] can be mobilized from *E. coli* to *Agrobacterium* and *Rhizobium* spp. when mobilization functions (*ColE1 mob*) and transfer functions (*tra*) are provided *in trans*.[39,40] In the protocol described here a strain of *E. coli* (GJ23), harboring the mobilization helper plasmids pGJ28 (*mob* function) and R64*drd*11 (*tra* functions),[36] is used to transfer pBR322–X::Tn*5* or pACYC184–X::Tn*5* (=pMX::Tn*5*) derivatives to the recipient *Rhizobium* sp. from which region X was originally cloned.

These pMX::Tn*5* plasmids are efficiently transferred to the recipient *Rhizobium* sp. but they and the helper plasmids cannot replicate there. Thus, selection for Tn*5*-encoded antibiotic resistance results in forced homologous recombination between wild-type X and X sequences flanking the Tn*5* on pMX::Tn*5*, resulting primarily in cointegrate formation. Resolution of this cointegrate (the second crossover event), and the resulting perfect gene replacement, is screened for by looking for the loss of resistance to marker M on pMX::Tn*5* (see Fig. 5) while maintaining selective pressure for Tn*5*-encoded resistances.

The two gene replacement protocols are outlined in detail in Fig. 5 and

[33] G. B. Ruvkun and F. M. Ausubel, *Nature* (*London*) **289,** 85 (1981).

[34] G. Ditta, S. Stanfield, D. Corbin, and D. R. Helinski, *Proc. Natl. Acad. Sci. U.S.A.* **77,** 7347 (1980).

[35] P. R. Hirsch and J. E. Beringer, *Plasmid* **12,** 139 (1984).

[36] E. Van Haute, H. Joos, M. Maes, G. Warren, M. Van Montagu, and J. Schell, *EMBO J.* **2,** 411 (1983).

[37] F. Bolivar, R. L. Rodriguez, P. J. Greene, M. C. Betlach, H. L. Heynecker, H. W. Boyer, J. H. Crosa, and S. Falkow, *Gene* **2,** 95 (1977).

[38] A. C. Y. Chang and S. N. Cohen, *J. Bacteriol.* **134,** 1141 (1978).

[39] G. J. Warren, A. J. Twigg, and D. J. Sherratt, *Nature* (*London*) **274,** 259 (1978).

[40] J. Finnegan and D. Sheratt, *Mol. Gen. Genet.* **185,** 344 (1982).

described below. For the sake of simplicity a *Rhizobium* sp. (*Rh*) is used as "recipient" in this scheme but the techniques are applicable to various gram-negative bacteria.

Experimental Protocol A

Strains and media are listed in Table I.

Steps A1,2,3 (Modified from Ref. 33). To transfer the *X*::Tn*5* segment to the bacterial species of origin, it must first be recloned into pRK290 (Tc^R [34]). *Eco*RI ("A") is a convenient enzyme to use, since it does not cut Tn*5* and has a single restriction site in pRK290. The ligation mixture is used to transform any desired auxotrophic strain of *E. coli* (QQ) and Tc^R Km^R transformants are selected, single colony purified, and checked for the presence of the desired pRK290–*X*::Tn*5* structure by minipreparation of plasmid DNA and restriction analysis (Fig. 2, Steps AB6,7).

Steps A4,5,6. To transfer pRK290–*X*::Tn*5* to a *Rhizobium* species (*Rh*), a "triparental mating" protocol is used. pRK290 derivatives can be transferred when *tra* functions are provided *in trans* by a helper plasmid pRK2013.[34] This helper plasmid is present in a separate *E. coli* host (*Ec* QQ/pRK2013 Km^R). Alternatively, *Ec* strains SM10 or S17-1, carrying RK2 transfer functions integrated in the chromosome can be used to provide mobilization helper functions.[15,16] In the triparental mating protocol, overnight cultures of QQ/pRK290–*X*::Tn*5* (containing Tc and Km), QQ/pRK2013 (Km) and *Rh* (with Y, an antibiotic resistance selectable marker of *Rh*) are washed 2 times to remove antibiotics and 100 μl of each culture is spotted on top of each other (or mixed first 1 : 1 : 1 and then spotted) on an LB or TY plate. The mating mixture is incubated at the optimal temperature for the *Rh* species for 24–28 hr. As controls, single spots of the three cultures used, and pairwise combinations are also prepared. The resulting bacterial patches are scraped off the plates with a sterile spatula, dilutions made whenever appropriate, and plated on selective plates containing Y (to select the recipient *Rh* species), Tc (to select pRK290) and Km(Neo), Str, or Bleo (to select for Tn*5*). Oftentimes a minimal medium is used for the selection plates in order to further counterselect the QQ auxotrophic donor and helper strains. The selection plates are incubated for 2–4 days, and a number of Y^R, Tc^R, Km^R (Neo^R, Str^R, $Bleo^R$) transconjugants are single colony purified.

Step A7. This procedural step is optional. To verify the presence of pRK290–*X*::Tn*5* in the *Rh* species a minipreparation of plasmid DNA can be made following the protocol of Fig. 2, Step AB6. This yields enough plasmid to retransform an *E. coli* strain, and one selects for Tc^R Km^R. The resultant transformants can then be analyzed for pRK290–*X*::Tn*5* plasmid DNA structure as described above (Fig. 2, Steps AB6,7).

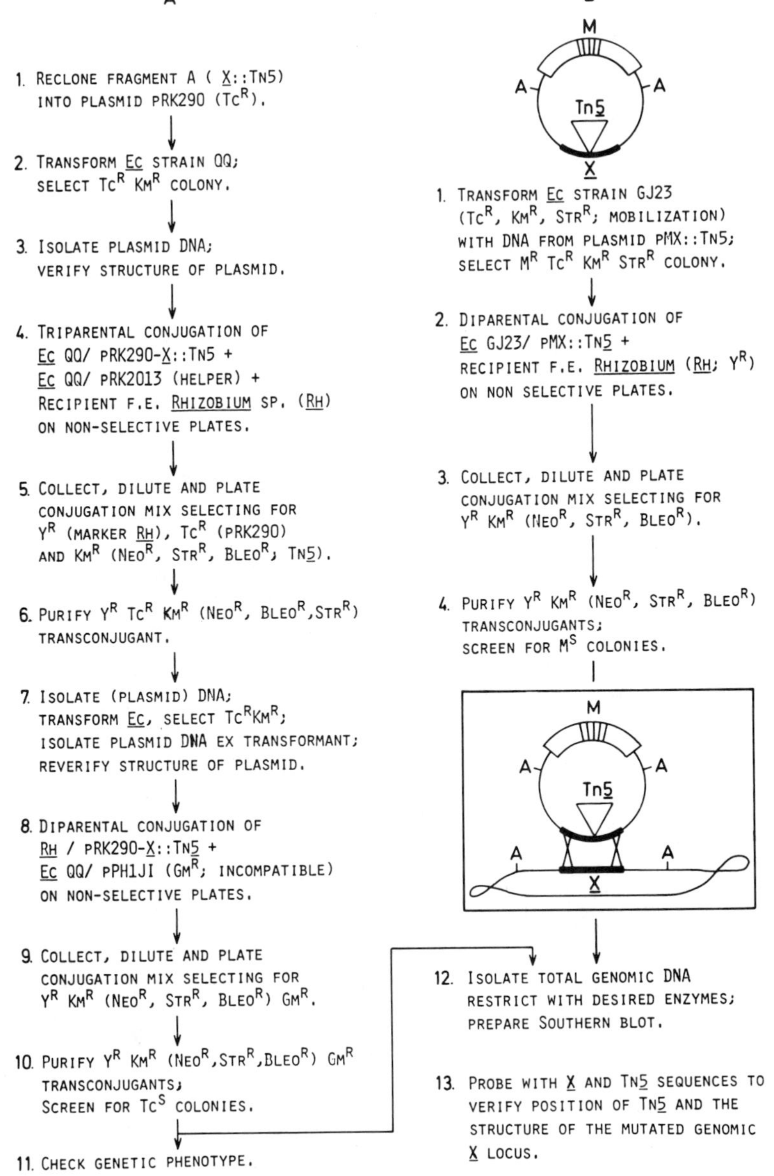

FIG. 5. Gene replacement with Tn5-mutated DNA segments. For further details see text.

Steps A8,9,10. To force recombination between *X*::Tn*5* sequences contained on pRK290 and the genomic wild-type copy of *X*, a "diparental" mating is carried out using cultures of *Rh*/pRK290–*X*::Tn*5* and a *E. coli* QQ strain harboring the Gm^R plasmid pPH1JI (see Steps A4,5,6). Gm^R Km^R (Neo^R, Str^R, $Bleo^R$) transconjugants are purified and checked for Tc^S. Tc sensitivity indicates loss of pRK290 vector sequences and is usually indicative of the desired double reciprocal crossover events in region X as opposed to cointegrate formation (integration of the entire pRK290–*X*::Tn*5* plasmid into region X via single crossover).

Step A11. Resultant *Rh* strains carrying the *X*::Tn*5* insertion mutations can now be examined for their genetic phenotype and used for further studies. For variations and additional details on this protocol, see Refs. 15, 16, 33, and 34.

Experimental Protocol B

Strains and media are listed in Table I.

Step B1. (modified from Ref. 36) *E. coli* strain GJ23, harboring plasmids pGJ28 (Km^R; carrying *mob* functions) and R64*drd*11 (Tc^R, Str^R; carrying *tra* functions) is transformed with pMX::Tn*5* plasmid DNA (where the pM is a *bom*$^+$ *ColE1*-type vector and the M marker is not Tc^R, Str^R). M^R, Km^R, Tc^R (Str^R) colonies are selected. Since both pGJ28 and Tn*5* encode Km^R, it is not possible to select for maintenance of pGJ28 in these M^R Km^R Tc^R (Str^R) transformants. But since pGJ28 is very compatible with incoming *ColE1*-type plasmids[36] (i.e., pMX::Tn*5*) and is rather stably maintained in GJ23 without selective pressure, this does not usually present a problem. To avoid complications due to spontaneous loss of pGJ28, one should not use a single GJ23/pMX::Tn*5* transformant as inoculum for making cultures but rather pool a number of colonies for this purpose.

Steps B2,3,4. A diparental conjugation is carried out using cultures of GJ23/pMX::Tn*5* and the recipient *Rhizobium* species (see Steps A4,5,6) and Y^R (*Rh* marker). Km^R (Neo^R, Str^R, $Bleo^R$, Tn*5*) transconjugants are screened for antibiotic sensitivity (M^S). This selects for those events where the vector sequences of pMX::Tn*5* have been lost due to double crossover events leading to perfect gene replacement (see Fig. 5, B4). The helper plasmids pGJ28 and R64*drd*11, although autotransmissable, are usually unable to replicate in species such as rhizobia and agrobacteria and therefore are rapidly segregated away. Only the loss of the R64*drd*1(Tc^R) markers can be checked since pGJ28 carries the same resistance marker as Tn*5* (Km^R) (see also Step B1).

Step AB12. To verify the structure of the genomic *X*::Tn*5* insertions, and compare it with the analogous *X*::Tn*5* insertions mapped on pMX::Tn*5*, total chromosomal DNA is isolated from *Rh X*::Tn*5* isolates (adapted from Ref. 41). A 1-ml saturated culture of the *Rh X*::Tn*5* strain grown in LB or TY is washed 2 times with 1 *M* NaCl in an Eppendorf tube and the cell pellet resuspended in 1 ml TE (10 m*M* Tris, 1 m*M* EDTA, pH 8). The cell suspension is incubated with 0.1 ml of 2 mg/ml lysozyme in TE with gentle mixing at 37° for 20 min. Then 0.125 ml of 10% sarkosyl, 5 mg/ml pronase in TE (predigested at 37° for 1 hr) is added and the mixture incubated at 37° for 1 hr. The viscous mixture is extracted with phenol (saturated with TE, pH 8) 2–3 times; caution is exercised so that one leaves the white interphase behind each time, and the final aqueous phase is extracted once with chloroform. Ammonium acetate is added to a final concentration of 0.3 *M* and the nucleic acids precipitated with 2.5× volume of ethanol. The DNA strands can be "spooled out" using a glass pipet or precipitated by centrifugation and resuspended in 0.2–0.5 ml of H_2O or TE.

These chromosomal DNA preparations are digested with the same enzymes used to characterize the original pMX::Tn*5* insertions (e.g., *Eco*RI and *Bam*HI; see Fig. 2, Steps 6, 7) and the agarose gels blotted as described.[29,42] The resulting Southern blots are hybridized to ^{32}P-labeled probes[29] of the *X* target DNA (and/or Tn*5*) to verify the position of Tn*5* within *X* and the absence of the wild-type X segment (see Fig. 6). In Fig. 6 *Eco*RI was used to digest the chromosomal DNAs of *X* (*gln*)::Tn*5* insertions of *Rhizobium meliloti* 1021[23,24] and the *Bam*HI fragment of pFB6162, carrying the wild-type *X* (*gln*) region (see Fig. 4) as a probe. One observes that in lanes 1, 2, 3, 5, and 6 the original *Eco*RI fragments, B or C of region X, have disappeared and given rise to new bands of B or C + *Tn5* which hybridize to the probe, suggesting they are true gene replacements. Lanes 4, 7, and 8 represent instances where cointegrates have formed since fragments A–D are present, but in addition a new D + *Tn5* band exists. In these cases both wild-type region X and *X*::Tn*5* are present. In lane 9 no new band can be observed originating from an A–D Tn*5* insertion. This may represent a secondary transposition of Tn*5* from the pRK290–*X*::Tn*5* to elsewhere in the genome which results in a Gm^R Km^R (Neo^R, Str^R, Bleo^R) Tc^S transconjugant (see Fig. 5, Step A10).

Thus, the protocols described in Figs. 2 and 5 represent examples of the "reverse genetics" approach which has been extremely useful in the molecular genetic analysis of genes via Tn*5* mapping. This approach has

41 H. M. Meade, S. R. Long, G. B. Ruvkun, S. E. Brown, and F. M. Ausubel, *J. Bacteriol.* **149,** 114 (1982).

42 E. Southern, this series, Vol. 69, p. 152.

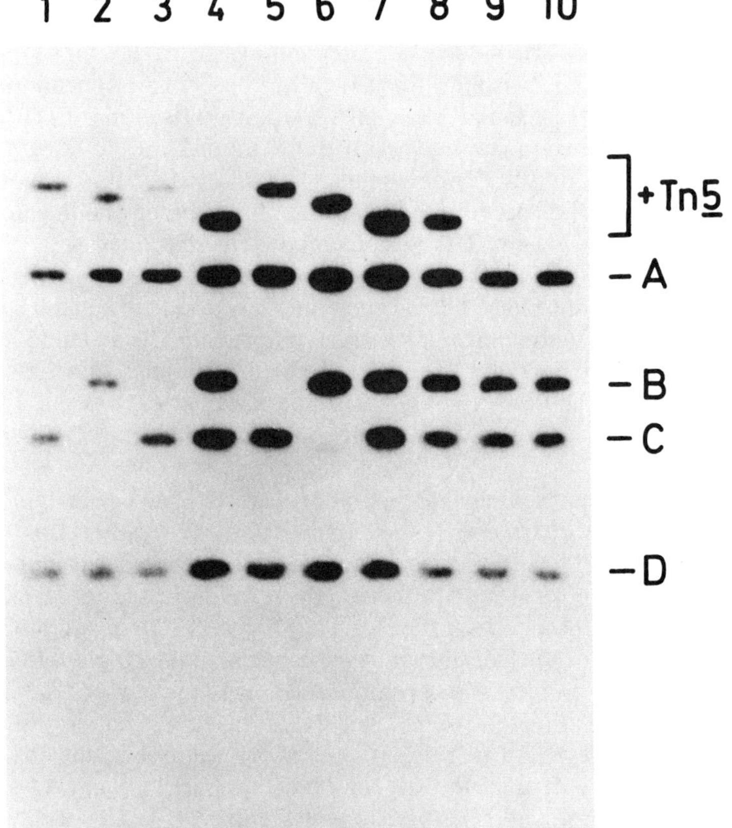

FIG. 6. Verification of the structure of the X::Tn*5* region in homogenotes. An autoradiogram of a Southern blot[42] of *Eco*RI-digested *R. meliloti* DNA of various presumptive homogenotes is shown. The blot was hybridized with ^{32}P-labeled[29] *Bam*HI-cut insert DNA from pFB6162 (pMX). A through D indicate the position of the original genomic *Eco*RI fragments of region X. The position of B or C fragments which have acquired Tn*5* are indicated by "+ Tn*5*." Lanes 1, 2, 3, 5, and 6 represent true gene replacements. Lanes 4, 7, and 8 represent cointegrate structures. The pattern observed in lane 9 is probably due to a secondary transposition of Tn*5* not into the intended region X. Lane 10 is a control of chromosomal DNA which was not subjected to mutagenesis with Tn*5*. For further details see text.

been of great utility for Tn*5* mapping of genes of rhizobia and agrobacteria. This is especially significant for genes initially identified by DNA–DNA hybridization with heterologous probes, by their vicinity to other regions involved in virulence (*Agrobacterium tumefaciens*), or symbiotic interaction with higher plants (*fix, nod; Rhizobium* sp.). These genes, for which no detectable phenotype exists in *E. coli,* could be mapped in this fashion (for examples, see Ref. 14 and Chap. [12] in this volume).

Discussion

Parameters of transposon Tn*5* which are pertinent to the mutagenesis protocols are described in Fig. 2: vectors used, transposition frequency, insertional specificity, polarity and stability of Tn*5* insertions have been discussed in detail.[13,14] One parameter is of particular interest when trying to generate a population of "random" Tn*5* insertions in a cloned DNA segment. That is, insertional specificity or the absence of "hot spots" of Tn*5* insertion.

Some "hot spots" for Tn*5* insertion have indeed been found. One of them in the promoter region of the Tc^R gene of the widely used cloning vector pBR322,[43] others in cloned DNA segments.[44,45] Whether the insertional specificity of Tn*5* found in these cases is due to a limited homology between the insertion target DNA and the Tn*5* ends,[44,45] to an apparent G + C cutting preference,[43] or to a preferential insertion site advantageous either to the target plasmid or the transposon itself, is not clear. It has been suggested that it may also depend on the nature of the vehicle used to deliver Tn*5*.[44] However, these findings do not detract from the general observation, based on the characterization of more than 800 Tn*5* insertions in more than 10 different cloned DNA segments,[14] that on a macroscale Tn*5* insertions are random enough to allow correlated physical and genetic maps to be constructed, and no distinct insertional specificity is found.

The utility of Tn*5* has recently been extended by the construction of Tn*5* derivatives carrying other selectable marker genes such as Tc^R, *trp*,[46] spectinomycin resistance (Sp^R [47]), or Gm^R/Km^R, Str^R/Sp^R.[48] Moreover, derivatives have been constructed which can function as "promoter probes" by fusing promoters at the site of Tn*5* insertion to promoterless "reporter" genes located within Tn*5*, such as Neo^R [49] or *lacZ*.[50] Last,

[43] D. E. Berg, M. A. Schmandt, and J. B. Lowe, *Genetics* **105,** 813 (1983).

[44] J. R. Lupski, P. Gershon, L. S. Ozaki, and G. N. Godson, *Gene* **30,** 99 (1984).

[45] L. Bossi and M. S. Ciampi, *Mol. Gen. Genet.* **183,** 406 (1981).

[46] D. E. Berg, C. Egner, B. J. Hirschel, J. Howard, L. Johnsrud, R. A. Jorgensen, and T. Tlsty, *Cold Spring Harbor Symp. Quant. Biol.* **45,** 115 (1980).

[47] K. M. Zsebo, F. Wu, and J. E. Hearst, *Plasmid* **11,** 118 (1984).

[48] G. F. De Vos, G. C. Walker, and E. R. Signer, *Mol. Gen. Genet.* **204,** 485 (1986).

[49] V. Bellofatto, L. Shapiro, and D. A. Hodgson, *Proc. Natl. Acad. Sci. U.S.A.* **81,** 1035 (1984).

[50] L. Kroos and D. Kaiser, *Proc. Natl. Acad. Sci. U.S.A.* **81,** 5816 (1984).

derivatives of Tn*5* have been constructed carrying other genetic units involved in the mobilization [Tn*5*(*mob*)[15,51]] of RP4 derivatives, which can be used to confer RP4 mobilization functions upon pMX::Tn*5*(*mob*) derivatives.

The parameters of the gene replacement protocols using Tn*5*-mutated fragments represented in Fig. 5, have been discussed in detail[33,36] and constitute a major part of Chap. [12] in this volume. One point that requires emphasis regardless of the protocol used (Fig. 5, A or B), concerns the frequency of obtaining double crossover (true gene replacement) versus single crossover (cointegrate formation) events. This is often variable. It appears to some extent to depend on the nature of the recipient organism. For example, while the technique described in Protocol A is highly efficient in generating true gene replacements in *R. meliloti* 1021,[33] in *R. sesbania* ORS571 often more than 500 Gm^R Km^R transconjugants must be screened to find Tc^S derivatives which have undergone double crossover recombination (Fig. 5, Step A10).[52] It is certainly dependent on the length of X sequences flanking the Tn*5* insertions, which are necessary as substrates for the homologous recombination events.

In conclusion "General methods for transposon Tn*5* mutagenesis" which work without problems for all gram-negative bacteria do not exist. It is hoped that the methods outlined here and in Chap. [12] in this volume will provide at least some methods to try.

Acknowledgments

I would like to thank Jim Lupski and J. Schell for helpful discussions and critical reading of the manuscript; Birgit Metz, Silvia Rossbach, Pascal Ratet, Katharina Pawlowski, and Nina Heycke for help in preparing the manuscript; and Eva Kath and Jutta Freitag for typing it.

[51] E. A. Yakobson and D. G. Guiney, Jr., *J. Bacteriol.* **160,** 451 (1984).
[52] K. Pawlowski, P. Ratet, J. Schell, and F. J. de Bruijn, *Mol. Gen. Genet.* **206,** 207 (1987).

[12] Site-Directed Tn5 and Transplacement Mutagenesis: Methods to Identify Symbiotic Nitrogen Fixation Genes in Slow-Growing *Rhizobium*

By JOHN D. NOTI, MITTUR N. JAGADISH, and ALADAR A. SZALAY

Introduction

The regulation of gene expression is best understood in those organisms for which convenient mutagenesis and screening techniques have been developed. The transposon Tn5 has been successfully used for *in vivo* mutagenesis in both prokaryotic[1,2] and eukaryotic organisms.[3,4] The transposition of Tn5 into a DNA sequence is generally random and usually results in the inactivation of the gene in which it is inserted.[5] Cells that contain Tn5 are resistant to kanamycin, a phenotype that enables cells with Tn5 insertions to be easily identified. The precise location of Tn5 within a region of DNA can be determined by restriction endonuclease analysis of the genomic DNA from cells that carry the Tn5 insertion.[6] By comparing the results of the functional analysis of the DNA regions that contain Tn5 insertions with the locations of these same insertions, a correlated physical and genetic map of a specific DNA region can be established.[6]

Tn5 mutagenesis is generally accomplished by one of two ways: (1) Tn5 can be directly introduced into the organism that is to be mutagenized[7]; (2) a genomic DNA fragment from the organism is introduced into an *Escherichia coli* strain that contains the transposon.[8] The former

[1] C. M. Berg, K. J. Shaw, J. Vender, and M. Borucha-Mankiewicz, *Genetics* **93,** 309 (1979).

[2] J. H. Miller, M. P. Calos, D. Galas, M. Hofer, D. E. Buchel, and B. Muller-Hill, *J. Mol. Biol.* **144,** 1 (1980).

[3] R. D. McKinnon, S. Bacchetti, and F. L. Graham, *Gene* **19,** 33 (1982).

[4] J. Ellis, L. S. Ozaki, R. W. Gwadz, A. H. Cochrane, V. Nussenzweig, R. S. Nussenzweig, and G. N. Godson, *Nature (London)* **302,** 536 (1983).

[5] D. E. Berg, *in* "DNA Insertion Elements, Plasmids and Episomes" (A. I. Bukhari, J. A. Shapiro, and S. L. Adhya, eds.), p. 205. Cold Spring Harbor Lab., Cold Spring Harbor, New York, 1977.

[6] G. E. Riedel, F. M. Ausubel, and F. C. Cannon, *Proc. Natl. Acad. Sci. U.S.A.* **76,** 2866 (1979).

[7] J. E. Beringer, J. L. Beynon, A. V. Buchanan-Wollaston, and A. W. B. Johnston, *Nature (London)* **276,** 633 (1978).

[8] M. N. Jagadish and A. A. Szalay, *Mol. Gen. Genet.* **196,** 290 (1984).

method is limited to those organisms in which Tn*5* transposition can occur. However, it has an advantage in that the entire genome is subject to mutagenesis. The latter method can be applied to any organism, but the DNA fragment containing the region of interest must be isolated first. The effect of Tn*5* insertion into a DNA sequence is then determined by screening for a selectable phenotype or for the loss of function of a specific gene. When a DNA fragment from another organism is mutagenized in *E. coli,* unless that organism's genes can be expressed in *E. coli* the Tn*5*-interrupted DNA fragment must be returned to the original organism for analysis. The ability of plasmids of the P-1 incompatibility group (i.e., RK2, RP4) to transfer to a wide variety of gram-negative bacteria has been used as the basis for the development of broad-host-range mobilization systems by Ditta *et al.*[9] and Simon *et al.*[10] Ruvkun and Ausubel[11] developed a method, referred to as site-directed Tn*5* mutagenesis, that enables genomic DNA fragments from fast-growing *Rhizobium meliloti* to be mutagenized in *E. coli* and then, via a broad-host-range plasmid, to be transferred back into *R. meliloti* where the Tn*5*-interrupted sequence on the plasmid is exchanged for the wild-type sequence in the genome by homologous recombination. Simon *et al.*[10] introduced a similar procedure for site-directed Tn*5* mutagenesis in *R. meliloti.*

The availability of a broad-host-range mobilization system for gram-negative bacteria facilitated the development of a method to introduce *in vitro* generated mutations including deletions, additions, inversions, and substitutions into *Rhizobium.* This method, referred to as site-directed transplacement and developed by Jagadish *et al.*,[12] utilizes Tn*5* as an unlinked selectable marker when transferring *in vitro* altered DNA sequences to *Rhizobium.*

In this chapter we show how site-directed Tn*5* mutagenesis is used in our laboratory to identify genes essential for symbiotic nitrogen fixation in slow-growing *Rhizobium* species. We have experimented with several methods for this kind of mutagenesis and will present the protocol that has worked most successfully in conjunction with these organisms. We also present a method for site-directed transplacement of *in vitro* altered DNA sequences that should be applicable to all gram-negative bacteria.

[9] G. Ditta, S. Stanfield, D. Corbin, and D. R. Helinski, *Proc. Natl. Acad. Sci. U.S.A.* **77,** 7347 (1980).

[10] R. Simon, U. Priefer, and A. Pühler, *Bio/Technology* **1,** 784 (1983).

[11] G. B. Ruvkun and F. M. Ausubel, *Nature* (*London*) **289,** 85 (1981).

[12] M. N. Jagadish, S. D. Bookner, and A. A. Szalay, *Mol. Gen. Genet.* **199,** 249 (1985).

Principle

The methods described here employ Tn5 directly as a means to generate mutations and localize genes in slow-growing *Rhizobium* by the insertion of Tn5 into a cloned genomic DNA fragment (site-directed Tn5 mutagenesis) or indirectly as a selectable marker to introduce a deletion, an addition, a base substitution, or an inversion (site-directed transplacement). Site-directed Tn5 mutagenesis is accomplished by first subcloning the DNA fragment of interest (target DNA) into a plasmid with the following properties: (1) It must be able to replicate in *E. coli* but not in slow-growing *Rhizobium;* (2) It must be able to be mobilized by conjugation. The target DNA is then subjected to random Tn5 mutagenesis in *E. coli.*[8] The transposition of Tn5 into the target DNA is verified by transformation of the plasmid into another *E. coli* strain and selecting for kanamycin resistance which is coded for by the NPTII gene of Tn5. The location of Tn5 within the target DNA is determined and the plasmid (containing the Tn5-interrupted target DNA) is mobilized by conjugation from *E. coli* to *Rhizobium.*

Within the *Rhizobium* cell homologous pairing can occur between the target DNA carried on the plasmid and the chromosome. A double-reciprocal crossover (one on either side of Tn5) will result in the exchange of the wild-type chromosomal DNA fragment for the Tn5-interrupted DNA fragment on the plasmid.[8,11] Because the plasmid vector (which now carries the wild-type fragment DNA) is unable to replicate in *Rhizobium,* it will be subsequently lost from the cell. Those *Rhizobium* cells in which a double-reciprocal crossover has occurred can be identified on the basis of their resistance to kanamycin. Inactivation of a gene that is essential for symbiotic nitrogen fixation by the insertion of Tn5 can be readily determined by inoculation of the strain containing the insertion onto the appropriate legume host plant.[8]

An *in vitro* altered DNA sequence (i.e., deletion, addition) is introduced into the *Rhizobium* genome[12] by exploiting the properties of the same plasmid vector. In order to substitute the altered DNA sequence for the wild-type genomic sequence, however, it is not essential to directly link a selectable genetic marker to the *in vitro* altered region of DNA. Instead, a selectable marker such as kanamycin resistance is required on the vector. The vector carrying the altered DNA fragment is transferred by conjugation from *E. coli* to *Rhizobium* where homologous pairing can occur. The entire plasmid is recombined into the chromosome via a single crossover between the altered DNA sequence on the plasmid and the wild-type chromosomal sequence. This event results in cells that contain both the wild-type and the altered DNA sequence which are separated by the vector DNA sequence. Cells that have integrated the entire plasmid into their chromosome can be identified on the basis of their resistance to kanamycin. The arrangement of the duplicated DNA sequences is unstable,[8,13,14] and cells with such duplications are maintained in culture only in the presence of kanamycin. Homologous recombination between the duplicated sequences mediated by a single crossover will lead to the excision of the vector along with one of the duplicated sequences. Subsequently, the vector along with one of these sequences will be lost from the cell because it is unable to replicate in *Rhizobium*. The resulting cells are kanamycin sensitive, and Southern analysis of genomic DNA from these cells will reveal whether the wild-type or the altered DNA sequence was retained.

The mobilizable plasmid pSUP202 was used to transfer DNA fragments from *E. coli* to *Rhizobium* following Tn*5* mutagenesis or *in vitro* alteration of the DNA sequence. This plasmid was constructed by Simon *et al.*[10] as part of a broad-host-range mobilization system for use with gram-negative bacteria. The important feature of this plasmid is that it contains a region of DNA from the broad-host-range plasmid RP4[15] (incompatibility P-1) referred to as the Mob site. The Mob site contains the origin of transfer which allows this plasmid to be mobilized from *E. coli* to other gram-negative bacteria. The plasmid also requires the transfer genes from RP4 for mobilization. The *E. coli* strain SM10, which contains a chromosomal integration of RP4, was constructed by Simon *et al.*[10] for use as a mobilizing strain in their system.

Materials

Bacterial Strains and Plasmids

The *Escherichia coli* strains used were HB101 (Str^r, *hsdM*$^-$, *hsdR*$^-$, *recA*, *proA*, *leuB6*, *thi*),[16] JA221 (*hsdM*$^+$, *hsdR*$^-$, *recA*, *leuB6*),[17] JN25 (Str^r, *hsdM*$^+$, *hsdR*$^-$, *trp*, *lacZ*, *thi*, *pro*::Tn*5*) (this study), SM10 (*hsdM*$^+$, *hsdR*$^+$, *recA*, *thi*, *thr*, *leu*, *supF*, *lac*, RP4-2-Tc::Mu),[10] MC1061 [Str^r, *hsdM*$^+$, *hsdR*$^-$Δ(*ara*,*leu*7697)Δ(*lac*)X74, *galU*, *galK*].[18] *Bradyrhizobium*

[13] A. Hinnen, J. B. Hicks, and G. R. Fink, *Proc. Natl. Acad. Sci. U.S.A.* **75,** 1929 (1978).

[14] J. L. Compton, A. Zamir, and A. A. Szalay, *Mol. Gen. Genet.* **188,** 44 (1982).

[15] C. M. Thomas, D. Stalker, D. Guiney, and D. R. Helinski, *in* "Plasmids of Medical, Environmental and Commercial Importance" (K. N. Timmis and A. Pühler, eds.), p. 375. Elsevier/North Holland, Amsterdam, 1979.

[16] H. W. Boyer and D. Roulland-Dussoix, *J. Mol. Biol.* **41,** 459 (1969).

[17] A. C. Chinault and J. Carbon, *Gene* **5,** 111 (1979).

[18] M. Casadaban and S. Cohen, *J. Mol. Biol.* **138,** 179 (1980).

sp. (*Vigna*) strain IRc78 (wild-type, Fix^+) was provided by A. Eaglesham. Plasmid pSUP202 (Amp^r, Tet^r, Cm^r)[10] was provided by A. Pühler. Plasmid pVK100 (Tet^r, Km^r)[19] was provided by E. Nester.

Growth Media

L-broth is 10 g of Bacto-tryptone, 5 g of Bacto-yeast extract, and 5 g of NaCl per liter of H_2O. RDM medium is 2.2 ml of 10% K_2HPO_4, ml of 10% $MgSO_4 \cdot 7H_2O$, 5 ml of 22% sodium glutamate, 1 ml of 1000× trace element solution, 1 ml of 1000× vitamin solution, and 10 g of mannitol per liter of H_2O. The pH of the medium is adjusted to 6.9. The 1000× trace element solution is 5 g of $CaCl_2$, 145 mg of H_3BO_3, 125 mg of $FeSO_4 \cdot 7H_2O$, 70 mg of $CoSO_4 \cdot 7H_2O$, 5 mg of $CuSO_4 \cdot 7H_2O$, 4.3 mg of $MnCl_2 \cdot 4H_2O$, 108 mg of $ZnSO_4 \cdot 7H_2O$, 125 mg of Na_2MoO_4, and 7 g of nitrilotriacetate per liter of H_2O (the pH is adjusted to 5.0 before adding nitrilotriacetate). The 1000× vitamin solution is 120 mg of inositol and 20 mg each of riboflavin, *p*-aminobenzoic acid, nicotinic acid, biotin, thiamin–HCl, pyridoxine–HCl, and calcium pantothenate per liter of 50 m*M* Na_2HPO_4, pH 7.0. RDY medium is RDM medium without mannitol and sodium glutamate and supplemented with 1 g of yeast extract per liter of medium. Solid media contained Bacto-agar at a final concentration of 2%.

Buffers

TE buffer is 10 m*M* Tris–HCl, 1 m*M* EDTA, pH 8.0.

Methods

Random Insertion of Tn5 into a Cloned Fragment

The general protocol for the insertion of Tn5 into a specific region of the *Rhizobium* chromosome is shown in Fig. 1. Specific DNA fragments isolated from the genome of a slow-growing *Rhizobium* species that will be the target for Tn5 mutagenesis are first excised from the original cloning vector and then ligated into pSUP202. The fragment can be conveniently subcloned into a unique restriction site within one of the three antibiotic resistant genes (Amp, Tet, Cm) carried on pSUP202. As an example, a fragment from the *Rhizobium* genome was ligated into a restriction site within the Tet gene of pSUP202 resulting in the plasmid pMJ10-4. The plasmid pMJ10-4 is then transformed into *E. coli* JN25

[19] V. C. Knauf and E. W. Nester, *Plasmid* **8,** 45 (1982).

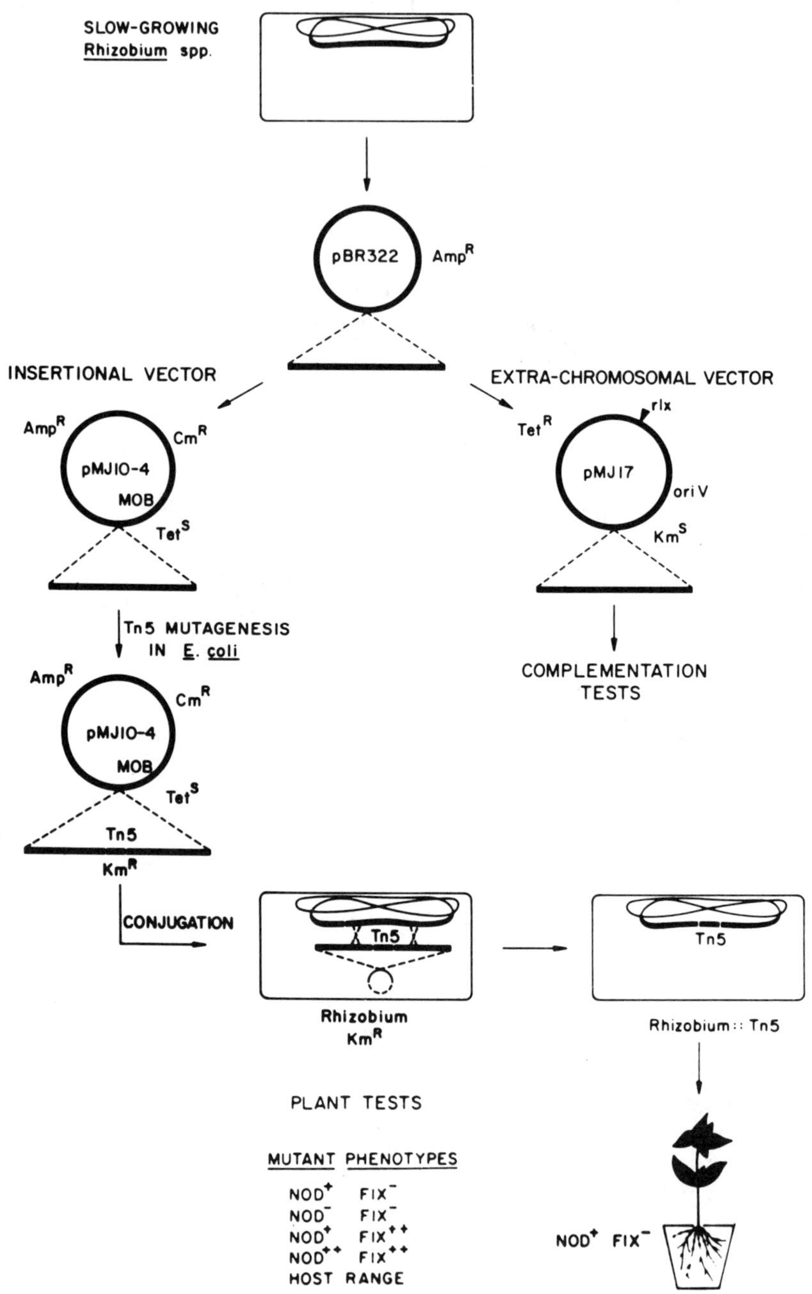
SLOW-GROWING
Rhizobium spp.
pBR322
Amp^R
INSERTIONAL VECTOR
EXTRA-CHROMOSOMAL VECTOR
Amp^R
Cm^R
pMJ10-4
MOB
Tet^S
Tet^R
rlx
pMJ17
oriV
Km^S
Tn5 MUTAGENESIS
IN E. coli
COMPLEMENTATION
TESTS
Amp^R
Cm^R
pMJ10-4
MOB
Tet^S
Tn5
Km^R
CONJUGATION
Tn5
Tn5
Rhizobium
Km^R
Rhizobium :: Tn5
PLANT TESTS
MUTANT PHENOTYPES
NOD+ FIX-
NOD- FIX-
NOD+ FIX++
NOD++ FIX++
HOST RANGE
NOD+ FIX-

which carries a chromosomal insertion of Tn5. It is within this strain that Tn5 mutagenesis will occur.

Transformants are selected on L-agar plates supplemented with kanamycin at a final concentration of 50 μg/ml and either ampicillin or chloramphenicol also at 50 μg/ml. Single colonies are then inoculated into 10 ml of L-broth supplemented with 250 μg/ml of kanamycin and 50 μg/ml of either ampicillin or chloramphenicol and incubated at 37° until the culture reaches stationary phase (usually 4–5 hr). Incubation of the culture in the presence of a high concentration of kanamycin leads to an enrichment of cells within the culture that carry Tn5 on the plasmid. This is probably because cells that contain multiple copies of plasmids with Tn5 insertions are more resistant to higher concentrations of kanamycin and thus have a selective advantage over cells that only contain a single chromosomal copy of Tn5.

Small-scale preparations of plasmid DNA are then prepared from 0.5 ml of each culture according to the procedure described by Birnboim and Doly.[20] The plasmid DNA from each 0.5 ml culture is resuspended in 30 μl of TE buffer, and 1–5 μl is then used to transform *E. coli* JA221. This step is performed to select for those individual plasmids in the total plasmid pool that now contain a Tn5 insertion. Transformants of strain JA221 are selected on L-agar plates supplemented with 50 μg/ml of kanamycin and 50 μg/ml of either ampicillin or chloramphenicol. The transposition of Tn5 from the chromosome of strain JN25 to either another part of the chromosome or to the plasmid it contains occurs with a transformation frequency of approximately 1×10^{-5}. Generally, approximately 10^3 kanamycin-resistant transformants can be expected from each transformation experiment. Single colonies of the transformants are then inoculated into

[20] H. C. Birnboim and J. Doly, *Nucleic Acids Res.* **7,** 1513 (1979).

FIG. 1. The general scheme used to identify and characterize symbiotic genes using site-directed Tn5 mutagenesis. A DNA fragment from *Rhizobium* is subcloned into a vector (insertional vector) that can be mobilized into a variety of gram-negative bacteria including *Rhizobium*. Tn5 is randomly inserted into this DNA fragment after it is transformed into an *E. coli* strain that contains Tn5. The plasmid (pMJ10-4) carrying the mutagenized fragment is transferred by conjugation into wild-type *Rhizobium*. A single crossover on either side of Tn5 will result in the exchange of DNA fragments between the plasmid and the chromosome. The effect of each Tn5 insertion on symbiotic nitrogen fixation is determined by inoculation of the *Rhizobium* strain onto the host legume plant. Some of the mutant phenotypes that are detectable with the plant tests are listed. For complementation analysis the same DNA fragment is subcloned into a vector (extrachromosomal vector) that can be mobilized and can replicate in *Rhizobium*.

2 ml of L broth containing the appropriate antibiotics, and after 5–6 hr of incubation the cells are harvested and small-scale preparations of plasmid DNA prepared.

The location of Tn*5* within each plasmid is determined by digestion with any one of the restriction endonucleases listed in Table I followed by agarose gel electrophoresis of the digested plasmid. Digestion of the plasmids with a restriction endonuclease that does not cleave Tn*5* can be most useful in determining the approximate location of Tn*5* with respect to the target DNA fragments generated by a particular restriction endonuclease. A more exact location can be determined by digestion of the plasmid DNA with a restriction endonuclease that cleaves within Tn*5* (Table I). For example, if Tn*5* is within a *Hin*dIII fragment of the target DNA, then digestion of the plasmid with *Hin*dIII will result in the appearance of three new *Hin*dIII fragments. One fragment corresponds to the internal *Hin*dIII fragment (3.35 kb) of Tn*5* and the other two are the junction fragments that consist of one end of Tn*5* and the adjoining target DNA. Since the two *Hin*dIII sites of Tn*5* are within the inverted repeats at the ends of the transposon, the length of a junction fragment minus 1195 bp (the distance of each *Hin*dIII site to the end of Tn*5*) corresponds to the distance of Tn*5* from one end of the target DNA.

Since the transposition of Tn*5* is a relatively frequent event, the number of cells within each culture of the transformed strain JN25 that have Tn*5* inserted in different locations can be quite high. Therefore, plasmid DNA isolated from these cultures and transformed into strain JA221 can result in transformants that contain plasmids with Tn*5* inserted into different locations. However, we have often found that the majority of transfor-

TABLE I
A LIST OF USEFUL ENDONUCLEASE RESTRICTION SITES PRESENT IN Tn*5*[a]

Restriction endonuclease	Number of sites	Approximate location
*Hin*dIII, *Pst*I, *Xho*I, *Bgl*II, *Hpa*I	2	Within the inverted repeat
*Sal*I, *Sma*I, *Pst*I, *Bam*HI, *Xho*I	1	Central region
*Kpn*I, *Pvu*I, *Eco*RI, *Cla*I, *Sst*I, *Bal*I	0	—

[a] For the precise location of these sites, see R. A. Jorgensen, S. J. Rothstein, and W. S. Reznikoff, *Mol. Gen. Genet.* **177,** 65 (1979).

mants contain plasmids with Tn*5* insertions in the same location. This is probably because transposition events that occur early during the growth of the culture result in an enrichment of the total plasmid population with plasmids containing the same insertion. In order to assure that each transformant of strain JA221 arises from an independent transposition event, only a single kanamycin-resistant colony should be isolated from among the approximately 10^3 transformants obtained with the plasmid DNA from each culture of strain JN25.

Site-Specific Insertion of Tn5 into the Rhizobium Genome

The plasmid pSUP202 contains the mobilization region (Mob) from the broad-host-range plasmid RP4 which enables it to be transferred by conjugation between various species of gram-negative bacteria. Individual plasmids containing a Tn*5* insertion can be transferred to *Rhizobium* strains from *E. coli* strain SM10 which contains a chromosomally integrated RP4 plasmid that provides essential transfer functions. The recipient *Rhizobium* strain is inoculated into 10 ml of RDY medium from a recently streaked culture growing on an RDM agar plate. The culture is incubated at 30° with vigorous shaking for 2–3 days until the density of the culture is $\sim 2 \times 10^8$ cells/ml ($OD_{600} \simeq 0.4$). The cells are harvested in a desk-top centrifuge at 2000 *g* for 10 min, washed once with 5 ml of RDY medium, and resuspended in 0.1 ml of RDY medium. The donor cells (*E. coli* SM10 cells containing a plasmid with a Tn*5* insertion) are inoculated into 2 ml of L broth supplemented with 50 μg/ml of kanamycin and 50 μg/ml of ampicillin or chloramphenicol and allowed to grow overnight.

In the morning one drop of the culture is inoculated into 2 ml of the same medium and allowed to grow for 1.5–2 hr until the density of the cells is $\sim 4 \times 10^8$ cells/ml ($OD_{600} \simeq 1.0$). The cells are harvested by centrifugation, washed once with L broth, and resuspended in 100 μl of RDY medium. Fifty microliters of each of the donor and recipient cell suspensions are then mixed and transferred to a sterile nitrocellulose filter [0.45 μm, 2.5 cm, GA6 (Gelman Sciences, Inc.)], placed on an RDY agar surface, and incubated for 1–2 days at 30°. The cell-laden filters are placed in 17 × 100 mm Falcon polypropylene test tubes, 5 ml of RDM medium containing 0.01% Tween 80 is added to each tube, and the cells are removed from the filters by vortexing. The cells are harvested by centrifugation at 2000 *g* for 10 min in a desk-top centrifuge, and resuspended in 0.5 ml of RDM medium. The undiluted samples (50–100 μl) are plated onto RDM-agar supplemented with kanamycin and streptomycin each at a final concentration of 100 μg/ml. Since Tn*5* also confers resistance to streptomycin, this also is included in the medium to select against the growth of spontaneous kanamycin-resistant cells. The selection plates are incubated for 7–10 days at 30°, and the transconjugants are purified by single colony isolation on the same medium.

Physical Analysis of Genomic DNA from the Transconjugants

The plasmid pSUP202 can only be stably maintained in *E. coli* and not in *Rhizobium* cells because it does not have a broad-host-range origin of replication. Therefore, kanamycin-resistant *Rhizobium* cells can be generated by any of three ways: (1) spontaneous mutation; (2) transposition of Tn*5* from the unstable plasmid DNA into the chromosome; (3) homologous recombination between the genomic DNA on the plasmid and the chromosome. The spontaneous mutation frequency is dependent on the particular strain being used, but in general is 10^{-8} to 10^{-9}. The mutation frequency is considerably higher when cells are selected for spontaneous resistance to a second antibiotic simultaneously. The transposition frequency is usually $1–2 \times 10^{-5}$, the same as that observed in *E. coli.*[5] Kanamycin-resistant cells that arise because of the integration of Tn*5* into the chromosome by a recombinational event occurs at a frequency of $1–3 \times 10^{-3}$.

The integration of Tn*5* into the chromosome can occur by either of two recombinational events. A single crossover between the chromosome and a homologous portion of the target DNA carried on pSUP202 leads to the integration of the entire plasmid.[8] This type of recombinational event results in the presence of a Tn*5*-interrupted target DNA fragment and an uninterrupted or wild-type fragment separated by the vector pSUP202 in the chromosome of the *Rhizobium* strain. Two single crossovers, one on each side of the Tn*5* insertion in the target DNA, results in the exchange of the Tn*5*-interrupted fragment on the plasmid for the wild-type fragment in the chromosome.[8] This is referred to as marker exchange. Because of the inability of PSUP202 to replicate in *Rhizobium,* the plasmid, which now carries the wild-type fragment, will be diluted out as the division of the cell continues.

Southern hybridization analysis of the genomic DNA from the kanamycin-resistant colonies is performed in order to determine the type of transconjugants that are obtained. Total genomic DNA from *Rhizobium* is isolated by a procedure slightly modified from Dhaese *et al.*[21] One milliliter of a late log-phase culture is centrifuged for 2 min in an Eppendorf centrifuge, and the pellet is resuspended in 200 μl of TE buffer containing 1.0% sarkosyl, 0.5 mg/ml of pronase B, and 100 μg/ml of ribonuclease A.

[21] P. Dhaese, H. DeGreve, H. DeCraemer, J. Schell, and M. Van Montagu, *Nucleic Acids Res.* **7,** 1837 (1979).

The mixture is incubated at 37° for 1 hr. The viscosity of the lysate is reduced by rapidly passing it through a 1 ml Pipetman tip 10–15 times. The sheared lysate is extracted twice with an equal volume of a 1 : 1 mixture of phenol (saturated with 0.5 *M* Tris–HCl, pH 8.0) and chloroform. The nucleic acids are precipitated with 0.1 vol of 3 *M* sodium acetate and 2.5 vol of ethanol at −70° for 1 hr. The pellet is washed with 70% ethanol, dried in a vacuum dessicator, and dissolved in 40 μl of TE buffer.

For Southern analysis, 8 μl of the total genomic DNA is digested with *Hin*dIII, and the DNA fragments are separated by agarose gel electrophoresis. The genomic DNA is then hybridized with ^{32}P-labeled pMJ10-4 plasmid DNA. If a double-crossover event occurred in the transconjugants, three DNA fragments in the genomic DNA would be visible. The three fragments correspond to the 3.35-kb internal Tn*5* DNA fragment and the two junction fragments containing one end of Tn*5* and the adjoining target DNA. No hybridization should be seen with ^{32}P-labeled vector DNA. If a single-crossover event occurred, two additional fragments would be visible that correspond to the uninterrupted wild-type fragment and the 7.3-kb pSUP202 vector DNA. If a transposition event occurred in the transconjugants, the sizes of the junction fragments in the genomic DNA would most probably be different from those of the plasmid DNA.

Functional Characterization of Rhizobium Strains Carrying Tn5 Insertions

In order to determine if a Tn*5* insertion in a specific site of the *Rhizobium* genome results in an alteration of its symbiotic nitrogen fixation properties, the phenotypes of the transconjugants generated as a result of the marker exchange event can be ascertained on an appropriate host plant. The procedure that is routinely used to test *Rhizobium* strains on plant hosts and which minimizes cross-contaminations of the plants is as follows.

Styrofoam pots (15.2 cm in diameter) and saucers are dipped in 70% ethanol and placed in plastic bags. The plastic bags form a tight-fitting sleeve around the saucer and pot, leaving only the top of the pot exposed. After drying, charcoal chips are added to a depth of 0.5 cm and the pots are then filled with fine silica sand. Approximately 500 ml of UV-sterilized nitrogen-free nutrient solution[8] is added to the sand. Seeds are surface sterilized using 0.1% acidified $HgCl_2$ (3 min in sterile water, 2 min in $HgCl_2$, and 10 rinses in sterile water). Three to six seeds are planted per pot, and each seed is inoculated with at least 10^6 cells of a 5–7-day-old rhizobial culture before covering with sand. The sand is covered to a depth of 2.5 cm with styrofoam beads to reduce the risk of airborne rhizobial contamination. At least eight plants of each cultivar used in each trial are left uninoculated. Each treatment is performed in triplicate. Ten days after planting all but two seedlings are removed from each pot. Plants are watered by adding nitrogen-free solution to the saucers through a hole in the plastic bag. Plants are grown in a light room (approximately 800 $\mu E\ m^{-2}\ sec^{-1}$) with a 13-hr photoperiod, and a 30°/25° day/night temperature regime. Plants are usually harvested 42 days after planting. At harvest time, shoots are removed, dried, and weighed. Acetylene reduction assays are carried out on the roots as described by Hardy *et al.*[22]

In order to be certain that the Tn*5* insertion is stably maintained in the genome of the *Rhizobium* strain after plant passage, the rhizobia can be isolated from the root nodules. Nodules are picked from the plants, immersed in 95% ethanol for 90 sec and then in 0.1% acidified $HgCl_2$ for 3 min, and washed with 10 changes of sterile water. The surface-sterilized nodules are crushed aseptically onto the surface of RDM agar supplemented with the appropriate antibiotics. After 7–10 days of incubation at 30°, single colonies are grown in culture and total DNA is prepared as previously described. The presence of Tn*5* can be determined by Southern hybridization analysis.

Complementation Analysis

Site-directed Tn*5* mutagenesis may reveal that more than one gene essential for symbiotic nitrogen fixation is present within a particular region of the genome. It is possible to determine whether or not these genes are arranged in more than one operon by organizing the Tn*5* insertions into complementation groups. Figure 2 shows the overall concept of how complementation analysis is performed. In a hypothetical example, a Tn*5* insertion in a particular DNA fragment reveals the presence of a gene that is required for nitrogen fixation. A *Rhizobium* strain containing the Tn*5* insertion can nodulate (Nod^+) the roots of the plant but is unable to fix nitrogen (Fix^-). The corresponding wild-type DNA fragment is introduced into this strain via a broad-host-range plasmid. The origin of replication (*oriV*) of plasmid pMJ17 (the broad-host-range plasmid containing the wild-type *Rhizobium* DNA fragment) enables this plasmid to be stably maintained in *Rhizobium*. In addition, this plasmid carries tetracycline resistance so that *Rhizobium* cells containing this plasmid are readily identified. Functional complementation of the Tn*5*-inactivated gene occurs *in trans* by the wild-type gene carried on pMJ17.

[22] R. W. F. Hardy, R. D. Holsten, E. K. Jackson, and R. C. Burns, *Plant Physiol.* **43,** 1185 (1968).

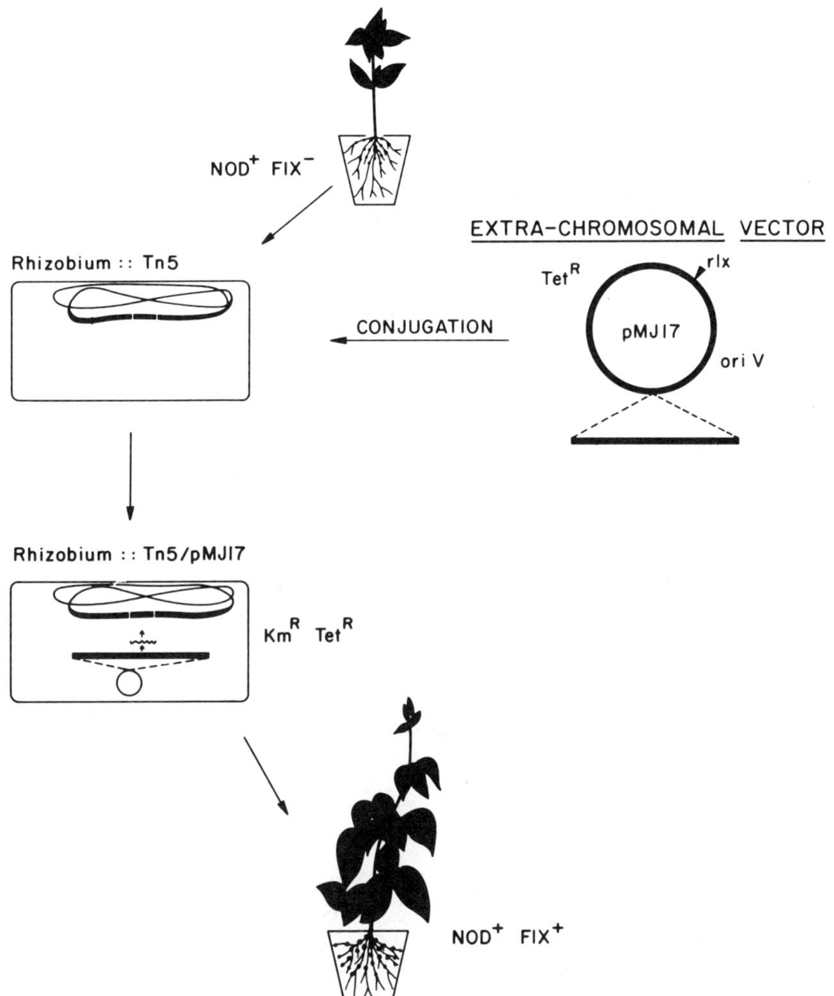

FIG. 2. The general scheme for complementation of a mutant phenotype in *Rhizobium*. The wild-type fragment that corresponds to the Tn5-interrupted fragment in *Rhizobium* is cloned into an extrachromosomal vector. The vector used in this study was pVK100 which contains a locus required for mobilization (*rlx*) and a broad-host-range origin of replication (*oriV*). In this scheme the wild-type DNA fragment was cloned into pVK100 which resulted in pMJ17.

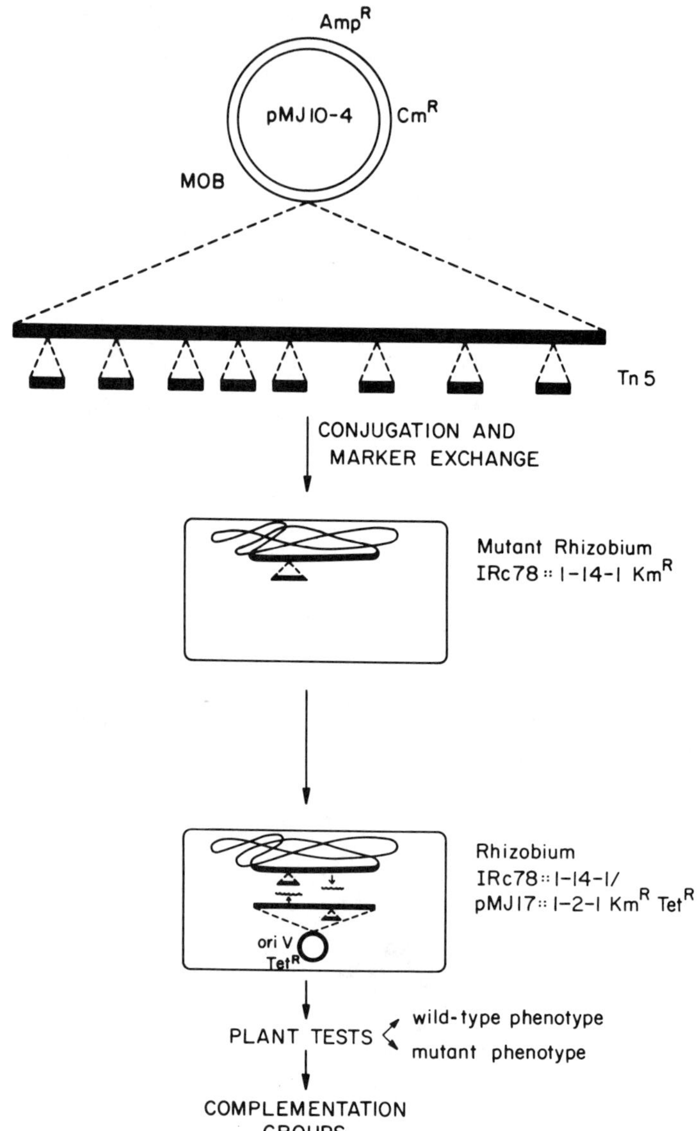

FIG. 3. An example of how two different Tn*5* insertions are determined to be in the same or in different complementation groups. The DNA fragment from wild-type *Rhizobium* was subjected to random Tn*5* mutagenesis. The location of eight different Tn*5* insertions within this DNA fragment is shown. Tn*5* insertion 1-14-1 was introduced into the wild-type *Rhizobium* chromosome by site-directed Tn*5* mutagenesis. The *Rhizobium* DNA fragment containing Tn*5* insertion 1-2-1 was subcloned from pMJ10-4 into a broad-host-range vector. The two Tn*5* insertions are not in the same transcript (〰), and, thus, the function inactivated by Tn*5* insertion 1-14-1 is provided *in trans* from the corresponding region present on the chromosome. If the two Tn*5* insertions are within the same operon, the RNA transcripts from both operons would be incompletely synthesized and complementation cannot occur.

Figure 3 shows how complementation groups are established. In one experiment, several Tn*5* insertions in distinct location within a large DNA fragment from the genome of *Rhizobium* strain IRc78 revealed the presence of several genes that are required for nitrogen fixation. In Fig. 3 the analysis of two different Tn*5* insertions, 1-14-1 and 1-2-1, is outlined. Tn*5* insertion 1-14-1 was transferred and recombined into the genome of *Rhizobium* strain IRc78 by the methods already described. Tn*5* insertion 1-2-1 was recombined into plasmid pMJ17 (see next section), the broad-host-range plasmid containing the same DNA fragment from *Rhizobium* strain IRc78 that was cloned into pMJ10-4 (see Fig. 1), to generate plasmid pMJ17::1-2-1. Plasmid pMJ17::1-2-1 was then mobilized into *Rhizobium* strain IRc78 that contained Tn*5* insertion 1-14-1, and the resulting strain IRc78::1-4-1/pMJ17::1-2-1 was inoculated onto the host plants. If the two Tn*5* insertions were in two different operons (and, thus, interrupt two different transcriptional units), the missing *Rhizobium* gene functions on the chromosome will be provided *in trans* from the plasmid. The result of this experiment is shown in Fig. 4. The plant inoculated with strain IRc78::1-4-1/pMJ17::1-2-1 (C) fixed nitrogen as well as the plant inoculated with the wild-type strain IRc78 (A). Since complementation of the mutant *Rhizobium* phenotype occurred, the conclusion was that these two Tn*5* insertions were not in the same operon. If the two Tn*5* insertions were within the same operon (and, thus, interrupted the same transcriptional unit), then complementation would not have occurred.

Transferring the Tn5 Insertions from pMJ10-4 to pMJ17

For complete complementation analysis the *Rhizobium* DNA fragments containing a single Tn*5* insertion have to first be cloned into the broad-host-range plasmid pVK100. This can be done by ligating each Tn*5*-interrupted fragment from pMJ10-4 into pVK100. However, convenient restriction sites may not be present to enable an intact DNA fragment to be removed from pMJ10-4. Additionally, this can be very tedious if several Tn*5* insertions are to be analyzed.

An alternative protocol is outlined in Fig. 5. In this example, a *Hin*dIII restriction fragment from the *Rhizobium* genome had been cloned into plasmid pSUP202 resulting in plasmid pMJ10-4. This DNA fragment was subcloned from pMJ10-4 into pVK100 to generate pMJ17 in the following manner. One microgram each of plasmids pMJ10-4 and pVK100 were digested to completion in the same Eppendorf tube with the restriction endonuclease *Hin*dIII. The restriction endonuclease was inactivated by heating to 65° for 10 min. The digested plasmids were ethanol precipitated, resuspended in 50 μl of TE buffer, and ligated using standard liga-

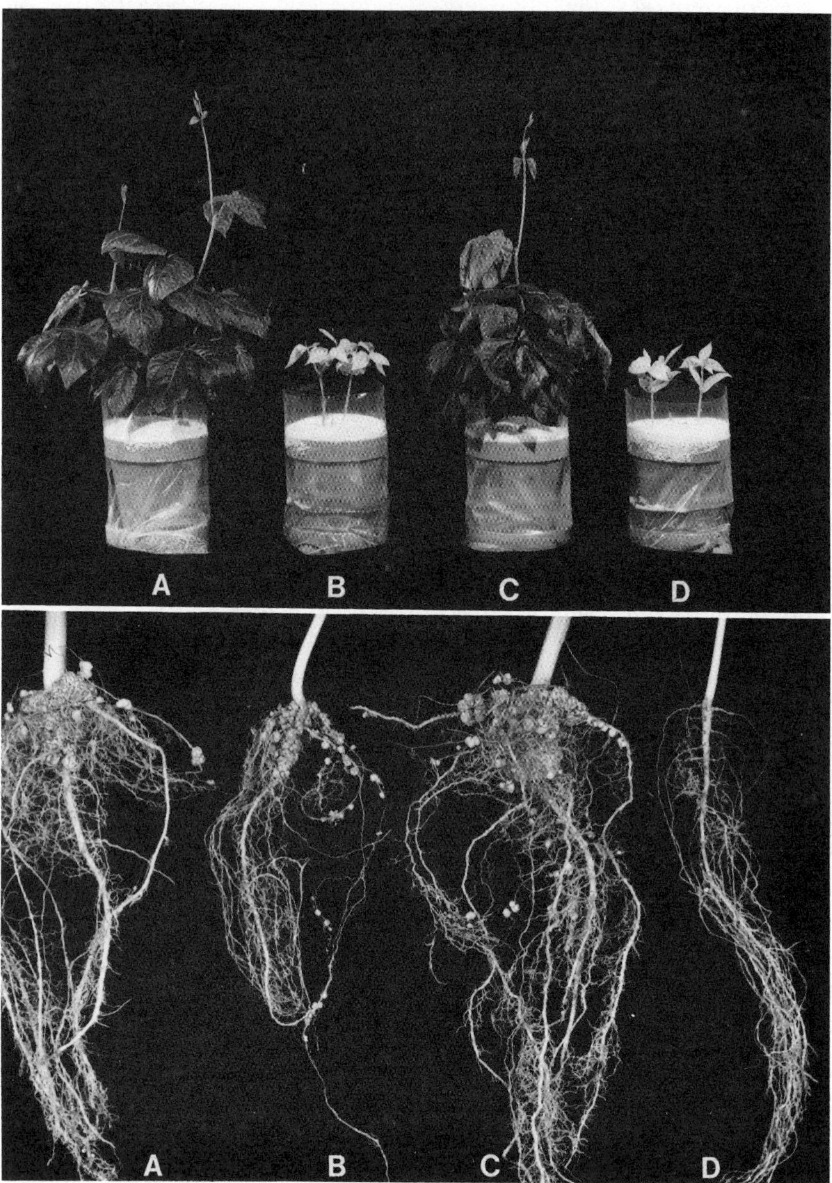

FIG. 4. Complementation analysis of a *Rhizobium* strain carrying two different Tn5 insertions. California black-eyed pea plants were inoculated with (A) wild-type *Rhizobium* IRc78, (B) *Rhizobium* IRc78::1-4-1, (C) *Rhizobium* IRC78::1-4-1/pMJ17::1-2-1, or (D) were uninoculated. The root systems are shown below each plant.

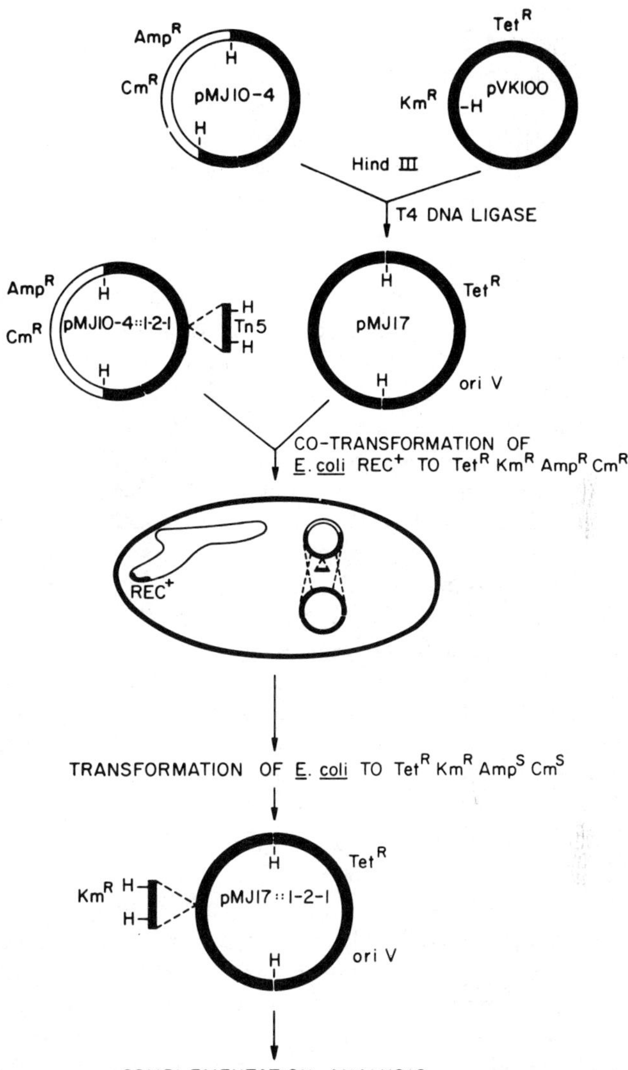

FIG. 5. The general scheme used to transfer Tn*5* from an insertional vector to an extrachromosomal vector. The blackened portion of pMJ10-4 is the wild-type DNA fragment that was isolated from *Rhizobium*. This fragment is removed by *Hin*dIII digestion and then ligated into the *Hin*dIII site in pVK100 resulting in pMJ17. A Tn*5* insertion (1-2-1) is transferred from the *Rhizobium* DNA fragment in pMJ10-4 to the same DNA fragment in pMJ17 by a reciprocal exchange of the homologous regions.

tion conditions. The ligated DNAs were transformed into *E. coli* JA221, and transformants were selected on L agar medium with 30 μg/ml of tetracycline. Plasmid DNA from transformants that are tetracycline resistant but kanamycin sensitive are screened by the method of Birnboim and Doly.[20]

The Tn*5* insertion was transferred from pMJ10-4 by homologous recombination into pMJ17. Approximately 0.1 μg of pMJ10-4 and pMJ17 were co-transformed into 0.1 ml of competent *E. coli* cells. Competent cells were prepared from a culture of strain MC1061 that was grown to early log phase in L broth. The cells were harvested by centrifugation, resuspended in an amount of 0.1 *M* $CaCl_2$ equal to one-half the original culture volume, held on ice for 2–4 hr, and concentrated 5-fold in 0.1 *M* $CaCl_2$. The transformed cells were plated onto L agar medium with 30 μg/ml of tetracycline, 50 μg/ml of kanamycin, 50 μg/ml of ampicillin, and 25 μg/ml of chloramphenicol. Because *E. coli* MC1061 is recombination proficient (Rec^+), recombination can occur between the homologous portions of the two plasmids. Two crossovers, one on either side of Tn*5*, will result in a reciprocal exchange of homologous *Rhizobium* DNA between the plasmids. A small number of plasmids in these transformants are expected to have undergone this double reciprocal crossover. Therefore, several colonies were scraped from a selective plate, pooled, and a small-scale preparation of plasmid DNA was prepared from the cells. Approximately 1–2 μg of this plasmid DNA was transformed into *E. coli* JA221, and transformants were selected on L agar medium with 30 μg/ml of tetracycline and 50 μg/ml of kanamycin. These transformants were screened for ampicillin and chloramphenicol sensitivity.

Site-Directed Transplacement

The general concept for introducing alterations into the *Rhizobium* is shown in Fig. 6. The basic approach is similar to that used for site-directed Tn*5* mutagenesis. The DNA region to be mutagenized is cloned into pSUP202. Any type of alteration can be made to the cloned region, the most common being an addition of a sequence, a deletion, a base substitution, or an inversion. The only requirement is that part of the original DNA region must remain unchanged to enable the plasmid to be integrated into the chromosome. Alternatively, the DNA region can be first cloned into another vector, mutagenized *in vitro,* and then subcloned into pSUP202. Additionally, a selectable marker is required on the vector. Only the tetracycline gene of pSUP202 is effectively expressed in slow-growing *Rhizobium* cells. If the *Hin*dIII site within this gene is used to clone the *Rhizobium* DNA fragment, then Tn*5* is inserted into pSUP202 by random Tn*5* mutagenesis as described in this chapter to serve as a marker.

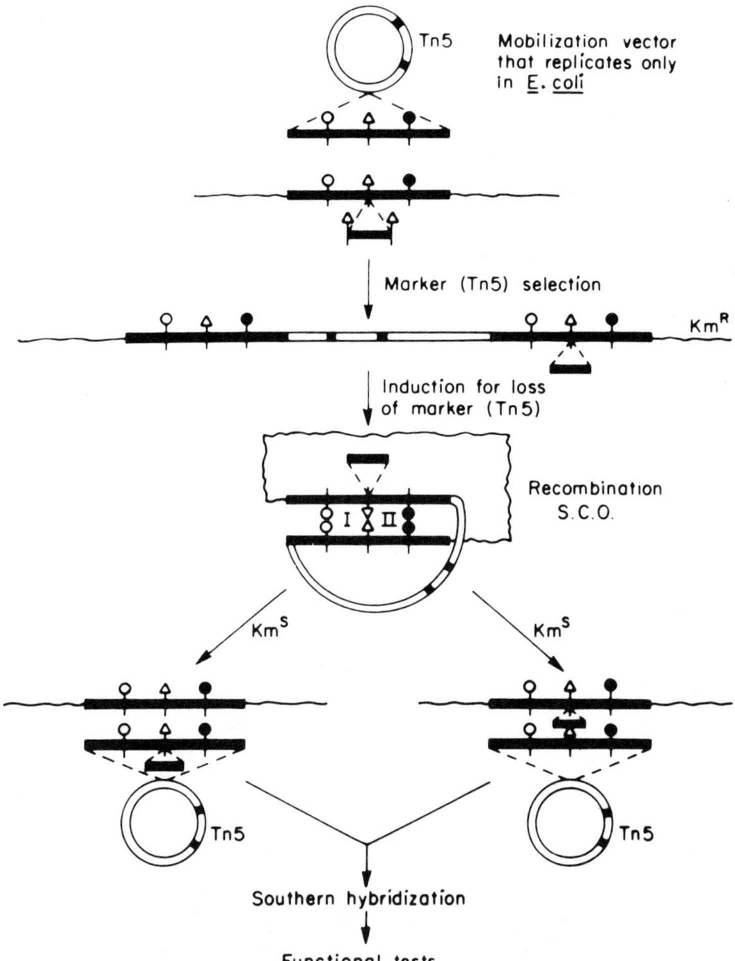

FIG. 6. Site-directed transplacement of an *in vitro* altered DNA fragment for the corresponding wild-type DNA fragment in *Rhizobium*. Small circles and triangles represent restriction endonuclease recognition sites shown here for the purpose of visualizing the *in vitro* alteration; ~~, flanking chromosomal DNA. A deletion was made in the *Rhizobium* DNA fragment present in the mobilization vector. The entire plasmid is integrated into the chromosome by a single crossover (S.C.O.) between the homologous DNA sequences on the plasmid and chromosome. A subsequent single crossover results in the loss of the vector and one of the duplicated sequences. Approximately 50% of the Km^s segregants are expected to contain the deleted sequence.

The entire plasmid containing the altered DNA sequence is mobilized from *E. coli* SM10 into the wild-type *Rhizobium* strain. *Rhizobium* cells resistant to both kanamycin and streptomycin are selected. A single crossover between the chromosome and the homologous region on the plasmid will result in the integration of the entire plasmid into the chromosome. *Rhizobium* cells that are kanamycin resistant because of this recombinational event, however, are not stably maintained in culture. A single crossover between various regions of the duplicated sequences in the chromosome of kanamycin-resistant cells will result in the excision of the vector and one of the duplicated sequences. To isolate kanamycin-sensitive segregants, a single kanamycin-resistant *Rhizobium* colony is inoculated into 10 ml of RDM medium without kanamycin which leads to an enrichment in the culture for cells that have lost the vector. Severalfold dilutions of the culture are made, and aliquots are plated on RDM agar. Single colonies are isolated and tested for either a Km^r or Km^s phenotype on RDM agar plates supplemented with 100 μg/ml of kanamycin. In a typical experiment, 75% or more of the tested colonies will be kanamycin sensitive. This indicates that DNA duplications in a chromosome are highly unstable.

Concluding Comments

Theoretically, any size DNA fragment can be mutated with Tn*5* and subsequently returned to the wild-type strain by marker exchange. In practice, however, DNA fragments smaller than 4 kb are not readily exchanged for the chromosomal sequence. For example, when Tn*5* was inserted approximately in the middle of a 4-kb DNA fragment and then this fragment was returned by conjugation to the wild-type *Rhizobium* strain, only 1 out of 250 kanamycin-resistant transconjugants was the result of a double-reciprocal crossover. The rest were a result of a single crossover that mediated the integration of the entire plasmid into the chromosome. In contrast, when Tn*5* was inserted in the middle of a 13-kb DNA fragment, 1 out of 12 of the kanamycin-resistant transconjugants were the result of a double-reciprocal crossover (marker exchange) event. Increasing the length of the target DNA increases the probability that a double crossover will occur. However, if Tn*5* is inserted near the extremities of a large fragment, the probability of a single crossover on both sides of Tn*5* is reduced.

Plasmid pSUP202 carries resistance to chloramphenicol, ampicillin, and tetracycline. However, only the tetracycline-resistance gene is expressed in the slow-growing *Rhizobium*. In order to facilitate screening for double crossovers in *Rhizobium,* the DNA fragment of interest should be cloned in sites present in the chloramphenicol or ampicillin gene. Then the kanamycin-resistant transconjugants are scored for resistance to tetracycline (loss of the vector).

Following random Tn*5* mutagenesis in *E. coli* JN25, plasmids that have incorporated Tn*5* are selected by transforming *E. coli* JA221 to kanamycin resistance. The plasmids are then transferred from *E. coli* JA221 to *E. coli* SM10 where they can be mobilized into *Rhizobium*. The low frequency of transformation observed with *E. coli* SM10 ($\sim 10^2/\mu g$) required that *E. coli* JA221, which has a high transformation capability, be used to initially select for the rare Tn*5*-containing plasmids.

Tn*5* is stably maintained in the chromosome throughout the life cycle of the *Rhizobium* cell in the nodules of the plant. Maintenance of an extrachromosomal plasmid such as pVK100 under these same nonselective conditions, however, is sometimes a problem. Moreover, false-positive complementation results can occur between two Tn*5* insertions that are actually within the same transcriptional unit as a result of marker exchange between the homologous *Rhizobium* DNA contained on the plasmid and in the chromosome.

Genes of interest can be linked to Tn*5* and recombined into the chromosome by marker exchange. For example, the *lacZ* gene of *E. coli* was fused to the promoter regions for the structural genes of nitrogenase, the key enzyme in nitrogen fixation. A region of the *Rhizobium* chromosome, previously determined to be unessential for nitrogen fixation, was cloned into a mobilizable plasmid and these nitrogenase promoter–*lacZ* fusions were inserted into this unessential region using standard ligation techniques.[23] Tn*5* was also inserted into this region to serve as a selectable marker. The fusions were then transferred to *Rhizobium* and inserted into the chromosome by marker exchange. The expression of the nitrogenase gene promoters are monitored easily by assaying for β-galactosidase activity. A similar approach was used to insert a nitrogenase promoter–bacterial luciferase gene fusion into *Rhizobium*.[24]

Although the advancement of the genetics of the slow-growing *Rhizobium* species has proceeded more slowly than that of the fast-growing *Rhizobium* species, the techniques for mutagenesis described in this chapter should facilitate progress in this field.

Acknowledgments

The authors wish to express their appreciation to Dr. A. Pühler for providing plasmid pSUP202 and strain SM10; to Ms. Deborah Bridwell for secretarial assistance; to Dr. J. Telford for illustrative work. This research was supported by a grant from the Allied Corporation.

[23] A. Yun, J. D. Noti, and A. A. Szalay, *J. Bacteriol.* **167,** 784 (1986).

[24] R. P. Legocki, M. Legocki, T. O. Baldwin, and A. A. Szalay, *Proc. Natl. Acad. Sci. U.S.A.* **83,** 9080 (1986).

Section III

Chemical Synthesis and Analysis of Oligodeoxynucleotides

[13] Simultaneous Synthesis and Biological Applications of DNA Fragments: An Efficient and Complete Methodology

By RONALD FRANK, ANDREAS MEYERHANS, KONRAD SCHWELLNUS, and HELMUT BLÖCKER

General Introduction

At this time, many a scientist has to decide how he can best obtain synthetic DNA. There is no general answer to the question of which method would be best, either synthesizing it in-house or having it done outside. One must first analyze the specific situation and consider factors of confidence, speed, flexibility, budget, available know-how, and annual demand. Suppose the decision was made to synthesize oligonucleotides in-house and to use them for the construction of genes or mutagenesis cassettes, or just for a variety of smaller projects. Two other questions are then raised: Aren't there "gene machines" on the market which work reliably for a reasonable price? Is there any additional advantage to hand-making oligonucleotides, mutagenesis cassettes, or complete genes?

The so-called "gene machines," which should be classified as "synthesis automats," "autosynthesizers," or simply "synthesizers," have been constantly improved over the past few years, especially for the synthesis of longer oligonucleotide chains. However, there are now manual methods available, which, on the one hand, enable the scientist to make more oligonucleotides per unit time and at much lower costs than with machines, but, on the other hand, usually require more human labor.

However, since the chemical synthesis of oligonucleotide chains is just one step on the way from nucleotides to probes, primers, adapters, and DNA double strands, the decision either to buy a good synthesizer or to follow a reliable procedure for manual or semiautomated synthesis is less important than some people still believe. More important are positive answers to questions such as the following:

Is our method economical, and are the running costs low? Can we, for example, recover building block excesses?

Do we have the option of producing either very small amounts (just enough for a few cloning experiments) or relatively large amounts (e.g., for biophysical studies)?

Does the applied chemical procedure provide DNA of such a high purity that, for example, after cloning no (or at least very few) unwanted mutations occur?

Do we have tools (computer programs) that help select the best possible primers and probes, and is there a simple way to design synthetic DNA double strands and to assemble them from oligonucleotides?

Do we circumvent error-prone steps in our methods? Are they foolproof?

All of these questions, and a few more, can be satisfied by a simple and complete methodology. In the following pages such a methodology is described.

Outline of the Methods

The basis of our methodology is the segmental support approach for the simultaneous synthesis of large numbers of oligonucleotides (''filter method,'' ''disk method,'' ''bulk method'') which was developed by us in the early 1980s and which has since been constantly improved.[1] The basic idea of this approach is rather simple. Instead of using small beads or fibers, as in the classical approaches to peptide and nucleic acid synthesis, we employ ''segmental supports,'' noninterchangeable entities, as the immobilizing support (Fig. 1). This type of support allows one to elongate simultaneously those oligonucleotide chains which require the same nucleotide.

We introduced paper disks for this purpose because it had been shown before that cellulose fibers could be used for support in chemical DNA synthesis.[2] To our knowledge, practically all oligonucleotides synthesized so far using the segmental support approach has been produced on paper disks, although other materials, such as glass, may also be used. Unfortunately, paper is chemically more reactive than, for example, inorganic carriers, but it is inexpensive, mechanically stable, flexible, and the starting nucleoside can be easily attached to it. The filter disks are rendered noninterchangeable—another advantage of paper—simply by numbering them unambiguously with a pencil. The numbers refer then to the different oligonucleotides to be synthesized.

Many oligonucleotides can be made simultaneously by stacking the filters into only four different reaction vessels for the addition of the usual four A, C, G, and T building blocks. After each coupling cycle, the filters are reshuffled into the correct vessels for the next chain elongation cycle.

[1] R. Frank, W. Heikens, G. Heisterberg-Moutsis, and H. Blöcker, *Nucleic Acids Res.* **11,** 4365 (1983).

[2] T. Horn, M. P. Vasser, M. E. Struble, and R. Crea, *Nucleic Acids Res. Symp. Ser.* **7,** 225 (1980).

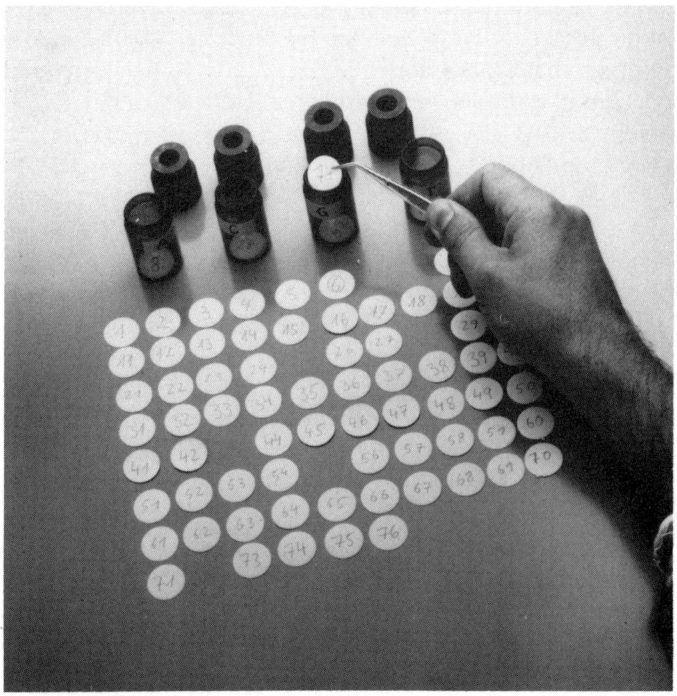

FIG. 1. Cellulose paper disks as support for the simultaneous chemical synthesis of large numbers of different oligonucleotides.

Through this procedure the time required for the synthesis is practically independent of the number of oligonucleotides, whereas the longest oligonucleotide determines the end point of the whole synthesis. In other words, ten additional oligonucleotides of a chain length not greater than the others' chain lengths would just make the piles in the vessels a little higher and require a few more seconds during the resorting step. However, one additional oligonucleotide which is only one nucleotide longer than that which was previously the longest, would require a complete extra reaction cycle.

It is obvious that the segmental support approach shows its striking advantage more convincingly the higher the number of oligonucleotides is. Advantages are also seen, however, when the oligonucleotides are either medium sized or even short in length (of the order of 10–25 base units). Thus, for the synthesis of a gene of about 500 base pairs, one can reduce the number of reaction cycles by more than 90%, when compared to conventional approaches, if the gene were constructed from relatively short oligonucleotides (for example, 50 20-mers: 20 cycles in each of the four reaction vessels = 80 cycles versus 20 × 50 = 1000 cycles).

The filter synthesis does not necessarily have to be carried out simultaneously in four reaction vessles but can also be done, for example, consecutively in one vessel. Such a one-vessel operation in the above example requires 80 successive cycles (80 × 1 = 80) instead of 20 synchronous cycles in each of the four reaction vessels (20 × 4 = 80). This will of course increase the labor time by a factor of 4. However, by an advantageous planning of the sequence of the elongation cycles one can to some extent reduce the number of cycles (depending on the oligonucleotide sequences of about 40–60 in the above example). Whether one will use fewer than four reaction vessels will have to be decided by what is considered most practical and adequate for the individual situation. Of course, the cycle time for any operational mode of the filter synthesis, which always includes the sorting of the filters, is much longer than that of the autosynthesizers. However, if one divides the cycle time by the number of oligonucleotides which are synthesized simultaneously, it drops below 1 min for the segmental support approach.

The filter method can also help to meet quite different demands with respect to quantities of DNA synthesized. Given the same loading of filter paper per gram and volume, same filter thickness, and same chemistry, the output of synthetic DNA should depend only on the diameter of the paper disks. Since we have shown that the filter method works equally well for disk diameters of 5–20 mm, one can set the yield range according to demand simply by selecting the appropriate disk diameter. For example switching from 5 to 20 mm will increase the yield by a factor of 16. An additional or alternative way of raising the amount would be to use more than one filter disk for the same sequence.

Initially, the filter method had been performed by us using phosphate triester procedures.[1] Later, Ott and Eckstein[3] were first to publish that the filter method also works nicely with phosphoramidites. In general the phosphoramidites reportedly give slightly higher coupling yields. However, we decided to continue using phosphotriester methods, mainly because of economical reasons. The building blocks, which are usually applied in 10- to 20-fold excesses, are fairly expensive for both DNA chemistries. Triester building blocks can be recovered.[4] We have recently established a rather simple protocol for recycling 60–80% of the excesses of the triester building blocks.[5]

[3] J. Ott and F. Eckstein, *Nucleic Acids Res.* **12,** 9137 (1984).

[4] G. R. Gough, M. J. Brunden, and P. T. Gilham, *Tetrahedron Lett.* **22,** 4177 (1981).

[5] K. Schwellnus, H. Blöcker, and R. Frank, unpublished results.

We have further shown that for removal of phosphate protecting groups, oximate reagents (very elegant reagents from a chemist's point of view, but unfortunately rather inconvenient in practice when compared to volatile reagents) can be replaced by concentrated ammonia, practically without affecting yield or quality of the synthetic DNA. This simplification of the workup has become possible because we introduced additional protecting groups on G[6] and T[7,8] to significantly reduce base modification reactions. Previously, these side reactions could only be reversed in a two-step deprotection employing oximate reagents. A one-step alkaline deprotection is also possible in the amidite chemistry since Sinha *et al.*[9] have introduced β-cyanoethyl phosphoramidites. Oligonucleotides produced with the usual methoxy phosphoramidites always require a two-step deprotection, and a certain number of base modification reactions have been described in the literature.[10] Hence, quite a few laboratories are currently testing additional base protecting groups in conjunction with the amidite chemistry.

A novice to DNA synthesis, even if convinced of the general advantages of the segmental support approach and the quality of the applied chemistry, might presume that simultaneous handling of more than 100 filters during the reshuffling steps might prove error prone or might at least require meticulous planning. However, we have written a simple computer program which does the planning within seconds. The program is called SYNPRO and is part of a menu-driven package, GENMON, designed for work with proteins and nucleic acids.[11] Like all other programs mentioned in this chapter it runs on VAX computers; personal computer versions (IBM, DEC) are also available.

SYNPRO accepts sequence inputs either from the keyboard, from sequence files stored on electronic medium, or directly from SYNGEN, a program to cut DNA double strands into favorable fragments for oligonucleotide synthesis (see below). SYNPRO also accepts degenerate bases, an option which proved to be very useful for shotgun synthesis of DNA double strands (see below) or for the preparation of highly degenerate

[6] T. Kamimura, M. Tsuchiya, K. Koura, M. Sekine, and T. Hata, *Tetrahedron Lett.* **24,** 2775 (1983).

[7] T. Kamimura, T. Masegi, K. Urakami, S. Honda, M. Sekine, and T. Hata, *Chem. Lett.* **1983,** 1051 (1983).

[8] C. J. Welch and J. Chattopadhyaya, *Acta Chem. Scand.* **B37,** 147 (1983).

[9] N. D. Sinha, J. Biernat, J. McManus, and H. Köster, *Nucleic Acids Res.* **12,** 4539 (1984).

[10] X. Gao, B. L. Gaffney, M. Senior, R. R. Riddle, and R. A. Jones, *Nucleic Acids Res.* **13,** 573 (1985).

[11] D. N. Lincoln, U. Leuner, W. Lehnberg, R. Frank, and H. Blöcker, submitted for publication.

SynPro << DNA synthesis protocol >>

a
```
A:XVARA     B:SYNSEQ   C:XHLSEQ    D:VAR11    E:VAR12    F:VAR13     G:X1LIN     H:X2LIN     I:X3LIN    J:PSRFRIM  K:MUT132G  L:PL1PRIM
M:PXR2PRIM  N:CAVVAR   O:EKL1PRIM  P:U11LIN   Q:KRALIN   R:PMSH1LIN  S:PMSH2LIN  T:PMSH3LIN
```

b

		Len	Mobil T	Mobil A	Extinc coeff	Alk depr	Cart	de-trit	HPLC	de-salt	lyo	OD	Vol	Gel	nmol totl
1- 1A	---------AGGATCTYTTATGTG	24	0.0	0.0	158815										
2- 2A	-------TTCATGGGTY--GTTTT	24	0.0	0.0	145215										
3- 1B	CGCTCCAGGTTTCCCCGGG	19	17.9	20.1	177260										
4- 2B	CTGGAGCGAACTGGCAAAAAA	21	19.8	22.2	253860										
5- 3B	CCCACCGTCCTTTTTTGCCAGTT	23	21.6	24.2	205480										
145- 1T	GATCAGAGATGGCGC	15	14.4	16.3	170810										
146- 2T	ATGATCAATGCGCCATCTCT	20	18.6	20.9	204730										
147- 3T	ATTGATCATTCCCCTCGAT	19	17.6	19.8	185920										
148- 4T	CGATCGAGGGGA	12	11.7	13.4	140160										

c Base usage: A=602 C=666 G=604 T=579 Y=2

d
```
Cycle 0
A:   4   6  10  13  14  17  19  25  34  36  60  61  68  72  73  74  78  79  83 100 101 104 105 108 110 112 114 132 136 137
   138 139 143 144 148 (35/0)
C:   7   8  11  18  23  27  29  30  40  42  43  44  46  48  50  52  55  56  58  59  62  63  64  67  69  70  71  80  85  86
    87  88  89  90  91  92  93  94  95  97  98 118 120 122 125 126 130 134 145 (49/0)
G:   1   3  12  15  20  24  33  37  38  39  41  54  66  75  76  77  84  96  99 103 107 109 111 113 117 124 128 129 133 140
   141 142 (32/0)
T:   2   5   9  16  21  22  26  28  31  32  35  45  47  49  51  53  57  65  81  82 102 106 115 116 119 121 123 127 131 135
   146 147 (32/0)

Cycle 1
A:   4  13  18  19  20  25  27  29  34  37  40  43  59  60  61  67  68  82  88  89  95  96 100 101 104 105 110 112 114 147 (30/0)
C:   6   9  10  11  15  24  28  32  44  56  62  64  65  70  81  86  87  90  99 103 107 111 115 118 126 129 130 133 134 146 (30/0)
G:   3  14  17  31  36  38  41  42  54  58  66  69  75  79  80  83 124 127 140 144 145 148 (22/0)
T:   1   2   5   7   8  12  16  21  22  23  26  30  33  35  39  45  47  49  84  92  93  97  98 102 106 108 109 113 117 128 (30/0)
-:  46  48  50  51  52  53  55  57  63  71  72  73  74  76  77  78  85  91  94 116 119 120 121 122 123 125 131 132 135 136 137 138 139 141 142 143 (36)

Cycle 7
A:   6   8  20  23  34  39  40  44  55  59  60  66  67  68  86  88  90  92  93  97  98 105 108 110 114 115 117 118 123 125 (30/0)
C:  13  14  15  16  21  22  27  30  31  32  37  41  42  58  61  62  64  65  70  81  82  84 100 130 134 146 147 (27/0)
G:   4   7   9  10  11  17  25  29  33  53  57  74  89  91  94  96 101 104 109 112 113 116 126 127 129 133 148 (27/0)
T:   3   5  12  18  19  24  26  28  35  36  43  54  56  69  72  76  80  95  99 102 103 106 107 111 124 138 139 141 143 145 (30/0)
Y:   1   2 (2/0)
-:  38  45  46  47  48  49  50  51  52  63  71  73  75  77  78  79  83  85  87 119 120 121 122 128 131 132 135 136 137 140 142 144 (32)

Cycle 22
A:  47  64 121 128 129 130 131 132 133 134 135 136 137 143 (14/9)
C:   5   9  10  12  16  22  24  25  30  32  34  35  40  41  42  43  44  46  48  50  52  53  57  61  62  67  68  69  70  76 (30/30)
G:  37  38  91 124 142 145 (6/6)
T:  33  56  58  66 109 113 125 126 127 144 (10/10)
-:  83  85  88  89  90  92  93  99 100 101 102 103 104 105 106 107 108 110 111 112 114 115 118 120 122 123 141 148 (28)

Cycle 23
A: 129 130 133 134 (4/4)
C:  64  83  85  88  89  90  92  93  99 100 101 102 103 104 105 106 107 108 110 111 112 114 115 118 120 122 123 141 148 (29/29)
```

probes/primers. In addition to degenerate or modified bases, blanks are also accepted. This option proved useful for optimal lining up of rare bases that are scattered in different oligonucleotides, thus minimizing extra manipulations for building in these uncommon bases. Any SYNPRO output can be seen on the screen, printed as hard copy (Fig. 2), or stored on disk. A synthesis protocol generated with SYNPRO writes a listing of all projects that will be covered by the synthesis, a listing of all oligonucleotides ordered according to the project order, and a status report for each elongation cycle. An additional option of SYNPRO allows the operator to set the maximal number of filters in the reaction vessels. This is of great experimental value because it helps not only to establish a constant flow in the flowthrough systems, but also allows standardized packs of reagents to be used. We have used SYNPRO over the last few years, and have synthesized thousands of oligonucleotides, but have made no mistakes during the resorting steps.

Recently, Seliger *et al.* reported on "simultaneous/asynchronous" filter synthesis.[11a] Instead of four columns the authors use only one. A computer program has been used to reduce the number of consecutive condensations by adjusting the optimal sequence of A, C, G, and T elongation reactions. We have included a column selection option in SYNPRO which enables the operator to select between one, two, three, four, or even more columns. The program is thus flexible enough to adapt to local conditions.[11]

A standard application of synthetic DNA fragments in biological projects lies in their use as mixed probes or as mixed primers. Here the segmental support approach offers additional options as already outlined in our original publication.[1] For example, all of the oligonucleotides in a mixed probe (as immediately selected by MIXPROBE, another subprogram of GENMON), can be synthesized simultaneously but on different filters. One may then combine bits of each of the filters and use them as

[11a] H. Seliger, K. Ballas, A. Herold, U. Kotschi, J. Lyons, F. Eisenbeiss, N. D. Sinha, and G. P. Talwar, *Chem. Scr.* **26,** 569 (1986).

FIG. 2. Protocol printout generated by SYNPRO. The different projects are listed in alphabetical order under (a). In (b) the numbers for all oligonucleotides and the sequences are listed together with sequence-specific data like length, mobility compared to A and T standards in a sizing gel, and extinction coefficient. In (c) for each phosphate component the total number of filter elongations is specified (base usage). In (d) the data for all cycles are listed: filters (represented by their numbers) to be coupled with the same building block. At the end of each line in parentheses the total number of filters for each elongation reaction and the total number of disks which are ready after this cycle are given. Underlined numbers represent sequences complete after this particular elongation cycle.

the usual mixed probes. If the target gene cannot be isolated with this mixed probe, one may work up other filter bits of certain groups of oligonucleotides: for example, A–T-rich or G–C-rich groups, or those containing the most probable codons. Finally, there would still be the option of working up all oligonucleotides separately and of using them separately or in well-adjusted stoichiometric amounts for probing.

Synthetic oligonucleotides can be used for the construction of double-stranded DNA as in regulatory sequences, structural genes, or mutagenesis cassettes. Essentially, there are two approaches. One orders the oligonucleotides like two rows of bricks in a wall; no gaps remain, and the nicks are sealed with joining enzymes (compare a in Fig. 3). The other approach uses incompletely overlapping oligonucleotides (usually very long ones) and applies polymerases to complete the double strands. Each of the approaches possesses inherent advantages and problems. Because of the power of the segmental support approach for the synthesis of many relatively short oligonucleotides we take special interest in the complete-overlap route, which derives from the pioneering work of Khorana and co-workers.[12] The synthesis of shorter oligonucleotides is certainly less risky. These oligonucleotides are more easily purified (which, in our experience, helps to lower the rate of unwanted mutations after cloning), and one can generate more flexible systems for cassette mutagenesis, since short oligonucleotides are easy to replace by their mutagenic counterparts. However, the purification of many short oligonucleotides usually takes much longer as compared to fewer but longer ones. This general drawback of short oligonucleotides for the construction of DNA double strands is virtually compensated for by procedures which allow simultaneous purification involving gels, thin-layer plates, or disposable cartridges. Even automated workup systems are under discussion.

A second disadvantage used to be that one had to work through cascades of joining and subjoining reactions before a fairly long DNA double strand had finally been made with short oligonucleotides. There are two main reasons why double-stranded DNA synthesis is now much easier than before. The ligation reactions are carried out at higher temperatures (up to 40°) to increase the selection against unwanted hybridization. Beside proper reaction conditions, this selection is only possible if the oligonucleotides, which are designed to form the double strand, have been carefully defined. We found it necessary to do this with the aid of a computer program. SYNGEN, an interactive subprogram of GENMON,

[12] H. G. Khorana, K. L. Agarwal, H. Büchi, M. H. Caruthers, N. K. Gupta, K. Kleppe, A. Kumar, E. Ohtsuka, U. L. RajBhandary, J. H. van de Sande, V. Sgaramella, T. Terao, H. Weber, and T. Yamada, *J. Mol. Biol.* **72,** 209 (1972).

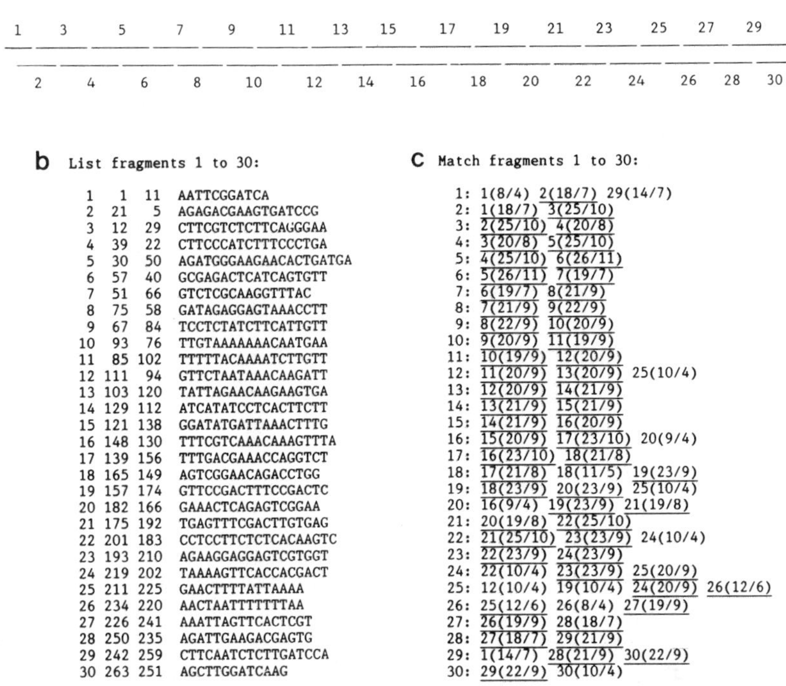

FIG. 3. Definition of oligonucleotides from a DNA strand by SYNGEN. In (a) the double strand is shown schematically by lining up the oligonucleotides. In (b) the oligonucleotides are listed with their numbers (first row). The second and third row indicate their 5′ and 3′ end positions in the double strand. In (c) the fragments are checked for their overlaps. Underlined numbers represent the desired matches. The first numbers in parentheses indicate the relative strengths of the overlaps, the second numbers give the lengths of the overlaps.

allows the operator to define oligonucleotides in a DNA double strand and move each of them along itself and all others, thus detecting any competing hybridizations above a certain threshold (Fig. 3). The oligonucleotide positions can be altered, interesting overlaps can be shown, and useful information can be stored on disk. We have constructed DNAs longer than 300 bp from more than 40 oligonucleotides in a one-step, one-pot reaction applying high temperatures for the ligation of fragments which had been designed by SYNGEN and produced with segmental supports.

More recently the ability of SYNGEN to accept degenerate bases enabled us to carry out reliable oligonucleotide-directed shotgun mutagenesis via construction of mixed double-stranded DNA ("mixed cassette mutagenesis," "shotgun cassette mutagenesis"). For example, we have been able to construct a pool of nearly 8000 predefined gene variants from a preparation of 90 mixed oligonucleotides in only a few ligation steps. The oligonucleotides were only 13 bases long on average, and the mixtures were designed to contain up to four components. Such a design allows us to isolate single oligonucleotides from the mixtures by chromatography or gel electrophoresis. This offers the possibility of constructing a single variant of interest, thus avoiding the search for a needle in a haystack.

We can state, in summary, that over the past years the methods detailed on the following pages proved economic and successful in many molecular biology and protein engineering projects.

Materials and General Methods[13]

Chemical Section

Equipment, Reagents, and Solvents

All parts used to build the manual synthesizer are from Omnifit. Chemically inert syringes and replaceable stainless steel needles are from Hamilton. Rubber septa are from Janssen. Ten-milliliter polyethylene tubes and 4-ml polystyrene tubes are from Greiner. Speed Vac Concentrators are from Savant. Disposable octadecyl columns (1 ml) and disposable filtration columns with 20-μm frits are from Baker, and Teflon solvent filters from LKB. Anion-exchange HPLC columns (SAX-10) are from Whatman. Si-60 F_{254} TLC plates (0.2 mm) are from E. Merck.

All reagents are of the highest available purity and are obtained from various commercial sources; solvents are of p.a. grade. Deoxynucleosides are from Pharma Waldhof. Triple-distilled pyridine and dimethoxytrityl chloride (DMTrCl) are from Cruachem. *o*-Chlorophenyl phosphorodichloridate and mesitylenesulfonyl-3-nitro-1,2,4-triazolide (MSNT) are from Biosyntech. The quality of MSNT is checked by its melting point (130–132°[14]) prior to use, because it decomposes slowly even if stored at −20°. DEAE–Sephadex A25 is from Pharmacia. PTFE

[13] Symbols for nucleotides and protecting groups are used according to the IUPAC–IUB Recommendations [*J. Biol. Chem.* **245,** 5171 (1970)]. Other abbreviations include the following: TEAB, triethylammonium bicarbonate (pH 7.5); DCA, dichloroacetic acid; PTFE, polytetrafluoroethylene; TLC, thin-layer chromatography; HPLC, high-performance liquid chromatography; RP C_{18}, reverse-phase silica gel with octadecyl coating; T4 PNK, T4 polynucleotide kinase; DTE, dithioerythritol; EDTA, ethylenediaminetetraacetic acid; XC, xylene cyanol; BPB, bromophenol blue; bp, base pair.

[14] S. S. Jones, B. Rayner, C. B. Reese, A. Ubasawa, and M. Ubasawa, *Tetrahedron* **36,** 3075 (1980).

powder (Fluon L 169) is from ICI. 1,2,4-Triazole is from Vega. For the preparation of TEAB and TEAP buffers, triethylamine is distilled first from toluenesulfonyl chloride and second from KOH. Dioxane is distilled from KOH and stored under nitrogen in a brown glass flask.

Preparation of Building Blocks

Protected Deoxynucleosides. 5′-*O*-Dimethoxytrityl-*N*⁶-(benzoyl)-2′-deoxyadenosine [(DMTr)bzA$_d$] and 5′-dimethoxytrityl-*N*⁴-(anisoyl)-2′-deoxycytidine [(DMTr)anC$_d$] are prepared according to Ti *et al.*[15] 5′-*O*-Dimethoxytrityl-*N*⁶-(*N*′,*N*′-di-*n*-butylformamidine)-2′-deoxyadenosine [(DMTr)fdA$_d$] is prepared according to Froehler and Matteucci.[16] 5′-*O*-Dimethoxytrityl-*N*²-(propionyl)-*O*⁶-(diphenylcarbamoyl)-2′-deoxyguanosine [(DMTr)pro,dpcG$_d$] is prepared according to Kamimura *et al.*[6]

Preparation of 5′-*O*-dimethoxytrityl-*N*³-(anisoyl)thymidine [(DMTr)-anT]: 50 mmol (12.1 g) thymidine is suspended in 50 ml dry pyridine; 55 mmol (16.6 g) dimethoxytrityl chloride is added, and the solution is stirred at room temperature until completion of the reaction. Two hundred milliliters pyridine and 250 mmol (31.5 ml) trimethylchlorosilane are added, and the solution is stirred for 30 min (a thick white precipitate forms). Then 250 mmol (33.8 ml) anisoyl chloride and 250 mmol (44 ml) diisopropylethylamine are added while stirring. The color of the solution changes to dark red. When the reaction is complete (about 2 hr), excess reagent is hydrolyzed with 50 ml water at 0°, and kept at room temperature for 30 min. The solution is concentrated to 300 ml and added dropwise to 4 liters of stirred petroleum ether (60–80). The precipitate is filtered off and dried under reduced pressure. The crude material (60 g) is suspended in 500 ml of $CHCl_3$ and filtered to remove residual anisic acid.

All compounds are purified by short column chromatography[17,18] on silica gel 60 (E. Merck) and are eluted with a stepwise gradient of methanol in chloroform.

Phosphate Components. Triethylammonium salts of the 3′-*O*-(*o*-chlorophenyl) phosphodiesters of the protected nucleosides are prepared by phosphorylation with *o*-chlorophenyl phosphoroditriazolide following the procedure of Duckworth *et al.*[19] with modifications as described by

[15] G. S. Ti, B. L. Gaffney, and R. A. Jones, *J. Am. Chem. Soc.* **104,** 1316 (1982).

[16] B. C. Froehler and M. D. Matteucci, *Nucleic. Acids Res.* **11,** 8031 (1983).

[17] B. J. Hunt and W. Rigby, *Chem. Ind.* (*London*) **1967,** 1868 (1967).

[18] B. S. Sproat and M. J. Gait, *in* "Oligonucleotide Synthesis: A Practical Approach" (M. J. Gait, ed.), p. 199. IRL Press, Oxford, England, 1984.

[19] M. L. Duckworth, M. J. Gait, P. Goelet, G. F. Hong, M. Singh, and R. C. Titmas, *Nucleic Acids Res.* **9,** 1691 (1981).

Frank *et al.*[1] If necessary (TLC analysis) the products are purified by short column chromatography on silica gel 60, with a stepwise gradient of methanol in chloroform/0.5% triethylamine. The following phosphodiester building blocks are used for the assembly of oligonucleotides: (DMTr)bzA_dp(oCP)$^-$TEAH$^+$ (MW 949.4), (DMTr)anC_dp(oCP)$^-$TEAH$^+$ (MW 955.4), (DMTr)pro,dpcG_d(oCP)$^-$TEAH$^+$ (MW 1112.7), (DMTr)-anTp(oCP)$^-$TEAH$^+$ (MW 970.4).

Nucleoside Succinates. 3′-*O*-Succinates of the protected nucleosides are prepared according to Chow *et al.*[20] The products are purified by short column chromatography on silica gel 60 with a stepwise gradient of methanol in chloroform/0.5% pyridine. The following succinates are used: (DMTr)fdA_d(Suc) (MW 809), (DMTr)anC_d(Suc) (MW 764), (DMTr)pro,dpcG_d(Suc) (MW 1213), (DMTr)anT(Suc) (MW 778).

Recycling of Excess Phosphate Components

Throughout the synthesis, excesses of coupling mixtures are collected separately in four bottles and are immediately quenched with water as described below. For recycling, most of the pyridine and water is removed in a rotary evaporator connected to an oil pump (do not try to concentrate to dryness!). The residue is partitioned between *n*-butanol/diethylether (2 : 1) and 0.2 *M* TEAB. After reextraction with 0.2 *M* TEAB the organic layer is dried with anhydrous sodium sulfate and again concentrated in a rotary evaporator; the remaining oil is diluted with dichloromethane, and the nucleotide material is precipitated in petroleum ether (60–80)/ether (9 : 1). The crude product is purified by column chromatography as already described for the phosphate components. Approximately 60–80% of the material used for the synthesis is usually recovered. ^{31}P and ^{1}H NMR show no differences from freshly synthesized material.

Enzymatic Section

Materials

Buffer salts and other chemicals are of p.a. quality and obtained from commercial suppliers. [γ-^{32}P]ATP (3000 Ci/mmol) is purchased from Amersham and diluted to lower specific radioactivity with ATP as required. Norit A is from Serva Feinbiochemica and is suspended in water to give 20% of settled bed volume. GF/C glass fiber filters (2.5 cm diame-

[20] F. Chow, T. Kempe, and G. Palm, *Nucleic Acids Res.* **9,** 2807 (1981).

ter) are from Whatman. Polynucleotide kinase (T4 PNK), T4 DNA ligase, and restriction endonucleases are from Bethesda Research Laboratories, and bacterial alkaline phosphatase (BAP C, suspended in ammonium sulfate) is from Worthington. Inorganic pyrophosphatase (yeast, lyophilized powder, from Sigma) is dissolved in water to 1 mg/ml and stored at $-20°$. Polyacrylamide slab gel electrophoresis is performed in a LKB Macrophore system. Bovine pancreas ribonuclease is from Boehringer; pGV sequencing vectors are from Amersham.

Norit Assay

For the yield determination of kinase and ligase reactions, the relative distribution of the ^{32}P label in a mixture of [γ-^{32}P]ATP, oligonucleotide [5'-^{32}P]phosphomonoesters, and [^{32}P]phosphodiesters is rapidly determined by selective degradation to inorganic [^{32}P]phosphate of first ATP by inorganic pyrophosphatase and second phosphomonoesters by bacterial alkaline phosphatase. Organically bound label is selectively adsorbed to activated charcoal (Norit A).[21,22]

All buffers and solutions are kept in ice if not otherwise indicated. A small aliquot (0.1–1 μl) of the reaction mixture (kinase or ligase reaction) is diluted with 100 or 150 μl of 50 m*M* Tris–HCl (pH 6.5)/2 m*M* $ZnCl_2$. Aliquots (45 μl) of this mixture are transferred to 4-ml polystyrene tubes (A, B, and C). Tube C is omitted if only kinase reactions are analyzed. Five microliters of inorganic pyrophosphatase solution is added to tubes B and C, and the mixtures are incubated at 40° for 15 min. Tube C is further incubated at 60° for 30 min after addition of 40 μl 0.2 *M* Tris–HCl (pH 9) and 1 μl of BAP C suspension. All tubes are then further treated separately in the same way. Two-tenths milliliter of a buffer containing 0.15 *M* ammonium sulfate, 25 m*M* potassium phosphate (pH 7), 2 m*M* sodium pyrophosphate, and 5 mg/ml BSA is added, followed by 0.2 ml of Norit A suspension and careful mixing. One milliliter 0.1 *N* HCl is added, and the solution is filtered through a GF/C glass fiber filter (soaked in 0.1 *N* HCl) which is placed on an equally sized glass sinter connected to a vacuum line. The filters are washed 3 times with 1 ml 0.1 *N* HCl, aspirated dry, and placed into a scintillation vial containing 5 ml toluene-based scintillation cocktail. Radioactivity is measured in a liquid scintillation counter. The filter from tube A carries the total organic bound label, the filter from tube B carries the total oligonucleotide bound label, and the filter from tube C carries the phosphodiester bound label. The radioactiv-

[21] C. L. Harvey, R. Wright, A. F. Cook, D. T. Maichuk, and A. L. Nussbaum, *Biochemistry* **12,** 208 (1973).

[22] H. Blöcker, R. Frank, and H. Köster, *Liebigs Ann. Chem.* **1978,** 991 (1978).

ity on the filter from tube B multiplied by the molar excess of ATP over oligonucleotide, divided by the radioactivity on the filter from tube A, gives the phosphorylation efficiency. The radioactivity on the filter from tube C divided by the radioactivity on the filter from tube B gives the ligation efficiency.

Oligonucleotide Synthesis on Cellulose Disks

Chemical Synthesis

Principle

The first step of the synthesis is the loading, that is, the coupling of a nucleoside via a succinyl linkage to cellulose. This nucleoside represents the 3′ end of the growing nucleotide chain. The chain elongation, which proceeds from 3′ to 5′, is carried out by repeating the following steps:

1. Detritylation: the acidic removal of the dimethoxytrityl group from the 5′-hydroxyl group
2. Condensation: coupling of the appropriate nucleotide building block to the growing chain
3. Capping: acylation of unreacted 5′-hydroxyl groups

All of these reactions are carried out in a continuous flow system with the filter disks stacked in column reactors.[23] We use a flexible manual synthesizer. Up to seven columns can be connected in parallel if mixed or modified nucleotides are also used in addition to the usual building blocks. At the end of each cycle the columns are taken apart and the filter disks are sorted for the next cycle. A pure oligonucleotide product is obtained at the end of the synthesis after several deprotection and purification steps.

The overall cycle time (including drying and sorting of the filter disks) is approximately 2.5 hr. If the capping reaction is omitted the cycle time can be reduced to less than 2 hr; however, a separation of 5′-hydroxyl and 5′-tritylated oligonucleotides with the aid of a reverse-phase cartridge (RP C_{18}) after the alkaline deprotection is no longer feasible because all truncated oligonucleotides also carry a trityl group at their 5′ ends.

The final amount of oligonucleotide product obtainable from a synthesis depends on the filter size used and on the loading of the first nucleo-

[23] H. W. D. Matthes, W. Zenke, T. Grundström, A. Staub, M. Wintzerith, and P. Chambon, *EMBO J.* **3,** 801 (1984).

side. Low loading (up to 25 μmol/g cellulose) is preferable because when too much nucleoside is loaded the flow of the solvent through the filter stack is impaired and problems arise because of incomplete washing steps.

The appropriate building blocks, commercially available for all four nucleotides, are base protected at the exocyclic amino functions of A, C, and G residues. Nevertheless, during the repetitive condensation reactions, base modifications can occur which are partly reversed in the normally applied oximate deprotection step. Modifications can be minimized if additional protecting groups at guanosine and thymidine are introduced. As a consequence, the oximate treatment may be omitted. Another side reaction in the elongation cycle is the acid-promoted depurination which preferably occurs at 3′-terminal A residues. A formamidine protected A nucleoside at this position has a stabilizing effect. The G nucleoside is protected with the propionyl group at N^2 and the diphenylcarbamoyl group at O^6 and T with the anisoyl group at N^3; 3′-terminal A residues are protected with the *N,N*-di-*n*-butylformamidino group at N^6.

Preliminary Details

Figure 4 shows a flow diagram of the multicolumn device. It is based on an Omnifit DNA bench synthesizer[24] and is supplemented by an 8-way valve (column selector) to connect up to seven columns, four 3-way valves at the outlets of the column reactors which allow control of the flow rates of the individual columns and which separate the excesses of the different phosphate components, and waste bottles to collect the excesses of phosphate components. Teflon sinters in the commercial columns are replaced by homemade perforated Teflon plates. The columns are marked with tape of different colors, one for each of the bases. The system is operated at 1300 torr (pressure limit for solvent bottles) to avoid formation of gas bubbles. Dichloromethane and pyridine are kept over molecular sieves (4 Å). The outlets of the solvent bottles are equipped with Teflon solvent filters (LKB). Flow rates are adjusted at the outlets of column reactors by the 3-way valves and should be 1.5–2 void volumes/min (see Table I). Sufficient washing of the support is required (pass at least 7 void volumes of solvent through each column). Backpressure of the columns depends markedly on the number of disks as well as on the overall charge with nucleotidic material (by initial loading and by chain elongations). If flow rates drop below 1.5 void volumes/min, the washing steps have to be prolonged proportionally.

[24] B. S. Sproat and M. J. Gait, *in* "Oligonucleotide Synthesis: A Practical Approach" (M. J. Gait, ed.), p. 101. IRL Press, Oxford, England, 1984.

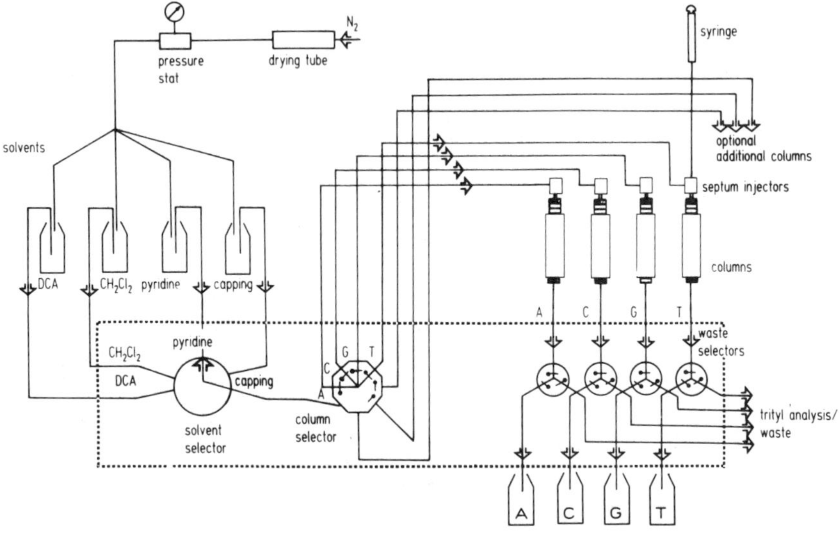

FIG. 4. Schematic flow diagram of a manual synthesizer.

To achieve high coupling yields, moisture has to be excluded from the reaction mixture. All glassware is dried in an oven prior to use (100°). The building blocks (phosphate components, nucleoside succinates) are freeze-dried from dioxane. This very effective drying step is considered

TABLE I
DEPENDENCE OF THE VOID VOLUME OF COLUMN REACTORS ON COLUMN SIZE AND FILTER STACK

Number of disks	Void volumes (ml)		
	6.6-mm column	10-mm column	15-mm column
1–10	0.20	0.36	0.85
10–15	0.23	0.48	1.10
15–20	0.27	0.60	1.30
20–25	0.30	0.70	1.55
25–30	0.35	0.80	1.80
30–35	0.40	0.90	2.05
35–40	0.45	1.00	2.25
40–45	0.50	1.10	2.50
45–50	0.55	1.20	2.75
100	1.00	2.25	5.00

much more convenient and safer than pyridine coevaporations. The nucleotidic material is dissolved in dioxane (about 1 ml per 50 mg), the solution cooled in liquid nitrogen, and freeze-dried overnight with the flasks kept in an ice bath. Aliquots of MSNT are dried in a desiccator overnight. Flasks containing the dried chemicals are tightly stoppered with rubber septa. Solvents and solutions are transferred with the help of gas-tight syringes, which are thoroughly washed with the respective dry solvent immediately before use. The color code of the column reactors is applied throughout all manipulations.

Coupling of Nucleoside Succinates to the Filter Disks (Loading)

Loaded filter disks are prepared in large amounts (up to 100) for the four standard nucleosides and can be stored at −20° for long periods (over 1 year). Loading of the filter disks is carried out in the manual synthesizer. Disks are cut out individually from Whatman 3MM paper with a punch. They must be a little wider than the inner diameter of the column reactors (6.8, 10.4, and 15.8 mm for 6.6, 10.0, and 15.0 mm columns, respectively) and, thus, will fit in tightly (important!). The disks are pressed carefully, one on top of the other, into the column reactors. Thereby they adopt a smooth U-shaped form which they will maintain throughout the synthesis. The bottom plunger of the columns is then adjusted to leave a small distance (~1 mm) between the upper plunger and the top filter. A fifth column is charged with disks in the same way and used to prepare unloaded capped filters (blank filters needed during elongation cycles).

A, C, G, T, and capping columns are flushed with pyridine until all air bubbles between the filters have disappeared (takes up to 20 min). However, it is possible to do this faster by injection of pyridine with a syringe through the septum injector (valves of the column selector have to be closed before injection!). The freeze-dried nucleoside succinates (0.25 mmol/g cellulose) are each dissolved in 1 void volume of pyridine and transferred to flasks each containing 1.75 mmol MSNT/g cellulose. Three millimoles *N*-methylimidazole per gram of cellulose is added to each flask, and the solutions are diluted with pyridine to 2 void volumes. One-half of these mixtures is injected within 10 min into the respective A, C, G, and T columns and left to react for 1 hr. The second half is then injected slowly (over about 10 min). After an additional hour columns are flushed with pyridine. (If higher loadings than usual are desired, up to 80 μmol/g cellulose can be achieved by repeating the procedure.) Then 2 void volumes of a 2:8:1 (v/v/w) mixture of acetic anhydride/pyridine/dimethylaminopyridine are injected slowly into each of the five columns. After 10 min the columns are flushed with pyridine and dichloromethane.

The filters are pressed from top to bottom out of the dissassembled columns into separate marked beakers, washed twice with ether and dried under reduced pressure. Two filter disks are taken from each of the batches and detritylated separately, each with 5 ml of 3% DCA in dichloromethane. After 20 min the absorbance is determined and the loading calculated (1 A_{500} = 0.0105 μmol). The usual yield is 25 μmol/g cellulose (1 μmol/15-mm disk; 0.2 μmol/6.6-mm disk).

Oligonucleotide Chain Assembly

All necessary data to prepare disks and reagents are displayed on the output generated by SYNPRO (Fig. 2). Loaded A, C, G, and T filters are assigned to the different oligonucleotide sequences by numbering them unambiguously (!) with a pencil according to the numbers on the A, C, G, and T lines under cycle 0 (loading cycle; see d in Fig. 2). Based on an average loading of 1 μmol/15-mm disk or 0.2 μmol/6.6-mm disk the amounts of phosphate components (10-fold excesses), MSNT (40-fold excess), and *N*-methylimidazole (50-fold excess) are calculated for each coupling reaction according to the number of disks (see parentheses at end of lines in Fig. 2, part d). Phosphate components are freeze-dried, and MSNT aliquots are dried in a desiccator.

For the chain elongations (see cycle 1 and the following cycles) the disks are sorted into the columns. Two capped blank filters are placed at the bottom and one on top of the nucleotide filters. The bottom plungers are adjusted to leave a small distance (about 1 mm) between the upper plungers and the top filters. Columns are flushed with dichloromethane until all air bubbles between the filters have disappeared. DCA, 3% in dichloromethane, is passed through the columns (double flow rate!) until the red color begins to disappear (but never longer than 3 min!). If necessary, DCA solution is injected with a syringe. The trityl eluate can be collected to monitor coupling yields.[25] After washing with dichloromethane and pyridine the columns are ready for injection of the coupling mixtures. The phosphate components are each dissolved in 1 void volume of pyridine and transferred to the respective MSNT flasks. *N*-Methylimidazole is added and the volume adjusted to 2 void volumes with pyridine. After closing the valves of the column selector and switching the waste selectors to "nucleotide excesses," the first halves of the mixtures are slowly injected (over 5 min) into the columns and left for 10 min. Then the second half is injected again within 5 min and left for an additional 10 min.

[25] Note that trityl yields may exceed 100% in the first elongation cycles if capping reactions are included.

Excesses of phosphate components are washed out with pyridine and collected in separate bottles containing 100 ml 0.2 *M* TEAB. Then the waste selectors are switched back to "general waste" and 20% acetic anhydride in pyridine is passed through the columns. After standing for 15 min the columns are washed subsequently with pyridine and dichloromethane and disassembled. The disks are pressed from top to bottom into a beaker, washed twice with ether and dried under reduced pressure for 15 min. Filter disks represented by underlined numbers in the SYNPRO output (see d in Fig. 2) are ready for workup after the respective cycle. In the last cycle capping may be omitted.

Deprotection, Purification, and Characterization

Principle

Various techniques are available for the workup of oligonucleotides. However, the production of large numbers of different oligonucleotides requires an efficient workup routine in order not to lose the particular advantage of the synthesis method. Time-consuming, tedious steps should be avoided, but parallel handling of many samples should be possible. These requirements strongly favor the one-step alkaline deprotection by the volatile reagent ammonia/pyridine.[26] For the same reasons we have adapted most of the necessary chromatographic purification steps to flash chromatography on small disposable, prepacked columns (cartridges). The separation media are as follows:

RP C_{18} silica gel (hydrophobic interactions) for prepurifications and for desaltings. In the current practice these materials give an eluate which is almost salt free, but small amounts of the solid phase may eventually be co-eluted. After concentration of very dilute oligonucleotide solutions these impurities may interfere, for example, with subsequent enzymatic reactions.

Sephadex A25 (ionic interactions) for desaltings. This material has a very high capacity, and the eluates contain high concentrations of salt which, however, can be removed later if volatile buffers are used.

PTFE powder for desaltings and for separation from radioactive ATP.[27] Like RP C_{18} this material binds oligonucleotides via hydrophobic interactions.[28] No salt is needed for elution, and a very clean product is obtained, but the capacity is rather low. Moreover, pressure has to be

[26] H. Blöcker, W. Heikens, and R. Frank, manuscript in preparation.

[27] K. Schwellnus, G. Kurth, and R. Frank, manuscript in preparation.

[28] S. Hjertén and U. Hellman, *H. Chromatrogr.* **202,** 391 (1980).

applied to pass any solvent through the PTFE powder, whereas cartridges filled with the other solid phases can easily be used in special devices (for example from Baker), which allow process of 10 samples at a time.

The workup path is selected according to the amount of material and the degree of purity that is required for different applications. If large amounts of only a few oligonucleotides are needed at highest purity, we suggest a two-step purification. First, tritylated oligonucleotides obtained after alkaline deprotection are separated from nontritylated material (truncated chains, protecting groups) over disposable cartridges. After detritylation the products are purified by anion-exchange HPLC. (Note that this two-step procedure is only sensible if capping reactions are included in the elongation cycles.) If, however, only small amounts of many oligonucleotides are needed (e.g., for the construction of DNA double strands) detritylation is carried out directly after alkaline deprotection and oligonucleotides are purified in only one step by either gel electrophoresis (see Fig. 5) or thin layer chromatography (TLC). The last methods allow simultaneous processing of many samples.

Routinely, we check the isolated oligonucleotides for homogeneity and identity by kinasing with [γ-^{32}P]ATP and analytical polyacrylamide gel electrophoresis (Fig. 6). Mobilities of product bands relative to coelectrophoresed chain length standards (A or T ladder) are compared with calculated values given in the SYNPRO output form (b in Fig. 2). This sizing takes into account chain length as well as base composition of oligonucleotides containing the four normal bases.[29] In case doubts arise from the sizing experiment, or if modified bases have been built in, identity of the product may be confirmed via sequence analysis by (1) fast atom bombardment–mass spectrometry (FAB–MS),[30] (2) two-dimensional fingerprint (''wandering spot'' analysis),[31] or (3) chemical degradation sequencing (Maxam/Gilbert). The first method is extremely fast and involves no enzymes, special chemicals, or radioactive materials. However, the method requires rather expensive equipment, consumes quite large amounts of product (up to 1 OD_{260}), and works to date only up to 13-mers. The second method gives detailed information about sequence and purity, and also about the nature of side products, for oligonucleotides up to about 25 nucleotides in length. The third method is the method of choice if large numbers of oligonucleotides or relatively long ones are to be analyzed. This applies especially when following the approach of Rosenthal *et al.*[32] Both methods 2 and 3 require small amounts of labeled

[29] R. Frank, and H. Köster, *Nucleic Acids Res.* **6,** 2096 (1979).

[30] L. Grotjahn, R. Frank, and H. Blöcker, *Nucleic Acids Res.* **10,** 4671 (1982).

[31] C.-P. D. Tu and R. Wu, this series, Vol. 65, p. 620.

[32] A. Rosenthal, R. Jung, and H.-D. Hunger, this series, Vol. 155 [20].

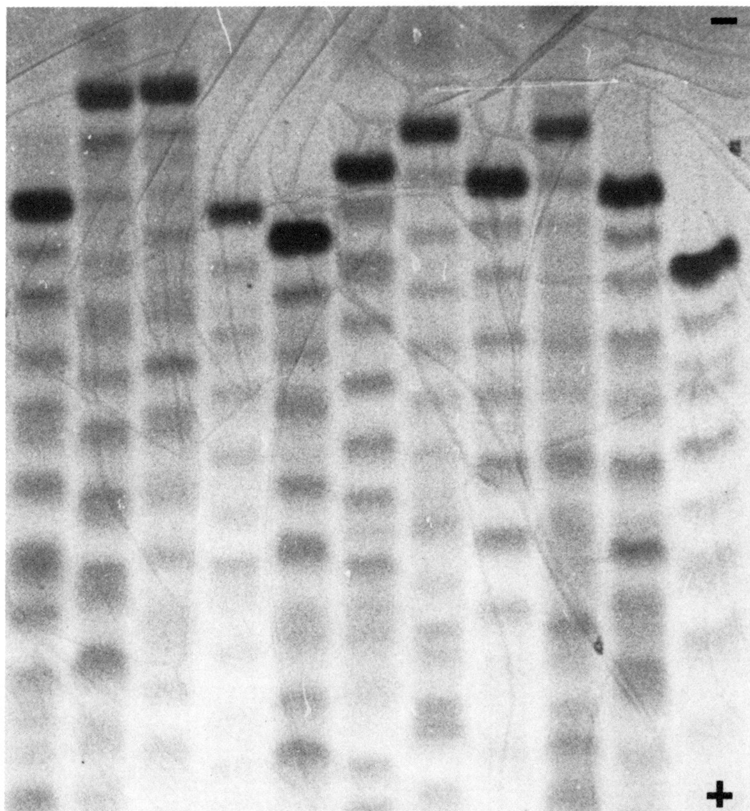

FIG. 5. Purification of oligonucleotides (17–21 nucleotides long) by preparative gel electrophoresis (conditions as described in the text).

material which can be separated conveniently from excess [γ-^{32}P]ATP, salts, and protein of the kinase reaction mixture by adsorption to a hydrophobic material such as PTFE. Thus, the labeled product is recovered salt free in yields greater than 90%.

Procedures

Alkaline Deprotection. Each filter is placed in a 10-ml plastic tube, and a mixture of aqueous ammonia (33%)/pyridine (9 : 1, v/v) is added (1 ml/6.6-mm disk, 4 ml/15-mm disk). Tubes are stoppered, sealed tightly with tape, and kept at room temperature overnight. After another 10–16 hr at 50°, the tubes are cooled on ice and opened carefully. Ammonia is removed by evaporation in a Speed Vac Concentrator connected to a water pump. The remaining liquid is transferred to another 10-ml tube. After

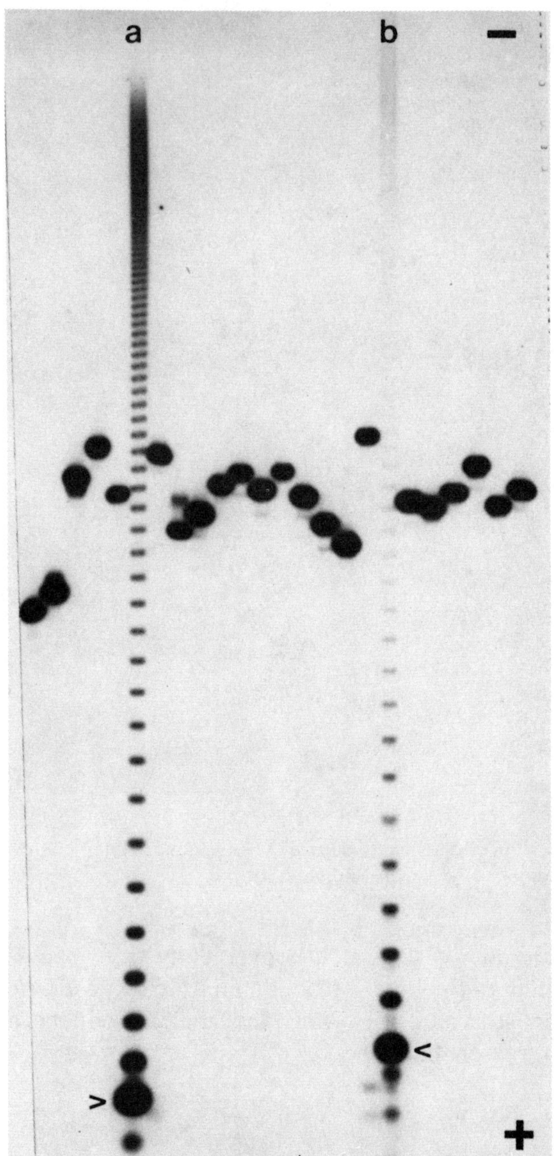

FIG. 6. Analytical gel electrophoresis (sizing) of 22 different oligonucleotides purified by preparative gel electrophoresis (for conditions see text). Lane a, dA homooligomers; lane b, T homooligomers (arrows indicate trimers).

washing the paper disk with 50 m*M* TEAB (about 3 ml for a 15-mm disk), both solutions are combined and evaporated to dryness.

RP C_{18} Cartridge. Each of the dry residues from alkaline deprotection is dissolved in 3 ml of 10% acetonitrile in 0.2 *M* TEAB and loaded on a disposable RP C_{18} cartridge equilibrated first with methanol and then with 10% acetonitrile in 0.2 *M* TEAB (2 ml each). After adsorption of the oligonucleotide, the cartridge is washed with 2 ml of the same buffer. The product is eluted with 2 ml of 80% methanol in 0.2 *M* TEAB. The eluate is evaporated to dryness, dissolved in 2 ml of water, and reevaporated.

Detritylation. The above dry residue is taken up in 80% acetic acid (0.5 ml for 6.6-mm filters, 2 ml for 15-mm filters). After 15 min at room temperature the solution is evaporated in a Speed Vac Concentrator. The residue is dissolved in 0.5 to 1 ml of water, extracted 3 times with an equal volume of ether, and reevaporated.

Anion-Exchange HPLC. After detritylation the residue is dissolved in 0.1–1 ml of 50 m*M* TEAB. One-quarter of the product from a 15-mm filter, or all of it from a 6.6-mm filter, may be separated in a single run on an analytical Whatmann SAX-10 column (4.6 × 250 mm). Conditions used for HPLC are identical to those described by Gait *et al.*[33] except for the cation of the buffer. For practical reasons (solubility) we prefer triethylammonium to potassium phosphate.[34] Pooled peak fractions are diluted with an equal volume of water and desalted on RP C_{18} cartridges as described above (omit acetonitrile). The fraction containing the product is evaporated to dryness, dissolved in 50 m*M* TEAB, and further purified on Sephadex A25.

Sephadex A25 Cartridge. One-tenth milliliter of pre-swollen Sephadex A25 is placed between the two polyethylene sinters of a 1-ml disposable filtration column. The column is washed with 1 ml 2 *M* TEAB and 2 ml 50 m*M* TEAB. After adsorption of the oligonucleotide, the cartridge is washed with 2 ml 50 m*M* TEAB. The oligonucleotide is eluted with 2 ml 2 *M* TEAB and the eluate evaporated to dryness. The residue is dissolved in 1 ml of water and lyophilized.

Preparative Gel Electrophoresis.[35] For purification of oligonucleotides on polyacrylamide gels the crude, dry products obtained after alkaline deprotection are directly detritylated (see above). The residue is then dissolved in 45% formamide, 10 m*M* EDTA, 0.02% XC, 0.02% BPB. The

[33] M. J. Gait, H. W. D. Matthes, M. Singh. B. S. Sproat, and R. C. Titmas, *Nucleic Acids Res.* **10,** 6243 (1982).

[34] J. Engels, personal communication.

[35] H. M. Hsuing, R. Brousseau, R. Michniewicz, and S. A. Narang, *Nucleic Acids Res.* **6,** 1371 (1979).

latter dye is omitted for oligonucleotides shorter than 14 bases. Depending on the amount of product obtained from the synthesis, up to 25% of the product from a 6.6-mm filter or 10% of it from a 15-mm filter can be applied on a 0.1 × 20 × 40 cm 20% acrylamide gel [acrylamide : methylene–bisacrylamide = 29 : 1, 7 *M* urea, 50 m*M* Tris–borate (pH 8.3)/1 m*M* EDTA, 7-mm-wide slots, 10 μl per slot]. Thus, 15 samples may be purified in parallel in a single gel run using the LKB Macrophore apparatus. The gel is preelectrophoresed for 1 hr at 800 V. After loading, the gel is run at 1000–1200 V and 55° until the blue dye has reached the bottom of the gel (5–7 hr). The lower half of the gel is placed onto a 20 × 20 cm TLC plate covered with SaranWrap. The product bands are cut out under UV light. Gel slices are crushed with a spatula in a 1.5-ml plastic tube and then soaked 3 times with 300 μl water for 20 min at 50° and centrifuged. The combined supernatants are subjected to purification over a PTFE cartridge (see below).

Analytical Gel Electrophoresis (Sizing). To check the purity, length, and base composition, about 2–10 pmol of the product oligonucleotide are labeled with 20 pmol [γ-^{32}P]ATP (about 50 Ci/mmol) and 0.2 U T4 PNK in a volume of 10 μl as described under ligation. To one-tenth of the kinasing reaction an equal volume of 90% formamide containing 0.05% XC, 0.05% BPB, 20 m*M* EDTA is added, and the mixture denatured by heating to 100° for 2 min. Co-electrophoresis with length standards is carried out on a 0.04 × 20 × 40 cm denaturing 20% polyacrylamide gel[29] at 1500 V and 50° until the BPB dye marker has moved 25 cm.

Thin-Layer Chromatography. Short oligonucleotides (up to 20-mers) can be obtained in sufficient purity on TLC as described by Wu *et al.*[36] After alkaline deprotection the crude product is detritylated without any intermediate prepurification. Up to 25% of the crude product from a 6.6-mm filter or up to 10% from a 15-mm filter are taken up in 10 μl ethanol/water (1 : 1) and applied as a narrow, 1.5-cm-wide band on a 20 × 20 cm TLC plate. The plates are developed and the oligonucleotides eluted as described. The products may be used either directly or after an additional purification on a PTFE cartridge. Thus, up to 10 different oligonucleotides can be purified simultaneously on a single TLC plate in amounts sufficient for many applications (0.1–2.0 OD_{260}).

PTFE Cartridge. PTFE powder (200 mg per 1 OD of oligonucleotide to be adsorbed) is placed between the two sinter plates of a disposable filtration column, first washed with ethanol and then with 2 ml 50 m*M*

[36] R. Wu, N.-H. Wu, Z. Hanna, F. Georges, and S. Narang, *in* "Oligonucleotide Synthesis: A Practical Approach" (M. J. Gait, ed.), p. 135. IRL Press, Oxford, England, 1984.

TEAB. After loading with the oligonucleotide solution, the cartridge is washed with 2 ml 50 m*M* TEAB, and the oligonucleotide is eluted with 2 ml of 50% ethanol in 50 m*M* TEAB. Oligonucleotides that have to be combined anyway for further application (e.g., for ligation reactions), can also be first combined in the required stoichiometric amounts and then purified together over PTFE. For separation of [γ-^{32}P]ATP from labeled oligonucleotides, the disposable cartridge (150 mg PTFE) is set up and equilibrated as described above. The kinasing mixture is diluted with 200 μl 50 m*M* TEAB, loaded on the cartridge, and [γ-^{32}P]ATP is washed off with 2 ml of 2% ethanol in 50 m*M* TEAB. The labeled oligonucleotide is eluted with 1 ml of 50% ethanol in 50 m*M* TEAB, evaporated in a Speed Vac Concentrator, dissolved in 500 μl water, and reevaporated.

Construction of DNA Double Strands

Principle

Enzymatic joining of synthetic oligodeoxyribonucleotides that are defined by SYNGEN to build up the two strands of a DNA duplex (Fig. 3) follows a rather simple procedure: phosphorylation of 5′-hydroxyl termini of the synthetic fragments with [γ-^{32}P]ATP catalyzed by T4 polynucleotide kinase and subsequent phosphodiester bond formation between 3′-hydroxyl and 5′-phosphate ends of neighboring fragments catalyzed by T4 DNA ligase. As T4 PNK and T4 DNA ligase have the same cofactor and buffer requirements, no workup of phosphorylation mixtures is necessary. Conveniently, all fragments to be ligated are combined, phosphorylated together, and joined in a one-pot reaction. Although this procedure is straightforward, some precautions have to be taken to avoid failures.

Ligation requires the formation of a double helical complex with the fragments to be joined held together end to end. The nicked double strand as shown in Fig. 3 is physically nonexistent under reaction conditions. Hybridization via short overlaps (5–10 bp) is a dynamic process, and the portion of double strands is dependent on the thermal stability of individual pairings. Temporarily formed duplices are then stabilized by complexation with the enzyme and joined. Careful design of the fragments ensures uniqueness of overlaps within a set of fragments to be joined. Hence, correct pairing between fragments provides the highest thermal stability of duplices. Therefore, ligation should be carried out at the highest possible temperature to avoid unwanted pairings to fragments other than the designed targets including modified sequences and contaminants of the chain length $n \pm 1$. T4 DNA ligase will endure even 40° for a short time.

The enzyme is known to catalyze ligation also at mismatch overlaps,[37,38] double strand breaks (blunt ends),[39] and small gaps.[40] Correct stoichiometry of the fragments combined is therefore crucial. If a fragment is underrepresented numerous side reactions may occur at the remaining ends of the truncated double strands. Even if correct amounts of oligonucleotides were combined, an incomplete 5′-phosphorylation will cause an insufficient amount of active components to be present. Completion of the phosphorylation reaction should be controlled before proceeding with the ligation. A simple and rapid assay is described under General Methods.

The final DNA double strands need suitable ends for recombination with restriction sites of the cloning vectors. Fragments forming these ends are commonly included without a 5′-phosphate to protect them from self-recombination. However, these ends are critical sites with respect to problems mentioned above. We often observed a much cleaner ligation by allowing self-recombination of ends and recutting of multimers with the corresponding restriction endonucleases. The experimental expenditure required for the complete assembly of a long DNA double strand greatly depends on the total number of subsegments that have to be constructed and isolated. To keep this work at a minimum it is desirable to include a maximum number of fragments in the first ligation step in order to increase the size of the subsegments. Considering all of the precautions mentioned above, we found that very long DNA double strands (up to 300 base pairs) can be obtained in high yields from up to 40 fragments in a quick one-step ligation reaction. Purity of the product is often high enough for direct cloning without any further purification. Figures 7 and 8 demonstrate the efficiency of the approach. Based on the improved chemistry and the computer-aided design of fragments the error rate presently observed with this cloned DNA is much lower than the previous rate of 1 in 500.[41]

DNA double strands constructed as described basically represent pieces of DNA built up from small cassettes (i.e., pairs of oligonucleotide fragments). New sequence variants are easily made, in high yields, by exchange, insertion, or deletion of two or more fragments. Most of the fragments can be reused many times. The simple ligation procedure allows simultaneous construction of large numbers of double strands. High

[37] R. Frank, Ph.D. thesis, Hamburg, Federal Republic of Germany, 1979.

[38] V. Sgaramella and H. G. Khorana, *J. Mol. Biol.* **72,** 427 (1972).

[39] V. Sgaramella, J. H. van de Sande, and H. G. Khorana, *Proc. Natl. Acad. Sci. U.S.A.* **67,** 1468 (1970).

[40] S. V. Nilsson and G. Magnusson, *Nucleic Acids Res.* **10,** 1425 (1982).

[41] E. Heron (Applied Biosystems), personal communication; W. H. McClain, K. Foss, K. L. Mittelstadt, and J. Schneider, *Nucleic Acids Res.* **14,** 6770 (1986).

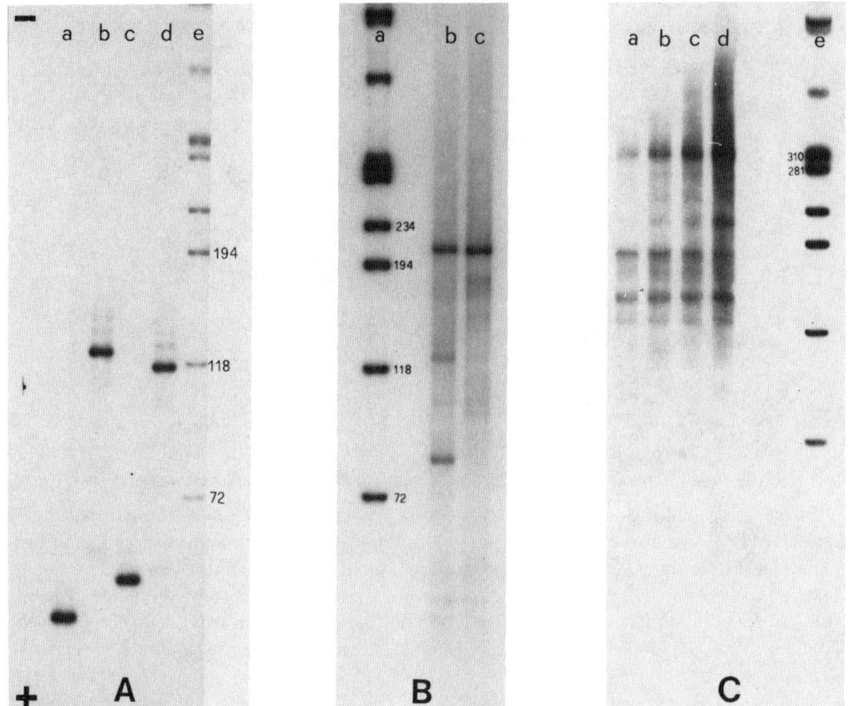

FIG. 7. Examples for one-step ligations of large numbers of short oligonucleotides. (A) Lane a, 53 bp from 7 fragments; lane b, 124 bp from 16 fragments; lane c, 58 bp from 8 fragments; lane d, 116 bp from 16 fragments; lane e, ^{32}P-labeled *Hae*III digest of ϕX174 RF DNA. (B) Lane a, as (A), lane e; lane b, 211 bp from 28 fragments, ligation temperature 30°, then 15°; lane c, as lane b, but ligation protocol as described in Procedure. (C) 288 bp from 38 fragments. Samples were analyzed after each temperature shift: lane a, 40°; lane b, 37°; lane c, 30°; lane d, 15°; lane e, as (A), lane e.

quality ligation products with very low error rate greatly facilitate the search for the correct sequences (sequence analysis of only one or two clones), especially if many DNAs are prepared at the same time.

Procedure

Ligation. Equimolar amounts (typically 10 to 100 pmol) of each oligonucleotide fragment are combined in siliconized plastic tubes (tube A, fragments comprising the upper strand; tube B, fragments comprising the lower strand). Those fragments that should not become phosphorylated (e.g., the ends of the duplex) are combined in a separate tube (tube C). The material is dried or lyophilized. The oligonucleotides of tube A and B

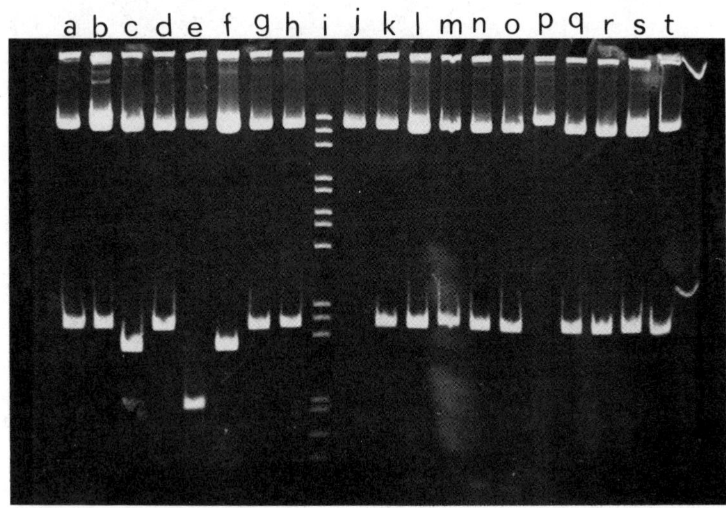

FIG. 8. Electrophoresis on a 5% nondenaturing polyacrylamide gel of *Hin*dIII- and *Sal*I-digested pGV451 vectors containing inserts deriving from the ligation mixture shown in Fig. 7C, lane d. Fourteen of 18 clones tested showed a band in the region of the expected length of 288 bp. One of two DNAs analyzed was found to have the correct sequence. Lane i, *Hae*III-digested PM2 DNA marker (from top to bottom, 1880, 1760, 1410, 860, 845, 672, 615, 525, 333, 295, 272, 167, 152, 120, 95 bp). Lane j, pGV451 digested with *Hin*dIII and *Sal*I.

are phosphorylated at 37° for 30 min with a 3-fold excess of [γ-^{32}P]ATP (about 2 Ci/mmol) over 5′-hydroxyl termini in reaction mixtures of 10–20 μl containing 60 m*M* Tris–HCl (pH 8), 6 m*M* $MgCl_2$, 10 m*M* DTE, and 2–4 units of T4 PNK. Efficiency of phosphorylation is determined with the Norit assay as described under General Methods. After quantitative phosphorylation the contents of tubes A and B are combined. (In case 5′-hydroxyl terminal fragments are used, T4 PNK is inactivated by heating tubes A and B to 95° for 2 min, and the contents are transferred to tube C.) Then 10 m*M* DTE, 0.2 m*M* ATP, and 2 units of T4 DNA ligase are added, and the mixture is incubated at 40° for 1 hr. After cooling to 37° another 2 units of ligase is added, and incubation is continued at 37° for 1 hr, at 30° for 3 hr and at 15° overnight. Small aliquots are removed after each temperature shift and ligation efficiency is determined with the Norit assay. Usually 60 to 80% efficiency is achieved. Then DNA ligase is inactivated by heating to 65° for 15 min. If multimers of the DNA double strand are produced, these are digested with the respective restriction endonucleases after increasing the volume of the reaction mixture by at least 100% and adjustment to the proper buffer conditions.

An aliquot of the mixture is then analyzed by electrophoresis on a

0.04 × 20 × 40 cm, 10% nondenaturing polyacrylamide gel [acrylamide : methylenebisacrylamide = 29 : 1, 50 m*M* Tris–borate (pH 8.3), 1 m*M* EDTA], thermostated to 20°, and run at 1000 V until the BPB has moved 25 cm. Samples for gel electrophoresis are loaded in a mixture containing 0.05% XC, 0.05% BPB, 10 m*M* EDTA, and 4 *M* urea and are not heated prior to loading. The original reaction mixture is stored at −20° until the result from gel electrophoresis indicates formation of a product of expected length. If the product has unphosphorylated ends, these may now be phosphorylated by addition of 2 units of T4 PNK and incubation at 37° for 30 min. The ligation product can be isolated from reaction mixtures by preparative gel electrophoresis on a 1-mm-thick gel under conditions as above or can be used directly for cloning after phenol extraction and ethanol precipitation.

Cloning and Sequencing. Approximately 200 ng (~0.17 pmol) of properly digested and dephosphorylated pGV vector[42] is ligated with at least a double molar excess of 5′-phosphorylated synthetic DNA fragment at 15° overnight. The ligation mixture (25 μl) contains 60 m*M* Tris–HCl (pH 8), 6 m*M* $MgCl_2$, 20 m*M* DTE, 0.25 m*M* ATP, and 4 U T4 DNA ligase. Competent cells of JM83[43] are transformed[44] with 5 μl of the ligation mixture (~40 ng vector). Usually 10–20 clones of the obtained transformants are picked and grown overnight in 2 ml LB medium[45] containing the appropriate antibiotics. A modified Triton X-100 lysis procedure[46] as recommended by Volckaert[42] is used to isolate plasmid DNA. Precipitated DNA is washed twice with 0.2 ml 70% ethanol before drying. The DNA is dissolved in 30 μl of bovine pancreas ribonuclease solution in water (3.3 mg/ml) and kept at room temperature for 15 min. One-tenth of the plasmid DNA is used for restriction enzyme digests and acrylamide gel electrophoresis.[45] Two DNAs with inserts of correct size are single-end labeled, left or right of the insert according to Volckaert.[42] This DNA is directly sequenced without further purification by the chemical degradation method, following the procedures of either Volckaert[42] or Rosenthal *et al.*[32]

[42] G. Volckaert, this series, Vol. 155 [17].

[43] J. Vieira and J. Messing, *Gene* **19,** 259 (1982).

[44] D. Hanahan, *J. Mol. Biol.* **166,** 557 (1983).

[45] T. Maniatis, E. F. Fritsch, and J. Sambrook, "Molecular Cloning: A Laboratory Manual." Cold Spring Harbor Lab., Cold Spring Harbor, New York, 1982.

[46] D. S. Holmes and M. Quingley, *Anal. Biochem.* **114,** 193 (1981).

[14] The Segmented Paper Method: DNA Synthesis and Mutagenesis by Rapid Microscale "Shotgun Gene Synthesis"

By HANS W. DJURHUUS MATTHES, ADRIEN STAUB, and PIERRE CHAMBON

Synthetic DNA has many uses and consequently is rapidly increasing in importance in gene technology and molecular biology. Cost, in terms of both time and materials, has been the most limiting factor for the use of large numbers of synthetic oligonucleotides. One of the major problems concerning cost has been the inability to scale down the use of expensive reagents without affecting overall yields. Most existing solid-phase synthesis protocols start with about a micromole or more of nucleoside attached to the solid support. They produce quantities of oligonucleotides (hundreds of micrograms) which are excessive for most purposes, and they result in a price which is too high for the general use of oligonucleotides in large numbers. The segmented microscale paper method,[1] by which over a hundred oligonucleotides can be synthesized simultaneously on cellulose paper support in 3 days, provides a powerful means for producing synthetic DNA at a low cost.

Several genes have been assembled from overlapping synthetic oligonucleotides.[2] The segmented microscale paper method in combination with "shotgun gene synthesis"[3] offers a particularly rapid procedure for the simultaneous construction of a series of genes. These genes can then be used for the *in vivo* synthesis of proteins or for the definition of functional DNA sequences. The classic approach of studying gene function is to create mutations and correlate the new phenotypes with these changes in the genetic material. Synthetic oligonucleotides provide an elegant means for the introduction of site-specific nucleotide changes in cloned genes, since the mutations are predetermined with respect to both their location in the DNA and the nature of the changes introduced.[4,5] The currently used protocols for oligonucleotide-directed *in vitro* mutagenesis

[1] H. W. D. Matthes, W. M. Zenke, T. Grundström, A. Staub, M. Wintzerith, and P. Chambon, *EMBO J.* **3,** 801 (1984).

[2] K. Itakura, *Trends Biochem. Sci.* **7,** 442 (1982) and references therein.

[3] T. Grundström, W. M. Zenke, M. Wintzerith, H. W. D. Matthes, A. Staub, and P. Chambon, *Nucleic Acids Res.* **13,** 3305 (1985).

[4] T. Harris, *Nature (London)* **299,** 298 (1982).

[5] M. Smith, *Trends Biochem. Sci.* **7,** 440 (1982).

are based on priming with a mismatching mutagenic oligonucleotide.[6–10] However, mismatch–primer mutagenesis has inherent limitations in the types and sizes of mutations which can be obtained. "Shotgun gene synthesis," using synthetic oligonucleotides,[3,11] offers an alternative approach to generate site-specific mutations. The segmented paper method for synthesis of oligonucleotides on a microscale[1] make this approach particularly appealing.

This chapter describes in detail the solid-phase phosphotriester and phosphoramidite methods (see Figs. 1 and 2) for the rapid and simultaneous synthesis of 20–150 oligonucleotides on a microscale (50–150 nmol scale), using a segmented paper support. It also describes the rapid "shotgun gene synthesis" technique, i.e., the experimental conditions under which a set of more than 15 oligonucleotides can be assembled on a microscale (<1 pmol) into phage M13, or plasmid-derived vectors without any purification step prior to transformation of *Escherichia coli* (see Fig. 8). Mutations in the assembled DNA segment can easily be introduced by exchanging the appropriate oligonucleotides within the "wild-type" set. As an example, the 75 oligonucleotides corresponding to a series of 30 mutations (15 oligonucleotides for the wild-type sequence, plus two oligonucleotides per mutation) within a DNA segment of 120 nucleotides can be prepared in one synthesis cycle in 2–3 days. Using this method we have successfully introduced a large number of different types of mutations flanked by two unique restriction endonuclease sites into the simian virus 40 (SV40) enhancer.[3]

The microscale "shotgun gene synthesis" technique has also been used by others to synthesize a gene encoding the minor A form of bovine intestinal calcium binding protein (ICaBP),[12] which was expressed in *E. coli,* and yielded the desired product at a level of 1–2% of total protein. In another study from our laboratory,[13] a series of insertion mutations of specific lengths were introduced into the SV40 early promoter to analyze the requirements for stereospecific alignments of the various elements of

[6] M. J. Zoller and M. Smith, *Nucleic Acids Res.* **10,** 6487 (1983).

[7] M. J. Zoller and M. Smith, this series, Vol. 100, p. 468.

[8] K. Norris, F. Norris, L. Christiansen, and N. Fiil, *Nucleic Acids Res.* **11,** 5103 (1983).

[9] Y. Morinaga, T. Franceschini, S. Inouye, and M. Inouye, *Bio/Technology* **2,** 636 (1984).

[10] W. Kramer, V. Drutsa, H.-W. Jansen, B. Kramer, M. Pflugfelden, and H.-J. Fritz, *Nucleic Acids Res.* **12,** 9441 (1984).

[11] K.-M. Lo, S. S. Jones, N. R. Hackett, and H. G. Khorana, *Proc. Natl. Acad. Sci. U.S.A.* **81,** 2285 (1984).

[12] P. Brodin, T. Grundström, T. Hofmann, T. Drakenberg, E. Thulin, and S. Forsen, *Biochemistry,* **25,** 5371 (1986).

[13] K. Takahashi, M. Vigneron, H. Matthes, A. Wildeman, M. Zenke, and P. Chambon, *Nature (London)* **319,** 121 (1986).

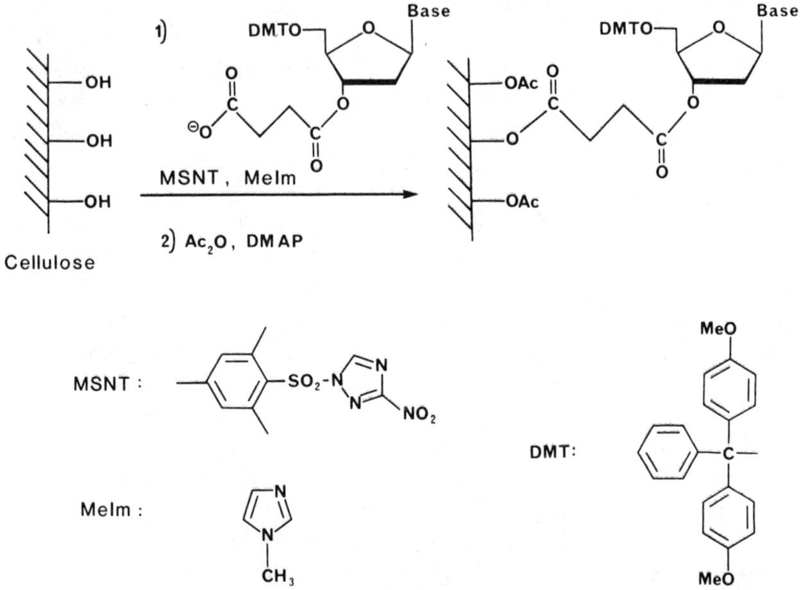

FIG. 1. Addition of the first nucleoside to the paper support.

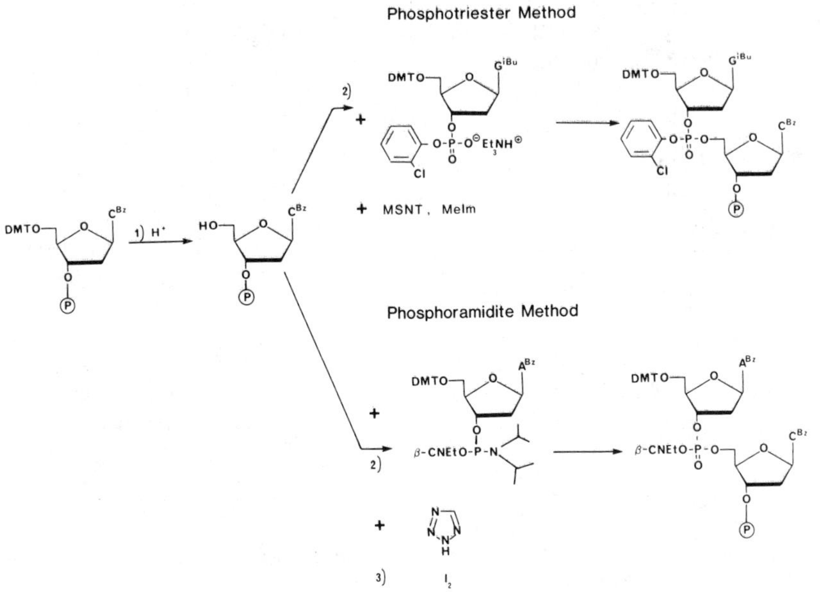

FIG. 2. Chemistry of phosphotriester and phosphoramidite methods. Ⓟ, Polymer support (cellulose paper or controlled pore glass).

this promoter. These mutations were introduced directly into the pBR322-derived vector which was used to study their effect on transcription *in vivo*. Direct "shotgun gene synthesis" into double stranded vectors such as pUC18 and pBR322 derivatives, followed by dideoxy sequencing of minipreparations of plasmids to characterize the recombinants, provides an extremely rapid method for extended mutagenesis in any DNA segment of interest.

I. The Segmented Paper Method and Its Application to Rapid Microscale "Shotgun Gene Synthesis" and Mutagenesis

A. Simultaneous Chemical Synthesis of Over One Hundred Oligodeoxyribonucleotides on a Microscale (the Segmented Paper Method for the Synthesis of Oligonucleotides)

Crea and Horn[14] have described the use of cellulose as a solid support for the phosphotriester method. Frank *et al.*[15] and Matthes *et al.*[1] used the phosphotriester method for the simultaneous synthesis of oligonucleotides on cellulose paper supports. Ott and Eckstein[16] have successfully applied the phosphite triester method for oligonucleotide synthesis on segmented paper supports. Here we describe these methods in more detail with some modifications to make them even more convenient for adaption to the current equipment and techniques used in an average molecular biology laboratory.

1. Phosphotriester Method (Figs. 1 and 2)[17]

The solid-phase phosphotriester method is currently the least costly method for oligonucleotide synthesis. Using this method we have synthesized oligonucleotides up to 25-mers on segmented paper supports in good yields (about 5%). Figure 3 shows the HPLC pattern of a 23-mer synthesized by this method. However, we generally keep an average length of about 18 nucleotides so as to obtain reproducibly high yields. When a segmented synthesis is underway, new demands can be added in the protocol at any sorting step. With a high demand for oligonucleotides it is convenient to have the simultaneous synthesis running continuously.

[14] R. Crea and T. Horn, *Nucleic Acids Res.* **8,** 2331 (1980).

[15] R. Frank, W. Heikens, G. Heisterberg-Moutsis, and H. Blöcker, *Nucleic Acids Res.* **11,** 4365 (1983).

[16] J. Ott and F. Eckstein, *Nucleic Acids Res.* **12,** 9137 (1984).

[17] M. J. Gait (ed.), "Oligonucleotide Synthesis: A Practical Approach." IRL Press, Oxford, England, 1984.

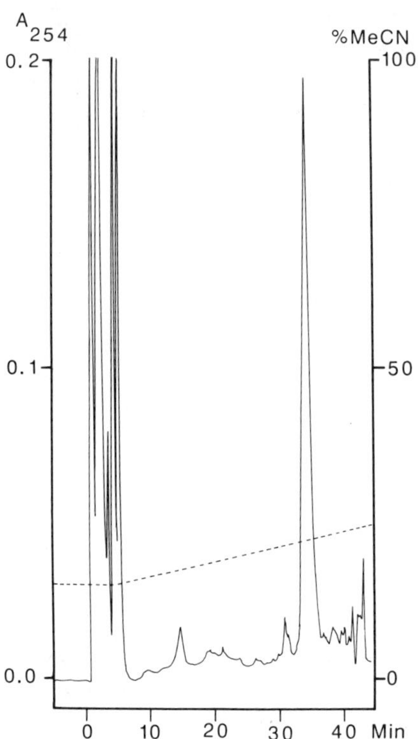

FIG. 3. Purification of 5′-*O*-DMT-oligonucleotides by reverse-phase high-performance liquid chromatography. Elution profile from the synthesis of a 23-mer (5′-DMT-TAACTGA-CACACATTCCACAGCT) synthesized by the phosphotriester method (10% of total injected). The early peaks correspond to failure sequences without 5′-protecting group. The 5′-DMT-23-mer elutes at 34 min.

2. *The Phosphoramidite Method (Figs. 1 and 2)*[17]

The solid-phase phosphite triester method[18,19] soon after its introduction in 1980 was considered to be a superior method for solid-phase oligonucleotide synthesis. However, the mononucleoside phosphites were too reactive to be conveniently used. It was only in 1983, when phosphites could be protected as the more stable morpholino- or diisopropyl phosphoramidites,[20,21] allowing purification by silica gel chromatography,

[18] M. D. Matteucci and M. H. Caruthers, *Tetrahedron Lett.* **21,** 719 (1980).

[19] M. D. Matteucci and M. H. Caruthers, *J. Am. Chem. Soc.* **103,** 3185 (1981).

[20] S. P. Adams, K. S. Kavka, E. J. Wykes, S. B. Holder, and G. R. Galluppi, *J. Am. Chem. Soc.* **105,** 661 (1983).

[21] L. J. McBride and M. H. Caruthers, *Tetrahedron Lett.* **24,** 245 (1983).

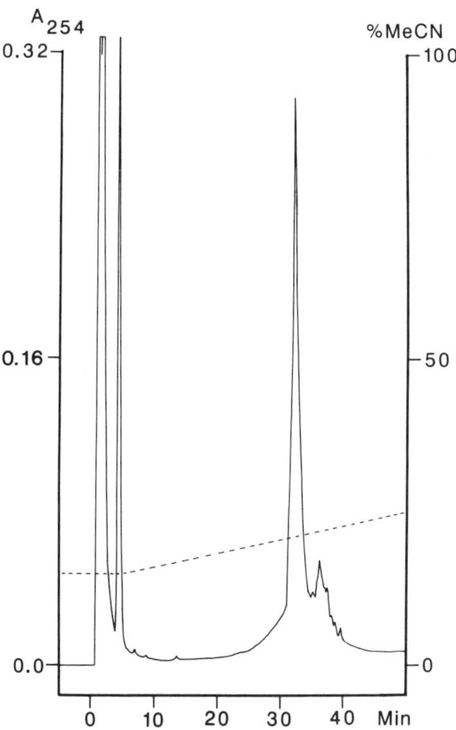

FIG. 4. Elution profile of a 33-mer (5′-DMT-*O*-TCCCAGTCACGACGTTGTAAAACG-ACGGCCAGT-3′) synthesized on paper support using the phosphoramidite method (10% of total injected).

that the method became as reliable as the phosphotriester method and superior with regard to yield and the maximal obtainable length of oligonucleotides. A disadvantage of the method is that it has a more complex reaction cycle and a higher cost for reagents. However, since the introduction of the β-cyanoethylphosphoramidites[22] the price has decreased substantially.

Ott and Eckstein reported the synthesis of eight decamers by the phosphoramidite method, using the paper support.[16] We have compared the β-cyanoethylphosphoramidite method on paper and controlled pore glass (CPG) supports simultaneously under identical conditions (see Figs. 4, 5, 6, and 7) without a capping step. We obtained good yields (25 and 89%, respectively, based on trityl analysis and 7.5 and 20% isolated yield after purification by HPLC) of 33-mers on both supports. The synthesis

[22] N. D. Sinha, J. Biernat, and H. Köster, *Tetrahedron Lett.* **24,** 5843 (1983).

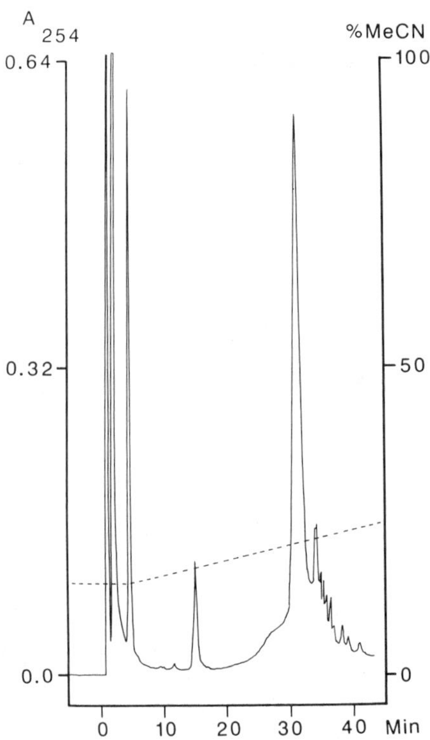

FIG. 5. Elution profile of a 33-mer (same as in Fig. 4) synthesized on a CPG support under identical conditions as the oligonucleotide shown in Fig. 4.

performed on CPG supports shows less chain termination in the early steps of the synthesis and therefore better overall yield, but the results obtained on paper support indicate that we can successfully synthesize 40-mers by the segmented method.

B. Sorting Step

There is a potential risk of error in oligonucleotide synthesis by the segmented method during the sorting of the paper disks before each coupling step. However, by carefully following predetermined protocols and by double-checking at each coupling step we have found no errors up to date in any of the oligonucleotides synthesized in our laboratory. A computer program OLIGO has been designed to facilitate the sorting steps and to minimize the error rate.[23] This program also includes options for

[23] C. M. Tolstoshev, H. W. D. Matthes, and P. Oudet, *Nucleic Acids Res.* **14,** 405 (1986).

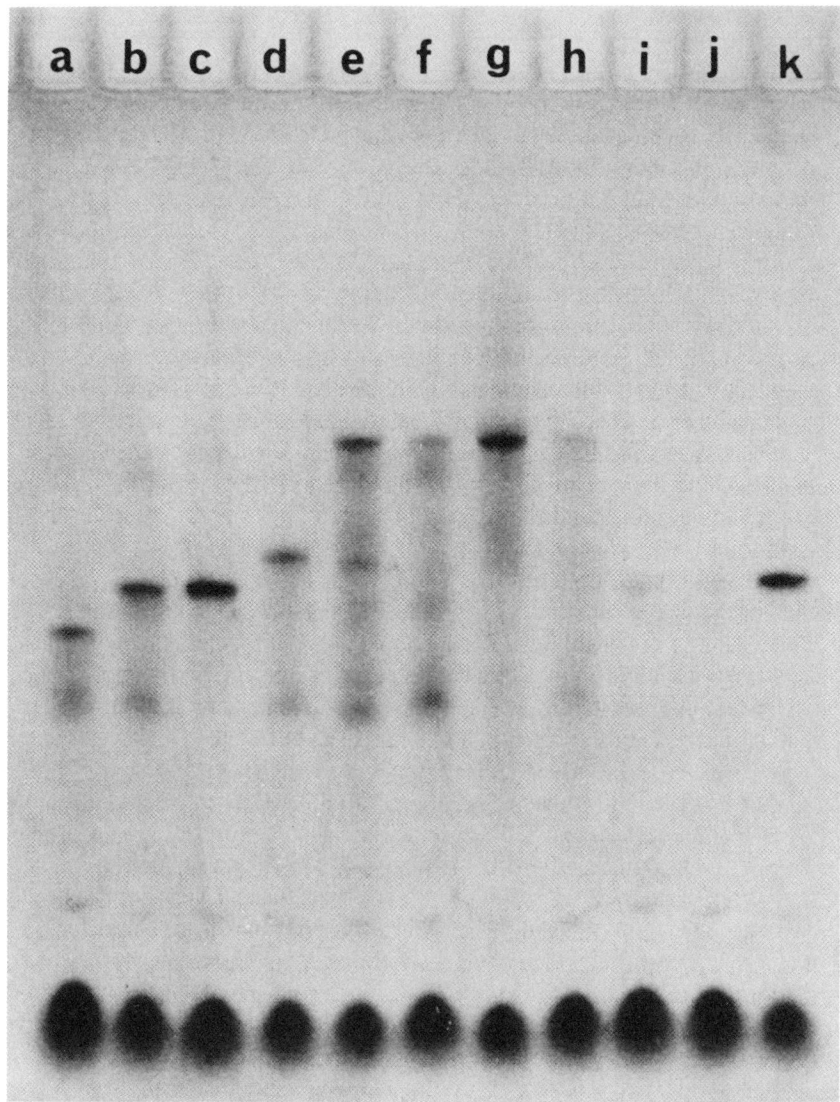

FIG. 6. UV-shadowing of oligonucleotides synthesized by the phosphoramidite methods on paper (0.15 μmol scale) or CPG (1.5 μmol scale) supports and separated by preparative (1.5 mm) polyacrylamide gel electrophoresis. A fraction (1–10%) of each synthesis was loaded. Lanes are as follows: a, 14-mer (paper), 5′-TCCCAGTCACGACG-3′; b, 17-mer (paper), 5′-GTAAAACGACGGCCAGT-3′; c, 17-mer (CPG), 5′-GTAAAACGACGG-CCAGT-3′; d, 20-mer (paper), 5′-GTTGTAAAACGACGGCCAGT-3′; e, 33-mer (paper), 5′-TCCCAGTCACGACGTTGTAAAACGACGGCCAAGA-3′; f, 33-mer (paper), 5′-TC CCAGTCACGACGTTGTAAAACGACGGCCAAGT-3′; g, 33-mer (CPG), 5′-TCCCAGT-CACGACGTTGTAAAACGACGGCCAAGT-3′; h, 33-mer (paper), 5′-TCCCAGTCAC-GACGTTGTAAAACGACGGCCAAGG-3′; i, 17-mer (paper), HPLC-purified; j, 17-mer marker, 1 μg; k, 17-mer marker, 10 μg.

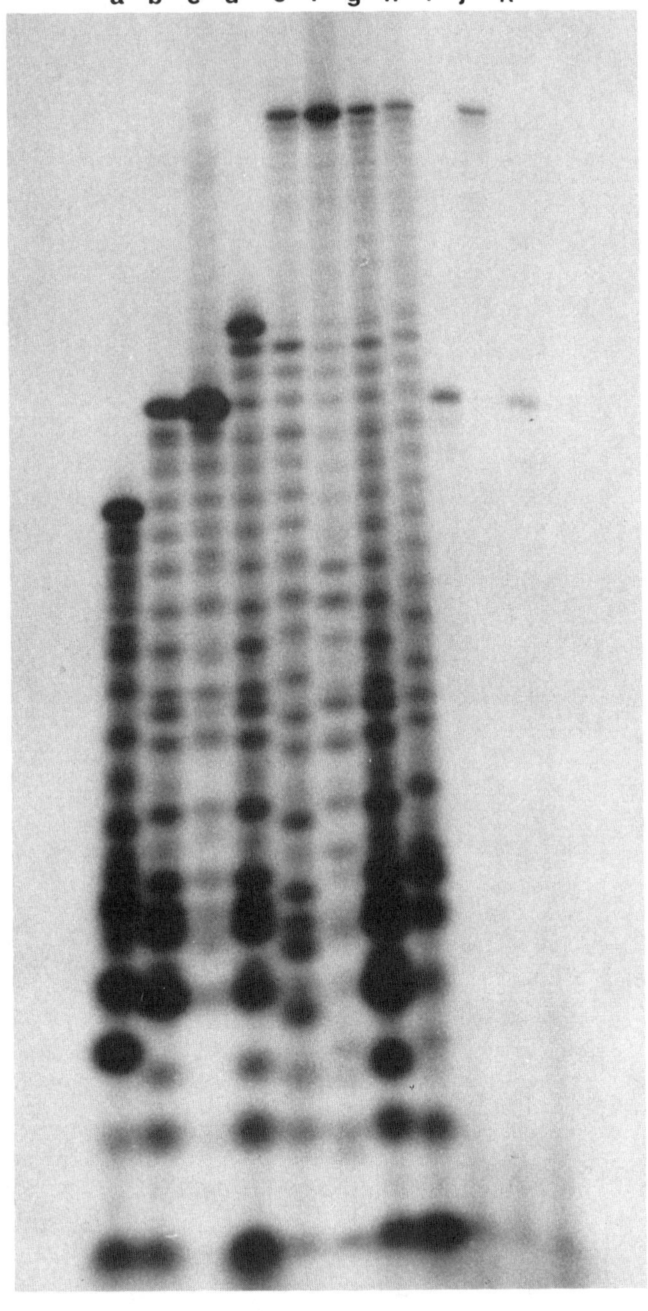
a b c d e f g h i j k

analyzing synthetic gene fragments for restriction endonuclease sites, sequence homologies, inverted and direct repeats, etc. Another option transfers the data input to storage files. With currently used word processing programs (e.g., PED, NOTIS, MS-WINDOWS), any filed oligonucleotide can be traced in a few minutes.

C. Purification of Oligonucleotides

The requirement of purity of oligonucleotides depends on the nature of the experiment it has to be used for. Below, we have given several methods for purification of oligonucleotides. HPLC is the method of choice for purifying large quantities of a few oligonucleotides. We are currently using polyacrylamide gel electrophoresis (PAGE) followed by ethanol precipitation for the purification of oligonucleotides for "shotgun gene synthesis." This procedure is the fastest for purification of large numbers of oligonucleotides and is sufficient to give high yields in "shotgun gene synthesis" as described (in Sections I,D and II,J).

D. Strategy of DNA Synthesis and Mutagenesis by Rapid Microscale "Shotgun Gene Synthesis"[3]

"Shotgun gene synthesis" (SGS) offers an alternative to the mismatch–primer approach for the generation of site-specific mutations. Currently available methods for gene synthesis require large amounts of synthetic oligonucleotides and employ time-consuming successive cycles of 5′-end phosphorylation, ligation, and purification of oligonucleotide blocks. The rapid microscale "shotgun gene synthesis" method[3] takes advantage of the fact that the transformation of *E. coli* with linear DNA is much less efficient than with circular DNA, and uses this as a biological selection for the direct cloning of a large number of correctly assembled genes using overlapping oligonucleotides. Only those synthetic DNA molecules which contain both the appropriate 5′ and 3′ sticky ends will allow circularization of the vector DNA in the presence of T4 DNA ligase and will create transformants in *E. coli*. The strategy is outlined in Fig. 8.

FIG. 7. Autoradiogram of the crude mixture (a–h) of the same oligonucleotides as in Fig. 6 after ^{32}P 5′-end labeling and gel electrophoresis (20% polyacrylamide, 8 *M* urea). Lanes a–h correspond to the same sequences as in Fig. 6 except that f and g are reversed. Lane i shows an HPLC-purified 17-mer synthesized on paper (same as in lane b), lane j is an HPLC-purified 33-mer (Fig. 4) synthesized on paper. Lane k is a 17-mer marker.

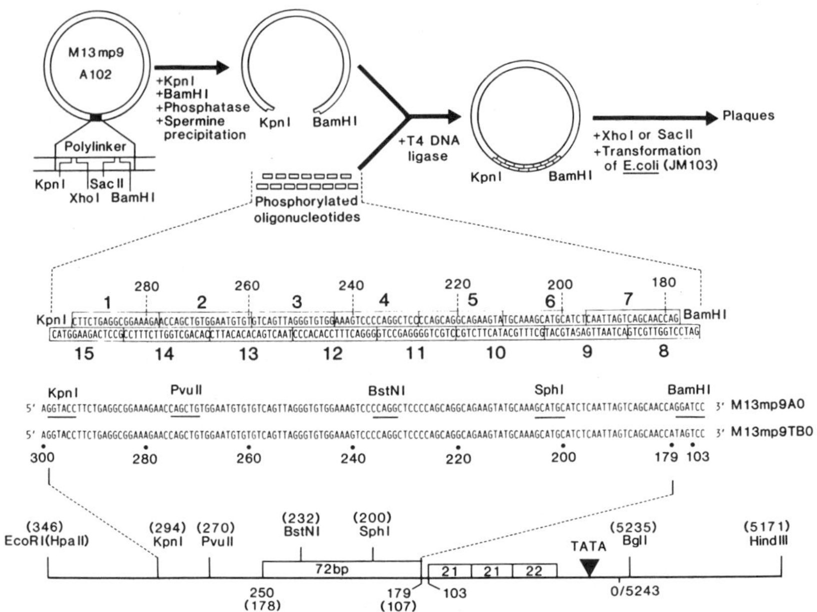

FIG. 8. Outline of oligonucleotide-directed mutagenesis by "shotgun gene synthesis" in an M13 phage vector. The lower part of the figure shows the SV40 promoter and enhancer region from *Hpa*II (nucleotide number 346; BBB system coordinates) to *Hin*dIII (nucleotide number 5171) containing only one copy of the 72-bp sequence (nucleotide number 179–250) cloned in the *Eco*RI and *Hin*dIII sites of M13mp9.[3] The vector M13mp9A102[3] was used for cloning of 15 overlapping synthetic oligonucleotides (1–15, in the middle part of the figure) which reconstruct the SV40 enhancer region from nucleotide numbers 102 to 298 (wild-type SV40 sequence, but containing only one copy of the 72-bp sequence and a *Bam*HI site at position 101).

For gene synthesis by SGS it is generally an advantage to use long oligonucleotides (≥35-mers) to minimize the number of purification and kination steps. SGS is also an excellent means to introduce a large variety of mutations in a particular gene fragment. However, for this purpose it is advantageous to use short oligonucleotides (~15-mers). For example, for the construction of a single point mutation in a gene fragment using n overlapping oligonucleotides (see upper part of Fig. 8) only two new oligonucleotides (one on each strand) are required to create the mutation while $n - 2$ oligonucleotides are kept wild type. A single wild-type set preparation (n oligonucleotides) can be used for hundreds of SGSs. Therefore, the shorter the oligonucleotides to be exchanged in the SGS set the shorter the length of new DNA that has to be synthesized. This is an important point, especially for extensive mutagenesis in a given gene

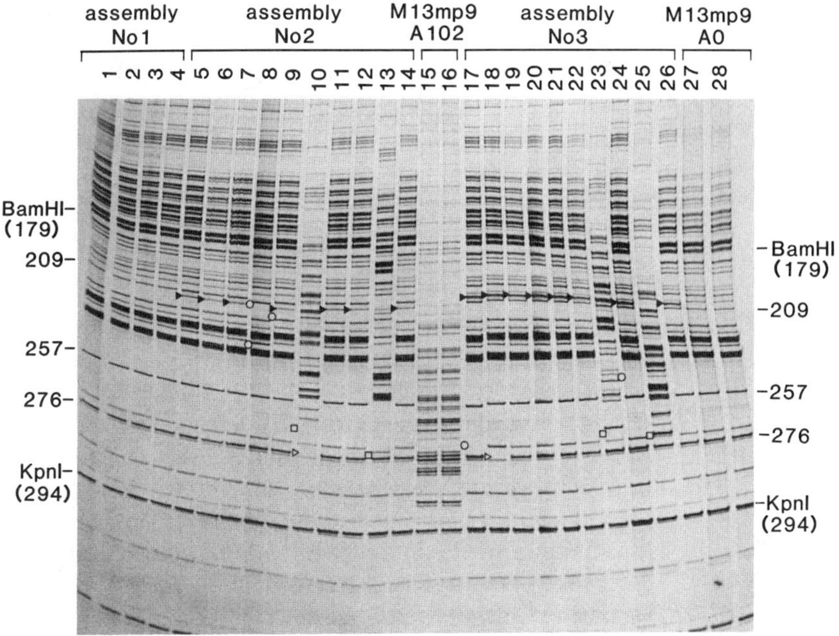

FIG. 9. DNA sequence analysis by the dideoxy method of M13 phage ss templates from shotgun ligation mutagenesis in the SV40 enhancer region (see Fig. 8). The C-reaction only is shown. (▶) Desired mutation; (▷) single nucleotide substitution; (○) single nucleotide deletion; (□) approximate start point of extensive deletion. About 40% of the clones gave the desired sequence.

fragment. A combination of long and short fragments may also be useful: If a flanking region has to be conserved as wild type in the mutagenesis study, then it can be constructed from long oligonucleotides while for the region to be mutated it is most economical to use short oligonucleotides. The yield of correct synthetic inserts by SGS is high. Screening is performed by dideoxy sequencing, using either double-stranded (ds) or single-stranded (ss) templates. An example of screening by the dideoxy method using ss templates is shown in Fig. 9. The rapid alkaline lysis method or a simplified rapid boiling method (see Section II,K,2,a) gives sufficiently good sequencing results to be used for screening for correct inserts by SGS (see Fig. 10).[24] For inserts of a length of 120 base pairs the success rate is about 40%.[3]

[24] R. G. Korneluk, F. Quan, and R. A. Gravel, *Gene* **40,** 317 (1985).

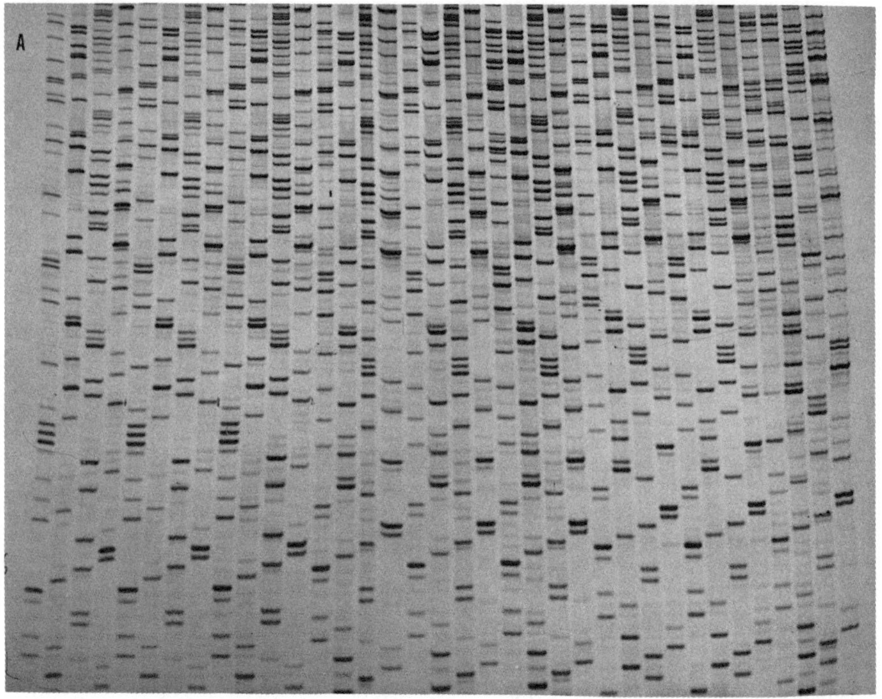

FIG. 10. (A) Sequence analysis of by the dideoxy method of ds pUC18 plasmid DNA from minipreparations prepared by the alkaline lysis method.[27] (B) Sequence analysis by the dideoxy method of ds pBR322 derived plasmid DNA from minipreparations by the modified rapid boiling method as described in the text.

II. Materials and Methods

A. Special Equipment for Oligonucleotide Synthesis

1. The bench DNA synthesizer—four-column model, Ref. No. 3603—is supplied by Omnifit, Ltd., Norfolk Street, Cambridge CB1 2LE, England (see Fig. 12). Similar synthesizers can be purchased from Cruachan Chemicals Ltd., 11 Napier Square, Livingston EH54 5DG, Scotland, or Cambridge Research Biochemicals Ltd., Button End, Harston, Cambridge CB2 5NX, England.
2. One gas cylinder with gauge. Any dry inert gas in a cylinder under pressure can be used for the solvent delivery system. Helium has an advantage since no bubbles are formed in the columns because of the pressure release. Argon is convenient due to its high density

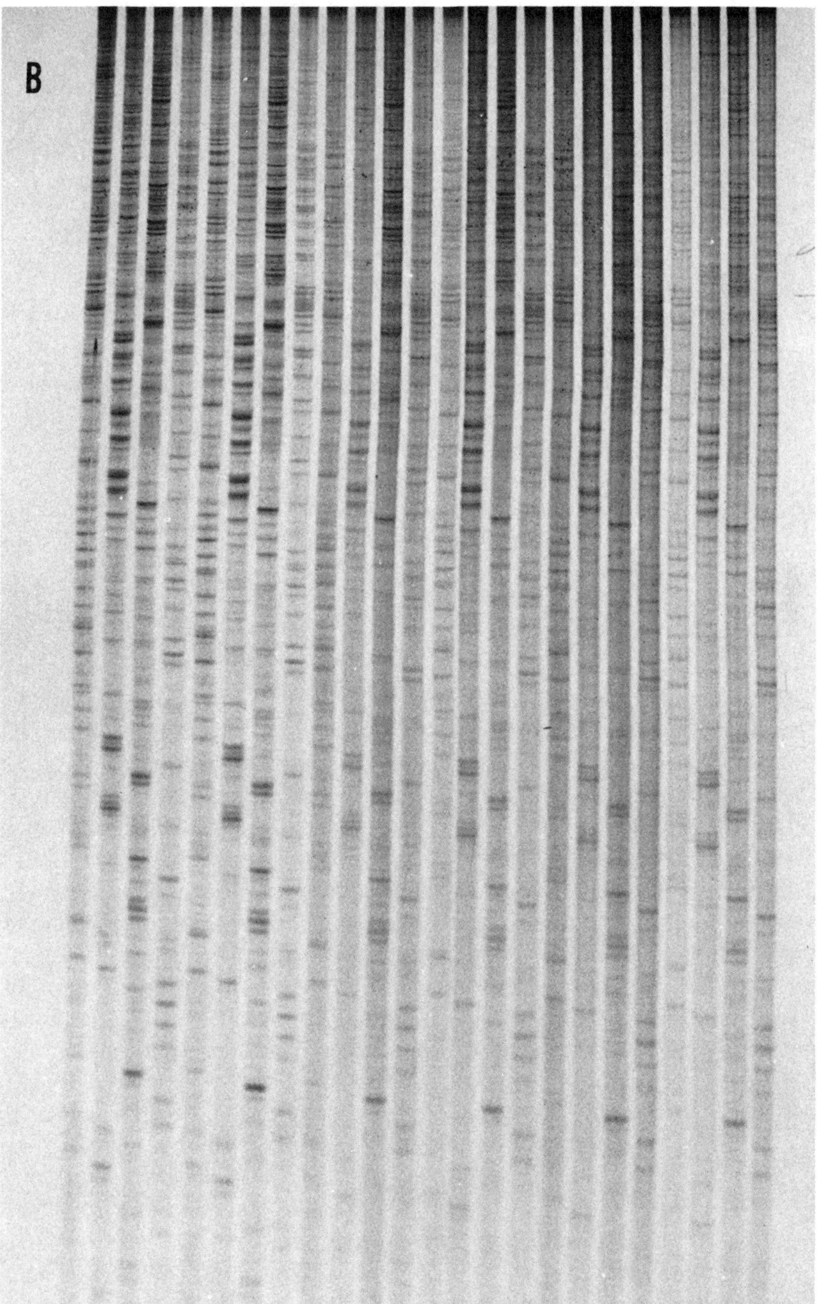

FIG. 10. (*continued*)

for filling up desiccators or the Speed Vac under reduced pressure.

3. A connector from the gas cylinder to the 3 mm (1.5 mm i.d.) Teflon tubing can be obtained from Omnifit (Ref. No. 1105).
4. The solid support is Whatman 3MM, Whatman 1 Chr, or Whatman ream paper (ordinary paper delivered in reams of 7 kg).
5. Six chemically inert, gas-tight syringes [e.g., Hamilton 1750 RN (81230), 0.5 ml, or Hamilton 1001 RN (81330), 1 ml] with two packs of spare needles (Ref. No. 80725, 22 gauge, 2 inches).
6. One Savant Speed Vac concentrator with oil pump.
7. An ultrasonic bath (not necessary if helium is used as driving gas).
8. Two 3-way glass valves.
9. A big calcium chloride or silica gel drying tube.
10. Four stainless steel HPLC filters, e.g., Alltech No. 05-0141, Alltech Associates, Inc., 2051 Waukegan Road, Deerfield, IL 60015.
11. Fifty silicone rubber septa, Suba-seal No. 30, Aldrich No. Z12,471-0.

B. Solvents

Warning: Pyridine, acetonitrile, and tetrahydrofuran are very toxic solvents and should be handled under a fume hood. Contact with the skin should be avoided. 1,2-Dichloroethane has been replaced by the less toxic dichloromethane. All solvents should be of analytical grade. Pyridine and acetonitrile should be anhydrous (water content below 20 ppm). This is achieved by keeping the solvents over activated molecular sieves (3 or 4 Å). Some qualities of molecular sieves create problems when used with pyridine due to leaking out of metal ions which can inhibit coupling reactions.[17] However, we have encountered no problems using molecular sieves from SDS, BP 4—Valdonne, F-13124 Peypan, France. Other solvents used are tetrahydrofuran (THF), dichloromethane (DCM), 2,6-lutidine, dichloroacetic acid (DCA), anhydrous diethylether, triethylamine.

C. Reagents

1. General Reagents for Both Methods

4-Dimethylaminopyridine (DMAP), Aldrich
Acetic anhydride
1-Mesitylene sulfonyl-3-nitro-1,2,4-triazolide (MSNT), Cruachem No. 1085
1-Methylimidazole, Merck-Schuchardt No. 805852

5′-*O*-(4,4′-Dimethoxytrityl)thymidine 3′-*O*-succinate, Cruachem No. 7400 (SPS grade)

5′-*O*-(4,4′-Dimethoxytrityl)-N^2-isobutyryl-2′-deoxyguanosine 3′-*O*-succinate, Cruachem No. 7410 (SPS grade)

5′-*O*-(4,4′-Dimethoxytrityl)-N^6-benzoyl-2′-deoxyadenosine 3′-*O*-succinate, Cruachem No. 7420 (SPS grade)

5′-*O*-(4,4′-Dimethoxytrityl)-N^4-benzoyl-2′-deoxycytidine 3′-*O*-succinate, Cruachem No. 7430 (SPS grade)

2. *Reagents for the Phosphotriester Method*

syn-2-Nitrobenzaldoxime, Cruachem No. 1210

1,1,3,3-Tetramethylguanidine, Merck No. 821897

5′-*O*-(4,4′-Dimethoxytrityl)thymidine 3′-(2-chlorophenyl)phosphate, triethylammonium salt, SPS grade, Cruachem No. 7300

5′-*O*-(4,4-Dimethoxytrityl)-N^2-isobutyryl-2′-deoxyguanosine 3′-(2-chlorophenyl)phosphate, triethylammonium salt, SPS grade, Cruachem No. 7310

5′-*O*-(4,4′-Dimethoxytrityl)-N^6-benzoyl-2′-deoxyadenosine 3′-(2-chlorophenyl)phosphate, triethylammonium salt, SPS grade, Cruachem No. 7320

5′-*O*-(4,4′-Dimethoxytrityl)-N^4-benzoyl-2′-deoxycytidine 3′-(2-chlorophenyl)phosphate, triethylammonium salt, SPS grade, Cruachem No. 7330

3. *Reagents for the Phosphoramidite Method*

Tetrazole, sublimed, Cruachem No. 20-1125-18, Biosearch No. NU 6288, Pharmacia No. 27-2875-01

Iodine (I_2), Aldrich No. 20,777-2

5′-*O*-(4,4′-Dimethoxytrityl)thymidine 3′-*O*-(β-cyanoethyl-*N*,*N*-diisopropylamino)phosphoramidite, Cruachem No. 20-8100-19, Biosearch No. NU 6103, Pharmacia No. 27-1736-01

5′-*O*-(4,4′-Dimethoxytrityl)-N^2-isobutyryl-2′-deoxyguanosine 3′-*O*-(β-cyanoethyl-*N*,*N*-diisopropylamino)phosphoramidite, Cruachem No. 20-8110-19, Biosearch No. NU 6102, Pharmacia No. 27-1734-01

5′-*O*-(4,4′-Dimethoxytrityl)-N^6-benzoyl-2′-deoxyadenosine 3′-*O*-(β-cyanoethyl-*N*,*N*-diisopropylamino)phosphoramidite, Cruachem No. 20-8120-19, Biosearch No. NU 6100, Pharmacia No. 27-1730-01

5′-*O*-(4,4′-Dimethoxytrityl)-N^4-benzoyl-2′-deoxycytidine 3′-*O*-(β-cyanoethyl-*N*,*N*-diisopropylamino)phosphoramidite, Cruachem

No. 20-8130-19, Biosearch No. NU 6101, Pharmacia No. 27-1732-01

D. Set-up of Four-Column Synthesis Apparatus

A schematic diagram of the apparatus is shown in Fig. 11 (see also Fig. 12). Use thick Teflon tubing (1.5 mm i.d.) for all helium-filled lines and solvent lines inside the solvent bottles. Use 0.8 mm i.d. Teflon tubing for the solvent delivery lines from the tops of the solvent reservoirs to the rotary solvent selector valve and also for the delivery line from the rotary solvent selector valve to the column inlet. Use about 0.5 m of 0.3 mm i.d. tubing from the column outlet to the waste bottle in order to generate a

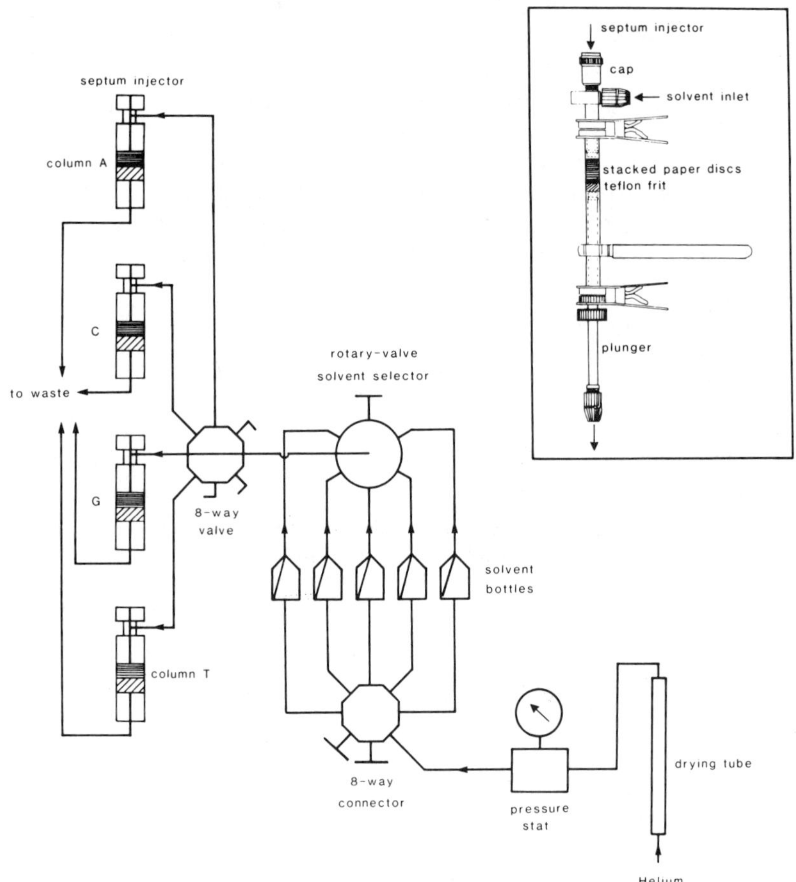

FIG. 11. Schematic illustration of oligonucleotide assembler.

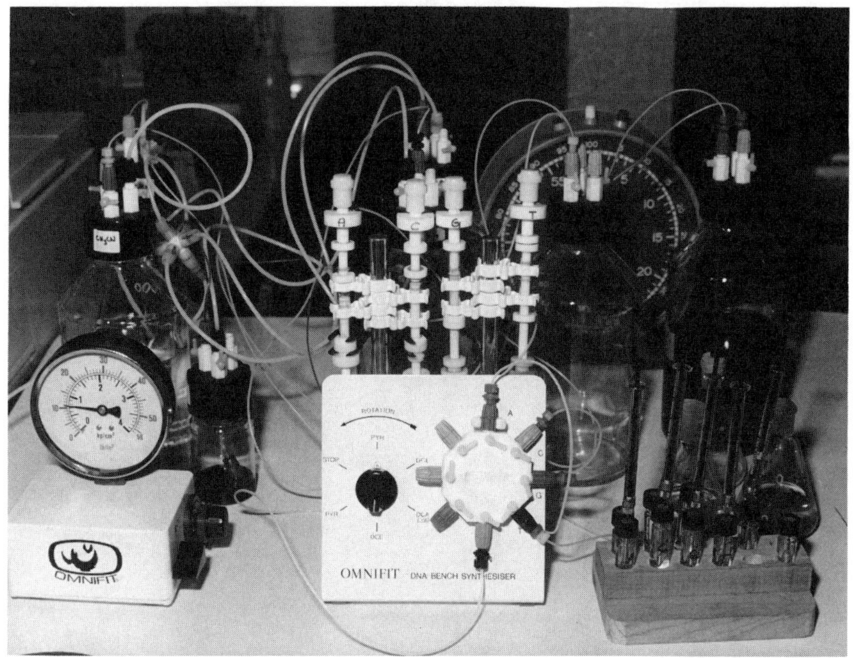

FIG. 12. Photograph of oligonucleotide assembler as used in the segmented method.

slight back pressure. This hinders bubble formation in the columns. First cut all tubing to the desired length and then slip on the threaded bolts [Omnifit 210(0) or 220(0) as included] plus the appropriate gripper fittings (Omnifit 2310 or 2312 as included; these are Teflon washers fitted with a 316 grade stainless steel backing ring). It is necessary to heat and stretch the end of the tubing before slipping on the gripper fitting. Note that all liquid flow fittings must be Teflon or equivalent, whereas the gas flow connectors can be made with Viton O rings.

The general setup procedure is as follows.

1. Use the long column to make a silica gel drying tube and connect this between the helium supply and the inlet valve of the pressure regulator (pressure stat; this has an on/off switch and a screw-in needle valve).
2. Connect the outlet valve of the regulator to one position on the 8-way connector (Omnifit 2404), then connect five other positions of this connector to one terminal on each of the five bottle tops (four reservoirs are 1 liter, while one is 200 ml). Block off the remaining two positions with the plug seals supplied, Omnifit 2320. One outlet

terminal on each bottle top will be used as a vent to enable the head space above the solvents to be flushed with inert gas when the bottles are filled.

3. Next, connect one terminal from each bottle to the rotary solvent selector valve using a 2-way connector and 0.8 mm i.d. Teflon tubing and fit a length of 1.5 mm i.d. tubing to the underside of these terminals such that the tubing reaches the bottom of the

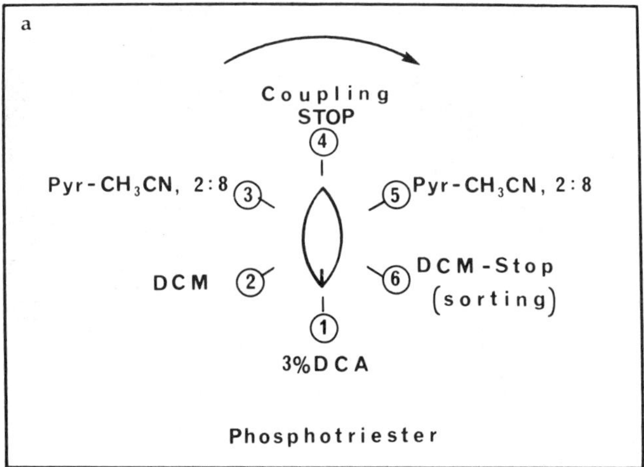

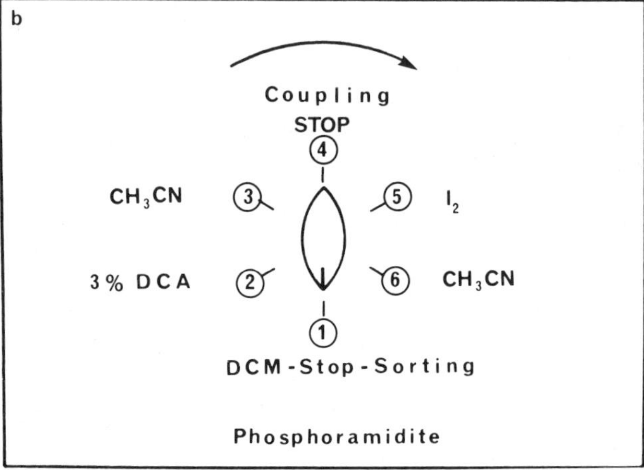

FIG. 13. Eight-way rotary solvent selector valve. Schematic illustration of connections (a) for the phosphotriester method and (b) for the phosphoramidite method.

reservoirs. Fit a stainless steel HPLC filter onto the Teflon tubing in the bottles (except for the acid bottle).

4. Connect the 6-way rotary solvent selector such that the solvent order is as shown on Figs. 13a and b, depending on the method you choose. The stop position is blocked off with a 2-way connector and a plug seal (Omnifit No. 2320, as included in the kit).
5. Connect the outlet (in the middle) of the rotary valve to the 8-way solvent distribution valve (Omnifit No. 1103) and use four other positions of the 8-way valve as outlets for connections to columns A, C, G, and T. Connect these with 0.8 mm i.d. Teflon tubing to the column inlets. Use a color code.
6. Finally, connect the outlet from the columns with 0.5 m of 0.3 mm i.d. Teflon tubing to the waste bottle(s).
7. Fit the columns with Teflon-faced silicone rubber septa in the screw caps on the injection end piece such that the Teflon face is downward and place a Teflon frit into each column to avoid blockages in the outlet lines.
8. Check the system for any leaks by putting some DCM in each bottle and pass it through one column at a time, leaving no air space above the Teflon frits in the columns. Empty the bottles, dry them in an oven at 120°, and then fill them with solvents according to Fig. 13a or b and Sections II,E,1 or II,E,2.

E. Solvent Systems

1. Solvent System for the Phosphotriester Method

1. 3% DCA/DCM, v/v
2. DCM
3. Acetonitrile–pyridine, 8 : 2, v/v (<20 ppm H_2O)
4. Stop position (plugged)
5. Acetonitrile–pyridine, 8 : 2, v/v (<20 ppm H_2O)
6. DCM

2. Solvent System for the Phosphoramidite Method

1. DCM
2. 3% DCA/DCM, v/v
3. Acetonitrile (<20 ppm H_2O)
4. Stop position (plugged)
5. 2.6 g Iodine + 80 ml THF + 20 ml 2.6-lutidine + 2 ml H_2O
6. Acetonitrile (<20 ppm H_2O)

F. Assembly of Oligonucleotides

1. Preparation of Paper Disks

Cut round disks of Whatman 3MM, Whatman 1Chr, or Whatman ream paper (6.5 mm diameter). Use a 6 mm punch on triply folded paper. Take enough care that the disks fit tightly into the columns so that the flow is through the paper and not around them. Disks of 3MM paper of the correct size weigh 6.3 mg, while 1 Chr or ream paper disks weigh only 2.8 mg.

2. Addition of the First Nucleoside (Loading of the Paper Disks, See Fig. 1)

The protected 2-deoxynucleoside succinates should be without any traces of free succinate. This can be achieved by carefully purifying them by "short column" silica gel chromatography.[17] 2-Deoxynucleoside succinates from Cruachem (SPS grade) are of this high purity. Traces of free succinate severely inhibit oligonucleotide synthesis on cellulose paper support (our unpublished observation; the CPG supports as used in most automatic machines are less sensitive to the purity of the 2-deoxynucleoside succinates for loading).

Sets of paper disks (15-20 Whatman 3MM disks or 50 Whatman 1Chr or Whatman ream paper disks), all properly numbered with a soft pencil on both sides, are stacked into each of four glass columns with a pair of tweezers (philatelist type). The columns are flushed with anhydrous acetonitrile–pyridine (8 : 2, v/v) or anhydrous acetonitrile (5 min), using the helium pressure-controlled solvent delivery system at a flow rate of 1 ml/min. Each of the four protected 2-deoxynucleoside succinates (40 mg each, about 50 μmol) is coevaporated twice with pyridine in different colored Eppendorf tubes (e.g., green for A, blue for C, red for G, and white for T) fitted in a Savant Speed Vac concentrator. Air is let in through a big calcium chloride drying tube and a 3-way valve. An even better solution to avoid humidity from the air is to let in argon from a balloon mounted on an additional 3-way valve so that it can be refilled without being disconnected. Anhydrous pyridine (300 μl) is added with a syringe to dissolve each of the four protected 2-deoxynucleoside succinates. The succinate solutions (300 μl), 200 μl (339 μmol) of a freshly prepared solution of MSNT in anhydrous pyridine (500 mg/ml), and 30 μl (72 μmol) of 1-methylimidazole are mixed with syringes in the color-coded Eppendorf tubes. A total of 530 μl of each activated succinate is injected slowly in parallel onto the respective columns over a 10-min period (330, 100, and 100 μl) and then left without flow for a further 2–3 hr. After flushing with pyridine–acetonitrile (2 : 8, v/v) or acetonitrile (1 ml/min, 5 min), the flow is stopped and the capping reagent (0.5 ml), acetic acid anhydride–pyridine–DMAP (2 : 8 : 1, v/v/w), is injected into each column and left there for 10 min. The reagent is removed by washing with pyridine–acetonitrile (2 : 8, v/v) or acetonitrile (5 min). After further washing with DCM (2 min), the columns are taken apart and the paper disks pushed with the plunger into a beaker. They are washed by swirling with ether and dried in the air on a sheet of Whatmann 3MM paper. At this stage the disks can be stored at −20° in closed vials. One disk from each column is routinely subjected to trityl analysis.

3. Trityl Analysis

Put a disk in a stoppered tube with 10 ml 3% DCA in DCM. Leave the filter for 10 min and swirl it with intervals to extract all color from the paper disk. Measure the absorbance at 503 nm in a 1-cm cuvette.

$$\text{Filter loading} = (A_{503} \times 10^7)/71000 \qquad \text{nanomoles of deoxynucleoside}$$

Average loading of 3MM disk: 100–150 nmol ≈ 20 μmol/g.

4. Sorting Protocols

a. Manual Sorting Protocol. Write down oligonucleotide sequences in a protocol (see Fig. 14). Write the sequences from the 5′ to the 3′ end, aligning all sequences at the 3′ end. The 3′-end nucleotide is the one to be attached directly to the paper support (loading, see Section II,F,2). Label each oligonucleotide with a code number (see Fig. 14) on each paper disk. Spread the paper disks on a clean sheet of 3MM paper. Line up four vials marked A, C, G, and T next to the sheet. Pick up each disk individually with the help of the tweezers. Take care by reading the code number on both sides of the disk, since disks may stick together, and match it with the protocol. Draw a circle on the protocol sheet around the following nucleotide to be attached and place the disk in the corresponding vial. When all the disks are sorted, make sure that all oligonucleotides in the protocol have been encircled, corresponding to the following round of couplings. Also recheck the disks in each vial. These precautions avoid any doubts about the oligonucleotide sequences. An extra precaution is to recheck the disks from each column after each coupling round, i.e., before sorting.

b. Computer-Directed Sorting. A computer program OLIGO has been designed in our laboratory[23] to facilitate data entry, analysis of oligonucleotides, sorting of disks, and preparation of files for storage. This program is available on request. Figure 15 shows the OLIGO program menu.

AC - SERIES

AC1	5'-CCTCGAGGAGTC-3'	12-mer
AC2	5'-AAGATGGCCGATCA-3'	14-mer
AC3	5'-GAACCAGAACACC-3'	13-mer
AC4	5'-TGCAGAGGT-3'	9-mer
AC5	5'-CTAGACCTCTGCAGGTGT-3'	18-mer
AC6	5'-TCTGGTTCTGATCG-3'	14-mer
AC7	5'-GCCATCTTGACTC-3'	13-mer
AC8	5'-CTCGAGGGTAC-3'	11-mer
AC9	5'-CTATTACACGATCA-3'	14-mer
AC10	5'-TGTAATAGGACTC-3'	13-mer
AC11	5'-CCTCGAGCTGCAG-3'	13-mer
AC12	5'-CAGCTGGCAGGAAGCAG-3'	17-mer
AC13	5'-GTCATGTGGCAAGGCTA-3'	17-mer
AC14	5'-TTTGGGGAAGGGAAAAT-3'	17-mer
AC15	5'-AAAACCACTAGGTAAAC-3'	17-mer
AC16	5'-TTGTAGCTGTGGTTTG-3'	16-mer
AC17	5'-AAGAAGTGGTTTTGAA-3'	16-mer

FIG. 14. Protocol for manual sorting of paper disks.

Options 1–3 are used for the input and checking of oligonucleotide sequences. Option 4 is used to analyze oligonucleotides (Fig. 16). This option is used to check a series of oligonucleotides for restriction enzyme sites or sequence homologies which can cause problems during ligation reactions. Option 5 is the sorting menu used for the assembly of the oligonucleotides. Figure 17 shows an example of a sorting step. Option 6 is used to transfer input data to storage files. An example of such a file is shown in Fig. 18.

```
0. Exit oligo synthesis system.
1. Input data for new project.
2. Verify oligo sequences.
3. Modify existing project data.
4. Analyse oligos.
5. Continue existing synthesis project.
6. Store oligo data in text file.


Type option number now :
```

FIG. 15. OLIGO program menu.

```
0. Exit analysis mode.
1. Search for inverted repeats.
2. Search for direct repeats.
3. Search for homologies with other sequences.
4. Search for restriction enzymes.

Choose the option number :
```

FIG. 16. Option 4 from OLIGO menu: Analyze oligos.

5. *The Elongation Cycle*

a. The Elongation Cycle for the Phosphotriester Method (Fig. 2) (for 50–90 3MM disks or 90–150 1Chr or ream paper disks). Each of the four protected 2-deoxynucleotide TEA salts (40 mg each) is coevaporated twice with pyridine in a dry Savant Speed Vac concentrator in color-coded Eppendorf tubes (a stock solution of 80 mg/ml can be made, enough for all cycles, and 0.5 ml is pipeted to each tube). It is convenient to prepare tubes enough for 1 day's work. The Speed Vac concentrator has a capacity for tubes sufficient for 10 cycles. Let in argon or dry air and close the tubes as quickly as possible. All tubes are kept in a desiccator

```
-------------------------------------------------------------------
                    *** STEP  12 ***
                        --------

 The following oligonucleotides were finished during the previous step :
       8,  31,  76,  77,

 Following is the oligo list for the next set of tubes.

*A* -->  25 filters.

     3,   9,  16,  17,  19,  20,  32,  35,  36,  40,  42,  44,  45,  48,
    49,  59,  60,  61,  62,  63,  65,  72,  73,  78,  79,

*C* -->  17 filters.

     1,   5,   7,  11,  15,  22,  26,  28,  29,  30,  39,  43,  52,  56,
    57,  58,  81,

*G* -->  14 filters.

     2,  10,  12,  13,  14,  23,  27,  47,  51,  54,  66,  67,  74,  80,


*T* -->  15 filters.

     6,  18,  24,  25,  33,  34,  37,  38,  41,  46,  50,  53,  64,  68,
    75,
```

FIG. 17. Example of sorting step from option 5 from OLIGO menu.

AC - SERIES

AC1	5'-CCTCGAGGAGTC-3'	12-mer
AC2	5'-AAGATGGCCGATCA-3'	14-mer
AC3	5'-GAACCAGAACACC-3'	13-mer
AC4	5'-TGCAGAGGT-3'	9-mer
AC5	5'-CTAGACCTCTGCAGGTGT-3'	18-mer
AC6	5'-TCTGGTTCTGATCG-3'	14-mer
AC7	5'-GCCATCTTGACTC-3'	13-mer
AC8	5'-CTCGAGGGTAC-3'	11-mer
AC9	5'-CTATTACACGATCA-3'	14-mer
AC10	5'-TGTAATAGGACTC-3'	13-mer
AC11	5'-CCTCGAGCTGCAG-3'	13-mer
AC12	5'-CAGCTGGCAGGAAGCAG-3'	17-mer
AC13	5'-GTCATGTGGCAAGGCTA-3'	17-mer
AC14	5'-TTTGGGGAAGGGAAAAT-3'	17-mer
AC15	5'-AAAACCACTAGGTAAAC-3'	17-mer
AC16	5'-TTGTAGCTGTGGTTTG-3'	16-mer
AC17	5'-AAGAAGTGGTTTTGAA-3'	16-mer
AC18	5'-ACACTCTGTCCAGCCC-3'	16-mer
AC19	5'-CACCAAACCGAAAGTCC-3'	17-mer
AC20	5'-AGGCTGAGAGGT-3'	12-mer
AC21	5'-CTAGACCTCT-3'	10-mer
AC22	5'-CAGCCTGGACTTTCGG-3'	16-mer
AC23	5'-TTTGGTGGGGCTGGAC-3'	16-mer
AC24	5'-AGAGTGTTTCAAAACCAC-3'	18-mer
AC25	5'-TTCTTCAAACCACAG-3'	15-mer
AC26	5'-CTACAAGTTTACCTA-3'	15-mer
AC27	5'-GTGGTTTTATTTTCC-3'	15-mer
AC28	5'-CTTCCCCAAATAGCCTT-3'	17-mer
AC29	5'-GCCACATGACCTGCTT-3'	16-mer
AC30	5'-CCTGCCAGCTGCTGCAG-3'	17-mer
AC31	5'-CTCGAGGGTAC-3'	11-mer
AC32	5'-CTATTACAAGGAAGCAG-3'	17-mer
AC33	5'-GTCTATTACAAAGGCTA-3'	17-mer
AC34	5'-CGGAGTAGTAGGTAAAC-3'	17-mer
AC35	5'-TTGTAGCTCTACTCCG-3'	16-mer
AC36	5'-AAGAACTACTCCGGAA-3'	16-mer
AC37	5'-AGAGTGTTTCCGGAGTAG-3'	18-mer
AC38	5'-TTCTTCGGAGTAGAG-3'	15-mer
AC39	5'-CTACTCCGATTTTCC-3'	15-mer
AC40	5'-TGTAATAGACCTGCTT-3'	16-mer

FIG. 18. Example of storage file created using option 6 in OLIGO menu.

over silica gel until needed. A set of four tubes is taken out before each coupling step, and 300 μl of anhydrous pyridine is added with a syringe by piercing the needle through the cap. Two hundred microliters of MSNT in pyridine (500 mg/ml) and 30 μl of 1-methylimidazole are added, each tube is mixed on a vortex, keeping the syringe in the tube, and then injected onto their respective columns (530 μl total). The amounts can be decreased to 20 mg monomer in 250 μl pyridine, 100 μl MSNT solution, and 15 μl 1-methylimidazole (365 μl total) for as many as 50 Whatman 3MM disks or 90 Whatman 1Chr or ream paper disks.

Before starting in the morning, degas all solvents in the bottles by opening a valve and releasing the pressure and then placing them one after another in a sonicating bath. This will prevent spontaneous bubble formation in the columns. Any bubbles in columns can be removed with the aid of a syringe (by suction with all valves closed). This precaution can be omitted by using helium as a driving gas for the solvent delivery system.

The elongation cycle is schematically represented as follows:

1. Sort the disks (sorting time is 10–30 min). Pull down the plunger (see inset in Fig. 11) to the lower part of the column and push down the Teflon frit to touch the plunger. Place the disks in the bottom part of the four columns (marked A, C, G, and T). Fill up with DCM. Introduce the injection end piece at an angle to avoid introduction of any bubbles and press it down firmly to chase any bubbles trapped between the disks. Then press up the disks with the plunger so that there is only 2 mm of liquid-filled space above the disks. Open up for the flow and check that no bubbles are trapped in the system. If any bubbles are seen, first release the pressure on the septum by turning the cap one turn counterclockwise to avoid damaging the septum when introducing the syringe. Tighten the cap back one turn, open up for the flow and suck up a little solvent with the syringe. Close the flow at the 8-way solvent distributor valve and suck up the bubbles.
2. Flush with DCM (rotary valve in position 6, Fig. 13a, 1 min) by opening the four outlet valves (mark them A, C, G, and T) of the 8-way valve which distribute the solvents to the columns.
3. Flush with 3% DCA in DCM (rotary valve in position 1, Fig. 13a, 2 min).
4. Flush with DCM (position 2, Fig. 13a, 1 min).
5. Flush with pyridine–acetonitrile (2:8, v/v) (position 3, Fig. 13a, 4 min). During this step add MSNT and 1-methylimidazole to the four mononucleotides as described above.
6. Stop the flow (position 4, Fig. 13a, also close all the valves on the 8-way valve solvent distributor). Unscrew the septum cap of the column one turn and slowly insert the syringe containing activated monomer solution and tighten the cap back one turn. Inject one-half of the activated nucleotides to the proper columns, leaving the

syringes on the columns. After 2 min inject another one-quarter of the mixtures, wait another 2 min and inject the rest. Leave for 11 min (15 min total).
7. Flush with pyridine–acetonitrile (2 : 8, v/v) (rotary valve in position 5, 3 min).
8. Flush with DCE (rotary valve in position 6, 1 min).
9. Stop the flow by closing all the outlet valves on the 8-way valve (A, C, G, and T) and press the disks into a beaker containing ether. Decant the ether and let the disks dry on a piece of filter paper.

b. The Elongation Cycle for the Phosphoramidite Method (For 40–90 Whatman 3MM disks or 90–150 1Chr or ream paper disks). Three 100-ml conical flasks with No. 1 Quick Fit seal (N/S 14/23) are dried in the oven at 120° and cooled in a desiccator over silica gel. Fill with anhydrous acetonitrile and about 5 g of activated molecular sieves and seal with a Suba-seal (silicone rubber) septum No. 30. Dissolve 0.5 g of each of the four protected β-cyanoethylphosphoramidites in 6 ml anhydrous acetonitrile (about 0.1 *M*). The solutions are distributed to dry, color-coded Eppendorf tubes (400 μl each) and evaporated to dryness in a Speed Vac concentrator. Argon is let in from a balloon and the tubes closed quickly and stored in a desiccator over silica gel at 4° or at −20° for longer periods. Sublimed tetrazole (1.40 g) is dissolved in 40 ml anhydrous acetonitrile (0.5 *M*). Before each coupling step, a set of four Eppendorf tubes with dry phosphoramidites is removed from the desiccator and 400 μl anhydrous acetonitrile injected by piercing through the cap with the syringe needle. Leave the needle in the tube. Add 400 μl tetrazole solution and mix on a vortex or by pumping with the syringe. Inject the activated mononucleotides onto the appropriate columns (800 μl total). If the protected nucleoside phosphoramidite solutions are to be used up in a few days, they can be kept as such in septum-sealed bottles (they are generally packed in such bottles). Before each coupling step, fractions of 400 μl are transferred to color-coded Eppendorf tubes and used immediately without drying down. Add 400 μl 0.5 *M* tetrazole (see above) in acetonitrile and mix. Inject immediately onto the appropriate columns. (For less than 40 3MM disks (10 disks/column) or 75 1Chr or ream paper disks the volumes injected can be halved.)

It is possible to include a capping step in the cycle by injecting a capping reagent (mix equal volumes of a solution of 6% DMAP in acetonitrile and a solution of 20% acetic anhydride in a mixture of *sym*-collidine–acetonitrile, 3 : 5) by syringe after step 8 in the scheme shown below. Leave the capping reagent for 1 min and rinse it out by a 1-min acetonitrile wash. In our hands this step has proved unnecessary but is generally included when using the phosphoramidite method.[17]

The elongation cycle is schematically represented as follows (see Fig. 13b):

1. Sort the disks and place them in the four columns (marked A, C, G, and T; see Section II,F,5,a,1) (time for sorting, 10–30 min).
2. Flush with DCM leaving the rotary valve in position 1 and opening the four outlet valves (marked A, C, G, and T) on the 8-way valve, which distributes the solvents to the four columns (1 min).
3. Flush with 3% DCA in DCM by turning the rotary valve for solvent selection to position 2 (2 min).
4. Flush with acetonitrile (rotary valve position 3, 4 min).
5. Stop the flow by turning rotary solvent selector valve to position 4. Also close valves A, C, G, and T on the 8-way solvent distribution valve to avoid pushing the coupling mixtures back into the distribution system. Inject one-half of the activated coupling mixtures onto the appropriate columns and leave the syringes on the columns. After 1 min inject another one-quarter of the mixtures, wait another 30 sec, inject the rest, and wait for 3 min (total 4.5 min).
6. Open the valves A, C, G, and T on the 8-way valve and turn the rotary valve counterclockwise back to position 3. Flush with acetonitrile (1 min).
7. Flush with the iodine solution by turning the rotary valve to position 5 (1 min).
8. Flush with acetonitrile (rotary valve in position 6, 1 min).
9. Flush with DCM (rotary valve in position 1, 1 min).
10. Stop the flow by closing the valves on the 8-way distribution valve. Leave the rotary valve solvent selector in position 1. Loosen the solvent inlets to the columns and pull out the upper end piece. With the plunger, push the disks into a small beaker with dry ether. Decant the ether and dry the disks by leaving them on a clean piece of Whatmann 3MM paper.

G. Deprotection of Oligonucleotides

1. Deprotection of Oligonucleotides Assembled by the Phosphotriester Method

a. Deprotection by the Ammonia Procedure. The procedure described in Section II,G,2 for the phosphoramidite method can also be used for the phosphotriester method. This procedure is the simplest and gives sufficiently good purity for most applications, e.g., for oligonucleotides that are to be used as primers. We have compared a large number of oligonucleotides assembled by the phosphotriester method and deprotected by both the ammonia method (as described below in Sections II,G,2 and II,G,1,b,ii) and the oximate method (Section II,G,1,b) and have found no differences in HPLC patterns. This means that internucleotide cleavage under the conditions given above seems to be at the same level as for the oximate method. However, the percentage of unwanted substitution mutations using the SGS technique (see Section II,J) is higher with oligonucleotides deprotected with the ammonia method. A reasonable explanation is that there are less base modifications in oligonucleotides deprotected by the oximate method.[25] For this reason we recommend the oximate method for deprotection of oligonucleotides to be used for the "SGS" method.

b. Deprotection by the Oximate Method (method of choice for oligonucleotides to be used for SGS). (i) After the last synthesis cycle (Section II,5,a), place the carefully marked disks in 0.5-ml Eppendorf tubes. Add 50 μl of deprotection solution [*syn*-2-nitrobenzaldoxime (0.56 g) in dioxane–water (1 : 1, v/v, 8 ml) and 1,1,3,3-tetramethylguanidine (0.40 ml)], vortex the tubes, centrifuge, and leave overnight (16 hr) at room temperature. Place the tubes in a Speed Vac concentrator and evaporate to near dryness (30 min approximately; the disks are kept in the tubes). Add concentrated ammonia (25%, 100 μl), vortex the tubes, and place them at 50° for 5 hr or overnight at 37°.

(ii) After cooling, divide each sample into two aliquots (50 μl each in 0.5-ml tubes; there is no need to remove the paper disks—they serve as an "internal control" if the number on the tube is lost by spilling ether or ethanol). These aliquots are then purified either by HPLC or by PAGE. Place all the tubes in a Speed Vac concentrator, evaporate for 30 min using a water pump, and then evaporate to near dryness with an oil pump (15 min). When aliquots are to be purified by HPLC, 0.1 *M* triethylammoniumacetate (TEAA), pH 7.0 (100 μl), is added and the solutions washed by extraction with ether (3 × 300 μl). Such samples may be stored indefinitely at −20° and, when required, injected directly onto an HPLC column (see Section II,H,1). Whereas the purification by HPLC (see Section II,H,1) is carried out at the level of the 5′-*O*-DMT-protected oligonucleotide and followed by detritylation, the purifications by preparative gel electrophoresis are carried out after detritylation.

[25] C. B. Reese and L. Zard, *Nucleic Acids Res.* **9,** 4611 (1981).

2. Deprotection of Oligonucleotides Assembled by the Phosphoramidite Method

After the last synthesis cycle [Section II,F,5,b (or II,F,5,a; see Section II,G,1,a)], place the paper disks in appropriately marked 0.5-ml Eppendorf tubes. Add 100 μl of concentrated ammonia (25–30%), leave overnight at room temperature, and then incubate 5 hr at 50° (or 1 hr at room temperature and then overnight at 37°). Continue as above (Section II,G,1,b,ii).

3. Detritylation

Dissolve the vacuum-dried deprotected oligonucleotide mixtures (Section II,G,1,b,ii) in 80% (v/v) acetic acid (50 μl) and keep them at room temperature for 20 min. Water (25 μl) is added, and the solution is extracted with ether (3 times, 200 μl each) and evaporated in a Speed Vac concentrator. Add water (25 μl) and reevaporate.

H. Purification of Oligonucleotides

1. Purification by HPLC (see Figs. 3, 4, and 5)

The 5′-*O*-DMT-protected oligonucleotide mixture (Section II,G,1,b,ii, before detritylation, Section II,G,3) in TEAA buffer (40 μl) is injected onto an analytical C_{18} reverse-phase column (4 × 125 mm; 5 μm spherical particles). The elution program is 5 min with 15% acetonitrile in 0.1 *M* TEAA, pH 7.0, at a flow rate of 1 ml/min, and then a linear gradient of 15–30% acetonitrile is applied over 60 min. The pure 5′-*O*-DMT-oligonucleotide, visible as a major peak, elutes after 32–45 min. Short oligonucleotides and oligonucleotides with a T at the 5′ end elute late, after approximately 45 min. The eluted oligonucleotide solution is collected in a 1.5-ml Eppendorf tube and evaporated to dryness in a Speed Vac concentrator. After detritylation (see Section II,G,3) the oligonucleotide is ready for use.

2. Purification using C_{18} Reverse-Phase Cartridges

If no HPLC is available, a relatively crude purification on a large scale (at least microgram scale) can be achieved using reverse-phase cartridges (Sep-Pak, Millipore No. 51910 or Bond Elut C_{18}, Analytical International, Harbor City, CA 90710). This will in general give oligonucleotides of sufficient purity to be used as primers. This purification method is also very convenient as an alternative to ethanol precipitation, before loading on a polyacrylamide gel (see Section II,H,3,b) or after elution from a gel slice. The advantage is that there is no loss of short oligonucleotides. However, this method is not recommended for purification of small amounts (nanogram scale) since in this case the major part is lost by irreversible absorption to the reverse-phase material. The C_{18} reverse-phase cartridge consists of a small cylinder containing reverse-phase silica gel. Solutions are pumped through it by attaching a syringe with a Luer-end fitting to the longer end of the cartridge (Sep-Pak) or using a syringe piston (Bond Elut). When polar solvents are pumped through, the less polar components of a mixture are retained by absorption on the reverse-phase material while the more polar components are eluted. A gel elution solution contains the synthetic oligonucleotide in an aqueous solution of buffer and urea. When this is passed through a reverse-phase cartridge, the oligonucleotide is retained while the aqueous salts and urea pass through it. The oligonucleotide is then eluted by using a less polar solvent mixture.

Procedure. A C_{18} reverse-phase cartridge is fitted to the Luer-end fitting of a 5-ml syringe and fixed on a stand with a clamp without tightening too firmly. The plunger is removed and 5 ml acetonitrile (chromatography grade) is added to the syringe to prewash and prewet the reverse-phase material. After 5 min, the remaining acetonitrile is pushed out with the plunger. When removing the plunger for refilling, the syringe should be disconnected from the cartridge.

Purification before Detritylation of 5′-O-DMT-Protected Oligonucleotide Mixture in 0.1 M TEAA Buffer (after procedure in Section II,G,1,b,ii). The cartridge is rinsed with 3 × 5 ml 0.1 *M* TEAA buffer (pH 7.0), using the plunger to give a medium flow rate (5 ml/min). The 5′-*O*-DMT-protected oligonucleotide mixture in 0.1 *M* TEAA (pH 7.0) (40 μl from the procedure in Section II,G,1,b,ii is diluted to 1 ml and added to the syringe fitted on the cartridge. With a slight pressure, the solution is slowly pushed onto the cartridge. The cartridge is washed with 15% acetonitrile in 0.1 *M* TEAA (pH 7.0) (2 × 5 ml). Then a solution of 30% acetonitrile in 0.1 *M* TEAA (pH 7.0) is applied and ten 1-ml fractions taken in 1.5-ml Eppendorf tubes. The fractions are checked for UV absorption at 260 nm. The 5′-*O*-DMT-oligonucleotide elutes in the first three fractions with 90% in the second fraction. The product-containing fractions are evaporated to dryness in a Speed Vac concentrator and detritylated (see Section II,G,3). A similar protocol can be used for entirely deprotected oligonucleotides using 20% acetonitrile in water as eluant. In this case the columns are prewashed and washed with water. This is an alternative protocol for preparing short oligonucleotide mixtures prior to their fractionation by PAGE (see Section II,H,3,b).

3. Purification by Polyacrylamide Gel Electrophoresis

a. Preparation of Samples. The oligonucleotide mixtures are contaminated with residues of chemicals which when applied to a gel in large quantities interfere with the electrophoretic run. A simple ethanol precipitation substantially improves the separation. If less than one-tenth of the detritylated mixture (procedure, Section II,G,3) is applied to a preparative polyacrylamide gel without a prior purification step, the separation and purity obtained is adequate. For use for SGS we currently load 10% of the detritylated mixture (Section II,G,3) directly onto the gel (see Fig. 6). However, when larger amounts of oligonucleotides are required we recommend one of the following three (Sections II,H,3,a,i, ii, or iii) prepurification procedures. These methods are equally useful for purification of oligonucleotides after PAGE to remove urea, acrylamide, salt, and other impurities, which otherwise would inhibit enzymatic reactions.

i. Ethanol precipitation. Dissolve the vacuum-dried detritylated aliquots (see Section II,G,3) in 50 μl of water. For medium length (>12 nucleotides) to long oligonucleotides, add 3 *M* sodium acetate (pH 6.5) (5 μl) and ethanol (150 μl) and place the mixtures at $-20°$ overnight. After centrifugation in a microfuge (15 min, approximately 13,000 rpm, 4°) wash the pellets with ethanol (100 μl) and lyophilize. For smaller oligonucleotides (<12 nucleotides), the following procedure is recommended. Add 3 *M* potassium acetate (very soluble in ethanol, 10 μl) and 6 volumes of ethanol (360 μl). Leave overnight at $-20°$. The isolation is as above.

ii. Drop dialysis on Millipore membranes

1. Pour 10 ml sterile water in a sterile, plastic Petri dish.
2. Float a 25 mm diameter, type VS Millipore membrane (Millipore Cat. No. VSWP02500, pore size 0.025 μm) with the shiny side up on the water surface. Allow the floating filter to wet completely before proceeding.
3. Pipet the oligonucleotide droplet carefully (20–100 μl) onto the center of the filter.
4. Cover the Petri dish. Dialysis can be done at room temperature or in the cold. Do not move the dish after depositing droplet. Dialyze for 1 hr.
5. Carefully retrieve the droplet with a micropipet.

iii. Microscale ion-exchange chromatography. A rapid microscale ion exchange chromatography (DEAE–Sephadex A25, Pharmacia, Uppsala, Sweden), can be performed for small oligonucleotides (e.g., linkers) for which ethanol precipitation is not efficient. Siliconized Eppendorf tubes (0.5 ml) are punctured in the bottom, tightly plugged with siliconized glass wool, and filled with 60 μl of DEAE–Sephadex A25 (prewashed with 3 *M* TEAA, pH 9, and reequilibrated with 0.1 *M* TEAA, pH 9). Remove most of the liquid from the columns by placing them in 1.5-ml Eppendorf tubes and centrifuge 10 sec in a microfuge. The oligonucleotides are bound to the columns by resuspending the resins in the oligonucleotide solutions (50 μl) and incubating 3–5 min at room temperature. Remove most liquid by a 10 sec centrifugation (see above). Resuspend resins in 150 μl water and 150 μl 0.1 *M* TEAA (pH 9) (twice each), followed by a 10-sec centrifugation after each washing. Oligonucleotides up to 8 nucleotides in length are efficiently eluted with 1 *M* TEAA (pH 9) whereas longer oligonucleotides (up to 22-mers) require 3 *M* TEAA (pH 9) for quantitative elution. Remove the buffer by evaporation in a Speed Vac concentrator at 45°. Dissolve the pellets in 100 μl water and dry again.

b. Gel Electrophoresis. For purification by PAGE, the pellets after detritylation (Section II,G,3) (or after one of the prepurifications, Section II,H,3,a,i, ii, or iii) are dissolved in 10 μl of 90% formamide, 0.2% xylene cyanol, 25 m*M* Tris–borate (pH 8.5), 0.6 m*M* EDTA and electrophorezed at 25 mA on 20% polyacrylamide (acrylamide–bisacrylamide, 19 : 1) 8 *M* urea gels (1.5 mm thickness). The gels are removed from the plates, wrapped in SaranWrap and placed on plates with a fluorescence indicator (Merck Silicagel 60F 254, No. 5735). The bands are visualized by UV irradiation (254 nm) from above (a photo can be taken at this point using a yellow filter), sliced out, and extracted with water (3 times the volume of the gel pieces) overnight at 4°. The supernatants containing oligonucleotides are freed from the polyacrylamide pieces by centrifugation through siliconized glass wool in Eppendorf tubes punctured at the bottom. Desalting and concentrating the purified oligonucleotides are done either by ethanol precipitation, by drop dialysis (see above), or by binding to DEAE–Sephadex A25 (all as described above).

I. 5′-End Phosphorylation of Oligonucleotides

For Maxam–Gilbert DNA sequence analysis[26] the oligonucleotides are 5′-end phosphorylated using standard methods.[27] Linkers and adapters for cloning and overlapping oligonucleotides for synthesis of DNA segments (SGS, see Section II,J), are 5′-end phosphorylated according to the following protocol: 0.05–0.5 nmol of oligonucleotide is incubated with 1.0 nmol ATP, 3 pmol of [γ-^{32}P]ATP (5000 Ci/mmol) and 5

[26] A. Rosenthal, R. Jung, and H.-D. Hunger, this series, Vol. 155 [20].

[27] T. Maniatis, E. F. Fritsch, and J. Sambrook, "Molecular Cloning: A Laboratory Manual," pp. 122–126. Cold Spring Harbor Lab., Cold Spring Harbor, New York, 1982.

units T4 polynucleotide kinase in 50 m*M* Tris–HCl (pH 7.5), 10 m*M* $MgCl_2$, 5 m*M* DTT, 0.1 m*M* spermidine in 20 μl final volume for 30 min at 37°.

J. "Shotgun Gene Synthesis" (SGS)[3]

Twenty micrograms of the vector to be used for the insertion of synthetic DNA fragments is digested with appropriate restriction enzymes to fit with the flanking regions of the synthetic fragments, dephosphorylated with calf intestine alkaline phosphatase using standard procedures,[27] and carefully extracted with phenol and chloroform. The DNA is precipitated with ethanol using standard conditions, lyophilized, and resuspended in 40 μl of TE buffer (0.1 pmol/μl).

For SGS 0.5 pmol of each of the different phosphorylated oligonucleotides are mixed, concentrated by ethanol precipitation (see Section II,H,3,a,i for special conditions) resuspended in 10 μl of 50 m*M* Tris–HCl (pH 7.5), 10 m*M* $MgCl_2$, 20 m*M* DTT, 1 m*M* ATP, 0.1 m*M* spermidine, and hybridized for 1 hr at 37°. In cases where the oligonucleotides contain self-complementary sequences, the hybridization can be done at higher temperatures, depending on the degree of selfcomplementarity. In this case ATP should be added after cooling of the hybridization mixtures to room temperature. After addition of 0.1 pmol of vector fragment and 1 unit of T4 DNA ligase (we use 1 unit of ligase from Promega or 200 units of ligase from New England Biolabs), the ligation reaction is performed overnight at room temperature. An aliquot is used for transformation in an appropriate bacterial strain using competent cells. The method by Hanahan[28] generally gives excellent yields and is recommended.

K. Screening of DNA by Dideoxy Sequencing

DNA sequence analysis can be achieved either on single- or double-stranded DNA, using standard methods,[24,29–33] with an appropriate primer ideally hybridizing 20–30 nucleotides 5′ to the region containing the mutation.

[28] D. Hanahan, *J. Mol. Biol.* **166,** 557 (1983).

[29] F. Sanger, S. Nicklen, and A. R. Coulson, *Proc. Natl. Acad. Sci. U.S.A.* **74,** 5463 (1977).

[30] A. T. Bankier and B. G. Barrell, "Techniques in the Life Sciences," Vol. B(5), pp. 1–34. Elsevier, Ireland, 1983.

[31] M. D. Biggin, T. J. Gibson, and G. F. Hong, *Proc. Natl. Acad. Sci. U.S.A.* **80,** 3963 (1983).

[32] R. B. Wallace, M. J. Johnson, S. V. Suggs, K. Migoshi, R. Bhatt, and K. Itakura, *Gene* **16,** 21 (1981).

[33] G. F. Hong, *Biosci. Rep.* **2,** 907 (1982).

1. Single-Stranded DNA

If SGS is performed in M13 phage vectors, the screening is by dideoxy sequencing on ss DNA minipreparations as described.[29–31]

2. Double-Stranded DNA

If SGS is performed in plasmid vectors (e.g., pBR322, pUC), screening is most rapidly done on minipreparations of ds DNA. We currently use the alkaline lysis method, very similar to published methods.[27] We also recommend a modification of the rapid boiling method as described below:

a. Preparation of Minilysates by the "Rapid Boiling" Method (A modification of the method of Holmes and Quigley[27,34]). This procedure gives readable sequences with pBR322 derivatives using the dideoxy method. Using pUC derivatives, the autoradiograms of the sequencing gel are practically without background. The rapid alkaline lysis method[27] gives similar results.

1. Grow 4 ml culture overnight without amplification in 15-ml Falcon tubes. It is convenient to prepare 24 or 48 lysates simultaneously. For plasmids harboring Amp resistance (pBR, pUC), add 0.1 mg/ml ampicillin.
2. Keep a fraction (100 μl) of each culture at 4° (0.5-ml tubes) for growing up positive clones or for glycerol stocks.
3. Spin at 3000 rpm for 5 min.
4. Remove carefully the medium by aspiration and resuspend in 350 μl STET buffer by shaking all tubes together in the stand on a vibrating mixer (Bioblock, Ref. No. E94326) for 10 min. Alternatively, the pellets can be resuspended by vortexing.
5. Pour suspension into 1.5-ml Eppendorf tubes.
6. Add 25 μl freshly prepared lysozyme (10 mg/ml in STET buffer, see below). Mix by vortexing.
7. Incubate 5 min at room temperature.
8. Float tubes in boiling water bath for 40 sec.
9. Centrifuge immediately for 15 min in microfuge.
10. Pour supernatant carefully (at small angle) into a clean tube.
11. Add 80 μl 5 *M* ammonium acetate and 400 μl propan-2-ol. Leave at room temperature for 10 min and spin for 15 min.
12. Remove supernatant carefully by aspiration using a drawn out Pasteur pipet; spin briefly and remove the remaining liquid.

[34] H. Pelham, *Trends Genet.* **1,** 6 (1985).

13. Dissolve pellet in 50 μl TE buffer.
14. Load 2 μl on a 1% agarose gel to check DNA for deletions or inserts. Remember a size control.

For restriction enzyme digestions or sequencing, the DNA should be treated as follows:

15. Add 50 μl 5 *M* LiCl.
16. Leave for 10 min at −20°.
17. Spin for 10 min in microfuge.
18. Transfer supernatant to a clean Eppendorf tube.
19. Add 250 μl of ethanol. Leave for 30 min at −20°.
20. Spin for 15 min in microfuge.
21. Remove liquid and wash pellet with 70% ethanol.
22. Dry pellet in a Speed Vac concentrator.
23. Redissolve in 50 μl TE buffer.

This preparation gives DNA which can be digested with restriction endonucleases and sequenced by the dideoxy method. It contains some t-RNA which may disturb the analysis of restriction enzyme digestions if you have fragments of the same length (about 70 nucleotides). In this case proceed as follows:

24. After endonuclease treatment of a fraction (e.g., 10 μl reaction mixture) add a mixture of 1 μl 0.2 M EDTA, 1μl RNase (0.5 mg/ml) in 3μl gel loading solution. Digest for 15 min at 37° before loading on a gel.

STET BUFFER

Ingredient	Amount for 100 ml
8% Sucrose	8 g sucrose
5% Triton X-100	5 ml Triton X-100
50 m*M* EDTA	10 ml 0.5 *M* EDTA (pH 8.0)
50 m*M* Tris (pH 8.0)	25 ml 2 *M* Tris–Cl (pH 8.0)

b. Dideoxy Sequencing of Double-Stranded DNA[24,33,34]

1. Mix together
   i. 0.5 μg ds plasmid DNA (~0.2 pmol).
   ii. 2 pmol of primer as described above.
   iii. 2 units of appropriate restriction enzyme, which cuts outside the region to be sequenced.

iv. 1 μl 10× annealing buffer (see below).
v. Water to 10 μl.
Leave for 30 min at 37°.

2. Denature DNA by floating samples for 2 min in a boiling water bath.
3. Transfer quickly to liquid nitrogen or ice. Leave for 1 min.
4. Leave for 30 min at room temperature.
5. Add DNA polymerase (Klenow fragment, 1 unit) and [α-^{35}S]dATP (0.5 μl, 1000 Ci/mmol) to each tube (2 μl total).
6. Distribute 2 μl to each of four wells of a microtiter plate with conical bottom wells.
7. Add dideoxy termination mixtures (2 μl) to each well (see below.[24,30,31]
8. Spin microtiter plates in a suitable centrifuge, e.g., Jouan GR2000 SX with adapters (Ref. no. 11179477, 11179466, and 11179475). Leave 20 min at room temperature.
9. Add chase mix (2 μl 0.25 m*M* dNTP, N = A + C + G + T). Leave 15 min at room temperature.
10. Add gel loading solution (4 μl, see below). Spin down to mix and heat 5 min at 120° in an oven. Place microtiter plate on ice.
11. Load on 8% (7 *M* urea) polyacrylamide gel and run at 30–50 Watts.

Termination Mixtures	
ddA:	50 μ*M* ddATP
	125 μ*M* dCTP
	125 μ*M* dGTP
	125 μ*M* dTTP
ddC:	10 μ*M* ddCTP
	5 μ*M* dCTP
	125 μ*M* dGTP
	125 μ*M* dTTP
ddG:	25 μ*M* ddGTP
	5 μ*M* dGTP
	125 μ*M* dCTP
	125 μ*M* dTTP
ddT:	250 μ*M* ddTTP
	5 μ*M* dTTP
	125 μ*M* dCTP
	125 μ*M* dGTP

10× Annealing Buffer
100 m*M* Tris (pH 8.5)
100 m*M* $MgCl_2$

Gel Loading Solution
9 ml deionized formamide
250 μl 10× TBE
750 μl H_2O
20 mg bromophenol blue
20 mg xylene cyanol blue

Acknowledgments

We are grateful to Thomas Grundström, Martin Zenke, and Marguerite Wintzerith for their fruitful collaboration during the development of these methods. Jia Hao Xiao kindly provided the autoradiogram used in Fig. 10A. We also thank Glenn Albrecht, Meera Berry, Irwin Davidson, Mary Macchi, Geoff Richards, and Christian Schatz for useful comments during the preparation of the manuscript. The excellent assistance of Bernard Boulay for photography is highly appreciated. Work was supported by grants from the CNRS, the INSERM, the MRT, from the Association pour le Développement de la Recherche sur le Cancer, and from the Foundation pour la Recherche Médicale.

[15] Chemical Synthesis of Deoxyoligonucleotides by the Phosphoramidite Method

By M. H. CARUTHERS, A. D. BARONE, S. L. BEAUCAGE, D. R. DODDS, E. F. FISHER, L. J. MCBRIDE, M. MATTEUCCI, Z. STABINSKY, and J.-Y. TANG

The phosphite triester approach to DNA synthesis using deoxynucleoside phosphoramidites as synthons has become the method of choice for the preparation of deoxyoligonucleotides. The general synthetic strategy involves adding mononucleotides sequentially to a deoxynucleoside attached covalently to a silica-based insoluble polymeric support. Reagents, starting materials, and side products are then removed simply by filtration. At the conclusion of the synthesis, the deoxyoligonucleotide is chemically freed of blocking groups, hydrolyzed from the support, and purified to homogeneity by either polyacrylamide gel electrophoresis (PAGE) or high-performance liquid chromatography (HPLC). In sections which follow we outline the synthesis methodology; detailed protocols for preparing the silica support, the phosphoramidites, and deoxyoligonucleotides; and the purification of synthetic DNA. As part of each detailed procedure, we also highlight potential problems and their solutions.

Outline of the Synthesis Method[1]

General Outline. The polymer support is derived from silica-based materials such as Vydak, Fractosil, and controlled pore glass. These resins appear to be ideal matrices for DNA synthesis since silica is chemically inert toward reagents used in DNA synthesis and does not swell in

[1] M. H. Caruthers, *Science* **230,** 281 (1985).

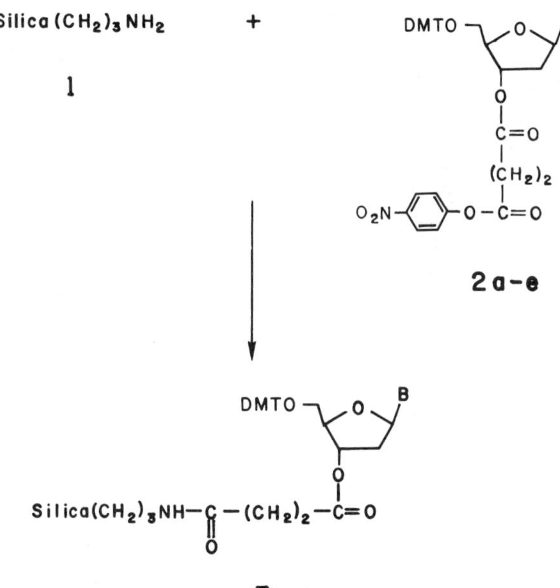

FIG. 1. Synthesis of deoxynucleosides linked to silica. **a**, B = Thymine; **b**, B = 6-*N*-benzoylcytosine; **c**, B = 6-*N*-benzoyladenine; **d**, B = 2-*N*-isobutyrylguanine; **e**, B = 6-*N*-[1-(dimethylamino)ethylidene]adenine; DMT = dimethoxytrityl.

common organic solvents. It can therefore be packed into a column and reagents pumped through the matrix without problems of compaction or swelling. Moreover these supports as originally formulated for HPLC were designed for efficient mass transfer during column chromatography. This latter characteristic allows DNA synthesis reagents to rapidly diffuse throughout the matrix which leads to high yields and minimal entrapment of reagents within the support.

The appropriately protected deoxynucleoside is linked to the matrix using the chemistry[2–4] as outlined in Fig. 1. Thus the initial step involves forming **1** by reacting 3-aminopropyltriethoxysilane with the support in ethanol for 4 hr. The next step, the synthesis of **2a–e**, proceeds by reacting the appropriate base-protected 5′-*O*-dimethoxytrityl deoxynucleoside with succinic anhydride. After an aqueous extraction against citric acid, the 3′-succinylated deoxynucleoside is converted to the *p*-nitrophenyles-

[2] M. D. Matteucci and M. H. Caruthers, *Tetrahedron Lett.* **21,** 719 (1980).

[3] M. D. Matteucci and M. H. Caruthers, *J. Am. Chem. Soc.* **103,** 3185 (1981).

[4] M. H. Caruthers, *in* "Chemical and Enzymatic Synthesis of Gene Fragments" (H. G. Gassen and A. Lang, eds.), p. 71. Verlag Chemie, Weinheim, Federal Republic of Germany, 1982.

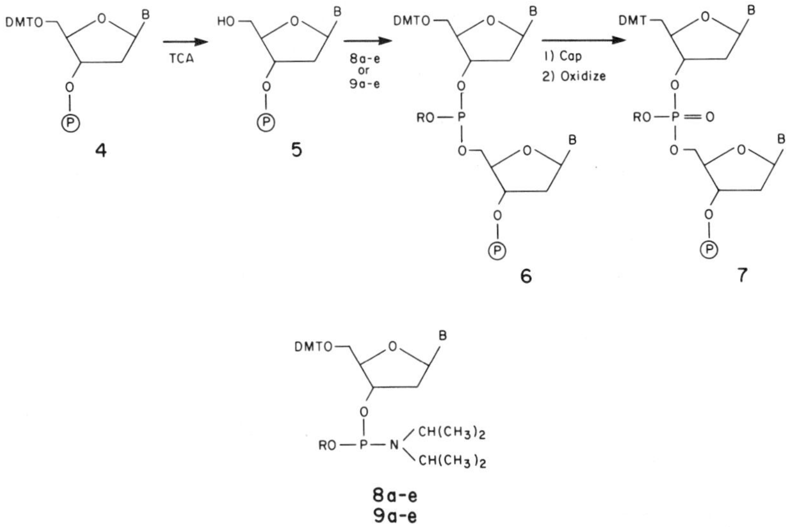

FIG. 2. Steps in the synthesis of a dinucleotide. Ⓟ, Silica support; TCA, trichloroacetic acid; **8a–e**, R = CH_3; **9a–e**, R = $NCCH_2CH_2$. **8a–e** and **9a–e** are differentiated by the base, B, as defined in Fig. 1.

ter using *p*-nitrophenol and dicyclohexylcarbodiimide (DCC). This deoxynucleoside is then added to **1** in a mixture of dimethylformamide, dioxane, and triethylamine. An intense yellow color forms rapidly, indicating the elimination of *p*-nitrophenol and the formation of **3a–e**. The ratio of reagents is usually adjusted so that 10–100 μmol deoxynucleoside per gram support is obtained. These substitution levels are convenient for synthesizing either the minimal quantities of DNA required for biological studies or the milligram amounts needed for biophysical studies.

The key intermediates for sequentially adding mononucleotides to the polymer support are appropriately protected deoxynucleoside 3′-phosphoramidites (Fig. 2, compounds **8** and **9**). These compounds are ideal synthons for preparing DNA since they are stable toward hydrolysis and oxidation but can easily be activated to form internucleotide linkages in very high yield. Since the methyl and β-cyanoethyl phosphoramidites (**8** or **9**) are currently used routinely, the use of both in DNA synthesis will be described. Although originally synthesized from diisopropylaminomethoxychlorophosphine[5,6] or diisopropylamino-2-cyanoethoxychlorophosphine,[7] these compounds are now prepared from bis(diisopropyl-

[5] S. L. Beaucage and M. H. Caruthers, *Tetrahedron Lett.* **22,** 1859 (1981).
[6] L. J. McBride and M. H. Caruthers, *Tetrahedron Lett.* **24,** 245 (1983).
[7] N. D. Sinha, J. Biernat, and H. Koster, *Tetrahedron Lett.* **24,** 5843 (1983).

amino)methoxyphosphine[8] and bis(diisopropylamino)-2-cyanoethoxyphosphine.[9] This is because the bis(diisopropylamino)alkoxyphosphines are very stable toward oxidation and hydrolysis, whereas the chlorophosphines are pyrophoric and must be stored anhydrous under an argon atmosphere. The synthesis of **8** and **9** therefore proceeds in acetonitrile by mixing an appropriately protected deoxynucleoside, the bis(diisopropylamino)alkoxyphosphine, and diisopropylammonium tetrazolide, a weak acid which activates the synthesis. This pathway is attractive since only stable, easily prepared reagents are used, and the deoxynucleoside 3′-phosphoramidite synthons, when isolated by a simple aqueous extraction, are free of 3′–3′ dinucleotides and 3′-phosphinic acid side products. These synthons can then be stored indefinitely after drying under reduced pressure.

Addition of a mononucleotide to **4** requires the following four steps (Fig. 2): (1) removal of the dimethoxytrityl protecting group with acid to form **5**; (2) condensation with a protected deoxynucleoside 3′-phosphoramidite; (3) acylation or capping of unreactive deoxynucleoside; and (4) oxidation of the phosphite triester to the phosphate triester to form **7**. Thus the synthesis proceeds stepwise in a 3′ to 5′ direction by the addition of one deoxynucleotide per cycle until the required DNA fragment has been prepared. A relatively new approach has also been developed which involves an *in situ* synthesis of deoxynucleoside 3′-phosphoramidites (Fig. 3).[8,10] A protected deoxynucleoside is first reacted with bis(diisopropylamino)alkoxyphosphine using diisopropylammonium tetrazolide as catalyst to form **8** or **9** *in situ* which in turn is activated with tetrazole and condensed with a deoxynucleoside covalently joined to silica. Other steps in the cycle are identical to those shown in Fig. 2.

The manual synthesis of deoxyoligonucleotides using sintered glass funnels as reaction vessels will be described.[11,12] However, precisely the same procedure can be used with automatic[13] and semiautomatic[2,11] machines or other manual methods where test tubes[14] and syringes[15] are the

[8] A. D. Barone, J.-Y. Tang, and M. H. Caruthers, *Nucleic Acids Res.* **12,** 4051 (1984).

[9] R. Kierzek, D. W. Kopp, M. Edmonds, and M. H. Caruthers, *Nucleic Acids Res.* **14,** 4751 (1986).

[10] S. L. Beaucage, *Tetrahedron Lett.* **25,** 375 (1984).

[11] M. H. Caruthers, *Recomb. DNA, Proc. Cleveland Symp. Macromol. 3rd, 1981* p. 261.

[12] M. H. Caruthers, S. L. Beaucage, C. Becker, W. Efcavitch, E. F. Fisher, G. Gallappi, R. Goldman, P. deHaseth, F. Martin, M. Matteucci, and Y. Stabinsky, *Genet. Eng.* **4,** 1 (1982).

[13] M. Hunkapiller, S. Kent, M. Caruthers, W. Dreyer, J. Firca, C. Griffin, S. Horvath, T. Hunkapiller, P. Tempst, and L. Hood, *Nature (London)* **310,** 105 (1984).

[14] P. L. deHaseth, R. A. Goldman, C. L. Cech, and M. H. Caruthers, *Nucleic Acids Res.* **11,** 773 (1983).

[15] T. Tanaka and R. L. Letsinger, *Nucleic Acids Res.* **10,** 3249 (1982).

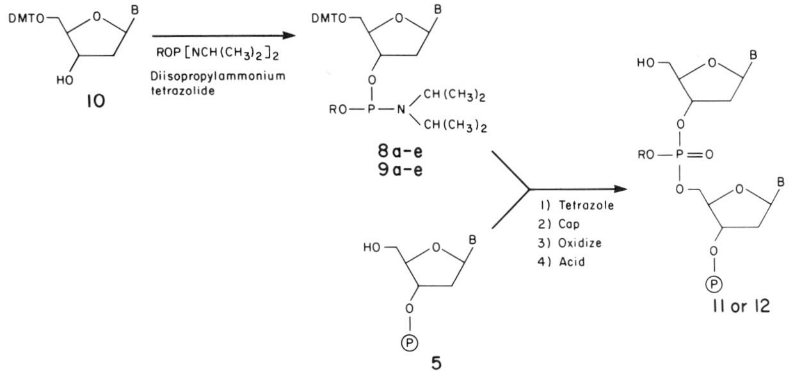

FIG. 3. Steps in the synthesis of a dinucleotide using *in situ* prepared deoxynucleoside 3′-phosphoramidites. **11**, R = CH_3; **12**, R = $NCCH_2CH_2$.

reaction flasks. Additionally, the *in situ* synthesis of deoxynucleoside 3′-phosphoramidites is included as part of this protocol. *In situ* synthesis methodologies are preferred where automatic or semiautomatic DNA synthesizers are still beyond the means of a small laboratory or uneconomical for a research group requiring only one or two deoxyoligonucleotides every few weeks. The same manual procedure can also be used with preformed deoxynucleoside 3′-phosphoramidites. When preformed synthons are used, they are simply dissolved in acetonitrile and substituted for the same *in situ* prepared compounds using the same concentrations and reaction times. These preformed synthons are especially attractive for preparing DNA on automatic or semiautomatic DNA synthesizers or for those who plan to manually synthesize a large number of deoxyoligonucleotides.

Detailed Description. Manual synthesis begins by loading each funnel with a silica support which has been derivatized with the protected deoxynucleoside corresponding to the 3′ terminus of the deoxyoligonucleotide to be synthesized in that funnel (**3a–e**). Generally a quantity of support to which has been linked 1 μmol of the desired deoxynucleoside is used in each synthesis. This is usually between 20 and 30 mg of derivatized support. The 5′-hydroxyl function of this deoxynucleoside is blocked by the acid-labile dimethoxytrityl group. The first step in the synthetic cycle is therefore the removal of this group with 3% trichloroacetic acid (TCA) in dichloromethane. In an acidic aprotic solvent, the dimethoxytrityl cation is bright orange (γ_{max} = 498 nm, $\varepsilon = 7.2 \times 10^7$). The completeness of each condensation may therefore be determined by measuring the amount of this cation recovered after the subsequent detritylation step. This parameter is generally called the trityl yield. Three percent TCA gives quantitative detritylation in approximately 90 sec. Since the protected purines are susceptible to depurination under acidic conditions, exposure to TCA should not exceed 2–2.5 min in order to insure complete detritylation of the 5′-hydroxyl group. After detritylation, all traces of acid must be removed to prevent detritylation of the phosphoramidite introduced during the following condensation step. Failure to do so leads to polycondensation, and the production of $N + 1$, $N + 2$, etc., deoxyoligonucleotides. This is especially important for deoxyguanosine which is the most labile toward detritylation.

Condensation of the next deoxynucleotide may proceed after the support and growing deoxyoligonucleotide chain have been rinsed free of any acid and dried by washing with anhydrous acetonitrile under a positive pressure of dry argon. This removes water which would compete for the reactive phosphoramidite to be introduced. A solution of tetrazole in dry acetonitrile is added to the support first, followed by a solution of the desired phosphoramidite (**8a–e** or **9a–e**). A 20 molar excess of phosphoramidite (relative to the loading of the first base on the support) and a further excess of tetrazole are used to ensure activation of the phosphoramidite. Coupling, which is complete in 2 min, forms a phosphite triester 3′–5′ linkage. This is a relatively labile linkage and must be oxidized to the stable phosphate triester form. However, a small amount of the phosphoramidite condenses with the enol tautomers of some of the carbonyl groups in the nucleoside bases, notably guanosine and thymidine. Although these phosphite adducts are very susceptible to hydrolysis, when oxidized to the phosphate form they become stable and can lead to the formation of modified, branched DNA. Thus, the hydrolysis of these unwanted side products must take place prior to the oxidation step. Fortunately this hydrolysis occurs during the capping reaction.

A small amount of the 5′-hydroxyl groups available for condensation with the added phosphoramidite fail to react. It is thus necessary to irreversibly block these groups to prevent the growing deoxyoligonucleotide chain from propagating in the next condensation step with a base deletion. Such remaining hydroxyl groups are acetylated with solutions of acetic anhydride and dimethyl aminopyridine (DMAP). At this time, the required hydrolysis of the ring adducts takes place, presumably via nucleophilic attack of acetate anions present in the capping solution. After the capping step is complete, strictly anhydrous conditions are no longer necessary until the dry acetonitrile wash immediately before the next condensation step. The capping step is followed by a wash with THF/lutidine/water to further ensure that phosphite adducts have been removed from the purine and pyrimidine bases.

Oxidation of the 3′–5′ phosphite triester linkage to the phosphate form is realized using a solution of iodine in aqueous lutidine. The iodine–

lutidine complex forms an adduct with the trivalent phosphorus, which is displaced by a water molecule leading to the release of a proton and oxidation of the phosphorus to the desired pentavalent form. The oxidation is complete within 30 sec. The support is then thoroughly washed to remove all of the iodine. The cycle begins again with detritylation of the 5′-hydroxyl of the deoxynucleoside just added to the growing DNA segment and continues until the synthesis is complete.

The deoxyoligonucleotide is next freed of protecting groups and isolated by PAGE, reverse-phase HPLC, or ion-exchange chromatography. For most applications such as DNA sequencing, cloning, or probing gene libraries, purification via PAGE is sufficient. The reaction mixture is simply loaded onto a polyacrylamide gel and fractionated by electrophoresis.[12] The product is then visualized with UV light and eluted from the gel by standard procedures.[16] When synthetic DNA is to be used directly for biochemical experiments, however, additional column chromatography is necessary in order to rigorously free the DNA of side products.

Typical results are illustrated in Figs. 4 and 5 for DNA purified by PAGE and reverse-phase HPLC, respectively. Although a large number of examples could have been selected, including many that are published,[1,8,11,12,14,17] the results shown in Fig. 4 are particularly interesting since the DNA is rich in deoxyguanosine. In the 36-mer illustrated in Fig. 4A, synthesis was completed using a capping step prior to oxidation (see above). As a consequence a very good yield of the product was obtained (other than dye markers, it was the only band observed when irradiated with UV light). Even a deoxyoligonucleotide containing essentially all deoxyguanosine (Fig. 4B) can be synthesized in high yield if guanine ring adducts are removed prior to oxidation. However if oxidation follows immediately after condensation, the synthesis of deoxyguanosine-rich DNA is more difficult using this procedure unless the O^6 position is protected.[18] We therefore recommend, as was also the case in the original[5] and subsequent publications,[1,4,11,12,14] that a hydrolytic wash and/or a capping step precede the oxidation step. Then O^6 protection is unnecessary.

The HPLC profile shown in Fig. 5 is also typical of the results obtained using this purification procedure. The major peak contains essentially pure product free of the impurities found in the trailing region of the product peak. This is the preferred purification procedure for DNA that is to be used directly for biochemical experiments. Unfortunately resolution is limited to compounds usually containing less than 20 mononucleotides.

[16] A. M. Maxam and W. Gilbert, this series, Vol. 68, p. 499.

[17] L. J. McBride, R. Kierzek, S. L. Beaucage, and M. H. Caruthers, *J. Am. Chem. Soc.* **108,** 2040 (1986).

[18] R. T. Pon, M. J. Damha, and K. K. Ogilvie, *Nucleic Acids Res.* **13,** 6447 (1985).

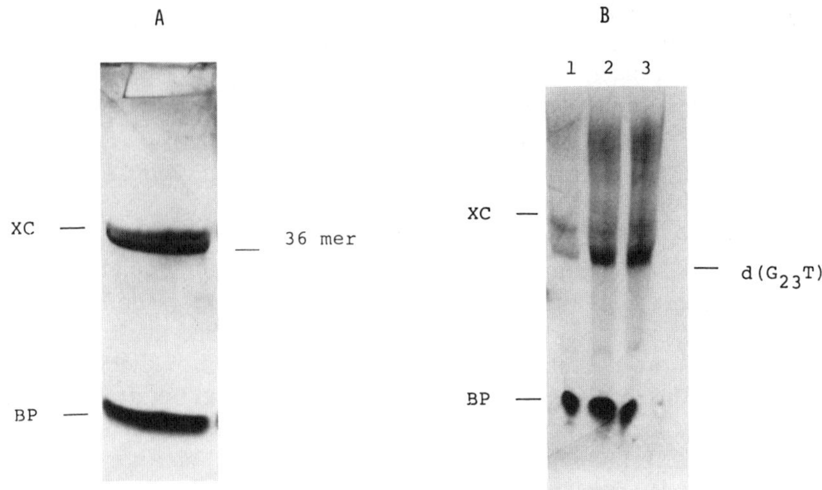

FIG. 4. Purification of guanine-rich DNA by PAGE. (A) Synthesis of d(CGCGGGGTGGAGCAGCCTGGTAGCTCGTCGGGCTCA). An aliquot of the crude reaction mixture after complete removal of protecting groups was loaded onto the 15% polyacrylamide gel. (B) Synthesis of $d(G_{23}T)$ where increasing concentrations of crude reaction mixture were loaded onto lanes 1, 2, and 3, respectively, after removal of all protecting groups. The sample of $d(G_{23}T)$ in 90% formamide was loaded at 95° onto the hot, 15% polyacrylamide gel. The electrophoresis was performed at high voltage in order to maintain a hot gel (approximately 65°) and therefore to minimize aggregation of the deoxyguanosine-rich DNA. The streaking above the more concentrated samples, however, is due to aggregation of the deoxyguanosine-rich DNA even at these high temperatures. Both deoxyoligonucleotides were enzymatically degraded to the normal mononucleotides. XC, xylene cyanol; BP, bromophenol blue. These dye markers absorb UV light and therefore also appear as dark bands. (The detailed visualization procedure is described in the section on DNA purification by PAGE.)

Experimental Procedures

Solvents

Since polynucleotide synthesis requires highly purified and anhydrous solvents in order to maximize yields, a first step is usually distillation. Dichloromethane is distilled from calcium hydride. 2,6-Lutidine, pyridine, and triethylamine are distilled first from toluenesulfonyl chloride and then calcium hydride. *N,N*-Diisopropylamine and morpholine are distilled from calcium hydride. Acetonitrile, required for anhydrous reaction conditions, must be distilled under argon, first from phosphorus pentoxide, and then from calcium hydride. Acetonitrile distilled only from phosphorus pentoxide is certainly dry, but is also acidic which affects both the preparation of phosphoramidites and the synthesis of DNA. The

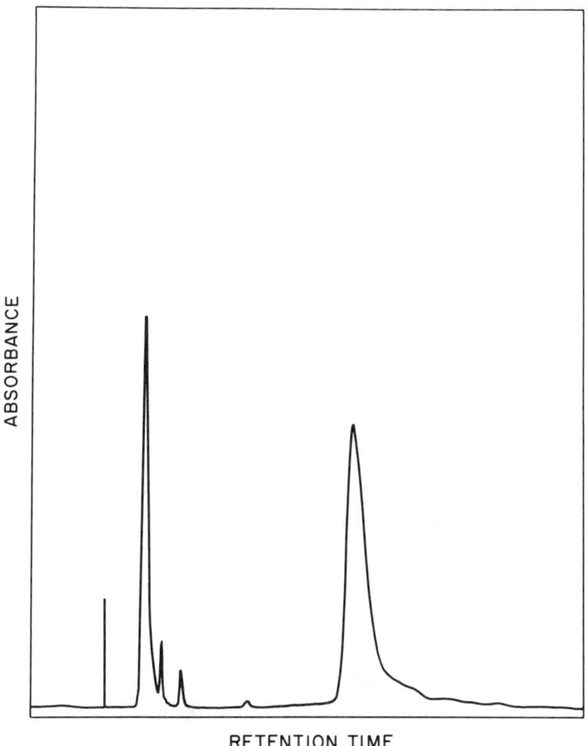

FIG. 5. Purification of synthetic DNA by reverse-phase HPLC. The unpurified but completely deprotected reaction mixture from the synthesis of DMTd(CCTTC-TAACAAGAAAACCTAGG) was fractionated on a C_{18} reverse-phase column. The eluant was a mixture of 24% acetonitrile and 76% triethylammonium bicarbonate (0.1 *M* v/v). The major peak retained on the column was the product. Early peaks not retained on the column are hydrolyzed protecting groups.

distillation from calcium hydride removes these acidic impurities. Dry acetonitrile should be stored in predried receiver flasks under septa and without any drying agent. If acetonitrile is stored over molecular sieves, dust from the molecular sieves eventually will clog the funnel frits used for DNA synthesis. Acetonitrile purchased from Burdick and Jackson (see commercial sources list) has been used without purification to give satisfactory yields of DNA. The acetonitrile for nonanhydrous washes can be used directly from the reagent bottle. Tetrahydrofuran (THF) is dried over sodium benzophenone and distilled. Acetic anhydride is distilled from phosphorus pentoxide. Methanol is refluxed over magnesium filings for 2 hr and distilled. *N,N*-Dimethylformamide (DMF) is distilled under reduced pressure below 50° first from phthalic anhydride and then calcium hydride. 3-Hydroxypropionitrile is distilled at less than 1 mm Hg. Pentane, hexane, and chloroform are dried by passage through a column of basic, chromatography grade alumina. All distillations, either of solvents or reagents, should be completed in a hood, behind a protective shield and under normal, safe laboratory conditions.

Reagents and Equipment

^{31}P-NMR spectra are recorded on a Bruker WM-250 spectrophotometer (101.2 MHz) and the chemical shifts (ppm) are reported relative to an external capillary standard of 85% H_3PO_4. HPLC is performed on a Waters Associates apparatus equipped with a solvent programer.

Dichloromethoxyphosphine. Dry methanol (133.5 ml, 3.3 mol) is added dropwise over 2 hr to a vigorously stirred solution of phosphorus trichloride (454 g, 3.3 mol) maintained at $-10°$ under an atmosphere of dry argon. The reaction mixture is then allowed to attain room temperature while being continuously stirred under the reduced pressure of a water aspirator (30 mm Hg). This step removes hydrochloric acid. The product is then purified by distillation as a single fraction using a water aspirator (30 mm Hg). The resulting product is next fractionally distilled under reflux using a glass helices packed column (2 $\times$ 20 cm) to give dichloromethoxyphosphine as a colorless liquid (bp 85–86°, 62.9 mm Hg) in 40–50% yield (100–130 ml). ^{31}P NMR ($CDCl_3$) shows a peak at δ 180.8. Based on ^{31}P NMR, the product should be at least 99% pure or redistillation is necessary.

The preparation of any chlorophosphine or aminophosphine must be completed properly in order to avoid explosions. During distillation of methoxydichlorophosphine, an orange polymer forms which can be especially dangerous. The first step therefore is not to distill to dryness but to leave approximately 25% of the product in the distillation flask. Alternatively after distilling approximately one-half of the product, the remainder can be transferred to a clean flask and the distillation continued until approximately 75% of the remaining phosphine has been collected. The pot residues, impure fractions, and polymeric material should then be neutralized. These residues are first dissolved in dichloromethane or carbon tetrachloride (approximately 5% solution) and transferred to a three-necked flask equipped with a reflux condenser and a vent tube. A 10% solution of methanol in dichloromethane or carbon tetrachloride is then placed in an addition funnel and added dropwise to the phosphine at a rate that maintains reflux. When the generation of heat ceases, the solution is neutral. However to insure complete reaction, an additional 100 ml of the 10% methanol solution should be added and the neutralized phosphine stirred for 12 hr. This solution can then be discarded in the proper manner with other organic waste. All glassware used to prepare these phosphines should also be rinsed thoroughly with a 10% methanol solution in dichloromethane or carbon tetrachloride before cleansing with water. Phosphines react violently with water.

2-Cyanoethoxydichlorophosphine. Phosphorus trichloride (32 ml, 0.36 mol) and triethylamine (59 ml, 0.44 mol) are dissolved in anhydrous ether (500 ml) and the solution maintained at $-12°$ with stirring under an atmosphere of dry argon. 3-Hydroxypropionitrile (25 ml, 0.36 mol) is dissolved in 250 ml anhydrous either and added dropwise over 2 hr while the temperature is maintained at $-12°$. After stirring overnight under argon at room temperature, the reaction mixture is filtered to remove triethylamine hydrochloride and the ether is removed with the aid of a water aspirator. The product is then fractionally distilled twice using a glass helices packed column (2×20 cm) to give 2-cyanoethoxydichlorophosphine as a colorless liquid (bp 78°, 0.6 mm Hg) in 52% yield. ^{31}P NMR (CH_2Cl_2) shows a characteristic peak at δ 175.98. Based on ^{31}P NMR the product should be at least 99% pure or redistillation is necessary. Various impure fractions and undistilled product should be neutralized in a manner described for the synthesis of dichloromethoxyphosphine.

Bis(diisopropylamino)methoxyphosphine. Diisopropylamine (96 g, 0.95 mol) is added dropwise over 1 hr to dichloromethoxyphosphine (14 g, 0.11 mol) in 500 ml of anhydrous ether at $-10°$ under a nitrogen atmosphere. The reaction mixture is allowed to warm to room temperature and is stirred an additional 16 hr. After filtration to remove diisopropylammonium hydrochloride, the crude product is isolated as a pale yellow liquid by removing solvent on a rotary evaporator. The product is fractionally distilled twice from calcium hydride (to remove residual amine salts) to afford bis(diisopropylamino)methoxyphosphine as a colorless liquid (bp 40–42°, 0.050 mm Hg; $d = 0.89$) in 77% yield (21.5 g) [^{31}P NMR (CH_2Cl_2) $\delta = 130.1$]. The product should be stored in a brown bottle under argon or nitrogen in a refrigerator at 4–5°.

Bis(diisopropylamino)-2-cyanoethoxyphosphine. Diisopropylamine (86 g, 0.85 mol) is added dropwise over 1 hr to 2-cyanoethoxydichlorophosphine (15 g, 0.087 mol) in 200 ml of anhydrous ether at $-10°$ under a nitrogen atmosphere. The reaction mixture is allowed to warm to room temperature and is stirred an additional 16 hr. After filtration to remove diisopropylamine hydrochloride, the crude product is isolated as a pale yellow liquid by removing solvent on a rotary evaporator. The crude product is fractionally distilled twice to afford bis(diisopropylamino)-2-cyanoethoxyphosphine as a colorless liquid (bp 107°, 0.020 mm Hg; $d = 1.04$) in 82% yield (21 g) [^{31}P NMR (CH_2Cl_2) $\delta = 123.4$]. The product should be stored in a brown bottle under argon or nitrogen in a refrigerator at 4–5°.

1H Tetrazole and Diisopropylammonium Tetrazolide. 1*H*-Tetrazole (trivially called tetrazole) is sublimed in a sublimation apparatus at 110–115° and 0.05 mm Hg prior to use. Tetrazole can explode violently if heated to a temperature close to its melting point (156°). If the crude tetrazole melts during sublimation the oil bath temperature should be reduced as an explosion could result. Thus the temperature of the sublimation apparatus should be controlled using an oil bath attached to a Variac. A thermal sensing device which will automatically disconnect the electrical circuit should also be inserted into the oil bath to guard against temperature fluctuations. Sublimations of tetrazole should be done in a closed hood behind an explosion shield. Diisopropylammonium tetrazolide is prepared in quantitative yield by adding diisopropylamine (0.81 g, 8 mmol) to a stirred solution of tetrazole (0.28 g, 4 mmol) dissolved in anhydrous acetonitrile (10 ml). The product is collected by filtration, washed with dry acetonitrile, and dried in a vacuum oven (40°) to yield a white crystalline solid. Both tetrazole and diisopropylammonium tetrazolide should be stored in a desiccator over Drierite.

Synthesis of Deoxynucleosides Attached to Silica and Deoxynucleoside 3′-Phosphoramidites

Protected Deoxynucleosides. The protected deoxynucleosides which are currently used routinely for DNA synthesis are 5′-*O*-dimethoxytritylthymidine, 5′-*O*-dimethoxytrityl-4-*N*-benzoyldeoxycytidine, 5′-*O*-dimethoxytrityl-6-*N*-benzoyldeoxyadenosine, and 5′-*O*-dimethoxytrityl-2-*N*-isobutyryldeoxyguanosine. Since these intermediates are available commercially and their synthesis has been described previously,[19] they will be used as examples for the purpose of outlining the DNA synthesis procedures. There are, however, several recently described, protected derivatives of deoxyadenosine which are recommended over 6-*N*-benzoyldeoxyadenosine. These include the succinoyl[20] and various amidines.[17,21,22] All of these derivatives are approximately 20-fold more resistant to depurination than 6-*N*-benzoyldeoxyadenosine which, as a consequence, leads to fewer side products and a higher overall yield of

[19] S. A. Narang, R. Brousseau, H. M. Hsiung, and J. J. Michniewicz, this series, Vol. 65, p. 610.

[20] A. Kume, R. Iwase, M. Sekine, and T. Hata, *Nucleic Acids Res.* **12,** 8525 (1984).

[21] L. J. McBride and M. H. Caruthers, *Tetrahedron Lett.* **24,** 2953 (1983).

[22] B. C. Froehler and M. D. Matteucci, *Nucleic Acids Res.* **11,** 8031 (1983).

the desired deoxyoligonucleotide. Of these derivatives, 5′-*O*-dimethoxytrityl-6-*N*-[1-(dimethylamino)ethylidene]deoxyadenosine is recommended because the amide acetal used in its preparation is commercially available and the completely protected deoxynucleoside can be prepared without isolating or transiently protecting any intermediates.[17] Immediately before use in preparing deoxynucleoside phosphoramidites, protected deoxynucleosides should be freed of trace amounts of water. This can most easily be accomplished by dissolving each protected deoxynucleoside in anhydrous pyridine, reconcentrating several times to a thick gum, dissolving in anhydrous pyridine, precipitating into hexane, and drying the precipitate under reduced pressure over a desiccant such as Drierite.

Synthesis of Deoxynucleoside 3′-Phosphoramidites. The same general procedure is used to prepare any appropriately protected deoxynucleoside 3′-phosphoramidite. The synthesis of **8a** will be described. 5′-*O*-Dimethoxytritylthymidine (1.0 mmol, 544 mg) and diisopropylammonium tetrazolide (0.5 mmol, 86 mg) are dissolved in dry dichloromethane (5 ml) under a nitrogen atmosphere and bis(diisopropylamino)methoxyphosphine (1.1 mmol, 282 mg) is added with stirring. After 1 hr, the reaction mixture is diluted to 25 ml with dichloromethane and extracted twice with 2% sodium carbonate (aqueous, 25 ml) followed by brine (25 ml). After a back-extraction of the combined aqueous phases with dichloromethane (25 ml), the organic fractions are pooled and dried over anhydrous sodium sulfate for 1 hr at 4°. The sodium sulfate is then removed by filtration and washed with anhydrous dichloromethane (10 ml). The organic layers containing the product are combined and concentrated with the aid of a rotary evaporator to a dry, white foam. The product is dissolved in anhydrous dichloromethane (5 ml) and precipitated with stirring into 300 ml of cold hexanes (−78°). The resulting suspension is filtered cold and dried under reduced pressure in a desiccator containing Drierite to afford the product in 87% yield (700 mg) as an amorphous white solid. These amidites can be stored at −20° indefinitely in a desiccator containing Drierite and an argon atmosphere. Before opening the desiccator it should be allowed to attain room temperature. Otherwise moisture from the air will condense on the cold amidites and on the inside surfaces of the desiccator.

For **8b–d**, the reaction times for synthesis of the deoxynucleoside 3′-phosphoramidites should be 1 hr for **8b** and 2 hr for **8c** and **8d**. [^{31}P NMR (CH_2Cl_2): **8a**, δ = 148.5, 148.1; **8b**, δ = 148.5, 148.3; **8c**, δ = 148.3, 148.1; **8d**, δ = 148.6, 148.1. ^{31}P NMR (CH_3CN): **8e**, δ = 149.1, 148.9.] The same general procedure can be used to prepare the deoxynucleoside 3′-(2-cyanoethoxy)-*N*,*N*-diisopropylaminophosphoramidites (**9a–e**). However the reaction time should be 4 hr with final yields being 80–90%. [^{31}P NMR (CH_2Cl_2): **9a**, δ = 146.03; **9b**, δ = 146.3, 146.07; **9c**, δ = 145.83; **9d**, δ = 145.83, 145.43.]

Synthesis of Deoxynucleosides Attached to Silica Supports. The same general procedure can be used for covalently linking deoxynucleosides to a large number of silica supports. The most important criteria is that the support be of a quality generally used for high-performance liquid chromatography. Currently the most popular are Fractosil[4,11] and controlled pore glass.[23–25] The stepwise yields with controlled pore glass are somewhat higher (1–2%) which is especially important for the synthesis of compounds containing more than 50 mononucleotides. The following procedure can be used for all silica matrices.

The initial step is synthesis of the succinylated deoxynucleosides as the *p*-nitrophenyl esters. Since these derivatives are prepared using the same general procedure, only the synthesis of **2a** will be described. To a solution of 5′-*O*-dimethoxytritylthymidine (2.5 mmol, 1.36 g) in anhydrous pyridine (5 ml) is added 4-*N*,*N*-dimethylaminopyridine (2.5 mmol, 0.30 g) and succinic anhydride (2 mmol, 0.22 g). The progress of the reaction is monitored by thin layer chromatography (TLC) on Merck F-254 plates using acetonitrile : water (9 : 1, v/v) as eluting solvent. Reaction products are visualized by irradiation with a UV lamp and by exposing the plate to HCl vapors. Both unreacted deoxynucleoside (R_f = 0.8) and product (R_f = 0.3–0.5 as a streaked spot owing to carboxylate anion) are UV absorbing and become orange with exposed to acid. The reaction is usually complete after 16 hr but occasionally a second portion of succinic anhydride (0.11 g, 0.1 mmol) is added in order to convert all the deoxynucleoside to the succinate half-ester.

The reaction mixture is next quenched by addition of water (0.2 ml) and, after 1 hr, solvent is removed using a rotary evaporator. The viscous oil is reconcentrated three times from toluene (3 $\times$ 20 ml) to remove pyridine, dissolved in dichloromethane (30 ml), transferred to an extraction funnel, and washed by vigorous shaking with an aqueous, 10% citric acid solution (10 ml, w/v). After removal of the aqueous phase containing succinic acid, the organic layer is washed with water (2 $\times$ 10 ml), diluted with pyridine (0.5 ml) to prevent detritylation, dried over anhydrous sodium sulfate, and concentrated to a gum using a rotary evaporator. The

[23] G. R. Gough, M. J. Brunden, and P. T. Gilham, *Tetrahedron Lett.* **22,** 4177 (1981).

[24] V. A. Efimov, S. V. Reverdatto, and O. G. Chakhmakhcheva, *Nucleic Acids Res.* **10,** 6675 (1982).

[25] S. P. Adams, K. S. Kavka, E. J. Wykes, S. B. Holder, and G. R. Galluppi, *J. Am. Chem. Soc.* **105,** 661 (1983).

product is dissolved in dichloromethane containing 5% pyridine, precipitated into 200 ml pentane : ether (1 : 1, v/v), and dried under reduced pressure (70–85% yield). The succinylated deoxynucleoside (1 mmol) is next dissolved in anhydrous dioxane (4 ml) containing dry pyridine (0.3 ml) and *p*-nitrophenol (0.14 g, 1 mmol). A solution of dicyclohexylcarbodiimide (1 mmol, 0.22 g) in anhydrous dioxane (1 ml) is added and the mixture is shaken for 2 hr at room temperature. The progress of the reaction is monitored by TLC in acetonitrile : water (9 : 1, v/v) where the product (R_f = 0.8) is separated from starting material (0.3–0.5). The reaction is usually complete after 2 hr but can be allowed to proceed for longer times if necessary. Dicyclohexylurea is removed by a low speed centrifugation and the supernatant containing the desired product is used directly in the condensation reaction with the silica support.

Deoxynucleosides containing 3′-*p*-nitrophenylsuccinate esters are attached to silica using the following general procedure. HPLC-grade silica such as Fractosil 500 (25 g) and 3-triethoxysilylpropylamine (13 ml) are added to 250 ml ethanol : water (95 : 5, v/v). After 4 hr of shaking on a wrist action shaking device, the silica is collected by filtration and dried under reduced pressure for 16 hr at 110°. (Silica should not be stirred with a magnetic stirring bar as this process will pulverize the silica particles.) The dried silica is suspended in anhydrous pyridine (100 ml) and treated with trimethylsilyl chloride (15 ml) for 12 hr at 20°. The silica is collected by filtration, washed thoroughly with methanol and ether, and dried under reduced pressure. The coupling then proceeds by first adding the silica amine (5 g) to dry *N,N*-dimethylformamide (7 ml) and triethylamine (1 ml) which is used to neutralize the propylamine salt. The solution containing 3′-O-succinylated deoxynucleoside (1 mmol) is then added to the silica amine. An immediate yellow color indicating the elimination of *p*-nitrophenol and the formation of silica-linked deoxynucleoside can be used to monitor the reaction. (If an immediate yellow color is not observed, more triethylamine should be added.)

After shaking the suspension for 4 hr, an aliquot of silica (approximately 1 mg) is removed for analysis. After washing the aliquot with DMF (2×), methanol (3×), and ether (2×), 0.1 *M* toluenesulfonic acid in acetonitrile (1 ml) is added, and the red–orange color of the dimethoxytrityl cation is observed. This analysis can be completed quantitatively if desired (see below). If these results appear satisfactory (i.e., a positive trityl test), the bulk of the silica is washed with DMF (3 × 10 ml), dioxane (3 × 10 ml), methanol (5 × 10 ml) and ether (3 × 10 ml). Unreacted propylamino silyl groups are blocked by placing the silica in a solution of acetic anhydride (0.7 ml) and pyridine (5 ml) followed by overnight agitation. The silica is recovered by washing with methanol (4 × 10 ml), then with ether (2 × 10 ml), and finally by drying under reduced pressure to a constant weight. Care must be taken during the initial stages of the drying procedure. If the desiccator is evacuated too rapidly, the silica will explode throughout the interior of the desiccator.

The efficiency of the final acylation step for propylamino groups can be qualitatively tested as follows. Aliquots (1 mg each) consisting of underivatized silica, silica derivatized with 3-triethoxysilylpropylamine, and silica-linked deoxynucleoside which has been blocked with acetic anhydride (compounds **3a–e**), are treated with 250 μl of saturated sodium borate containing 0.2 mg/ml picryl sulfate. After centrifuging the reaction products, the underivatized silica should remain white, the aminopropyl silyl silica should appear bright orange–red, and the acylated silica containing deoxynucleoside will be a pale yellow. If the silica covalently linked with deoxynucleoside (**3a–e**) is orange–red, it should be treated a second time with acetic anhydride.

With some preparations, a contaminant consisting of succinylated *n*-propylamino groups may be present and must be blocked. The succinylation could result either from incomplete hydrolysis of succinic anhydride or from a failure to remove all of the succinic acid during aqueous extraction with citric acid. This free succinic acid would be further activated in the presence of DCC to reform succinic anhydride. If succinylated propylamino groups are present, they can be blocked using the following procedure. The protected silica containing succinylated deoxynucleoside (5 g) is suspended in a solution of dry pyridine (5 ml) containing DCC (0.22 g, 1 mmol) and *p*-nitrophenol (0.14 g, 1 mmol) and shaken overnight at room temperature. Morpholine (0.2 ml) is added and the suspension shaken for 1 hr. The silica is transferred to a filter, washed with methanol (4 × 10 ml), THF (3 × 10 ml), and ether (3 × 10 ml), and dried under reduced pressure to a constant weight. This procedure can also be used with an aliquot (5 mg) to test for the presence of succinate. A yellow color following addition of morpholine indicates that succinate is present and has to be blocked.

A quantitative assay for trityl cation and therefore for loading of deoxynucleoside on silica is as follows:

1. Weigh accurately approximately 1 mg of dry silica.
2. Add 1 ml of 0.1 *M* toluenesulfonic acid in acetonitrile.
3. Measure the absorbance at 498 nm. If the absorbance approaches 2.0, dilute before reading. The absorbance should be determined immediately since it fades with time. The loading can be calculated using the following equation:

Deoxynucleoside in micromoles/g =

$$\frac{\text{(absorbance at 498 nm)(dilution factor)}}{\text{weight silica in mg}} \times 14.3$$

Synthesis of Deoxyoligonucleotides

Assembly of Glassware and Equipment. During several synthesis steps, reagents must be maintained under anhydrous conditions. One source of water contamination is moisture that is adsorbed on glassware surfaces. This water can be removed and glass surfaces protected against further adsorption by a silation procedure. Thus all glassware to be used for DNA synthesis should first be rigorously cleaned using normal laboratory procedures and then soaked for 1 hr in a solution consisting of trimethylsilylchloride : carbon tetrachloride (5 : 95, v/v). The glassware is next rinsed with ethanol, ether, and dried in an oven at 110° for a minimum of 2 hr.

The equipment involved is minimal. Argon lines are required for each of the phosphoramidite solutions as well as for each of the two capping solutions and also the anhydrous acetonitrile flask. The synthesis takes place in a 2-ml medium frit funnel which has a Teflon stopcock inserted in its stem. This is held by a rubber stopper on a heavy-walled suction flask connected to an aspirator. The vacuum line from the aspirator is fitted with a 3-way valve to allow for breaking the vacuum in the suction flask. The funnel is fitted on top with a rubber septum for the anhydrous sections of the synthesis, and another argon line is necessary for each funnel/flask assembly. It is also necessary to arrange the argon delivery so that a positive pressure may be delivered through this line to the funnel to push out the reactant solutions during the anhydrous steps of the cycle instead of using the aspirator vacuum. In our laboratory, a stopcock is inserted in the main argon delivery line immediately before the outlet bubbler. This may be closed for short periods of time to provide the required positive back-pressure in the line going to each funnel. The number of different oligonucleotides which may be synthesized simultaneously is limited only by the number of funnels and flasks available, and by the patience of the person doing the work.

The solutions which are required to be anhydrous are delivered to the funnel via syringe and septa. The solutions not required to be rigorously dry may be delivered from a wash bottle. These are the detritylation solution, the aqueous lutidine/THF solution, the iodine/lutidine oxidation mixture, and the acetonitrile and dichloromethane solutions that are used between steps not requiring anhydrous conditions. Each of these wash bottle solutions should be prefiltered through a mesh finer than the one in the funnel being used in the synthesis in order to eliminate the plugging of this funnel by particulate matter.

Detritylation, Capping, Oxidation, and Wash Solutions. The reagent composition for various solutions is as follows:

1. Detritylation solution: 6.75 g TCA in 225 ml CH_2Cl_2.
2. Oxidation solution: 2.54 g I_2 in 20 ml H_2O, 40 ml lutidine, and 40 ml THF. These solutions should be prepared and then filtered through a very fine frit such as Millipore 0.5-μ FH filters. The detritylation and oxidation solutions are kept in wash bottles. The oxidation solution should be stored in a refrigerator at 5° and filtered every 3–4 days if not prepared fresh before each synthesis. This is important as a brown precipitate forms upon storage and will plug filters if not removed prior to usage.
3. Capping solutions
   a. Capping solution A: 3.25 DMAP in 50 ml dry THF.
   b. Capping solution B: 10 ml acetic anhydride plus 10 ml lutidine.
   These solutions are prepared in predried flasks under anhydrous conditions and are not filtered.
4. Aqueous hydrolytic wash solution: 40 ml THF, 40 ml lutidine, 20 ml water. This solution can be kept in a wash bottle.
5. Reagent grade dichloromethane and acetonitrile are used for washes that do not require anhydrous conditions. These solutions can be kept in wash bottles.

The quantities given for all of the above solutions are sufficient for approximately 75 condensation cycles.

Synthesis Reagents. The detailed synthesis of the following deoxyoligonucleotides using compounds **8a–d** prepared *in situ* will be considered.

5′-GCCAGACCAAAACAGCTAAGGACC-3′	24 bases
5′-CGAGGTCCTTAGCTGTTTTGGTCT-3′	24 bases

Two funnels are needed. One contains 1 μmol of silica-C (**3b**) and the other 1 μmol of silica-T (**3a**), the two deoxynucleosides at the 3′ ends of the deoxyoligonucleotides. There are 46 deoxynucleotide condensations which are broken down as G_{12}, A_{12}, C_{12}, T_{10}. This number excludes the deoxynucleotides linked to silica. Twenty micromoles of phosphoramidite is needed for each condensation, and it is advisable to make up 40% more phosphoramidite solution when working on a small scale to account for wetting the sides of the flasks and syringes, spilling, etc. One-half equivalent of the diisopropylammonium tetrazolide salt and 1.0 equivalent of the bis(diisopropylamino)methoxyphosphine are used to prepare the phosphoramidite from each protected deoxynucleoside. Therefore the amount of protected deoxynucleoside required for any total synthesis is determined by the following equation, where 1.4 accounts for the additional 40%:

$$\text{Deoxynucleoside in g} = (\text{number of condensations})(\text{moles per condensation})(1.4)(\text{molecular weight})$$

The molecular weights of the protected deoxynucleosides are DMTdG^{ib}, 639.7; DMTdA^{bz}, 671.7; DMTdC^{bz}, 633.7; DMTdT, 544.6. Thus for the preparation of the phosphoramidite of DMTdG^{ib}, the amount of protected deoxynucleoside is

$$0.215 \text{ g} = 12 \times 0.00002 \times 1.4 \times 639.7$$

For the entire DNA synthesis in the two examples given, the amounts of various reagents are as shown in the following tabulation:

	G_{12}	A_{12}	C_{12}	T_{10}
Deoxynucleoside, mmol	0.336	0.336	0.336	0.280
Deoxynucleoside, mg	215	226	213	153
Bis(diisopropylamino)methoxyphosphine, μl	29	29	29	24
Diisopropylammonium tetrazolide, mg	99	99	99	83
Dry acetonitrile, ml	3.4	3.4	3.4	2.8

For the synthesis of each deoxynucleoside 3′-phosphoramidite, the protected deoxynucleoside and diisopropylammonium tetrazolide are added to a 5-ml flask. The flask is then sealed with a rubber septum, and dry acetonitrile is added to give, after completion of the reaction, a final concentration of 100 m*M* deoxynucleoside 3′-phosphoramidite. The bis-(diisopropylamino)methoxyphosphine is next added through the rubber septum via a syringe, and the reaction is stirred under argon at room temperature. The formation of the phosphoramidite should be followed using TLC by noting the disappearance of starting material near the bottom of the plate and the appearance of a spot of higher R_f. A silica gel plate containing a UV indicator is used with an elution solvent of hexanes and acetone (1 : 1) made 5% in triethylamine. The TLC plate is prerun before the application of the samples, and the developed plate is observed under short-wave UV light. The R_f values of the desired phosphoramidite products are as shown in the following tabulation:

Base	R_f	Reaction time
G	0.44	2 hr
A	0.36	1.5 hr
C	0.69	1.0 hr
T	0.71	45 min

Streaking of the sample on the TLC plate will occur as the phosphoramidite reacts slowly with the silica gel, but this does not interfere with observing the synthesis. Some undissolved salt will remain in the phosphoramidite solutions after reaction is complete.

The amount of tetrazole solution required for DNA synthesis is determined by totaling the number of condensations, in this case 46, multiplying by the amount of solution used in each condensation (0.3 ml), and adding 20% for wastage. In this case

$$46 \times 0.3 \times 1.2 = 16.6 \text{ ml}$$

or approximately 17 ml. A 0.45 *M* tetrazole solution is required, and the molecular weight of tetrazole is 70; thus

$$0.17 \times 0.45 \times 70 = 0.536 \text{ g}$$

or approximately 540 mg of resublimed tetrazole is required to make up this solution. As with the phosphoramidite and capping solutions, the tetrazole solution must be made up in a predried flask, closed with a septum, and supplied with an argon line. The solutions shown in the following tabulation should now be at hand:

In wash bottles	In flasks under argon
Acetonitrile	Phosphoramidite solutions (4)
Dichloromethane	Tetrazole solution, 0.45 *M*
	Anhydrous acetonitrile

Synthesis Cycle. With 1 μmol of the first nucleoside linked to silica (**3a–e**) in each funnel, the synthesis cycle proceeds as outlined below. First, turn on the aspirator so that there is reduced pressure in the flasks. The stopcocks in the funnel stems should be closed. Unless otherwise specified, a "wash" means simply filling the funnel about two-thirds full, then opening the stopcock to drain off the wash solution, and finally closing the stopcock.

Detritylation

2 × dichloromethane wash.

3 × the following:

Add approximately 1.0 ml TCA solution.

Wait 45 sec.

Open stopcock and drain off TCA.

Dichloromethane wash.

3 × acetonitrile wash (from wash bottle).

Condensation

Use a 3-way stopcock in the vacuum line in order to break the vacuum in the flasks, then cover the funnels with septa and argon bleeds.

3 × the following:

Add 1.5 ml anhydrous acetonitrile.

Open funnel stopcock and push through with positive argon pressure.

Close stopcock.

Add 300 μl tetrazole solution.

Add 200 μl desired phosphoramidite solution.

Wait 2 min, gently shake funnel during this time.

Open stopcock, push through spent phosphoramidite solution.

Close stopcock.

Capping

Add 500 μl capping solution A.

Add 100 μl capping solution B.

Wait 2 min, gently shake funnel.

Remove septum from funnel, evacuate the flask, and open the funnel stopcock to remove spent capping solution.

Close stopcock.

1 × wash with acetonitrile (from wash bottle).

Aqueous wash

Add 1 ml of the THF/lutidine/water (2 : 2 : 1; v/v/v) solution.

Wait 3 min.

Open stopcock and drain off the solution.

Oxidation

Add approximately 500 μl of the oxidation solution.

Wait 30 sec.

Open stopcock and drain off oxidation solution.

4 × wash with acetonitrile (from wash bottle).

If an iodine color remains, wash with additional acetonitrile.

The cycle is now ready to be repeated. Following the last condensation, proceed to first the aqueous wash and then to the oxidation step as the capping step is unnecessary. It is advisable to proceed directly to the removal of protecting groups and the hydrolysis of the DNA from the silica matrix. Long-term storage on silica is not advised as silica-catalyzed degradation is possible. A convenient stopping point for an interrupted synthesis (i.e., overnight) is after an oxidation step and before detritylation. The silica can be stored as a suspension in acetonitrile. If a preformed deoxynucleoside 3′-phosphoramidite is to be used, the appropriate amount should be dissolved in dry acetonitrile at 100 m*M* and stored over argon in a flask equipped with a septum. Such a sample can be added directly during the condensation step and used in place of the *in situ* prepared phosphoramidite.

This same general procedure can be used with bis(diisopropylamino)-2-cyanoethoxyphosphine. Usually the synthesis *in situ* of these deoxynucleoside 3′-phosphoramidites requires somewhat longer reaction times but the reaction progress can be followed on TLC. Additionally for maximum yield, the condensation time should be 5 min and the time for the capping step should be reduced to 1 min.

Removal of Protecting Groups. Silica containing the fully protected DNA as isolated after the last oxidation step is transferred to a screw-cap test tube equipped with a Teflon-lined plastic cap. For syntheses containing *O*-methylphosphate protection, a solution (1 ml) containing dioxane/thiophenol/triethylamine (2 : 1 : 1; v/v/v) is added and the suspension agitated for 90 min at room temperature. The silica is then freed of the supernatant after a low speed centrifugation, washed at least 5 times with methanol and diethyl ether, and air dried. For best results, the solution containing thiophenol should be prepared fresh using reagent grade thiophenol and purified triethylamine. Decomposition occurs on prolonged storage, especially with unpurified triethylamine. If the β-cyanoethyl group is used to protect phosphorus, this step is unnecessary. Failure to remove the *O*-methyl phosphorus protecting groups with thiophenol leads to extensive methylation of the nucleoside bases, predominantly at N-3 of thymine.[26]

The next step is treatment with concentrated ammonium hydroxide to hydrolyze the ester linking the DNA to the support and to remove protecting groups from the purine and pyrimidine bases. The silica is first treated with concentrated ammonium hydroxide (1 ml) for 3 hr at room temperature in a sealed test tube in order to cleave the DNA from the support. After a low speed centrifugation, the supernatant is transferred to a clean, dry screw-cap test tube. The silica is then washed with concen-

[26] L. J. McBride, J. S. Eadie, J. W. Efcavitch, and W. A. Andrus, *Nucleosides Nucleotides* **6,** 297 (1987).

trated ammonium hydroxide (1 ml). The test tube should be kept tightly sealed during centrifugations in order to prevent ammonia from escaping. The two concentrated ammonium hydroxide supernatants are combined in the screw-cap test tube, sealed firmly, and warmed at 60° for 16 hr to remove base-protecting groups. The sample is cooled, concentrated to dryness in a Speed-Vac, and reconcentrated once from ethanol (1 ml). The sample is now ready to be purified either by reverse-phase HPLC or PAGE.

The above step involving removal of protecting groups must be carried out properly. Otherwise a successful synthesis will lead to DNA that is useless for any biological or biochemical application. The most critical step is storage of concentrated ammonium hydroxide. A small bottle of concentrated ammonium hydroxide (100 or 250 ml) is used for this deprotection step and is kept tightly sealed in a freezer at −20°. When needed, the bottle is opened quickly, an aliquot removed, and resealed immediately. After about 50% of the bottle's contents have been consumed, it is discarded. Additionally, during the 60° step, the screw-cap test tube must also be tightly sealed. If these precautions are not followed, most ammonia escapes and the base-protecting groups will not be removed completely. Another potential problem is excessive treatment of the silica with concentrated ammonium hydroxide. This leads to solubilization of silica which can create purification problems with either HPLC or PAGE. This step should therefore be limited to 3 hr as most of the DNA is cleaved from the silica during this time period (80%).

Purification by Reverse-Phase HPLC. The sample obtained after deprotection with ammonium hydroxide is dissolved in 0.35 ml of 0.1 *M* triethylammonium acetate (pH 7.0) and purified on a C_{18}-reverse-phase, HPLC column (Waters Associates). The eluting gradient is acetonitrile (22–26%) and 0.1 *M* triethylammonium acetate (78–74%) at pH 7.0. Alternatively elution can be at a constant mixture of acetonitrile and aqueous phase. The peak containing 5′-*O*-dimethoxytrityl deoxyoligonucleotide is concentrated under reduced pressure to dryness. Care should be taken so that fractions from the trailing region of the peak are not pooled. These fractions usually contain incompletely deprotected DNA and deoxyoligonucleotides of shorter length than the product which are derived from depurination during synthesis. the recovered deoxyoligonucleotide is next treated with acetic acid/water (4 : 1, v/v) at room temperature for 45 min (0.5 ml) in order to remove the 5′-*O*-dimethoxytrityl group. The completely deprotected deoxyoligonucleotide is dissolved in water (0.75 ml), extracted 3 times with ether (3 × 0.5 ml) and ethyl acetate (3 × 0.5 ml) to remove dimethoxytritanol, lyophilized to dryness, and dissolved in 10 m*M* Tris–HCl (pH 7.6) containing 1 m*M* EDTA. The amount of DNA recovered can then be determined spectrophotometrically at 260 nm using the molar extinction coefficients at pH 7.6 listed in the following tabulation:

Deoxynucleoside	Extinction coefficient
dA	1.54×10^4
dG	1.17×10^4
dC	0.75×10^4
dT	0.88×10^4

Deoxyoligonucleotides purified using this procedure can be used directly for various biochemical experiments.

Purification by PAGE. The lyophilized sample obtained after deprotection with concentrated ammonium hydroxide is first reconcentrated twice to dryness with ethanol/water (4 : 1, v/v) to remove trace amounts of basic salts. These salts buffer acetic acid used during detritylation and lead to incomplete removal of the dimethoxytrityl group. The residue is treated at room temperature for 45 min (0.5 ml) with acetic acid/water (4 : 1, v/v), lyophilized to dryness, and redissolved in water (0.75 ml). The solution of a completely deprotected deoxyoligonucleotide is then extracted 3 times with ether (3 × 0.5 ml) and ethyl acetate (3 × 0.5 ml) to remove dimethoxytritanol, lyophilized to dryness, and dissolved in water. The concentration is then determined spectrophotometrically using the extinction coefficients listed in the previous section. An aliquot of the unpurified DNA mixture (10–15 OD units at 260 nm) is concentrated to dryness and dissolved in 50 μl of deionized formamide/water (9 : 1, v/v) containing 0.01% xylene cyanol and 0.01% bromophenol blue dye markers. This sample is then purified on a 0.3 × 20 × 20 cm polyacrylamide slab gel using a solvent of 7 *M* urea, 90 m*M* Tris–borate (pH 8.3), and 2 m*M* EDTA; 20, 12, and 8% polyacrylamide gels are used to purify deoxyoligonucleotides containing up to 20, 30, and more bases, respectively. The total sample (50 μl) is loaded in a 2.5-cm well, and electrophoresis is carried out at 350 volts until the bromophenol blue dye marker migrates to the bottom of the gel. The gel is removed from the glass plates, enclosed in SaranWrap, and placed on a fluorescent, 20 × 20 cm silica gel plate.

The deoxyoligonucleotides are visualized by irradiation with long-wavelength UV light. Short-wavelength UV should be avoided as cross-linking of thymine bases to acrylamide occurs and, additionally, mutation lesion are generated. If a short-wavelength UV light is the only instrument available it can be used, but irradiation of the gel should be limited to the few seconds required for identifying the product band. When marking the bands of DNA, the UV lamp and the observer should be positioned directly over the gel. Otherwise the apparent displacement of the DNA absorbance band will lead to improper marking of the gel and a failure to excise the correct DNA sample.

The acrylamide gel containing the desired deoxyoligonucleotide, which is usually the slowest migrating, most pronounced band is carefully cut out from the total slab gel with a razor blade, crushed, and soaked for 8 hr at room temperature with gel elution solution (1.5 ml) containing 0.5 *M* ammonium acetate, 0.01 *M* magnesium acetate, 0.1% sodium dodecyl sulfate, and 0.1 m*M* EDTA. The polyacrylamide suspension is pelleted by a low speed centrifugation. The solution is decanted and washed with equal volumes of *n*-butanol (3×) and ether. In order to remove salts, the DNA is next passed through a Sephadex G50–40 column (1 × 10 cm) using 10 m*M* triethylammonium bicarbonate as eluant. The peak containing the desired product is lyophilized to dryness. The deoxyoligonucleotide is dissolved in 10 m*M* Tris–HCl (pH 7.6) and the concentration determined spectrophotometrically. The sample is now ready for use in various biochemical experiments. Synthetic DNA samples should be stored frozen at pH 7.5–8.0.

During purification by PAGE, several potential problems can arise. One that is particularly troublesome is poor resolution of DNA bands. This is usually caused by the presence of salt in the sample or by loading too much DNA in the well. Occasionally this problem is due to the presence of DNA secondary structure. For these cases, the best solution is to run the electrophoresis using more drastic denaturation conditions. These would include increasing the urea concentration to 8 *M*, boiling the sample just prior to loading, and pre-running the gel to at least 50° before loading the sample. Another related problem is the presence of more than one deoxyoligonucleotide, especially failure sequences containing one less deoxynucleotide, in the purified sample. This is a very serious problem for experiments where synthetic DNA is enzymatically joined using T4 DNA ligase and then cloned. The presence of failure sequences in these ligated samples will lead to various deletion mutants among cloned duplexes. This problem is usually caused by overloading the gel or by cutting too broad a gel band. Since gel purification usually yields more material than is needed for cloning experiments, we cut out and purify only the more slowly migrating one-half of the product band. Occasionally a parallax (see above) caused by visualizing the gel with a UV lamp not located perpendicular to the gel can cause this problem as well. If the lamp is positioned improperly, failure sequences might be the primary product cut from the gel. Another problem is the presence of curved bands (i.e., bands that smile). This problem is usually caused by thick, uneven glass plates that do not heat evenly or by using too high a voltage. Bands that smile have to be avoided, otherwise the isolation of pure DNA from such a gel is almost impossible. A final problem is a dark background absorption throughout the gel when it is illuminated with UV light. Although this problem can occasionally be corrected by using a new fluorescent silica gel plate, a more serious source is contamination of buffers or acrylamide with UV-absorbing material. This problem can be corrected by filtering acrylamide, solvents, and buffers through activated charcoal.

Purified synthetic deoxyoligonucleotides can then be characterized using standard procedures such as DNA sequencing,[16] two dimensional analysis,[27] and degradation with snake venom phosphodiesterase.

Commercial Sources of Reagents and Equipment

Below are listed manufacturers or commercial suppliers of chemicals, solvents, and equipment used in DNA synthesis procedures in this chapter. This does not imply that they are superior to others but only that we have found them acceptable. Following the supplier's name, product numbers are listed.

Acetic acid: EM, AX0073-9
Acetic anhydride: Fisher, Reagent Grade, A-10
Acetonitrile: Burdick and Jackson, 015
Acrylamide: Kodak, 5521
Alumina oxide: Woelm Pharma, 02084
Anhydrous ether: Mallinkrodt (American Scientific Products) 0848-5
Bisacrylamide: Fisher, 0-3586
Bromophenol blue: Fisher Scientific, B-392
Calcium hydride: Aldrich, 21, 326-8
Chloroform: EM, CX1055-9
Concentrated ammonium hydroxide: Fisher, A-669
Controlled pore glass: Pierce
2′-Deoxyadenosine: Sigma, D7400
2′-Deoxycytidine: Sigma, D3897
2′-Deoxyguanosine: Sigma, D7145
2′-Deoxythymidine: Sigma, D9250
Dichloromethane: EM, DX0835-3
Dicyclohexylcarbodiimide (DCC): Sigma, D3128
Diisopropylamine: Sigma, D3022
4-*N*,*N*-Dimethylaminopyridine (DMAP): Aldrich, 10,770
Dioxane, Reagent Grade: Fisher Scientific, D-111
Drierite: J. T. Baker, 8 Mesh, 7-L056

[27] C.-P. D. Tu and R. Wu, this series, Vol. 65, p. 620.

EDTA, disodium: EM, EX0539-1
Ethanol: Gold Shield, Midwest Solvent Co. of Illinois
Film, X-ray, for autoradiography: Eastman Kodak
Filters: Millipore, 0.5 μm, FMUP 04700
Fluorescent thin-layer plates: Merck, 60 F-254
Fractosil 500: Merck
Hexanes: EM, HX0299-5
3-Hydroxypropionitrile: Sigma, C6773
Iodine: Fisher Scientific, I-36
2,6-Lutidine: Aldrich, 24, 395-7
Medium frit funnel, 2 ml: Kimble, 7700-5
Methanol, Reagent Grade: EM, MX0485-7
Microcentrifuge, 15,000 rpm: Eppendorf, 5412
Molecular sieves, 4 Å: Davison-Fisher Scientific, M-513
Morpholine: Aldrich, 13, 423-6
p-Nitrophenol: Fisher, N-105
Pentane: Fisher, P-393
Phosphorus pentoxide: Fisher Scientific, A-245
Phosphorus trichloride: V.W.R., EM-PX1050-3
Pyridine: Fisher Scientific, P-368
Rubber septa: V.W.R., Aldrich
Screw-cap test tubes, 10–15 ml: Fisher 14-959-25C
Sephadex G50–40: Sigma, G-50-40
Sodium sulfate: EM, SX0760-1
Tetrahydrofuran (THF): Fisher Reagent Grade, T397
Tetrazole: Aldrich, L-390-0
Thiophenol: Aldrich, 24, 024-9
Toluenesulfonyl chloride: Eastman Kodak, 523
Trichloroacetic acid (TCA): EM, TX1045-1
3-Triethoxysilylpropylamine: Aldrich, 11, 399-5
Triethylamine: Kodak, 1073576
Trimethylchlorosilane: Aldrich, C7, 285-4
Tris base: Sigma, T-1503
Urea: Schwarz/Mann Biotech., 821527
Xylene cyanol: V.W.R., JTX539-5

Acknowledgments

The development of these procedures was supported by the United States Public Health Service Grants GM25680 and GM21120. M.H.C. was the recipient of a USPHS Career Development Award during the early stages of this work (1 KO4 GM00076). S. L. Beaucage and D. R. Dodds were the recipients of postdoctoral fellowships from the Natural Sciences and Engineering, and the Medical Research Councils of Canada, respectively. L. J. McBride acknowledges an Upjohn Graduate Fellowship.

[16] An Automated DNA Synthesizer Employing Deoxynucleoside 3′-Phosphoramidites

By SUZANNA J. HORVATH, JOSEPH R. FIRCA, TIM HUNKAPILLER, MICHAEL W. HUNKAPILLER, and LEROY HOOD

We have constructed an automated synthesizer that employs activated deoxynucleoside 3′-phosphoramidites as key intermediates in the solid-phase synthesis of DNA. This instrument can be used to synthesize long oligonucleotides (>50 bases) and can incorporate two, three, or four nucleotides (degeneracy) into any particular base position. Its mechanical design permits rapid, simple, and reliable operation at a relatively low cost.

One of the experimental techniques that has played a key part in the rapid development of genetic engineering is the chemical synthesis of deoxyoligonucleotides of defined structure. Such deoxyoligonucleotides are often an essential part of the process of isolating a single gene from libraries containing tens of thousands to millions of gene fragments. Because of the unique chemical structure of DNA, a piece of DNA with a particular sequence will recognize and bind only to another piece with a complementary sequence, ignoring all other pieces of DNA with other sequences in a complex mixture. Therefore, small pieces of chemically synthesized DNA can be used as unique molecular probes to isolate gene-sized pieces of DNA that contain the appropriate complementary sequence. The correct sequence for the DNA probe is often deduced from the amino acid sequence of the corresponding protein by translating, using the genetic code dictionary, protein sequence into DNA sequence. Since the genetic code is degenerate for most amino acids, often a set of DNA probes containing the alternative nucleotides at the third base positions must be synthesized in order to insure that the one sequence complementary to the designed gene fragment is present. This requirement for mixed or heterogeneous DNA probes places severe constraints on the methods used to synthesize DNA. Compared to other ways in which genes can be obtained, the synthetic method offers new potentials. Total genes can be produced by chemically synthesizing overlapping fragments from both strands of the double-stranded gene in such a way that its fragments can hybridize to one another. These double-stranded fragments then are linked together by enzymatic ligation. In order to be routinely useful, the synthetic procedure must be rapid, simple, reliable, and inex-

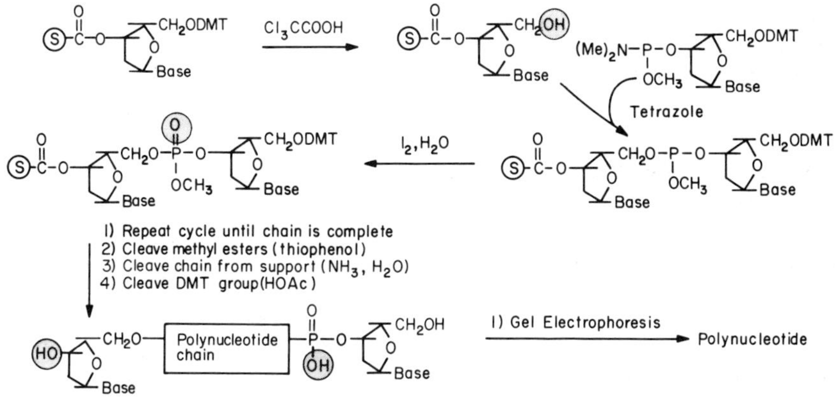

FIG. 1. Synthesis of DNA using the phosphoramidite chemistry. Taken from Hunkapiller *et al.*[11]

pensive, because often ten or more different DNA molecules must be synthesized for locating or assembling a single gene.

The chemistry of deoxyoligonucleotide synthesis has been under study for many years. The knowledge and experience which accumulated during these years by using Khorana's phosphodiester approach[1] followed by many modifications of the phosphotriester approach[2–6] and Letzinger's phosphite triester method[7] have led to the development of the first reliable, relatively easy manual process of deoxyoligonucleotide synthesis. Caruthers and co-workers, using a modified version of phosphite triester chemistry worked out a method for synthesizing oligonucleotides on polymer support.[8] In an extension of this work they also developed new phosphite reagents of high stability.[9,10] Our rapid, routine automated synthesis is based on Caruthers' solid-phase phosphoramidite chemistry (Fig. 1).[11]

[1] P. T. Gilham and H. G. Khorana, *J. Am. Chem. Soc.* **80,** 6212 (1958).

[2] R. L. Letsinger and V. Mahaderan, *J. Am. Chem. Soc.* **87,** 3526 (1965).

[3] C. B. Reese and R. Saffhill, *Chem. Commun.* **13,** 767 (1968).

[4] F. Eckstein and I. Risk, *Chem. Ber.* **102,** 2362 (1969).

[5] K. Itakura, C. P. Bahl, N. Katagiri, J. Michniewicz, R. H. Wightman, and S. A. Narang, *Can. J. Chem.* **51,** 3469 (1973).

[6] J. C. Catlin and F. Cramer, *J. Org. Chem.* **38,** 245 (1973).

[7] R. L. Letsinger and W. B. Lursford, *J. Am. Chem. Soc.* **98,** 3655 (1976).

[8] M. D. Matteucci and M. H. Caruthers, *J. Am. Chem. Soc.* **103,** 3185 (1981).

[9] S. L. Beaucage and M. H. Caruthers, *Tetrahedron Lett.* **22,** 1859 (1981).

[10] L. J. McBride and M. H. Caruthers, *Tetrahedron Lett.* **24,** 245 (1983).

[11] M. Hunkapiller, S. Kent, M. Caruthers, W. Dreyer, J. Firca, C. Giffin, S. Horvath, T. Hunkapiller, P. Tempst, and L. Hood, *Nature (London)* **310,** 105 (1984).

Ogilvie and co-workers[12] have described an automated device for synthesizing oligonucleotides up to 14 bases in length using deoxynucleoside 3′-phosphochloridites. However, because of their inherent instability, these reagents have generally been replaced by the more stable deoxynucleoside 3′-phosphoramidites.[9] We report here the design of an automated DNA synthesizer that can synthesize longer strands of DNA at lower cost. Moreover, two, three, or four bases can be incorporated at appropriate positions in accordance with the ambiguity of the genetic code for individual amino acid residues. The design and operation of such a synthesizer is described below.

Synthesizer Design

The DNA synthesizer has several features in common with the protein sequenators designed previously in our laboratories.[13] The general design of the instrument is shown in the schematic diagram (Fig. 2). The primary reactor is a flowthrough column, and the reagent and solvent deliveries are controlled by a series of zero-dead volume, pneumatically actuated diaphragm valves (Fig. 3). To move reagents and solvents from the reservoirs, positive argon pressure is used rather than a mechanical pump. The unique design of the valves and well-controlled argon pressure in each reservoir provides a simple, accurate, and reliable method of reagent delivery that allows for complete purging of previously delivered reagent from the reaction train prior to delivery of the next reagent. The valves are automatically operated by a control unit. A separate reactor, the activation vessel, is designed for the activation of each phosphoramidite prior to coupling. If mixed probes are being synthesized, the activation vessel is also used for the accurate premixing and the simultaneous activation of two, three, or four different phosphoramidites.

Primary Reaction Vessel. The vessel that houses the silica support on which the oligonucleotide is attached is similar to the one used in the gas-phase sequenator.[13] The cartridge is formed from three Pyrex cylinders housed in a metal jacket. The silica resides in the central cavity of the middle cylinder and is held in place by porous Teflon membranes [Zitex filter membranes, extra-coarse grade (Chemplast, Inc., Wayne, NJ)] that also provide the seals between the glass cylinders.

Activation Vessel. The deoxynucleoside 3′-phosphoramidites are activated by reaction with tetrazole in an activation vessel before they are

[12] G. A. Urbina, G. M. Sathe, W. C. Liu, M. F. Gillen, P. T. Duck, R. Bender, and K. K. Ogilvie, *Science* **214,** 270 (1981).

[13] R. M. Hewick, M. W. Hunkapiller, L. E. Hood, and W. J. Dreyer, *J. Biol. Chem.* **256,** 7990 (1981).

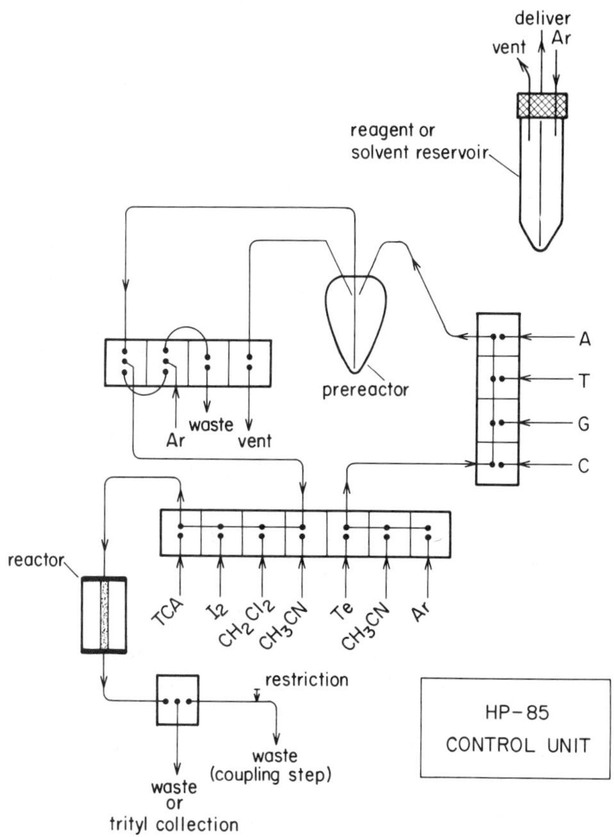

FIG. 2. Schematic diagram of the DNA synthesizer.

delivered into the primary reaction vessel. The prereaction vessel is a conical Pyrex flask similar to the conversion flask used in the gas-phase sequenator.[13] It has an internal volume of about 3 ml and is fitted with Teflon tubing connectors. The nucleoside solution is delivered into the flask, the tetrazole solution is added, and the solutions are thoroughly mixed by bubbling argon through it before it is transfered from the flask to the primary reaction vessel. In each synthetic cycle, the prereactor is automatically cleaned by solvent washes.

Delivery Valves. The valves (Fig. 3) that control liquid and gas movement through the instrument are zero-dead volume, pneumatically actuated diaphragm valves similar to those used in our gas-phase protein sequenator.[13,14] They have been modified in two ways: (1) FEP–Teflon

[14] M. W. Hunkapiller and L. E. Hood, U.S. Patent No. 4,242,769 (February 24, 1981).

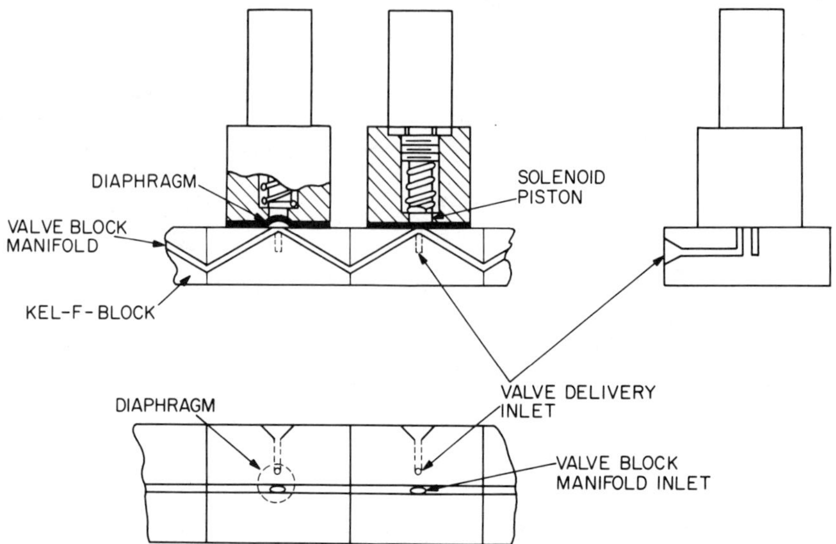

FIG. 3. Delivery valve assembly of DNA synthesizer.

diaphragms used in the earlier valves have been replaced by 0.02-inch-thick Kalrez diaphragms and (2) the diaphragms are held closed by the piston of a solenoid unit mounted as an integral part of the valve. When the solenoid is energized, the piston lifts away from the diaphragm and allows the reduced pressure inside the piston chamber (supplied by a remote vacuum pump) to lift the diaphragm away from the surface of the valve block.

Reagent/Solvent Storage. Reagents and solvents are stored in Pyrex bottles (reservoirs) protected from light. All tubing and bottle cap surfaces exposed to chemical vapors or liquid are made of either Teflon or Kalrez (E. I. du Pont de Nemours & Co., Wilmington, DE).

Argon Supply. Two 200-cubic-foot cylinders supply argon to the reagent/solvent reservoirs and the reaction vessel. This supply is passed through a molecular sieve filter (Model #461, Matheson Gas Products) to eliminate water vapor and distributed through a manifold where it is reduced by four low pressure, low flow gas regulators (Veriflo Corp., Richmond, CA) to the 3–8 psig level required.

Waste Collection. Effluent from either reaction vessel can be directed into a Pyrex collection bottle through the appropriate delivery valve. During the delivery of activated nucleoside from the activation vessel to the primary reaction vessel, the effluent from the primary vessel is directed first through an adjustable needle valve that precisely regulates the flow rate. This valve (Nupro Co., Willoughby, OH) is fitted with a Kalrez O ring. Effluent from the primary vessel during and immediately after the detritylation step is directed into a fraction collector to facilitate quantitative monitoring of each coupling reaction.

Control Unit. Valve operations are controlled by an HP-85 microcomputer (Hewlett-Packard, Palo Alto, CA) interfaced to the valves through a Hewlett-Packard Model 3497 system controller. The system software allows automatic operation with manual override of any valve function. The synthesizer programs can be customized to effect modification of the valve and timing sequencing of any or all steps in them. Normally, however, all the user must do is type in the base sequence of deoxyoligonucleotide to be synthesized. This sequence can include from one to four bases to be added at each cycle of the synthesis, allowing for the synthesis of heterogeneous or mixed probes.

Synthesizer Operation

In order to optimize conditions for routine synthesis, we have carefully selected and prepared solvents and reagents for each individual step of the synthesis cycle listed in Table I and studied reaction conditions in correlation with machine performance to minimize the cycle time that is required for each base addition with high coupling yield.

Synthesis Start-up. The synthesis of deoxyoligonucleotide with defined sequence begins by loading the proper silica-bound, 3′ end deoxynucleoside into the primary reactor. The derivatization of the polymer support (Fractosil-500, Merck) was made according to the published protocol of Matteucci and Caruthers.[8] The synthesis then automatically proceeds in the 3′ → 5′ direction by adding one deoxynucleotide per cycle. The synthesis cycle includes three reactions (Fig. 1, Table I): 5′-detritylation, coupling (activation and condensation), and oxidation.

Detritylation. The 5′-hydroxyl of the silica-bound deoxynucleoside (or

TABLE I
GENERAL PROTOCOL FOR DNA SYNTHESIS

Reagent or solvent	Function
CH_2Cl_2	Wash
3% TCA/CH_2Cl_2	*5′-Detritylation*
CH_2Cl_2	Wash
CH_3CN	Wash
Activated nucleoside (Nu + Te)	*Coupling*
CH_3CN	Wash
I_2/H_2O : 2,6-lutidine : THF	*Oxidation*
CH_3CN	Wash

deoxyoligonucleotide) is protected with the acid-labile dimethoxytrityl group. For automated synthesis, the 3% trichloroacetic acid (TCA)/dichloromethane solution proved to be an excellent reagent. The detritylation reaction is complete in less than a minute, and, because of the subsequent fast and efficient dichloromethane wash to remove excess reagent, the depurination is negligible. The dimethoxy cations give a bright orange color in solution with λ_{max} 498 nm, and they are automatically collected and used for quantitative monitoring of the coupling yield of each previous cycle.

Coupling. 5′- and base-protected deoxynucleoside 3′-*N*,*N*-diisopropylaminophosphoramidites are used for the sequential addition of the four bases to the silica-bound deoxynucleoside (or deoxyoligonucleotide). While the column is being washed with anhydrous acetonitrile, the selected deoxynucleoside 3′-phosphoramidite : acetonitrile solution is delivered to the activation vessel and mixed with 1*H*-tetrazole : acetonitrile solution. The activation is extremely rapid. In mixed base positions, the proper deoxynucleoside 3′-phosphoramidite solutions are delivered into the activation vessel one after the other in a well defined molar ratio and then mixed with the 1*H*-tetrazole solution. By applying positive argon pressure in the activation vessel and by opening the proper valves, the activated deoxynucleoside 3′-phosphoramidite solution passes through the column and the condensation is complete in 3 min 30 sec. The flow rate and thus the reaction time can be finely adjusted with a needle valve built in the outlet line of the column. Using only 10-fold molar excess of the deoxynucleoside 3′-phosphoramidites, the coupling yields are ≥97%. The carefully prepared, protected deoxynucleoside 3′-*N*,*N*-diisopropylaminophosphoramidites are stable in anhydrous acetonitrile solution more than a week. After coupling, both the activation vessel and the column are washed with acetonitrile.

Oxidation. The 3′–5′ phosphite triester bond formed in the previous coupling step is oxidized to phosphate triester by using I_2 in a water : 2,6-lutidine : tetrahydrofuran solution. This oxidation is complete in 30 sec. The I_2 solution can be removed from the column using a 55-sec acetonitrile wash.

The detailed protocol of our DNA synthesizer is given in Table II. One synthesis cycle involves 24 steps, and, as indicated, some of these steps have double functions in order to shorten the cycle time. The total cycle, the addition of one base, takes 9 min 24 sec.

Deprotection and Cleavage from the Support. Once the synthesis cycle described above is repeated the desired number of times, the silica-bound, protected deoxyoligonucleotide is removed from the primary reactor (column), cleaved from the silica support and freed of protecting

TABLE II
STEPS INVOLVED IN EACH CYCLE OF THE AUTOMATED DNA SYNTHESIS[a]

Step	Function	Duration (sec)
1	Deliver CH_2Cl_2 to reactor	20
2	Flush argon through reactor	6
3	Deliver 3% TCA/CH_2Cl_2 to reactor (5′ detritylation)	55
4	Flush argon through reactor	6
5	Deliver CH_2Cl_2 to reactor; Pressurize Nu and Te reservoirs	20
6	Flush argon through reactor	6
7	Deliver CH_3CN to reactor Deliver Nu to prereactor	13
8	Deliver CH_3CN to reactor; Flush argon through prereactor to push Nu into prereactor (clean the line)	3
9	Deliver CH_3CN to reactor; Deliver Te to prereactor	19
10	Deliver CH_3CN to reactor; Flush argon through prereactor to push Te into prereactor (clean the line)	3
11	Deliver CH_3CN to reactor; Bubble argon into prereactor to mix Nu and Te solution (activation)	8
12	Flush argon through reactor	5
13	Push activated nucleotide from prereactor to the inlet of reactor	2
14	Deliver activated nucleotide to reactor (condensation)	210
15	Flush argon through reactor	3
16	Deliver CH_3CN to prereactor (wash)	5
17	Push CH_3CN from prereactor to reactor	35
18	Deliver CH_3CN to reactor; Deliver CH_3CN to prereactor (wash)	4
19	Deliver CH_3CN to reactor; Push CH_3CN from prereactor to waste	37
20	Flush argon through reactor	5
21	Deliver I_2/H_2O : 2,6-lutidine : THF	30
22	Flush argon through reactor	9
23	Deliver CH_3CN to reactor	55
24	Flush argon through reactor	5

[a] Cycle time: 9 min 24 sec.

groups. The deprotection procedure starts with treating the silica-bound reaction product with a thiophenol : triethylammonium : dioxane solution at room temperature for 45 min. The thiophenoxide ions remove the methyl groups from internucleotide phosphotriester bonds. The silica then is extensively washed with methanol in order to completely remove the thiophenol reagent and dried under reduced pressure. This step is followed by treatment with concentrated ammonium hydroxide at room temperature for 3 hr to cleave the base-protected diester oligonucleotides from the support. After separating the silica and the supernatant by centrifugation, the supernatant containing the product is treated with additional concentrated ammonium hydroxide at 50° for up to 12 hr to remove the base-protecting groups, namely, *N*-benzoyl groups from deoxycytosines and deoxyadenosines and *N*-isobutyryl groups from deoxyguanosines. The ammonium hydroxide solution is removed by vacuum drying, and finally the last 5′-end dimethoxytrityl group is cleaved by using 80% acetic acid solution for 15 min. The reaction mixture containing the crude, unprotected deoxyoligonucleotide is lyophilized to dryness.

Purification. For purification of the crude deoxyoligonucleotides, either preparative polyacrylamide gel electrophoresis[15,16] or preparative reverse-phase HPLC (Axxiom C_{18} column; A: 0.1 *M* triethylammonium acetate, pH 7.0; B: 50% A and 50% CH_3CN, OD 260 nm) have been used. In practice, mixed probes, single probes, and gene fragments with high guanosine content and/or with long chains (≥30) are usually purified by preparative gel electrophoresis. Other oligonucleotide chains, such as primers for dideoxy DNA sequencing, are usually purified by HPLC (Fig. 4).[17]

Materials

The following protected nucleosides were obtained from American BioNuclear (Emeryville, CA): 5′-*O*-dimethoxytrityl (DMTr) deoxythymidine, 5′-*O*-DMTr-*N*-benzoyldeoxycytidine, 5′-*O*-DMTr-benzoyldeoxyadenosine, and 5′-*O*-DMTr-*N*-isobutyryldeoxyguanosine. Chloro-*N*,*N*-dimethylaminomethoxyphosphine and nucleoside 3′-*N*,*N*-diisopropylaminophosphoramidite of the protected deoxynucleosides of each of the four bases were prepared according to published protocols.[9] Solvents and reagents were purified or dried as follows: dichloromethane (reagent grade, Mallinckrodt), dried over 3 Å molecular sieves (Fisher) and passed through an activated alumina column prior to use; acetonitrile (HPLC grade, Burdick & Jackson), distilled over calcium hydride and stored over 3 Å molecular sieves; tetrahydrofuran (Aldrich), distilled over sodium spheres (1 mol/liter) in the presence of benzophenone (0.125 mol/liter);

[15] T. Maniatis, A. Jeffrey, and H. van de Sande, *Biochemistry* **14,** 3787 (1975).

[16] A. M. Maxam and W. Gilbert, this series, Vol. 65, p. 499.

[17] E. C. Strauss, J. A. Kobori, G. Siu, and L. Hood, *Anal. Biochem.* **154,** 353 (1986).

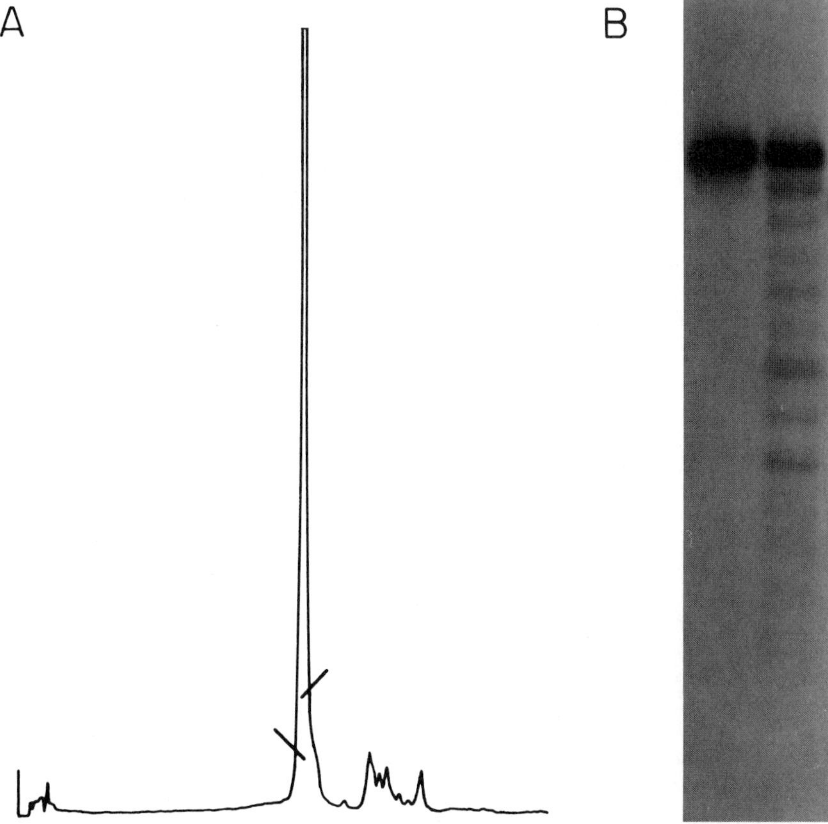

FIG. 4. Reverse-phase HPLC purification of a synthetic deoxyoligonucleotide (25-mer). (A) Preparative reverse-phase HPLC run of the crude, fully deprotected product indicating the portion of the main peak, which was collected. (B) Polyacrylamide gel electrophoresis: radiogram of ^{32}P-labeled HPLC-purified and crude deoxyoligonucleotide (25-mer) of (A).

2,6-lutidine (Aldrich), double-distilled over *p*-toluenesulfonyl chloride and calcium hydride, stored over 4 Å molecular sieves; iodine (Aldrich), resublimed; trichloroacetic acid (analytical reagent, Mallinckrodt); 1*H*-tetrazole (Aldrich), resublimed.

Results and Discussion

The unique instrument design and the carefully optimized machine performance have made the synthesis of deoxyoligonucleotides a reliable, easy, and routine procedure. Our automated DNA synthesizer utilizes a modified, highly efficient solid-phase phosphoramidite chemistry which

TABLE III
APPLICATIONS OF CHEMICALLY SYNTHESIZED DNA PRIMERS

Name	Sequence	Purpose
H-2L^d (third external domain)	CTTACCTCATCTCAG	Site-specific mutagenesis
H-2L^d (transmembrane domain)	CCTTTTCAACCTGTG	Site-specific mutagenesis
H-2L^d (transmembrane domain)	TGAATGAATAAGGTAAGCATCGATG CATCGATGCTTACCTTATTCATTCA	Oligonucleotide linker with stop codons in every reading frame
J558 V_H gene of IgG: 5′ end	GAGCTGGTGAAGCCTGGGGCTTCAGTCAAG	Screen cosmid library
3′ end	GCTCAACAGCCTGACATCTGAGGAC	
D gene flanking segment	ACAAAAACCC	Screen sperm genomic cosmid library
Acetylcholine receptor α subunit of *Torpedo californica*	GTTCAGAACCTAGTACCAG	Screen cDNA library
Human interleukin 2: N-Terminal	AGGTGCACTGTTTGTGACAAGTGCAAGACT	Screen genomic and cDNA library of human and mouse
Internal	TTCTGTGGCCTTCTTGGGCATGTAAAACTT	
C-Terminal	TCAAGTTAGTGTTGAGATGATGCTTTGACA	
$C_{56} \rightarrow G$ Mutagenic primer for yeast tRNA^{ARG}	GGGGGTCCAACCCAT	Site-specific mutagenesis
Universal primer R	GTATCACGAGGCCCTT	Dideoxy sequencing in pBR322

TABLE IV
APPLICATIONS OF CHEMICALLY SYNTHESIZED DNA PRIMER MIXTURES

Name	Sequence	Purpose
Complement C4 class III genes (β)	AAACCAGGACT G G G T T C C	Identify coding regions on plasmid clones
(γ)	GAAGCACCAAAAGT G G G G T T C C	
Lens main intrinsic protein (MP26)	ACCCAATGATTAGTAAAATT G G G G G T C	Screen cDNA library from bovine lens protein mRNA
Drosophila photoreceptor cell-specific antigen	GGATAATGAGTTTCTTCCAT G G G C C T C TGCATAACATTATA G G G T C	Screen genomic library
Transducin γ subunit	TGATCAACTTCCATTTT G G C C T C GTAACTTCTTTTTT G C C C T C	Screen cDNA library
Transforming growth factor	CCATGAAAACAATATTG G G G G C	Clone TGF

allows the synthesis of deoxyoligonucleotides with excellent yield (≥97%/cycle) and high speed (<10 min/cycle). From the large number of different deoxyoligonucleotide chains which have been synthesized on the machine, we selected a few; their sequences and applications are summarized in Tables III and IV. Mixtures of deoxyoligonucleotides having identical base sequences at most positions but having two or more bases represented in others can be prepared in a single synthesizer run (Table IV). In the activation vessel, two to four different deoxynucleoside 3′-phosphoramidites can be mixed, activated, and subsequently simultaneously coupled to the silica-bound deoxyoligonucleotide chains. In this way mixed probes with high degeneracy (up to 256) have been synthesized, and all have proved useful as hybridization probes. We have carried out extensive studies concerning the synthesis of DNA mixtures. Using color-coded 5′ protecting groups on each of the four bases[18] and computer analysis for multiple absorption spectrum analysis, we were able to prove the equimolar existence of all desired sequences in the final mixture.

[18] E. F. Fisher and M. H. Caruthers, *Nucleic Acids Res.* **11,** 1589 (1983).

Section IV

Site-Specific Mutagenesis and Protein Engineering

[17] Oligonucleotide-Directed Mutagenesis: A Simple Method using Two Oligonucleotide Primers and a Single-Stranded DNA Template

By MARK J. ZOLLER and MICHAEL SMITH

Introduction

This chapter presents a simple and efficient method for oligonucleotide-directed mutagenesis using vectors derived from single-stranded phase. It is a modification of our previously published procedure[1,2] and features the use of two primers, one of which is a standard M13 sequencing primer and the other is the mutagenic oligonucleotide.[3] Both primers are simultaneously annealed to single-stranded template DNA (SS-DNA), extended by DNA polymerase I (large fragment), and ligated together to form a mutant–wild-type gapped heteroduplex. *Escherichia coli* is transformed directly with this DNA; the isolation of covalently closed circular DNA (CCC-DNA) as in our previous report is not necessary. Mutants are identified by plaque lift hybridization using the mutagenic oligonucleotide as a probe. As an example of the method, a heptadecanucleotide was used to create a T to G transversion in the *MATa* gene of *Saccharomyces cerevisiae* cloned into the vector M13mp5. The efficiency of mutagenesis was approximately 50%. Production of the desired mutation was verified by DNA sequencing. The same procedure has been used without modification to create insertions of restriction sites as well as specific deletions of 500 bases. This chapter also describes the use of two new pUC vectors from Messing that can be interconverted between a double-stranded and single-stranded form.

In vitro mutagenesis of cloned DNA has become a standard tool in the functional analysis of nucleic acids and proteins. Many approaches exist which can be broadly grouped into (1) random and (2) site-directed methods. Two excellent reviews extensively cover these techniques and their uses.[4,5] Random techniques are used to identify the location and boundaries of a particular function or activity. Once an important region has been identified, site-directed mutagenesis can be employed to determine

[1] M. J. Zoller and M. Smith, this series, Vol. 100, p. 468.

[2] M. J. Zoller and M. Smith, *Nucleic Acids Res.* **10,** 6487 (1982).

[3] M. J. Zoller and M. Smith, *DNA* **3,** 478 (1984).

[4] D. Shortle, D. DiMaio, and D. Nathans, *Annu. Rev. Genet.* **15,** 265 (1981).

[5] M. Smith, *Annu. Rev. Genet.* **19,** 423 (1985).

the role of specific nucleotides (or amino acids). The construction of cloned DNA bearing a specified alteration is best accomplished by either gene synthesis or by oligonucleotide-directed mutagenesis. Both of these strategies have been successfully used to create DNA with a desired sequence.

Gene Synthesis

The availability of reliable procedures for manual oligonucleotide synthesis[6–11] along with an increasing access to DNA synthesizers have made gene synthesis a reasonable method to produce specific mutations. The gene is constructed using segments of oligonucleotides that are ligated together. By this procedure, an entire gene can be synthesized that contains a desired change. In addition, a simple construct can be produced that allows for the replacement of segments by a number of mutated "oligonucleotide cartridges" positioned between restriction endonuclease sites.[12–15]

Enzymatic Extension of a Mutagenic Oligonucleotide

Oligonucleotide-directed mutagenesis is similar in principle to mutagenesis by gene synthesis in that an oligonucleotide that bears the desired mutant sequence is inserted into a cloned gene. However, oligonucleotide-directed mutagenesis differs from gene synthesis in the manner in which the cloning is accomplished. In this case, an oligonucleotide consisting of the mutant sequence is hybridized to its complementary se-

[6] S. P. Adams, K. S. Kavka, E. J. Wykes, S. B. Holder, and G. R. Galuppi, *J. Am. Chem. Soc.* **105,** 661 (1983).

[7] M. H. Caruthers, S. L. Beaucage, C. Becker, J. W. Efcavitch, E. F. Fisher, G. Galuppi, R. Goldman, P. Dehaseth, M. Matteucci, L. McBride, and Y. Stabinsky, *in* "Gene Amplification and Analysis" (T. Papas, M. Rosenberg, and J. G. Chirikjian, eds.), Vol. 3, p. 1. Academic Press, New York, 1983.

[8] R. Frank, W. Heikens, G. Heister-Moutsis, and H. Blocker, *Nucleic Acids Res.* **11,** 4365 (1983).

[9] M. J. Gait, H. W. D. Mathes, M. Singh, B. S. Sproat, and R. C. Titmas, *Nucleic Acids Res.* **10,** 6243 (1982).

[10] H. W. D. Mathes, W. M. Zenke, T. Grundtrom, A. Staub, M. Winzerith, and P. Chambon, *EMBO J.* **3,** 801 (1984).

[11] M. Gait (ed.), "Oligonucleotide Synthesis." IRL Press, Oxford, England, 1984.

[12] M. Matteucci and H. L. Heyneker, *Nucleic Acids Res.* **10,** 3113 (1983).

[13] K. M. Lo, N. R. Hackett, and H. G. Khorana, *Proc. Natl. Acad. Sci. U.S.A.* **81,** 2285 (1984).

[14] J. Wells, *Gene* **34,** 315 (1985).

[15] M. Urdea, this series, in press.

quence in a clone of wild-type DNA, thereby forming a mutant–wild-type heteroduplex. The oligonucleotide serves as a primer for *in vitro* enzymatic DNA synthesis of regions that are to remain genotypically wild type. A double-stranded heteroduplex is formed, which is subsequently segregated *in vivo* into separate mutant and wild-type clones. The two can be distinguished by a number of screening procedures or the mutated DNA can be selected for.

The basic strategy of oligonucleotide-directed mutagenesis was developed using the single-stranded phage ϕX174.[16,17] The technique has been successfully applied to genes cloned into either plasmid[18–21] or phase vectors.[1–3,22–25] In an earlier report we described an efficient procedure for oligonucleotide-directed mutagenesis using M13-derived vectors. The major features of this method were the use of single-stranded clone DNA as template, the purification of *in vitro* synthesized CCC-DNA by alkaline sucrose gradient centrifugation, and the use of the mutagenic oligonucleotide to screen for mutants by hybridization. We report here a variation of our original method that is simpler, faster, and of equal efficiency.[3] It is based on the use of two primers, one of which is the mutagenic oligonucleotide and the other is a standard sequencing primer. This method obviates the need to isolate CCC-DNA in order to obtain mutants in high yield yet retains the convenience of the M13 system for template isolation, sequencing, and screening. Variations of two-primer mutagenesis have been described by Wallace and co-workers[26] and by Norris *et al.*[23] In addition, several improvements in the basic procedure have been

[16] C. A. Hutchison, S. Phillips, M. H. Edgell, S. Gillam, P. Jahnke, and M. Smith, *J. Biol. Chem.* **253,** 6551 (1978).

[17] A. Razin, T. Hirose, K. Itakura, and A. D. Riggs, *Proc. Natl. Acad. Sci. U.S.A* **75,** 4268 (1978).

[18] S. Inouye and M. Inouye, *in* "DNA and RNA Synthesis" (S. Narang, ed.). Academic Press, New York, 1985.

[19] E. D. Lewis, S. Chen, A. Kumar, G. Blanck, R. E. Pollack, and J. L. Manley, *Proc. Natl. Acad. Sci. U.S.A.* **80,** 7065 (1983).

[20] B. A. Oostra, R. Harvey, B. K. Ely, A. F. Markham, and A. E. Smith, *Nature (London)* **304,** 456 (1983).

[21] R. B. Wallace, M. Schold, M. J. Johnson, P. Dembek, and K. Itakura, *Nucleic Acids Res.* **9,** 3647 (1981).

[22] C. Montell, E. F. Fisher, M. H. Caruthers, and A. J. Berk, *Nature (London)* **295,** 380 (1982).

[23] K. Norris, F. Norris, L. Christiansen, and N. Fiil, *Nucleic Acids Res.* **11,** 5103 (1983).

[24] B. Wasylyk, R. Derbyshire, A. Guy, D. Molko, A. Roget, R. Toule, and P. Chambon, *Proc. Natl. Acad. Sci. U.S.A.* **77,** 7024 (1980).

[25] P. Carter, H. Bedouelle, and G. Winter, *Nucleic Acids Res.* **13,** 4431 (1985).

[26] M. Schold, A. Colombero, A. A. Reyes, and R. B. Wallace, *DNA* **3,** 469 (1984).

introduced by Carter *et al.*,[25] Kunkel,[27] and by Kramer *et al.*[28] These protocols eliminate hybridization screening by implementing genetic selections.

Our original method used M13-derived vectors since the procedures were modeled after the original ϕX174 mutagenesis studies. See Messing[29] for a description of these vectors. Dente *et al.*[30] introduced a set of pEMBL vectors that were based on pUC8/9 and included an origin of replication from the single stranded phage f1. By superinfection of plasmid bearing cells with a f1 helper phage, single-stranded plasmid DNA could be isolated and used for sequencing or mutagenesis. Since then other workers have constructed other dual vectors (see Mead and Kemper for a review[31]). Recently, Messing and co-workers (unpublished results) have introduced two new pUC vectors that contain M13 intergenic (IG) DNA and a new helper phage, M13KO7, with which single-stranded DNA can be isolated from these vectors. These dual vectors are very useful in conjunction with oligonucleotide-directed mutagenesis for three reasons: (1) the pUC vectors (3.2 kb) are much smaller than M13 vectors (7.2 kb), thus larger fragments can be cloned into the pUC vectors, (2) double-stranded DNA is easier to isolate than M13 RF and thus subsequent steps with the mutagenized clone are simpler, and (3) the helper phage M13KO7 does not replicate as well as wild-type M13 in the presence of the pUC vector and thus represents a small fraction of the isolate SS-DNA. The IG region from M13 can be added to almost any vector, therefore mutagenesis and functional analysis can be accomplished using the same vector. We have used the same mutagenesis procedure outlined here on these single-stranded pUC vectors with great success. The only difference is that hybridization screening is done on colonies not plaques as with M13 vectors. Use of these vectors will be described in this chapter as well.

Materials and Methods

Strains. Escherichia coli JM101 [Δ(*lac–pro*), *SupE, thil,* F′, *proAB*$^+$, *laci*$^+$, *lacZ* AΔ15 *tra*D*36*] was used for all work with M13 and was the

[27] T. Kunkel, *Proc. Natl. Acad. Sci. U.S.A.* **82,** 488 (1985).

[28] W. Kramer, V. Drutsa, H.-W. Jansen, B. Kramer, M. Plfugfelder, and H.-J. Fritz, *Nucleic Acids Res.* **12,** 9441 (1984).

[29] J. Messing, this series, Vol. 101, p. 20.

[30] L. Dente, G. Caesarini, and R. Cortese, *Nucleic Acids Res.* **11,** 1645 (1983).

[31] D. A. Mead and B. Kemper, *in* "Vectors: A Survey of Molecular Cloning Vectors and Their Uses." Butterworth, Massachusetts (in press).

generous gift of J. Messing (Waksman Institute, Piscataway, New Jersey). A streak of JM101 was maintained at 4° for approximately 1 month on a plate of supplemented M9 + B1. Overnight cultures were grown in liquid M9 (supplemented) + B1. *Escherichia coli* MV1193, Δ(*lac–proAB*), *thi*, *rpsi* (strep^r), *endA*, *sbcB15*, *hspR4*, Δ(*srl–recA*)306::Tn*10* (tet^r) [F′: *traD36*, *proAB*, *lacI*^{q}zdelM15], used with pUC118 and pUC119, was also supplied by J. Messing.

Vectors. For mutagenesis with M13 vectors, a 4.2-kb *Hind*III fragment containing the *MATa* gene from *Saccharomyces cerevisiae* was cloned into the vector M13mp5 as described by Messing.[29] A detailed description of the construction of this clone is presented in Zoller and Smith.[2] pUC118 and pUC119 were supplied by J. Messing. Mutagenesis using these vectors was conducted using a 6-kb fragment containing the *CDC25* gene from *S. cerevisiae* cloned into the *Sal*I site of pUC118.[32]

Materials. *Escherichia coli* DNA polymerase (large fragment) and T4 DNA ligase were obtained from Bethesda Research Laboratories. T4 polynucleotide kinase was obtained from New England Biolabs. [α-^{32}P]dATP and [γ-^{32}P]rATP were purchased from Amersham. Unlabeled deoxyribonucleic acids, rATP, and dideoxyribonucleic acids were obtained from PL Biochemicals and supplied in solid form. Nitrocellulose filters (BA85) were purchased from Schleicher and Schuell. All other materials were of standard quality for molecular biological work. The heptadecanucleotide used for mutagenesis (5′-CCGCAACAG-GAAAATTT-3′) was synthesized by a manual phosphite triester method similar to that of Adams *et al.*[6] The M13 sequencing primer used as the "second primer" (5′-CCCAGTCACGACGTT-3′) and the hexadecanucleotide used to sequence the mutation (5′-TAAACGTATGAGATCT-3′) were also synthesized manually according to Adams *et al.*[6] Purification of oligonucleotides following synthesis was carried out using a 20% polyacrylamide–7 *M* urea sequencing gel (40 × 20 × 0.05 cm) in TBE buffer. The desired product was eluted from the gel by a crush–soak method described by Maxam and Gilbert[33] using 0.5 *M* ammonium acetate. Isolation of the oligonucleotide was accomplished either by ethanol precipitation or by C_{18} Sep-Pak chromatography.[13,34]

Mutagenesis of the *CDC25* gene in pUC118 was accomplished using an 18-mer that produced a single base insertion. The second primer was a

[32] L. Levin (Cold Spring Harbor Lab., Cold Spring Harbor, New York), unpublished results (1986).

[33] A. M. Maxam and W. Gilbert, this series, Vol. 65, p. 499.

[34] P. R. Sanchez and M. S. Urdea, *DNA* **3,** 334 (1984).

15-mer that primed approximately 1 kb upstream from the mutagenic oligomer. Both oligonucleotides were synthesized on an Applied Biosystems 380A machine.

Solutions

10× kinase buffer:	1 *M* Tris–HCl, 0.1 *M* $MgCl_2$, 0.1 *M* dithiothreitol (DTT) (pH 8.3)
10× Buffer A:	0.2 *M* Tris–HCl, 0.1 *M* $MgCl_2$, 0.5 *M* NaCl, 10 m*M* DTT (pH 7.5)
10× Buffer B:	0.2 *M* Tris–HCl, 0.1 *M* $MgCl_2$, 0.1 *M* DTT (pH 7.5)
2 m*M* dNTPs:	2 m*M* dCTP, 2 m*M* dTTP, 2 m*M* dATP, 2 m*M* dGTP
20× SSC:	3 *M* NaCl, 0.3 *M* sodium citrate, 10 m*M* EDTA (pH 7.2)
100× Denhardt's:	2% Bovine serum albumin, 2% polyvinylpyrrolidone, 2% Ficoll
Prehybridization solution I:	3 ml 20× SSC, 1 ml 100× Denhardt's, 0.2 ml 10% SDS, 5.8 ml H_2O
Prehybridization solution II:	3 ml 20× SSC, 1 ml 100× Denhardt's, 6 ml H_2O
Probe solution:	10 ml prehybridization solution II, 10^6–10^7 cpm oligonucleotide (1–2 ml)
TBE:	50 m*M* Tris, 50 m*M* borate, 1 m*M* EDTA (pH 8.3)

Media

YT:	8 g tryptone, 5 g yeast extract, 5 g NaCl per liter
YT plates:	YT media plus 15 g agar per liter
YT top agar:	YT media plus 8 g agar per liter
2YT:	16 g tryptone, 10 g yeast extract, 5 g NaCl per liter
M9 supplemented + B1:	1× M9 plus 0.2% glucose, 1 m*M* $MgSO_4$, 0.0001% B1, 0.4% casamino acid, 0.1 m*M* $CaCl_2$

For pUC118/119 SS-DNA preparation: 2YT plus 0.001% B1 plus 150 μg/ml ampicillin

Transformation of E. coli JM101. Competent cells are prepared on the day of use by diluting 0.5 ml of an overnight JM101 culture into 50 ml YT media. Cells are grown at 37° to an A_{600} of 0.5, harvested, and resuspended in 25 ml ice-cold 50 m*M* $CaCl_2$. Cells are left on ice 30 min, harvested, and resuspended in 5 ml ice-cold 50 m*M* $CaCl_2$. Glycerol can be added to 15% final concentration, and the cells can be stored at −70° without significant loss of transformation efficiency.

Two hundred microliters of $CaCl_2$-treated cells are aliquoted into Falcon 2059 tubes and placed on ice. Appropriate amounts (under 10 μl) of DNA are added to the cells which were kept on ice for 30 min. The cells are placed at 42° for 2 min, after which 100 μl overnight JM101 and 2.5 ml YT top agar are added and the entire mixture poured onto a YT plate. Freshly prepared competent cells can be plated without the addition of the 100 μl overnight culture.

Transformation of E. coli MV1193. Competent cells are prepared as described above for JM101 except that cells are grown in LB and are concentrated 50× instead of 10× in 100 m*M* $CaCl_2$. An appropriate aliquot of DNA is added to 100 μl of cells in 2059 tubes. After 10 min on ice, tubes are placed at 42° for 2 min, after which 1 ml LB is added and tubes are incubated without shaking at 37° for 1 hr. Next, aliquots are spread onto LB–amp plates and the plates are incubated overnight at 37°. See below for appropriate dilution.

Preparation of Single-Stranded Template DNA

A. Preparation of Single-Stranded M13: SS-DNA is prepared as described by Sanger *et al.*[35] A single plaque is cored with a 100-μl glass micropipet and added to 1.5 ml 2YT to which 15 μl overnight JM101 has also been added. Growth is carried out in a Falcon 2059 tube with vigorous shaking at 37° for 5–6 hr. The cells are transferred to a 1.5-ml Eppendorf tube and are pelleted by a 5-min spin in a microcentrifuge. The supernatant is transferred to a fresh 1.5-ml Eppendorf tube. Fifty microliters of supernatant is removed and stored as a phage stock at 4°. Two hundred microliters 2.5 *M* NaCl/20% PEG 8000 are added to the remainder of the phage suspension, the tube is mixed, and incubated at room temperature for 15 min. The phage are pelleted by centrifugation for 5 min

[35] F. Sanger, A. R. Coulson, B. G. Barrell, A. J. H. Smith, and B. A. Roe, *J. Mol. Biol.* **143,** 161 (1980).

at room temperature, and the supernatant is removed by aspiration. At this point a small white pellet can be observed. The phage are resuspended in 100 μl TE (10 : 1) (i.e., 10 mM Tris–HCl, 1 mM EDTA, pH 8.0), extracted with 50 μl phenol, 50 μl phenol/chloroform (1 : 1), and finally 2 $\times$ 500 μl ether or chloroform. Ten microliters 3 M sodium acetate (pH 5.5) and 250 μl ethanol are added and the tube is placed at $-70°$ for 15 min. The DNA is pelleted by a 5-min centrifugation at 4°, washed with 0.5 ml 70% ethanol, dried under reduced pressure, and resuspended in 50 μl TE (10 : 1). Typical yields range from 2–5 μg. This is usually enough to do several mutagenesis experiments.

To prepare larger amounts of template DNA, a phage stock from a single plaque grown in 1.5 ml 2YT is prepared as described above. The cells are spun out, and the phage-containing supernatant is poured into a fresh 1.5-ml tube. For each milliliter of 2YT are added 10 μl overnight JM101 and 10 μl of phage stock. For example, to prepare SS-DNA from a 50-ml culture, 0.5 ml of overnight JM101 and 0.5 ml phage stock are added to 50 ml 2YT, incubated at 37° for 5–6 hr, after which the cells are harvested by centrifugation and the supernatant is treated as described above with the volumes scaled up proportionally.

B. Preparation of Single-Stranded pUC118/119 DNA. First make a stock of helper phage M13KO7. Make serial dilutions of phage in YT or TE. Add 10 μl of diluted phage to 2.5 ml YT top agar contianing 100 μl overnight MV1193 (or JM101) and plate onto YT plates. Once individual plaques are visible, core 5 plaques with a sterile micropipet and add them to 5 ml 2YT plus 70 μg/ml kanamycin plus 50 μl overnight MV1193 cells. Grow the phage at 37° for 10–14 hr with vigorous shaking. Pellet the cells by centrifugation. Use the supernatant as described below. Titer phage before using.

To make single-stranded DNA start from a single colony on a plate. Use a sterile loop to transfer the entire colony into a 2059 tube containing 2 ml 2YT supplemented with 0.001% B1 and 150 μg/ml ampicillin. To this add an aliquot of M13KO7 helper phage. (The amount added must be titrated to find optimal yields of SS-DNA. Usually this is around 1 μl of stock that is 5×10^{10} phage/ml.) Grow in a shaker for 1 hr at 37°, then add 4 μl 25 mg/ml kanamycin. Continue growing at 37° for 8–14 hr. Pour 1 ml of culture into a 1.5-ml Eppendorf tube. Spin 5 min then pour the supernatant into a fresh tube. Work up as for SS-DNA preparation as described above.

Preparation of M13-RF. Five milliliters of phage stock is prepared as described in Part B above. Fifty microliters overnight JM101 and 50 μl phage from a 1-ml stock are added to 5 ml 2YT. (Alternatively, one plaque can be added in place of phage stock.) The culture is incubated at 37° for 5–6 hr after which the cells are harvested and the supernatant is retained. Two hours after a phage culture is started 2.5 ml of JM101 overnight is added to 500 ml M9 (supplemented) and incubated at 37° with shaking. This culture is grown to an A_{600} of 0.7–0.8 then innoculated with 5×10^{12} phage (10^{10} phage/ml). Growth at 37° is continued for 2 hr after which the cells are pelleted and M13RF is isolated by alkaline lysis plasmid preparation method (Maniatis *et al.*, 1981, see Ref. 39). The DNA is resuspended in 10 ml TE (50 : 1), digested with 0.5 mg RNAse for 15 min at 23°, then subjected to cesium chloride–ethidium bromide isopycnic centrifugation using a Beckman VTi50 rotor for 24 hr at 45,000 rpm and 18°. A single fluorescent band of DNA is usually obtained. Typical yields range from 100 to 200 μg per 500 ml preparation. The success of this procedure depends on the proper multiplicity of infection (at least 10^{10} phage/ml culture). The phage stock can be prepared a day ahead, titered, then used the next day to prepare RF.

5′ Phosphorylation of the Mutagenic Oligonucleotide

A. For Mutagenesis. The mutagenic oligonucleotide (200 pmol) is mixed with 2 μl 10× kinase buffer, 1 μl 10 m*M* rATP, and 4 units T4 polynucleotide kinase in a total volume of 20 μl. The reaction is carried out in a 1.5-ml Eppendorf tube incubated at 37° for 1 hr and is terminated by heating at 65° for 10 min.

B. For Hybridization or Primer Extension. The mutagenic oligonucleotide (20 pmol) is phosphorylated as in A with 50 μCi [γ-^{32}P]ATP (3000 Ci/mmol) as the sole source of ATP.

Separation of Labeled Oligonucleotide from Unincorporated Label

Labeled oligonucleotide is removed from unincorporated ATP by DE-52, Sephadex G-50, or C_{18} Sep-Pak chromatography as described below. For preparation of hybridization probes, the DE-52 method is used. For preparation of a salt-free oligonucleotide solution for primer extension, the C_{18} Sep-Pak method is used. The effectiveness of labeling can be assessed by running an aliquot of the kinase reaction on a 20% sequencing gel. Alternatively, an aliquot can be chromatographed on Whatman DE-81 paper using 0.3 *M* ammonium formate (pH 8.0). Under these conditions, the oligonucleotide remains at the origin and ATP migrates with an R_f of approximately 0.7.

A. DE-52 Batch Elution. Twenty-five hundredths milliliter of DE-52 equilibrated in TE (10 : 1) is packed into a 1 ml Pipetman tip or a Dispocolumn. The sample is diluted by addition of 200 μl (TE (10 : 1), applied to the column, and washed with 5 × 1 ml of TE (10 : 1). When using the Pipetman tip, the washes can be forced through by attachment of the Pipetman to the end. Unincorporated ATP is eluted by 5 × 1 ml washes with 0.2 *M* NaCl in TE (10 : 1). The labeled oligonucleotide is eluted with 3 × 1 ml washes with 1 *M* NaCl in TE (10 : 1).

B. Sephadex G-50 Chromatography. Sephadex G-50 (superfine) is prepared according to the manufacturers directions, then equilibrated in either 10 m*M* triethylammonium bicarbonate (pH 8.0) or 10 m*M* ammonium bicarbonate (pH 8.0). The column is prepared in a 10-ml plastic syringe (bed volume is 10 ml). The buffer head is dropped to the top of the Sephadex and the sample is applied (total volume 200–500 μl). Allow the sample to enter the bed, then carefully apply 1 ml of buffer. Collect 1 ml fractions. (The unique characteristic of this column is that once the buffer reaches the top of the Sephadex, the flow essentially stops until more buffer is added). Continue to add 1-ml aliquots of buffer and collect fractions. Monitor each fraction using a hand-held Geiger counter. The labeled oligonucleotide elutes reproducibly between fractions 4 and 7. The sample is dried on a Speed Vac Concentrator. The dried sample can be resuspended in water and evaporated again.

C. C_{18} Sep-Pak Chromatography. A Sep-Pak (Waters) column is prepared for use by applying 10 ml acetonitrile (Baker, HPLC grade) to the matrix using a 10-ml polypropylene syringe. The organic solution is washed out with 10–20 ml water. The Sep-Pak is now ready. The oligonucleotide sample is diluted with 1–2 ml of TE (10 : 1) and applied through the syringe to the Sep-Pak. Wash the Sep-Pak sequentially with 10 ml 25 m*M* ammonium bicarbonate (pH 8.0), 10 ml 25 m*M* ammonium bicarbonate/5% acetonitrile, 2 × 10 ml 5% acetonitrile/95% H_2O, and then elute the oligonucleotide with 3 × 1 ml 30% acetonitrite/70% H_2O. Evaporate the samples to dryness on a Speed Vac. This is a modification of a procedure described by Lo *et al.*[13] and Sanchez and Urdea.[34]

Two-Primer Oligonucleotide-Directed Mutagenesis

A. Annealing. Single-stranded template DNA (0.5 pmol) is mixed with 5′-phosphorylated mutagenic oligonucleotide (10 pmol), nonphosphorylated M13 sequencing primer (10 pmol), and 1 μl 10× Buffer A in a total volume of 10 μl. The mixture is heated at 55° for 5 min then placed at room temperature (23°) for 5 min. During the annealing reaction, the "Enzyme/nucleotide" solution is prepared by addition of the following components: 1 μl 10× Buffer B, 1 μl 2 m*M* dNTPs, 1 μl 10 m*M* rATP, 6 units T4 DNA ligase, 2 units *E. coli* DNA pol I (large fragment), and H_2O to 10 μl. This is kept on ice until use. (Note that the concentration of dNTPs have been lowered from our previous report after a suggestion by Shi and Fersht.[36] Also the amount of ligase has been increased.)

B. Extension and Ligation. After 5 min at room temperature, 10 μl of the "Enzyme/nucleotide" solution is added to the annealed DNA. The contents are mixed then the tube is placed at 15° for 4–12 hr. (We generally use 6–8 hr.)

C. Transformation. For M13, the DNA sample is diluted 20×, 100×, and 500× with TE (10 : 1). Quantities of 1 and 5 μl of each dilution are used to transform JM101 as described above.

For pUC118/119, add 1 μl of the DNA sample to 100 μl of competent MV1193 in 2059 tubes. Follow transformation procedure above. Spread plate 200, 150, 100, and 50 μl onto LB–amp plates. For hybridization screening, it is convenient to plate directly onto nitrocellulose filters (BA85 S and S) on top of LB–amp plates. Try to obtain a filter with approximately 100–200 colonies on it. Therefore, it is probably a good idea to transform a second aliquot of cells with 5 μl of sample.

Plaque Hybridization (M13)

A plate with approximately 100–200 plaques is chosen and placed at 4° for 15 min. A dry piece of nitrocellulose is placed onto the plate for 5 min (orient by poking with a needle). The filter is carefully removed, dried plaque side up for 5 min, then baked under reduced pressure at 60–80° for 1 hr. The filter is placed in a boilable bag, 10 ml of prehybridization solution I is added, and the sealed bag is held at 65° for 1 hr. Following this, the prehybridization solution is removed, and 10 ml of probe solution is added. Hybridization is carried out at 37° for 4–16 hr. (For shorter oligomers or AT-rich oligonucleotides, the temperature of hybridization is 23°.) The probe solution is removed and stored at −20° for future use. The filter is washed at room temperature with 3 × 100 ml of 6× SSC, covered with SaranWrap, and autoradiographed for 6–12 hr at −70° using Kodak XAR-5 film and a Dupont screen. Subsequent washes at higher temperatures are carried out by placing the filter into 100 ml of preheated 6× SSC for 2 min at the desired temperature. Autoradiography is carried out as above for 6–12 hr depending on the strength of the hybridization signal.

Dot Blot Hybridization (M13)

Single-stranded DNA (2 μl out of 50 μl from a minipreparation) is spotted onto a dry sheet of nitrocellulose placed over a piece of Whatman

[36] J.-P. Shi and A. R. Fersht, *J. Mol. Biol.* **177,** 268 (1984).

3MM. The filter is baked under reduced pressure for 1 hr at 60–80°, then wetted with 10 ml of prehybridization solution II for 15 min at room temperature in a Petri dish. This is removed, and 10 ml of probe solution is added. Hybridization is carried out at 37° for 1 hr. All washes are carried out as described above. Autoradiography is conducted for 1 hr.

Colony Hybridization Screening (pUC118/119)

Follow the procedure for filter replication and colony screening with oligonucleotide probes as described by Woods[37,38] (also see Maniatis *et al.*[39]). Briefly, pick one or two filters with 100–200 colonies on each. Replicate filters, lyse colonies, neutralize, then bake filters for 1–2 hr at 80°. Prewash filters according to Wood with 3× SSC/0.1% SDS at 55° for 4 hr to overnight. Make sure that all of the colony debris is removed. Prehybridize for several hours then hybridize with phosphorylated probe for 4 hr to overnight. Wash filters as described above for plaque lift.

Plaque Purification of M13 Clones

Individual plaques are touched with a sterile toothpick or needle, and the phage are added to 100 μl of sterile TE (10 : 1). This is mixed well, diluted 10^4 and 10^5 with TE (10 : 1), and 10 μl of each dilution is plated with 100 μl overnight JM101 and 2.5 ml YT top agar. Single-stranded DNA is prepared from six plaques of each putative mutant.

Colony Purification of pUC118/119 Clones

As with M13 plaques, a positive colony may contain a small amount of wild-type molecules. Simply streaking for single colonies will not segregate mutant from wild-type molecules since the plasmid copy number is so high. Thus, it is important to make DNA and retransform in order to obtain a pure mutant clone. Innoculate 3 ml of LB plus ampicillin with a putative positive colony. Grow overnight at 37°. Harvest the cells from 1.5 ml and prepare minipreparation DNA by either an alkaline lysis or boiling procedure as described in Maniatis *et al.*[39] This DNA is then used to transform MV1193 (or JM101). SS-DNA is prepared from a number of colonies and is sequenced to verify the construction of the desired mutant (see below). The plasmid preparation can also be done immediately fol-

[37] D. E. Woods, A. F. Markham, A. T. Ricker, G. Goldberger, and H. R. Colten, *Proc. Natl. Acad. Sci. U.S.A.* **79,** 5661 (1982).

[38] D. E. Woods, *Focus* **6**(3) (1984).

[39] T. Maniatis, E. F. Fritsch, and J. Sambrook, "Molecular Cloning: A Laboratory Manual." Cold Spring Harbor Lab., Cold Spring Harbor, New York, 1982.

lowing the initial transformation with the heteroduplex DNA. Simply transform MV1193 and grow overnight instead of plating onto nitrocellulose. Prepare minipreparation DNA, transform with an aliquot of this DNA, plate onto nitrocellulose, and screen as described above.

DNA Sequencing

Chain-terminator DNA sequencing is carried out according to Sanger *et al.*[35] There is virtually no difference in the procedure whether using M13 or pUC118/119. Often the site of the mutation is too far from the site of the universal sequencing primer. In these cases, we usually synthesize a 15-mer that primes 50–150 nucleotides upstream of the mutation. We find that the oligonucleotide synthesized by the Applied Biosystems machine (and probably other machines) does not have to be gel purified to use as a sequencing primer.

Single-stranded DNA (5 μl out of 50 μl from a minipreparation) is mixed with the sequencing primer (0.2 pmol) and 1 μl 10× Buffer A in a total volume of 10 μl. The mixture is heated at 55° for 5 min, then placed at room temperature for 5 min. One microliter of [α-^{32}P]dATP (600–800 Ci/mmol, 10 μCi/μl) and 0.5 μl (2 units) of DNA pol I (large fragment) are added to the annealed DNA. Two microliters of this mixture is added to 2 μl of each of the four dideoxy stop solutions in a 1.5-ml Eppendorf tube, and the reaction is initiated by a quick spin in a microfuge. After 15 min at room temperature, 2 μl of 1 m*M* dNTPs is added to each of the tubes. The reaction is terminated by addition of 4 μl of a formamide/dye/EDTA solution. The tubes are heated at 90° for 3 min with the caps off. One microliter from each tube is electrophoresed on an 8% polyacrylamide–8 *M* urea gel at 32 watts constant power. Following electrophoresis the gel is dried and autoradiographed with Kodak XRP film overnight.

Preliminary Tests for Specific Priming by the Mutagenic Oligonucleotide

Twenty picomoles of mutagenic oligonucleotide is kinased as described above. The unincorporated. [α-^{32}P]ATP is removed by either G-50 or Sep-Pak chromatography. The labeled oligonucleotide is dried down using a Speed Vac then resuspended in 20 μl of water. Single-stranded template DNA (0.5 pmol) is mixed with 1 μl 10× Buffer A, 1 μl unlabeled mutagenic oligonucleotide (8 pmol), and 2 μl kinased oligonucleotide (2 pmol) in a total volume of 10 μl. The mixture is heated at 55° for 5 min then placed at room temperature for 5 min. During annealing, a mixture of DNA pol I (large fragment), buffer, and dNTPs is made up as described above for mutagenesis except that DNA ligase and the M13 sequencing primer are omitted. Ten microliters of this mixture is added to the primer–template mixture, and extension is carried out at 15° for 2 hr. Following this, the reaction is stopped by heating the tube at 65° for 10 min. After cooling, 10 units of the desired restriction endonuclease are added and digestion is conducted for 1–2 hr. The extension reaction conditions are compatible with enzymes that are active in 25 m*M* NaCl, 20 m*M* Tris, 10 m*M* $MgCl_2$ 5 m*M* DTT. For enzymes that cut under other conditions, the solution can be either diluted to decrease NaCl or the NaCl concentration can be increased. The endonuclease reaction is stopped by addition of an equal volume of formamide/dye/EDTA solution. The mixture is treated at 90° for 5 min then placed on ice until use. An aliquot (5 μl) is electrophoresed on a sequencing gel alongside the reactions from dideoxynucleotide sequencing using the mutagenic oligomer (see below).

DNA Sequencing using a Mutagenic Oligonucleotide

The mutagenic oligonucleotide is substituted for the sequencing primer in the standard dideoxynucleotide sequencing procedure described above. The oligonucleotide does not have to be kinased. The ratio of primer to template should be adjusted to 20 : 1 (or the ratio that will be used with the mutagenesis reaction).

Results

This chapter presents a simplified protocol of our original method for oligonucleotide-directed mutagenesis using M13-derived vectors. A single T to G transversion in the yeast *MATa* gene was constructed to illustrate the procedure. In addition, we describe the appropriate modifications of the procedure to be used in conjunction with the pUC118/119 vectors.

Experimental Rationale

Figure 1 shows the DNA sequence of the template in the region of the desired mutation hybridized with the mutagenic oligonucleotide. The heptadecanucleotide is perfectly matched with the template except for a single G–A mismatch that encodes the desired change. Prior to conducting the mutagenesis reaction, the two preliminary tests, primer extension and

```
                          G                       Mutagenic
             5' CCGCAACAG AAAATTT  3'             Oligonucleotide
3'-AAGTCGAAAGGCGTTGTC TTTTAAAATATTT -5'           Single-stranded
                          A                       Template DNA
```

FIG. 1. Mutagenic oligonucleotide annealed to single-stranded template DNA.

PRIMER EXTENSION

CHAIN TERMINATOR SEQUENCING

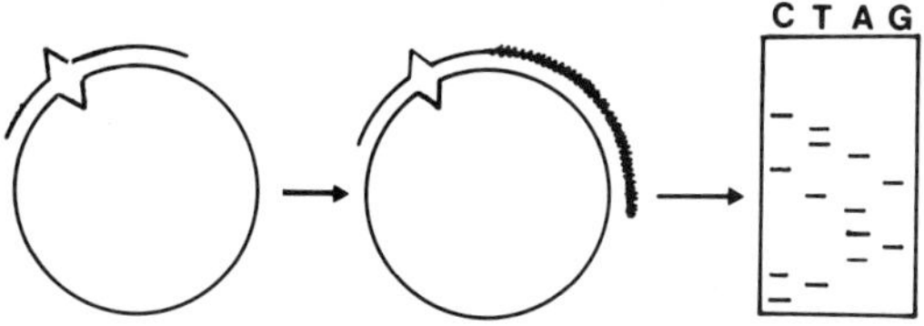

FIG. 2. Two preliminary tests used to demonstrate specific priming of the mutagenic oligonucleotide at the desired site.

chain terminator sequencing using the mutagenic oligonucleotide as primer, were carried out. The basis for these two experiments are summarized in Fig. 2.

Preliminary Tests

The purpose of these preliminary test is to determine the optimum conditions to get priming by the mutagenic oligonucleotide from the desired site. The primer to template ratio can be altered as well as the temperature of annealing and extension. Generally, we use a 10- to 30-fold molar excess of primer over template, the annealing is done at 55° for 5 min, and extension is carried out at 15°. For this particular experiment, we used a 20 : 1 primer to template ratio. The results (data not shown) indicated that the mutagenic oligonucleotide hybridized to the desired site. Primer extension followed by *Hae*III digestion yielded a major fragment 100 bases long that represented greater than 90% of the extended products observed on the autoradiogram. In the second test, use of the mutagenic oligonucleotide as a sequencing primer, the sequencing pattern was unique and clear, and showed the region immediately downstream from the desired priming site. The primer extension products were coelectrophoresed with the chain terminator sequencing reactions. The 100-base fragment comigrated with the second G in the sequence GGCC (*Hae*III) at the expected site.

Scheme for Two-Primer Mutagenesis

The scheme for two-primer mutagenesis is shown in Fig. 3. A universal squencing primer and a 5′-phosphorylated mutagenic oligonucleotide are annealed to single-stranded template DNA. Extension of each primer is carried out using *E. coli* DNA polymerase I (large fragment). The newly synthesized strand from the sequencing primer is ligated to the 5′ end of the mutagenic oligonucleotide by T4 DNA ligase. This forms a gapped heteroduplex bearing a mismatch in the middle. Covalently closed circular molecules are not formed since the sequencing primer is not phosphorylated. Extension and ligation is carried out for 4–12 hr at 15°. The DNA is diluted then used to transform *E. coli*. The resulting plaques (or colonies) are screened for mutants by filter hybridization using the mutagenic oligonucleotide as a probe.

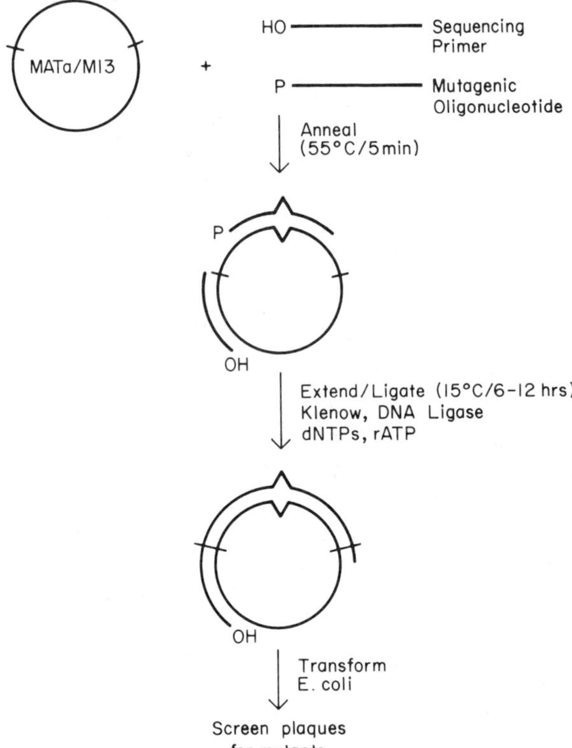

FIG. 3. Scheme for two-primer mutagenesis.

Screening for Mutants

The basis for detection of mutant DNA is shown in Fig. 4. Radioactively labeled mutagenic oligonucleotide is hybridized to a filter onto which mutant and wild-type phage DNA are bound. The principle of this procedure is that the mutagenic oligonucleotide will form a more stable duplex with a mutant clone with a perfect match than with a wild-type clone bearing a mismatch. Hybridization is carried out under conditions of low stringency where both mutant and wild-type DNA hybridize with the oligonucleotide (Figure 4A). On increasing the temperature at which the filter is washed, the mutagenic oligonucleotide remains hybridized to

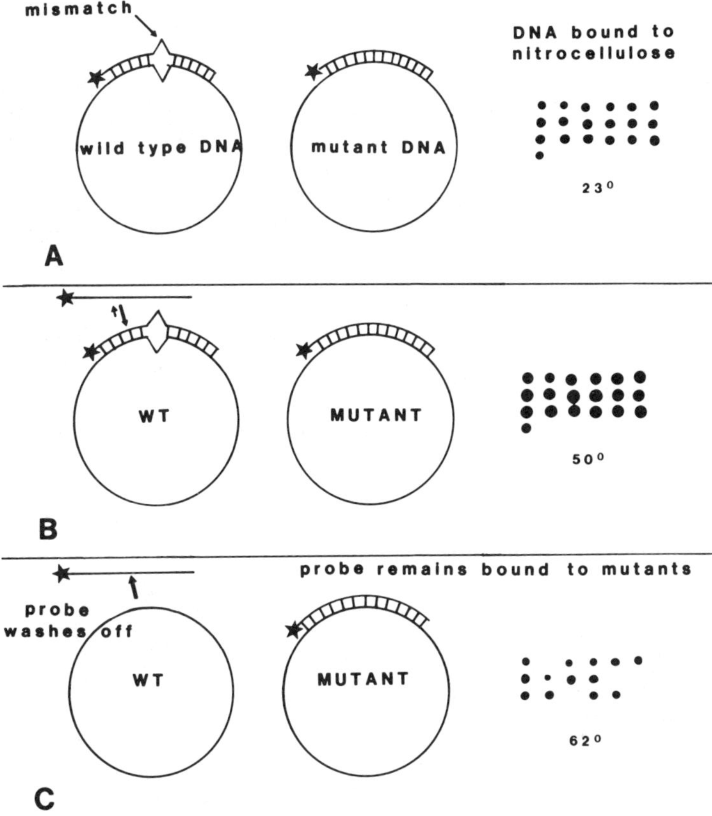

FIG. 4. Screening for mutants using the mutagenic oligonucleotide as a hybridization probe. Example of 36 clones assayed by dot blot. Hybridization was conducted at 37°, washes were carried out at (A) 23°, (B) 50°, and (C) 62°.

mutant DNA and dissociates from wild-type DNA (Fig. 4B). At a certain temperature, mutant DNA can be easily discriminated from wild-type DNA (Fig. 4C). The temperature at which mutant and wild-type DNA are discriminated depends on the particular mutation and the specific sequence of the mutagenic oligonucleotide. For a 17-mer with one or two mismatches, the wash temperature is increased in intervals of 10°. Deletions can be made using the same procedure with a large oligonucleotide that spans across the region to be deleted. In these cases, it is often convenient to synthesize a separate screening probe of 15–17 nucleotides that contains the sequence from the middle of the mutagenic oligonucleotide.

With M13 vectors, screening can be carried out by dot blot hybridization of phage DNA prepared from individually isolated plaques (Fig. 4 and Zoller and Smith[1,2]) or by plaque hybridization where 100–200 plaques can be screened at once.[3] It is generally easier to do a plaque lift to obtain a number of putative mutants than to grow up and screen individual plaques. Another advantage of the plaque lift is facilitation of screening several different mutagenesis experiments at the same time. Figure 5 shows autoradiographs of the plaque lift hybridization following a low (23°) and a high (62°) temperature wash. A number of the hybridization signals present at 23° were lost after the 62° wash. These are presumably wild-type whereas the plaques that still exhibit a signal after the 62° wash are putative mutants. On this basis, the efficiency of mutagenesis was 50%. Two putative mutants and one wild-type from this plate were plaque purified. SS-DNA was prepared from six individual plaques from each original clone. Aliquots were spotted onto nitrocellulose and hybridized

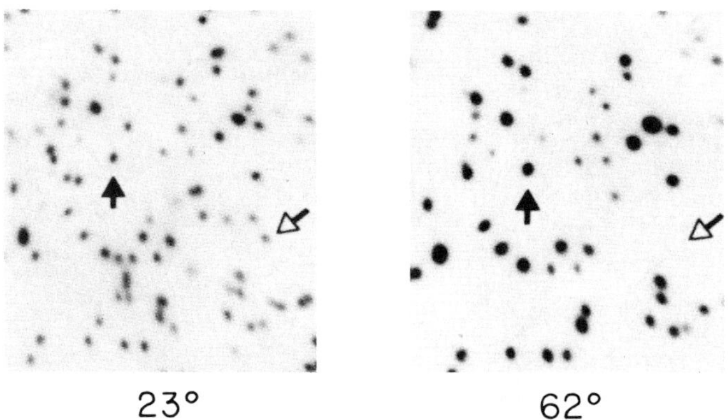

FIG. 5. Screening for mutants by plaque lift hybridization using the mutagenic oligonucleotide as a probe.

with the labeled mutagenic oligonucleotide (dot blot hybridization). All isolates from the suspected mutants hybridized after both the 23 and 62° wash. In contrast, the DNA from the suspected wild type hybridized only after the 23° wash (data not shown). DNA from one of the suspected mutants and one wild type were sequenced in the region of the desired mutation using an oligonucleotide that primed 65 bases away. As shown in Fig. 6, the correct T to G transversion was constructed in the putative mutant. No other base change was observed in this region. For functional analysis of this mutant, M13 RF was prepared and the *MAT* insert was isolated and cloned into a yeast vector (see Discussion).

It is significant to note that, theoretically, each original plaque should

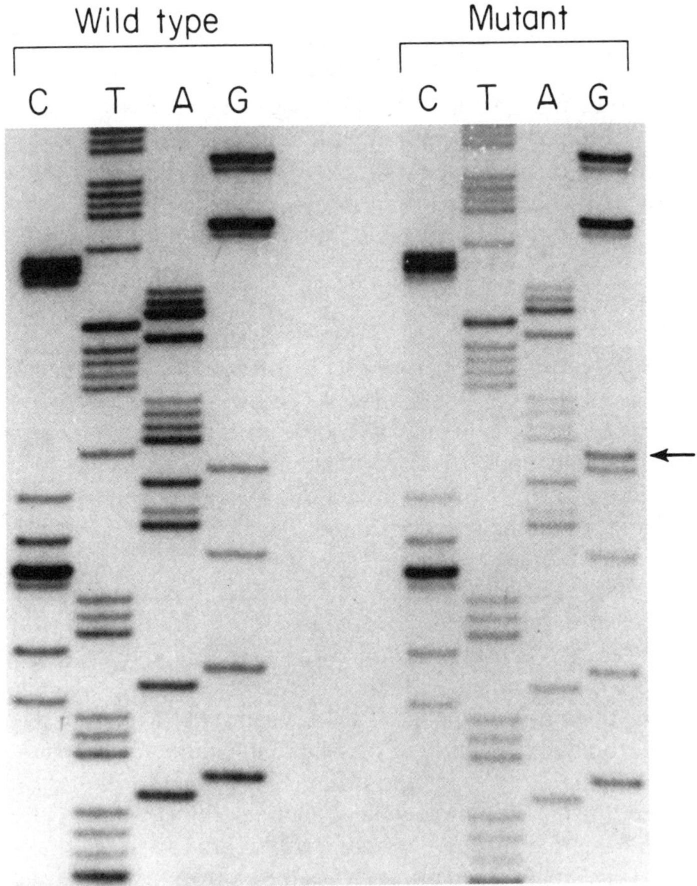

FIG. 6. DNA sequence of mutant and wild-type clones in the region of the mutation.

contain a 50:50 mixture of wild-type and mutant phage. In practice, however, we typically find that a suspected mutant plaque contains between 80 and 100% mutants by analysis of plaque purified clones. This may reflect a strand bias at an early stage of DNA replication or correction of the mismatch prior to DNA replication.

Mutagenesis of CDC25 Gene in pUC118

The procedure for mutagenesis using single-stranded pUC118 or pUC119 vectors is similar to that for M13 vectors. The procedure for transformation is the same as for any pUC vector. Screening is accomplished by hybridization of the mutagenic oligonucleotide to colony DNA bound to nitrocellulose filters. The basic concept and procedures are the same as plaque hybridization, only the preparation of the filter is different.

Using the two-primer method on the *CDC25* gene cloned into pUC118, the efficiency of mutagenesis was approximately 25% as judged by hybridization screening (data not shown). Single-stranded DNA from several putative positive colonies were prepared and sequenced in the same manner as M13. Subsequent isolation of the mutagenized *CDC25* gene fragment can be carried out from minipreparation plasmid DNA.

Discussion

This chapter presents an improved method for oligonucleotide-directed mutagenesis using M13-derived vectors and single-stranded pUC vectors. This procedure entails fewer steps than our previous method and results in an equally high efficiency of mutagenesis. Hybridization screening can be eliminated when the procedure is used in conjunction with the selection and enrichment procedures developed by Carter *et al.*,[25] Kunkel,[27] and Kramer *et al.*[28]

Two primers are used, one of which is the mutagenic oligonucleotide and the other is an oligonucleotide that primes some distance away. Each primer is extended on the same single-stranded template by *E. coli* DNA polymerase I (large fragment). The "upstream" primer extends toward the 5′ end of the mutagenic oligonucleotide and, on ligation of it, forms a gapped heteroduplex. Wild-type and mutant phage are propagated and segregated *in vivo* by transfecting *E. coli*. Screening for mutant phage is accomplished using the mutagenic oligonucleotide as a hybridization probe. The basis for this screen is the differential thermal stability of a matched versus mismatched duplex. As an example, a 17-mer was used to create a T to G transversion in the *MATa* gene of *S. cerevisiae*. The procedure, from primer extension to sequenced mutant, took about 5 days with less than 3 hr of hands-on work per day. The efficiency of mutagenesis was approximately 50%.

With pUC vectors, screening is accomplished by colony screening. The basic concept and procedures are the same as plaque hybridization, only the preparation of the filter is different. Using the two-primer method on the *CDC25* gene cloned into pUC118, the efficiency of mutagenesis was approximately 25%.

There are several differences between this procedure and out previously published protocol. The extension and ligation reaction has been shortened from 12–20 to 4–12 hr. We generally extend and ligate for 6–8 hr. Following this, the DNA is used directly for transfection. The isolation of CCC-DNA by alkaline sucrose gradient centrifugation, as in our previous procedure, has been omitted. In fact, the formation of covalently closed double-stranded molecules cannot occur since the "upstream" primer was not 5′ phosphorylated. The use of the "upstream" primer is to place the mismatch within the interior of the extended fragment, reminiscent of marker rescue experiments with ϕX174.[40]

The position of the "upstream" primer may be important. We have not investigated this aspect fully. In this experiment the "upstream" primer was an M13 sequencing primer that extends from a position 2.9 kb upstream of the mutagenic oligomer. Oligonucleotides other than the universal sequencing primer have been used which prime 1 kb from the mutagenic oligomer. In addition, a successful experiment has been completed in which the upstream primer was only 200 bases away.[41]

The features of the mutagenic oligonucleotide for use in this procedure are the same as described previously. Generally, in making small changes of 1–3 bases, 17- to 20-mers are used. The same procedure has been successfully used to make large deletions (0.5–1 kb) as well as insertions of restriction endonuclease sites. Deletions and restriction site insertions are generally made using 30- to 40-mers. In addition, mutations can be made randomly within a limited region by synthesizing a wild-type oligonucleotide that is "doped" with non-wild-type nucleotides. Prior to conducting the experiment, it is important to carry out the preliminary tests to demonstrate that the mutagenic oligonucleotide primes specifically.

Recently, several mutagenesis schemes have been developed that increase the efficiency of mutagenesis by genetic selection. The frequency

[40] C. A. Hutchison and M. H. Edgell, *J. Virol.* **8,** 181 (1971).

[41] R. Kostriken (Cold Spring Harbor Lab., Cold Spring Harbor, New York), unpublished results (1985).

of mutants is high enough so that screening is eliminated; one simply picks clones following transformation and sequences them. Refer to chaps. [18], [19], and [20] in this volume. Many mutagenesis procedures are relatively simple once they are set up in the lab. The choice of which method to use may depend simply on the familiarity of an individual with the particular vector system and enzymes utilized with the method.

Summary

The important features of the protocol described here are as follows: First, the procedure consists of a few simple steps and results in a reasonably high frequency of mutagenesis. Second, using two primers, there is no need to isolate covalently closed double-stranded molecules as in our previous method. Third, the use of vectors derived from single-stranded phage facilitates template preparation, mutagenesis efficiency, screening, and DNA sequencing. Fourth, the same basic steps can be directly applied when using the single-stranded pUC derivatives.

Acknowledgments

I thank Louisa Dalessandro for typing the manuscript, and I thank the members of my lab: Karen Johnson for conducting the M13/MATa experiment, Lonny Levin for conducting the pUC118/CDC25 experiment, and Richard Kostriken for developing shortcuts. In addition, thanks go to Jeff Vieira and Jo Messing for pUC118/119 vectors and M13KO7 helper phage. Research was funded by general laboratory grants from Cold Spring Harbor Laboratory and by the National Institutes of Health (GM 33986-01).

[18] Oligonucleotide-Directed Construction of Mutations via Gapped Duplex DNA

By WILFRIED KRAMER and HANS-JOACHIM FRITZ

Introduction

Oligonucleotide-directed construction of mutations has become the method of choice to introduce predetermined structural changes into DNA. The basic concept of the method, as originally outlined by Hutchison and Edgell,[1] rests on annealing a short (synthetic) oligonucleotide to a target region on a single-stranded replicon that is almost, but not entirely,

[1] C. A. Hutchison III and M. H. Edgell, *J. Virol.* **8,** 181 (1971).

complementary to the target region. A partial DNA duplex with at least one mispaired or unpaired nucleotide results, and the marker carried by the oligonucleotide can be salvaged into complete progeny replicions by a combination of *in vitro* and *in vivo* reactions on the heteroduplex DNA.

Several laboratories have evolved techniques to improve the versatility and efficiency of the process (see Ref. 2 for a review; also see Chaps. [17], [19], and [20] in this volume).

We have developed the gapped duplex DNA approach to oligonucleotide-directed mutation construction.[3–6] The key intermediate in this process is a partial DNA duplex of a recombinant M13 genome (gapped duplex DNA, gdDNA) which has only the target region of mutation construction exposed in single-stranded form and which, furthermore, carries distinguishable genetic markers in the two DNA strands in such a way that a rigorous selection can be applied in favor of phage progeny arising from the shorter strand (i.e., the minus strand of the M13 genome). Since the synthetic oligonucleotide becomes physically integrated into the latter DNA strand, an indirect selection for the synthetic marker is applied without the need of the constructed mutation itself being associated with a discernible phenotype. Marker scrambling is suppressed by the use of a transfection host that is deficient in DNA mismatch repair.

Two alternative variants of the gdDNA method are applicable: the "fill-in" protocol (route B, Fig. 1), which incorporates DNA polymerase/DNA ligase reactions *in vitro*, and the simplified "mix–heat–transfect" protocol (route A, Fig. 1), which bypasses these enzymatic manipulations (step 6, see below). The individual steps of the process are as follows (see Fig. 1):

1. Cloning of the target DNA fragment for the mutation construction into phage M13mp9 (I). This phage carries two amber codons in vital phage genes.[7]
2. Preparation of (single-stranded) virion DNA of this recombinant M13mp9 phage (III).
3. Cleavage of (double-stranded) replicative form (RF) DNA of phage M13mp9rev (II) with the same restriction enzyme(s) used for cloning the target DNA into M13mp9 (IV). In M13mp9rev the two

[2] M. Smith, *Annu. Rev. Genet.* **19,** 423 (1985).

[3] W. Kramer, K. Schughart, and H.-J. Fritz, *Nucleic Acids Res.* **10,** 6475 (1982).

[4] W. Kramer, V. Drutsa, H.-W. Jansen, B. Kramer, M. Pflugfelder, and H.-J. Fritz, *Nucleic Acids Res.* **12,** 9441 (1984).

[5] W. Kramer, A. Ohmayer, and H.-J. Fritz, *Nucleic Acids Res.*, in press (1986).

[6] H.-J. Fritz, J. Hohlmaier, W. Kramer, A. Ohmayer, and J. Wippler, *Nucleic Acids Res.*, in press (1986).

[7] J. Messing, R. Crea, and P. H. Seeburg, *Nucleic Acids Res.* **9,** 309 (1981).

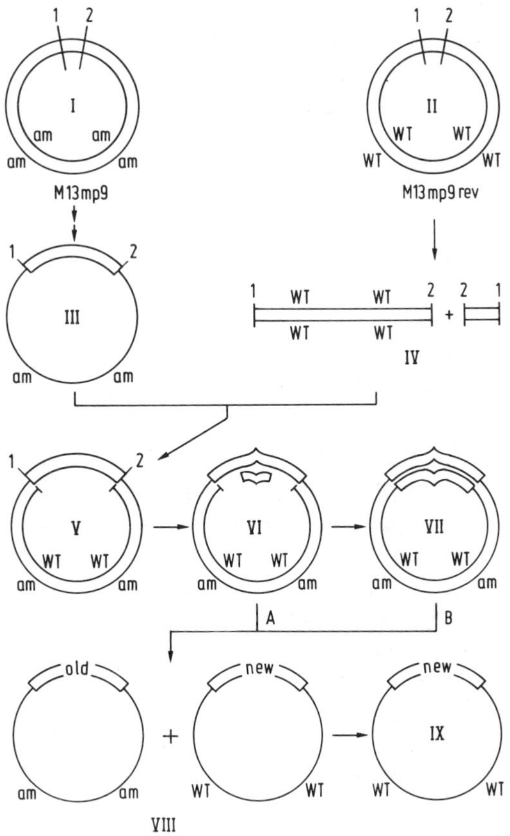

FIG. 1. Schematic representation of the gapped duplex DNA method of mutation construction. For details refer to the section Introduction. Symbols: am, amber codon; WT, corresponding wild-type codon; 1, 2, recognition sites for restriction endonucleases.

amber codons mentioned under step 1 are replaced by wild-type codons.[4]

4. Construction of gdDNA by *in vitro* DNA/DNA hybridization (V).
5. Annealing of the synthetic oligonucleotide to the gdDNA (VI).
6. Filling in the remaining gaps and sealing the nicks by a simultaneous *in vitro* DNA polymerase/DNA ligase reaction (VII, route B). This step can be omitted without drastic loss of marker yield (route A).
7. Transfection of a host deficient in mismatch repair and production of mixed phage progeny (VIII).
8. Elimination of plus strand progeny by reinfection (at very low multiplicity) of a host unable to suppress amber mutations. The syn-

thetic marker is enriched together with the minus strand progeny (IX).

9. Screening of the clones resulting from the reinfection for the presence of the desired mutation plus DNA sequence verification.

The gapped duplex DNA method combines the following favorable features:

1. Annealing of the synthetic oligonucleotide to inappropriate sites within the recombinant M13 genome is suppressed since only a relatively small window of single-stranded DNA is available for hybridization. This helps to minimize unwanted side reactions such as formation of deletions.
2. Since the entire vector part of the recombinant phage genome is already covered by the complementary minus strand in the gdDNA, the enzymatic fill-in reaction to yield fully double-stranded DNA is greatly facilitated and can be safely left to the natural mechanisms within the transfected cell.
3. As outlined above, an indirect and general selection can be applied in favor of the synthetic marker independent of any phenotype associated with this marker.

Following route B, marker yields of 88% have been achieved with a gap of 120 nucleotides in length, 65% with a larger gap (~1.64 kb).[5] Route A, tested with the larger gap, lead to marker yields of up to 53%.[6] Clearly, these yields are sufficiently high to make screening of a few randomly picked clones by DNA sequence analysis fast and economic. This combines two indispensable steps of analysis into one: identification of the correct mutant and verification of the integrity of the rest of the DNA segment under study. Starting from DNAs II and III (Fig. 1) plus oligonucleotides for mutation construction and DNA sequence analysis (chain termination method[8]), the entire procedure takes about 4–5 days.

The marker yields stated above were obtained in model experiments in which large samples of phage progeny were screened to ensure statistically meaningful results. In addition, the method has been used in a sizable number of biological studies in this laboratory and others. Results of these experiments suggest that yields in the range indicated above can generally be expected—possible exceptions being rare cases of special target DNA structure (see Scope, Limitations, and Practical Hints).

Route B, as outlined below, makes use of an enzyme combination[5] consisting of T4 DNA polymerase, *Escherichia coli* DNA ligase, and gp 32, the single-stranded DNA binding protein encoded by gene *32* of phage

[8] A. T. Bankier, K. W. Weston, and B. Barrell, this series, Vol. 155 [7].

T4. A similar combination was described by Craik *et al.*[9] Alternatively, *E. coli* DNA polymerase I (large fragment) and T4 DNA ligase can be used as described earlier.[4] The last procedure is slightly less efficient.

Materials

Chemicals and Buffers

Buffer A: 1.5 *M* KCl, 100 m*M* Tris–HCl, pH 7.5
Buffer B: 500 m*M* Tris–HCl, 600 m*M* ammonium acetate, 50 m*M* $MgCl_2$, 50 m*M* dithiothreitol, pH 8.0
100 m*M* $CaCl_2$, autoclaved
Chloroform/isoamyl alcohol: 24 : 1 (v/v) (analytical grade, Merck)
CsCl (analytical grade), solid
Diethyl ether (analytical grade)
10 (4 × 2.5) m*M* dNTPs (deoxynucleoside triphosphates): equimolar mixture of dGTP, dATP, dTTP, and dCTP (Boehringer Mannheim), each at a final concentration of 2.5 m*M*
DOC solution: 50 m*M* Tris–HCl, pH 8.0, 1% Brij 58 (polyethylene glycol monostearyl ether; Serva), 0.4% sodium deoxycholate, 62.5 m*M* EDTA
EDTA (ethylenediaminetetraacetate), pH 8, 500 and 250 m*M*
Ethanol (analytical grade), precooled to −20°
Ethidium bromide solution (10 mg/ml)
2 m*M* NAD (nicotinamide adenine dinucleotide; Boehringer Mannheim)
Phage buffer: 20 m*M* sodium phosphate, 0.002% gelatin, pH 7.2; autoclaved
Phenol (analytical grade, Merck), saturated with TE buffer
Phenol/chloroform: 1 : 1 (v/v) mixture of TE-saturated phenol and chloroform/isoamyl alcohol mixture (24 : 1, v/v)
3 *M* sodium acetate (analytical grade)
Solution A: 2.5 *M* NaCl, 20% polyethylene glycol 6000
Solution B: 25% sucrose (w/v), 50 m*M* Tris–HCl, pH 8.0
TE buffer: 10 m*M* Tris–HCl, 0.1 m*M* EDTA, pH 8.0
TES buffer: 50 m*M* NaCl, 10 m*M* Tris–HCl, 1 m*M* EDTA, pH 8.0
Tetracycline stock solution (2 mg/ml), filter sterilized

Microbiological Growth Media

Antibiotic Medium 3: 17.5 g Antibiotic Medium 3 (Difco; premixed), 1 liter H_2O; autoclaved

[9] C. S. Craik, C. Largman, T. Fletcher, S. Roczniak, P. J. Barr, R. Fletterick, and W. J. Rutter, *Science* **228,** 291 (1985).

EHA plates: 13 g Bactotryptone (Difco), 8 g NaCl, 10 g Bactoagar (Difco), 2 g sodium citrate·$2H_2O$, 1 liter H_2O; autoclaved; add 7 ml 20% glucose (sterile) after autoclaving; use 25–30 ml per petri dish (85 mm diameter)

M9 medium: 7 g $Na_2HPO_4 \cdot 2H_2O$, 3 g KH_2PO_4, 1 g NH_4Cl, 1 liter H_2O; autoclaved; add the following sterile solutions after autoclaving: 25 ml 20% glucose (w/v), 1 ml 100 m*M* $CaCl_2$, 1 ml 1 *M* $MgSO_4$, 5 ml 100 μM $FeCl_3$, and 1 ml thiamin (1 mg/ml)

Top agar: 10 g Bactotryptone (Difco), 5 g NaCl, 6.5 g Bacto-agar (Difco), 1 liter H_2O; autoclaved

Enzymes

Escherichia coli DNA ligase (New England Biolabs)
T4 DNA polymerase (New England Biolabs)
gp 32 (gene *32* protein from T4-infected *E. coli;* Pharmacia)
Lysozyme (Sigma)

Bacterial and Phage Strains

BMH 71-18: [Δ(*lac–proAB*), *thi*, *supE*; F′, *laci*^{q}, *Z*ΔM15, *proA*$^+$*B*$^+$]
BMH 71-18 mutS: (BMH 71-18, *mutS2*15::Tn*10*)
MK 30-3: [Δ(*lac–proAB*), *recA*, *galE*, *strA*; F′, *laci*^{q}, *Z*ΔM15, *proA*$^+$*B*$^+$].
M13mp9[10]
M13mp9rev[4]

Procedures

Procedure 1: Preparation of Virion DNA of Recombinant M13mp9 (Adapted from Walker and Gay[11])

1. Dilute a fresh overnight culture of strain BMH 71-18 1 : 100 with Antibiotic Medium 3, transfer 2.5 ml of the dilution to a sterile 15 ml culture tube, infect the culture with 3–5 μl of a phage stock (see Procedure 6) of the recombinant M13mp9 clone, and incubate overnight with agitation at 37°.
2. Transfer 1.4 ml of the culture to an Eppendorf tube and centrifuge for 5 min in an Eppendorf centrifuge. (The rest of the culture may be used for preparation of phage stock, see Procedure 6.)
3. Withdraw 1 ml of the supernatant with a Gilson or Eppendorf

[10] J. Messing, this series, Vol. 101, p. 20.
[11] J. E. Walker and N. J. Gay, this series, Vol. 97, p. 195.

pipet. To avoid transfer of cells, slide the pipet tip down along the wall of the tube opposite the cell pellet.

4. Centrifuge again for 5 min and transfer 800 μl of the supernatant to another tube as in step 3.
5. Add 200 μl of solution A to the 800 μl of supernatant, mix, and leave for 15 min at room temperature.
6. Centrifuge for 5 min and remove the supernatant carefully with a pasteur pipet; recentrifuge for 30 sec and remove all remaining traces of the supernatant with a drawn-out capillary pipette.
7. Add 110 μl TE buffer to the phage pellet (which should be visible), then 50 μl of TE-saturated phenol, vortex for 15 sec, shake the tube for 10 min at room temperature on an Eppendorf mixer, and vortex again for 15 sec.
8. Centrifuge for 3 min, transfer the aqueous phase to a fresh Eppendorf tube, and reextract with 50 μl phenol/chloroform then with 100 μl chloroform/isoamyl alcohol.
9. Transfer 90 μl of the aqueous phase after the chloroform extraction to a fresh Eppendorf tube, add 10 μl 3 *M* sodium acetate and 300 μl ethanol (−20°), and keep the mixture at −20° for at least 1 hr.
10. Centrifuge at 4°, 15,000 rpm (rotor type SS34, Sorvall, or equivalent) for 15 min and remove the supernatant with a drawn-out pasteur pipet.
11. Add 1 ml ethanol (−20°) to the tube, vortex for 10 sec, and centrifuge again as in step 10. Remove the supernatant with a drawn-out pasteur pipet and dry the pellet (which is normally not visible) in a Speed Vac concentrator (or vacuum desiccator).
12. Dissolve the pellet in 30 μl TE buffer and store the DNA at −20°.
13. Analyze 3 μl of the virion DNA solution by gel electrophoresis (1% agarose). Use M13 virion DNA of known concentration for comparison of band intensities. The typical yield of this procedure is 2–3 μg of virion DNA.

Procedure 2: Preparation of M13mp9rev RF-DNA[12] *(Adapted from Clewell and Helinski*[13]*)*

1. Inoculate 2 × 75 ml Antibiotic Medium 3 with 1 ml each of fresh overnight culture of strain BMH 71-18 and infect one of the cul-

[12] If RF-DNA of M13mp9rev is required for only very few experiments, a scaled-up procedure of the alkaline lysis method as described by T. Maniatis, E. F. Fritsch, and J. Sambrook ("Molecular Cloning: A Laboratory Manual," p. 368. Cold Spring Harbor Lab., Cold Spring Harbor, New York, 1982) may be used.

[13] D. B. Clewell and D. R. Helinski, *Proc. Natl. Acad. Sci. U.S.A.* **62,** 1159 (1969).

tures immediately with 100 μl of M13mp9 phage stock suspension (see Procedure 6). Shake both culture flasks overnight at 37°.

2. Centrifuge the infected culture at 15,000 rpm (rotor type SS34, Sorvall, or equivalent), 4°, for 10 min. Carefully decant the phage suspension and store at 4° until use.
3. Dispense 3 liters Antibiotic Medium 3 into three sterile 2-liter Erlenmeyer flasks, inoculate each flask with 25 ml of the uninfected overnight culture of BMH 71-18, and shake at 37° until the culture has reached an OD_{546} of 0.5–0.6.
4. Infect each culture with 25 ml of the phage suspension (step 2) and continue shaking at 37° for another 3–4 hr.
5. Collect the cells by centrifugation, decant the supernatant, and resuspend the cells at 0° in 22.5 ml solution B using a 10-ml pipet. (If several centrifuge buckets were used for centrifugation, add the 22.5 ml to the first one, resuspend the cells, transfer the suspension to the next bucket, and so on.)
6. After resuspending, add 6 ml lysozyme solution (5 mg/ml in 50 m*M* Tris–HCl, pH 8.0) and shake gently for 10 min at 0°.
7. Add 7.5 ml 250 m*M* EDTA, pH 8, and continue incubation at 0° for 10 min, again with gentle shaking.
8. Add 27.5 ml DOC solution and again gently shake at 0° for 10 min.
9. Transfer the mixture (which should now be viscous and cloudy) carefully to screw-cap tubes (for the rotor type 30, Beckman, or equivalent) and centrifuge at 30,000 rpm, 4° for 60 min.
10. Pool the supernatants ("cleared lysate") in a 100-ml graduated cylinder (be careful not to transfer any material from the viscous pellets), add 0.97 g CsCl and 20 μl ethidium bromide solution (10 mg/ml) per milliliter of supernatant, and distribute the solution equally into two 35-ml polyallomer Quick Seal centrifuge tubes (Beckman).
11. Fill up the tubes with taring solution [50 m*M* Tris–HCl, pH 8.0, with 0.97 g CsCl and 20 μl ethidium bromide solution (10 mg/ml) added per milliliter starting buffer], seal the tubes, and centrifuge for at least 14 hr at 45,000 rpm, 15°, in a rotor type VTi50 (Beckman) or equivalent.
12. After centrifugation, two (sometimes three) bands should be visible, corresponding to linear plus "open circle" DNA and to supercoiled DNA (the occasional third band most likely is single-stranded DNA). Puncture the top of the tube for aeration, pierce the wall of the tube with a wide-bore canula mounted to a 10-ml hypodermic syringe just below the upper band(s), and draw off these bands. Without removing the first syringe, repeat the procedure for the lowest band containing the supercoiled DNA (which

should be approximately in the middle of the tube). This band should be collected in a total volume of 5 ml per tube.

13. Transfer these two 5-ml samples to two 5-ml Quick Seal tubes (Beckman) and recentrifuge for at least 8 hr at 45,000 rpm, 15°, in a rotor type VTi65 (Beckman) or equivalent.
14. Collect the band containing the supercoiled DNA as in step 12 in a volume of 1–2 ml each. Add 1 volume TE buffer, mix, then add 4 times the starting volume ethanol (room temperature), mix again, and store in the dark for about 1 hr at room temperature.
15. Centrifuge for 30 min at 12,000 rpm, room temperature, in a rotor type HB 4 (Sorvall) or equivalent. Decant the supernatant and wipe the tube with a lint-free paper towel, being careful not to dislodge the pellet. Dry the pellet under reduced pressure and redissolve in approximately the starting volume of TE buffer.
16. Transfer the DNA solution to a 14-ml capped polypropylene tube (Greiner), add an equal volume of TE-saturated phenol, mix by shaking, and centrifuge until clear phase separation is achieved (~10 min). Transfer the aqueous (upper) phase to a new tube and repeat the extraction another 2 times.
17. Extract the aqueous phase 3 times with 2 volumes diethylether, evaporate residual ether by warming to 65° for 15–30 min, and dialyze 2–3 times against 1000 volumes of TES buffer (~6 hr each time).
18. Determine the DNA yield spectrophotometrically ($\varepsilon_{260} = 2 \times 10^{-2}$ $cm^2/\mu g$) using the last dialysis buffer as reference. The ratio of A_{260}/A_{280} should be close to 2.0. The yield is typically 100–400 μg RF-DNA per liter of culture.
19. Store DNA at −20°.

Procedure 3: Construction of Gapped Duplex DNA (gdDNA)

1. Cleave 0.5 μg (0.1 pmol) RF-DNA of M13mp9rev with the restriction enzyme(s) used for construction of the recombinant M13mp9 clone in a 1.5-ml Eppendorf tube. It is important that the cleavage reaction(s) be driven to completion.
2. Extract the DNA solution with phenol/chloroform and precipitate the DNA with ethanol (analogously to Procedure 1, steps 9 and 10). Wash the pellet with 70% ethanol and dry in a Speed Vac concentrator (or vacuum desiccator).
3. Add 0.5 pmol (~1.3 μg) of the recombinant M13mp9 virion DNA (Procedure 1), 5 μl buffer A, and water to a final volume of 40 μl. Dissolve the pellet by flicking the tube several times.

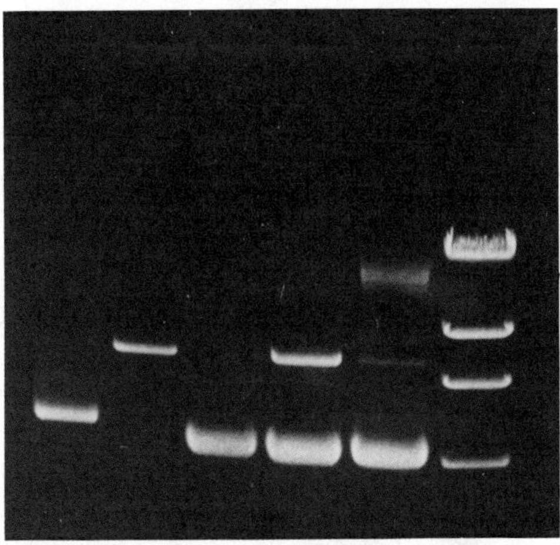

FIG. 2. Gel electrophoretic analysis of different intermediates in the construction of gdDNA (gap size, 120 nucleotides). Lane 1, RF-DNA; lane 2, linear RF-DNA fragment; lane 3, single-stranded virion DNA; lane 4, hybridization mixture (DNA as in lane 2 plus DNA as in lane 3) before denaturation; lane 5, same mixture as in lane 4, but after denaturation/renaturation; lane 6, λ DNA, cleaved with *Hin*dIII. Electrophoresis (field 3.5 V/cm) was carried out on a 1% agarose gel. The DNA was stained with ethidium bromide. The electrophoretic mobility of gdDNA molecules of small gap size as in the example shown here is practically indistinguishable from that of relaxed fully double-stranded DNA (compare uppermost bands of lanes 1 and 5). In lane 5 several minor bands beside remaining single-stranded virion DNA, reformed linear DNA, and gdDNA can be observed. Such bands appear occasionally upon hybridization but do not interfere with the mutation construction experiment.

4. Heat the reaction mixture (in a metal block) to 100° for 3 min, then incubate it at 65° for 5–10 min.
5. Electrophorese an aliquot (about three-fifths) on a 1% agarose gel to check formation of gdDNA. Apply to the same gel as markers 300 ng RF-DNA of M13mp9rev, 100 ng linearized RF-DNA of M13mp9rev, and 300 ng of the recombinant M13mp9 virion DNA (compare Fig. 2).
6. If necessary, the procedure can be interrupted at this point by storing the rest of the hybridization mixture at −20°.

Procedure 4: Hybridization of the Mutagenic Primer to the gdDNA and Optional DNA Polymerase/DNA Ligase Reaction

1. Mix 8 μl of the hybridization mixture and 2 μl of an aqueous solution of the mutagenic primer (2 pmol/μl) in a 1.5-ml Eppendorf

reaction tube and heat the mixture to 65° for 5 min. Keep the reaction tube at ambient temperature for approximately 10 min. (In route A, i.e., for mutation construction experiments without polymerase/ligase reactions, 90 μl of 100 m*M* $CaCl_2$ is added and the resulting mixture is immediately used for transfection; Procedure 5, step 4.)

2. Add 4 μl buffer B, 4 μl 2 m*M* NAD, 4 μl of 4 × 2.5 m*M* dNTPs, and 15 μl water, mix, and add 1 μl *E. coli* DNA ligase (5 U/μl) and 1 μl T4 DNA polymerase (1 U/μl).
3. Incubate for 15 min at 25°, add 2 μl gp 32 (2 mg/ml; if necessary dilute with water immediately prior to use), and incubate again for 90 min at 25°.
4. Stop the reaction by adding 1 μl 500 m*M* EDTA, pH 8.0, and heating to 65° for 10 min.
5. Extract the mixture once with 40 μl phenol/chloroform and 3 times with 200 μl diethyl ether. Evaporate residual ether at 65°, put on ice, add 60 μl 100 m*M* $CaCl_2$, and use immediately for transfection (Procedure 5, step 4).

Procedure 5: Transfection and Segregation (Transfection Adapted from Cohen et al.[14]*)*

1. Inoculate 50 ml of Antibiotic Medium 3 with 0.5 ml fresh overnight culture of BMH 71-18 mutS grown in M9 medium (optionally containing tetracycline at a concentration of 20 μg/ml).
2. Grow the culture with shaking at 37° to an OD_{546} of 0.4, collect the cells by centrifugation (6,000 rpm at 4°, rotor type SS34, Sorvall, or equivalent), and resuspend the cells in 20 ml ice-cold 100 m*M* $CaCl_2$.
3. Centrifuge as before, resuspend the cells in 10 ml ice-cold 100 m*M* $CaCl_2$, centrifuge again, resuspend in 2 ml ice-cold 100 m*M* $CaCl_2$, and keep on ice for at least 30 min.
4. Add 200 μl of this cell suspension to the 100 μl DNA mixture prepared by Procedure 4 and keep the resulting mixture on ice for 90 min.
5. Vortex briefly, heat the suspension for 3 min to 45°, and inoculate 25 ml of Antibiotic Medium 3 with (the bulk of) this suspension (see below). Shake overnight at 37°. (It is advisable to check the transfection efficiency. To this end, remove 20 μl of the 300 μl suspension, dilute 1 : 10 and 1 : 100 in 100 m*M* $CaCl_2$, and plate 100 μl of the dilutions analogously to steps 9–11.)

[14] S. N. Cohen, A. C. Y. Chang, and L. Hsu, *Proc. Natl. Acad. Sci. U.S.A.* **69**, 2110 (1972).

6. Inoculate 5 ml Antibiotic Medium 3 with strain MK 30-3 and incubate overnight with agitation at 37°.
7. Transfer a 1.4-ml aliquot of the transfected BMH 71-18 mutS culture (step 5) to a sterile 1.5-ml Eppendorf tube, centrifuge for 5 min, and transfer the supernatant into another sterile Eppendorf tube. (Store at 4°.)
8. Dilute the supernatant $1:10^6$, $1:10^7$, $1:10^8$, and $1:10^9$ in phage buffer.
9. Place five sterile 12-ml glass tubes in a 45° water bath, dispense 2.5 ml of thoroughly melted top agar into each tube. After the agar is cooled to 45°, add 3 drops of the overnight culture of strain MK 30-3 (step 6) to each tube.
10. Add 100 μl of each of the different phage dilutions (step 8) to four of these tubes, mix by vortexing, and pour the top agar onto EHA plates. The content of the fifth tube without phages is also plated as a control to determine whether the host strain is contaminated by phage.
11. Keep the plates faceup at room temperature for about 20 min until the top agar has solidified and then incubate the plates upside down at 37° for at least 7 hr.

Procedure 6: Preparation of M13 Phage Stocks and Template DNA for Nucleotide Sequence Analysis

1. Dilute a fresh overnight culture of strain BMH 71-18 1 : 100 with Antibiotic Medium 3, transfer 2.5 ml of the diluted cell suspension into a sterile culture tube.
2. Punch out a plaque resulting from a segregation step (Procedure 5) with a sterile glass tube (e.g., a 200-μl capillary pipet) and blow the agar disk from the pipet into the culture tube. Vortex the mixture and incubate with agitation for 4.5–5.5 hr at 37°.
3. The preparation of template DNA for nucleotide sequence analysis can now be carried out as described in Procedure 1, steps 2–13.
4. For the preparation of a phage stock, transfer 1 ml of the infected culture to a sterile Eppendorf tube. Centrifuge the sample for 5 min in an Eppendorf centrifuge and transfer the supernatant into another (sterile) Eppendorf tube. Carry-over from the bacterial pellet should be avoided.
5. Store at 4°. The tube can be sealed with moldable laboratory film (Parafilm M). Phages from this stock should be viable for at least 1 year.
6. The phage suspension resulting from step 4 may be lyophilized in a Speed Vac concentrator for transport. The lyophilized sample can

be redissolved by simply adding the starting volume of water. Prior to lyophilization the phages may be precipitated with polyethylene glycol (see Procedure 1, steps 5 and 6) and redissolved in phage buffer. The phage titer may drop rather drastically upon lyophilization. In case of a low titer (which is sometimes observed after lyophilization), step 1 of Procedure 1 may be repeated.

Procedure 7: Maintenance of Bacterial Strains (Adapted from Miller[15])

For storage of bacterial strains three methods are commonly used: plates, glycerol cultures, and stab cultures.

Agar Plates. Agar plates are good for strains that are used frequently. For strains BMH 71-18, BMH 71-18 mutS, and MK 30-3 minimal plates are used to select for the *pro* marker of the F′ episome.

1. For preparation of minimal plates, mix 7 g $Na_2HPO_4 \cdot 2H_2O$, 3 g KH_2PO_4, 1 g NH_4Cl with 400 ml H_2O and, in a second flask, 15 g Bactoagar (Difco) with 600 ml H_2O. Autoclave both mixtures separately, then combine them. Add the following sterile solutions: 25 ml 20% glucose, 1 ml 100 m*M* $CaCl_2$, 1 ml 1 *M* $MgSO_4$, 5 ml 100 μM $FeCl_3$, and 5 ml thiamin (1 mg/ml). Use about 35–40 ml per petri dish (85 mm diameter).
2. Prepare an overnight culture of the bacterial strain to be stored, plate 0.2 ml overnight culture per plate, and incubate the plates upside down at 37°, until a bacterial lawn is visible (1–2 days).
3. Seal the plates with moldable laboratory film (Parafilm M) and store at 4°. The strain should be viable for about 1 month.
4. To start a liquid culture, scrape bacteria with a sterile glass rod from the plate (one touch or a short streak) and swirl the rod in culture medium for a few seconds.

Glycerol Cultures. Glycerol cultures are convenient for less frequent use and for storage of large numbers of different bacterial clones. For preparation of glycerol cultures, buffered growth media are preferred over standard rich media since the low pH prevalent in an overnight culture leads to reduced shelf life.

1. A fresh overnight culture of the strain to be stored is diluted with sterile glycerol to a final concentration of 50% glycerol.
2. Transfer the mixture to a sterile 5-ml screw-cap tube and store the tube at −20°. Avoid repeated warming of the culture. The strain should be viable for at least 1 year.

[15] J. H. Miller, "Experiments in Molecular Genetics," p. 434. Cold Spring Harbor Lab., Cold Spring Harbor, New York, 1972.

3. To start a liquid culture, transfer a small aliquot of the glycerol culture into culture medium.

Stab Cultures. Stab cultures are chosen for backup storage and transport of strains. They cannot be used very often.

1. Add 6.5 g Bactoagar (Difco) to 1 liter Antibiotic Medium 3 (Difco) and melt the agar.
2. Fill 3-ml aliquots into 4-ml screw-cap glass vials (autoclavable). Autoclave and allow to cool to room temperature.
3. Dip a sterile glass rod, or (preferably) a platinum loop, into a fresh overnight culture of the strain to be stored and stab into the agar in the vial.
4. Close the vial, seal it with moldable laboratory film (Parafilm M) and keep at room temperature. After 2–3 days bacterial growth should be visible in the stab culture along the path of the loop.
5. Store the culture at room temperature in the dark. The strain should be viable for at least 1 year.
6. To start a liquid culture, remove some bacteria from the agar with a sterile glass rod or platinum loop and transfer to culture medium (swirl rod in medium for a few seconds).

Scope, Limitations, and Practical Hints

A large portion of the accumulated experience on oligonucleotide-directed mutation construction did not originate from systematic model investigations but, rather, was collected in a more or less incidental way while applying the method to a great number of biological problems. Thus, there exists a fair amount of "soft" but (from a practical point of view) useful information. The following section on experimental requirements, possible pitfalls, etc., rests in part on that body of experience with no attempt to make reference to the numerous and diverse individual cases.

Structure of the Target Region

Gap Size. Gap size is a significant, yet not overwhelming factor for the efficiency of the method. Thus, somewhat higher marker yields can be expected if the length of the DNA fragment cloned into M13mp9 is kept as small as possible. However, gaps in the size range of a few thousand bases are still good substrates and the difference in efficiency between the "fill-in" (route B) and the "mix–heat–transfect" protocols (route A) is smaller with larger gaps.

Sequence Redundancies. Sequence redundancies are a more serious concern: Direct repeats will result in equivocal choice of the primer hybridization target or bridging of two remote DNA sites and formation of deletions. Inverted repeats can form snapback structures and prevent primer annealing altogether, if the hybridization target of the mutagenic oligonucleotide is part of the double-stranded stem of such a structure. Hairpin/loop structures outside the hybridization target can block the *in vitro* fill-in reaction catalyzed by purified DNA polymerase. In this situation, route A offers a potential advantage.

In the gdDNA method, such sequence redundancies are rendered insignificant if occurring between insert and vector part because the latter is covered by the complementary DNA strand. If possible, however, the gap sequence should be checked for repetitions. In problematic cases possible countermeasures can be (1) cloning a subfragment which does not contain the repeated DNA sequence, (2) use of longer oligonucleotides for mutation construction to give the correct hybridization product an energetic advantage, or (3) covering the nontarget repeat by an additional DNA segment complementary to that part of the gdDNA (restriction fragment or synthetic oligonucleotide). Circumstantial evidence exists that in addition to the above-mentioned constraints there are *other* (poorly understood) *factors* of target DNA structure that can lower the efficiency of the method. Extremely high G + C content of the target DNA may be one such factor.

Structure of the Synthetic Oligonucleotide

Length of Primer. The mutagenic primer consists of a core sequence to be left unpaired after annealing to the M13 genome and two flanking sequences complementary to the target DNA which provide site-specific and stable hybridization. For a sequence of normal G + C content and an unpaired core of just one nucleotide these flanking sequences should be about 7–9 nucleotides long (route B). The flanking sequences should be longer if they have a high A + T content or if extended unpaired regions must be accommodated in the core. With the "mix–heat–transfect" protocol (route A) mutagenic primers about 10 nucleotides longer should be used to ensure sufficient hybrid stability during transfection.

Chemical Nature of the Primer Near its 5′ Terminus. A fraying 5′ end of the primer annealed to the template seems to invite loss of the mutagenic oligonucleotide by strand displacement after enzymatic elongation of the 3′ end of the gdDNA strand has reached this point. Whenever possible, the primer sequence at the 5′ terminus should therefore consist of one or several G or C residues. In route A, the gap filling and sealing reactions are left to the enzymes of the transfected cell. In this situation, the mutagenic oligonucleotide not only can be lost by strand displacement but also by nick translation. We have found it useful to protect the 5′ end of the primer against 5′ to 3′ exonucleolytic attack by one or several phosphorothioate internucleotidic linkages.[6] Such bonds can be prepared conveniently[16,17] by automated DNA synthesis.[6]

Sequence Redundancies within the Mutagenic Primer. Inverted repeats within the primer sequence (contiguous or interrupted) give the synthetic oligonucleotide (partial) self-complementarity. Because of oligonucleotide duplex formation, this can strongly interfere with annealing to the target site.

Direct repeats near one of the primer ends (as small as three nucleotides in length) can lead to primer–template slippage with loop formation on the primer side. This results in an unwanted insertion. When designing the primer sequence to be synthesized, attention should be paid to such local redundancies. Here too, synthesis of a longer primer can be helpful.

Purity of the Primer. Chemical homogeneity of the primer is essential for specific mutation construction. Two alternative purification schemes have proven successful: (1) two consecutive rounds of high-pressure liquid chromatography (HPLC) exploiting different separation principles in each run, and (2) a combination of HPLC and preparative gel electrophoresis. Current synthetic methods (for a comprehensive treatise, see Ref. 18) yield oligonucleotides with a 5′-terminal dimethoxytrityl group attached. This group provides a convenient handle for purification by HPLC on a reverse stationary phase[19] (bonded hydrocarbon phases with chain lengths ranging from C_4 to C_{18}).

After detritylation,[20] the oligonucleotide is enzymatically phosphorylated[20] before either of the two purification procedures outlined below is carried out: (1) rechromatography by HPLC using either an anion-exchange column[21] or a reverse-phase column as above but operated in the

[16] B. A. Connolly, B. V. L. Potter, F. Eckstein, A. Pingoud, and L. Grotjahn, *Biochemistry* **23,** 3443 (1984).

[17] W. J. Stec, G. Zon, W. Egan, and B. Stec, *J. Am. Chem. Soc.* **106,** 6077 (1984).

[18] M. J. Gait (ed.), "Oligonucleotide Synthesis: A Practical Approach." IRL Press, Oxford, England, 1984.

[19] H.-J. Fritz, R. Belagaje, E. L. Brown, R. H. Fritz, R. A. Jones, R. G. Lees, and H. G. Khorana, *Biochemistry* **17,** 1257 (1978).

[20] B. Kramer, W. Kramer, and H.-J. Fritz, *Cell* **38,** 879 (1984).

[21] L. W. McLaughlin and N. Pier, *in* "Oligonucleotide Synthesis: A Practical Approach" (M. J. Gait, ed.), p. 117. IRL Press, Oxford, England, 1984.

paired-ion mode,[22] or (2) preparative polyacrylamide gel electrophoresis as described.[23,24] Both methods separate by chain length *and* state of the 5′ terminus (phosphorylated/unphosphorylated) and yield 5′-phosphorylated oligonucleotides of high purity. If, however, the synthetic material, as isolated from the first round of HPLC, is significantly contaminated by products of lower chain length, method (1) is preferable.

Side Reactions

Errors of DNA Polymerase. The "fill-in" protocol (route B), like most techniques of oligonucleotide-directed mutagenesis used to date, requires a DNA polymerase reaction *in vitro.* It is known that a purified DNA polymerase has a high error frequency.[25] Thus, additional mutations outside the mutagenic primer may occasionally be expected. Such events have indeed been observed and, for this reason, it is important to sequence the entire DNA segment under study after mutagenesis. Conceivably, the "mix–heat–transfect" protocol (route A), which leaves all enzymatic reactions to the transfected cell, may be more accurate in that respect. At present, however, the body of experience with route A is not large enough for a meaningful comparison.

Mutator Phenotype of the Transfection Host. Escherichia coli strains deficient in DNA mismatch repair (such as the mutS strain used in this method) have an increased frequency of spontaneous mutations (mutator phenotype). It is possible, in principle, that additional and unwanted mutations may be introduced into the target DNA during propagation of the recombinant M13 phage. This, however, is of no practical significance, since the rate of spontaneous mutation is still lower by several orders of magnitude than the frequency of the constructed mutation.

Hydrolytic Deamination of Cytosine Residues. The construction of gdDNA involves heating of the DNA solution to 100°. Under these conditions, hydrolytic deamination of 2′-deoxycytidine residues to 2′-deoxyuridine may be a significant reaction. Such reactions would be detrimental because they would provide start points of unwanted repair synthesis when occurring in the double-stranded portion of the gdDNA. More seri-

[22] H.-J. Fritz, D. Eick, and W. Werr, *in* "Chemical and Enzymatic Synthesis of Gene Fragments" (H. G. Gassen and A. Lang, eds.), p. 199. Verlag Chemie, Weinheim, Federal Republic of Germany, 1982.

[23] T. Atkinson and M. Smith, *in* "Oligonucleotide Synthesis: A Practical Approach" (M. J. Gait, ed.), p. 35. IRL Press, Oxford, England, 1984.

[24] R. Wu, N.-H. Wu, Z. Hanna, F. Georges, and S. Narang, *in* "Oligonucleotide Synthesis: A Practical Approach" (M. J. Gait, ed.), p. 135. IRL Press, Oxford, England, 1984.

[25] A. R. Fersht and J. W. Knill-Jones, *J. Mol. Biol.* **165,** 633 (1983).

ously, a C/G to T/A transition would be induced at the site of such a lesion when it was located within the single-stranded gap. In one out of many dozens of cases studied (route B), we have found an additional C/G to T/A transition. It is, of course, not possible to decide whether this mutation was due to the described cytosine deamination or to an error of the DNA polymerase (see above). Again, route A may offer an advantage due to the expected *in vivo* killing of entering gdDNA molecules that carry a 2′-deoxyuridine residue in their single-stranded part.

Flexibility

The applicability of the gdDNA method reaches beyond oligonucleotide-directed mutation construction. Gapped duplex DNA molecules are also ideal substrates for other methods of directed mutagenesis such as forced gap misrepair[26] or attack by single-strand selective chemicals (e.g., bisulfite[27]). Studies in "reversed genetics" can be planned in such a way that a stock of gdDNA is prepared and the last two (or similar) methods are used for saturation of the gap sequence with (single) point mutations in order to identify functionally critical points within the cloned gene or regulatory DNA segment. Once these are known, the same original gdDNA can be used to ask specific questions via oligonucleotide-directed construction of predetermined mutations.

[26] D. Shortle, P. Grisafi, S. J. Benkovic, and D. Botstein, *Proc. Natl. Acad. Sci. U.S.A.* **79**, 1588 (1982).

[27] D. Shortle and D. Botstein, this series, Vol. 100, p. 457.

[19] Rapid and Efficient Site-Specific Mutagenesis without Phenotypic Selection

By THOMAS A. KUNKEL, JOHN D. ROBERTS, and RICHARD A. ZAKOUR

The deliberate alteration of DNA sequences by *in vitro* mutagenesis has become a widely used and invaluable means of probing the structure and function of DNA and the macromolecules for which it codes. In an ideal experiment, alterations are produced at high efficiency and with minimum effort; such features become especially important when sequence changes produce silent, unknown, or nonselectable phenotypes. To overcome some of the limitations that lead to low efficiency, several variations of *in vitro* mutagenesis techniques have been developed.[1–5] Each procedure has its own advantages, but each also requires additional time and/or technical expertise. An alternative method[6] is presented here that is simple, rapid, and efficient. This method takes advantage of a strong biological selection against the original DNA template which is preferentially destroyed on transfection. The use of this special template can be combined with many of the previously described *in vitro* mutagenesis methods. What we describe here is not, therefore, a new procedure for site-directed mutagenesis but is, rather, the use of standard and well-established procedures in conjunction with an unusual template. This combination permits flexibility in the choice of mutagenesis techniques and makes possible the highly efficient recovery of mutants.

Principle

The basis of this method is the performance of site-directed mutagenesis using a DNA template which contains a small number of uracil residues in place of thymine.[6] The uracil-containing DNA is produced within an *Escherichia coli dut$^-$ ung$^-$* strain. *Escherichia coli dut$^-$* mutants lack the enzyme dUTPase[7,8] and therefore contain elevated concentrations of dUTP which effectively competes with TTP for incorporation into DNA. *Escherichia coli ung$^-$* mutants lack the enzyme uracil *N*-glycosylase[9] which normally removes uracil from DNA.[10] In the combined *dut$^-$ ung$^-$* mutant, uracil is incorporated into DNA in place of thymine and is not removed.[11–13] Thus, standard vectors containing the sequence to be

[1] M. Smith, *Annu. Rev. Genet.* **19,** 423 (1985).

[2] M. Zoller, this volume [17].

[3] W. Kramer and H.-J. Fritz, this volume [18].

[4] P. Carter, this volume [20].

[5] G. Cesarini, C. Traboni, G. Ciliberto, L. Dente, and R. Cortese, *in* "DNA Cloning: A Practical Approach" (D. M. Glover, ed.), Vol. 1, pp. 137–149. IRL Press, Oxford, England, 1985.

[6] T. A. Kunkel, *Proc. Natl. Acad. Sci. U.S.A.* **82,** 488 (1985).

[7] E. B. Konrad and I. R. Lehman, *Proc. Natl. Acad. Sci. U.S.A.* **72,** 2150 (1975).

[8] S. J. Hochhauser and B. Weiss, *J. Bacteriol.* **134,** 157 (1978).

[9] B. K. Duncan, P. A. Rockstroh, and H. R. Warner, *J. Bacteriol.* **134,** 1039 (1978).

[10] T. Lindahl, *Proc. Natl. Acad. Sci. U.S.A.* **71,** 3649 (1974).

[11] B.-K. Tye and I. R. Lehman, *J. Mol. Biol.* **117,** 293 (1977).

[12] B.-K. Tye, J. Chien, I. R. Lehman, B. K. Duncan, and H. R. Warner, *Proc. Natl. Acad. Sci. U.S.A.* **75,** 233 (1978).

[13] D. Sagher and B. Strauss, *Biochemistry* **22,** 4518 (1983).

changed can be grown in a *dut⁻ ung⁻* host to prepare uracil-containing DNA templates for site-directed mutagenesis.

For the *in vitro* reactions typical of site-directed mutagenesis protocols, uracil-containing DNA templates are indistinguishable from normal templates. Since dUMP in the template has the same coding potential as TMP,[14] the uracil is not mutagenic, either *in vivo* or *in vitro*. Furthermore, the presence of uracil in the template is not inhibitory to *in vitro* DNA synthesis. Thus, this DNA can be used *in vitro* as a template for the production of a complementary strand that contains the desired DNA sequence alteration but contains only TMP and no dUMP residues.

After completing the *in vitro* reactions, uracil can be removed from the template strand by the action of uracil *N*-glycosylase.[10] Glycosylase treatment can be carried out with purified enzyme[6] but is most easily achieved by transfecting the unfractionated products of the *in vitro* incorporation reaction into competent wild-type (i.e., *ung*$^+$) *E. coli* cells. Treatment with the glycosylase, either *in vitro* or *in vivo,* releases uracil-producing apyrimidinic (AP) sites only in the template strand.[15] These AP sites are lethal lesions, presumably because they block DNA synthesis,[13,16,17] and are sites for incision by AP endonucleases which produce strand breaks.[15] Thus, the template strand is rendered biologically inactive, and the majority of progeny arise from the infective[6,18–20] complementary strand which contains the desired alteration. The resulting high efficiency of mutant production (typically >50%) allows one to screen for mutants by DNA sequence analysis, thus identifying mutants and confirming the desired alteration in a single step. This feature is particularly advantageous when no selection for the desired mutants is available.

Materials and Reagents

Bacterial Strains

The *E. coli* strains and their sources are listed in Table I. Their use and maintenance are described below.

[14] J. Shlomai and A. Kornberg, *J. Biol. Chem.* **253,** 3305 (1978).
[15] T. Lindahl, *Annu. Rev. Biochem.* **51,** 61 (1982).
[16] R. M. Schaaper and L. A. Loeb, *Proc. Natl. Acad. Sci. U.S.A.* **78,** 1773 (1981).
[17] T. A. Kunkel, *Proc. Natl. Acad. Sci. U.S.A.* **81,** 1494 (1984).
[18] P. Rüst and R. L. Sinsheimer, *J. Mol. Biol.* **23,** 545 (1967).
[19] J. E. D. Siegel and M. Hayashi, *J. Mol. Biol.* **27,** 443 (1967).
[20] E. L. Loechler, C. L. Green, and J. M. Essigmann, *Proc. Natl. Acad. Sci. U.S.A.* **81,** 6271 (1984).

TABLE I
Escherichia coli STRAINS

Strain designation	Genotype	Source
BW313	HfrKL16 PO/45 [*lysA*(61-62)], *dut1*, *ung1*, *thi1*, *relA1*	*a*
CJ236[b]	*dut1*, *ung1*, *thi1*, *relA1*/pCJ105 (Cm^r)	*c*
RZ1032	As BW313, but Zbd-279::Tn*10*, *supE44*	*d*
NR8051	[Δ(*pro–lac*)], *thi*, *ara*	*a*
NR8052	[Δ(*pro–lac*)], *thi*, *ara*, *trpE9777*, *ung1*	*a*
KT8051	[Δ(*pro–lac*)], *thi*, *ara*/F′ (*proAB*, $lacI_q^-Z^-$ ΔM15)	*e*
KT8052	[Δ(*pro–lac*)], *thi*, *ara*, *trpE9777*, *ung1*/F′ (*proAB*, $lacI_q^-Z^-$ ΔM15)	*e*
CSH50	[Δ(*pro–lac*)], *thi*, *ara*, *strA*/F′ (*proAB*, $lacI_qZ^-$ ΔM15, *traD36*)	*f*
NR9099	[Δ(*pro–lac*)], *thi*, *ara*, *recA56*/F′ (*proAB*, $lacI_q^-Z^-$ ΔM15)	*g*

[a] As described by T. A. Kunkel, *Proc. Natl. Acad. Sci. U.S.A.* **82,** 488 (1985), but see "Uses, Maintenance, and Characteristics of Bacterial Strains" in this chapter.
[b] The plasmid pCJ105 (Cm^r) was constructed as described by C. M. Joyce and N. D. F. Grindley, *J. Bacteriol.* **158,** 636 (1984).
[c] C. M. Joyce, Yale University, New Haven, Connecticut.
[d] See text.
[e] K. Tindall, National Institute of Environmental Health Sciences, Research Triangle Park, North Carolina.
[f] T. A. Kunkel, *Proc. Natl. Acad. Sci. U.S.A.* **81,** 1494 (1984).
[g] R. M. Schaaper, B. N. Danforth, and B. W. Glickman, *Gene* **39,** 181 (1985).

Growth Media

YT medium: Bactotryptone, 8 g; Bactoyeast extract, 5 g; NaCl, 5 g. Add to 1 liter of H_2O and sterilize in an autoclave.

2× YT medium: Bactotryptone, 16 g; Bactoyeast extract, 10 g; NaCl, 10 g; pH adjusted to 7.4 with HCl. Add H_2O to 1 liter and sterilize in an autoclave.

Soft agar: NaCl, 9 g; Difco agar, 8 g. Add H_2O to 1 liter and sterilize in an autoclave.

VB salts (50×): $MgSO_4 \cdot 7H_2O$, 10 g; citric acid (anhydrate), 100 g; K_2HPO_4, 500 g; $Na_2HPO_4 \cdot 2H_2O$, 75 g. Dissolve the above in 670 ml dH_2O, bring volume to 1 liter, and sterilize in an autoclave. After dilution, pH is 7.0–7.2.

Minimal plates: Add 16 g of Difco agar to 1 liter of dH_2O and sterilize in an autoclave. When the agar has cooled to 50°, add 0.3 ml of 100 m*M* IPTG, 20 ml of 50× VB salts, 20 ml of 20% glucose, and 5 ml of 1 mg/ml thiamine–HCl. [Each of these solutions was sterilized either by filtration (0.2-μm pore) or in an autoclave prior to their addition to the 50° agar.] The mixture is mixed well and dispensed into sterile petri dishes (30 ml/plate).

Enzymes and Reagents

T4 DNA polymerase and T4 polynucleotide kinase were from Pharmacia, Molecular Biology Division. T4 DNA ligase was from New England Biolabs or International Biotechnologies, Inc. Deoxynucleoside triphosphates (HPLC grade, 100 m*M* solutions) were purchased from Pharmacia, Molecular Biology Division, and used without further purification. 5-Bromo-4-chloroindolyl-β-D-galactoside (Xgal) was from Bachem Chemicals. Isopropylthio-β-D-galactoside (IPTG) was from Bethesda Research Laboratories. Phenol (ultrapure grade) was obtained from Bethesda Research Laboratories or International Biotechnologies, Inc., and used without further purification. All other chemicals were obtained from standard suppliers of molecular biological reagents.

Stock Solutions

Xgal: 50 mg/ml in *N,N*-dimethylformamide (DMF), stored at −20°. Avoid exposure to light.

IPTG: 24 mg/ml in dH_2O, stored at −20°.

PEG/NaCl (5×): Polyethylene glycol 8000, 150 g; NaCl, 146 g. Dissolve in dH_2O, adjust volume to 1 liter, and filter sterilize using a 0.2-μm filter.

Phenol extraction buffer (PEB): 100 m*M* Tris–HCl (pH 8.0); 300 m*M* NaCl; 1 m*M* EDTA.

Phenol: equilibrated versus multiple volumes of PEB until the pH of the aqueous phase is ~8.0, stored in a brown bottle at 4°.

TE buffer: 10 m*M* Tris–HCl (pH 8.0); 0.1 m*M* EDTA.

Kinase buffer (10×): 500 m*M* Tris–HCl (pH 7.5); 100 m*M* $MgCl_2$; 50 m*M* dithiothreitol.

SSC (20×): 3 *M* NaCl; 300 m*M* sodium citrate.

SDS dye mix (10×): 10% sodium dodecyl sulfate; 1% bromophenol blue; 50% glycerol.

TAE buffer (50×): Tris base, 242 g; glacial acetic acid, 57.1 ml; EDTA, 100 ml of a 500 m*M* solution (pH 8.0). Dissolve in dH_2O, adjust volume to 1 liter.

Methods

Uses, Maintenance, and Characteristics of Bacterial Strains

Uracil-containing DNA was first prepared as a template for *in vitro* mutagenesis as described by Sagher and Strauss[13] using *E. coli* strain BW313. This strain was chosen on the basis of three criteria which are crucial for the successful production of uracil-containing viral DNA templates: (1) susceptibility to infection by small filamentous bacteriophages (e.g., M13), which requires the F (sex factor) pilus; (2) the presence of the Dut$^-$ and Ung$^-$ phenotypes, which are required for the stable incorporation of uracil into phage DNA; and (3) a low rate of spontaneous mutation in the progeny phage, so that unwanted mutations are not introduced into the DNA target.

In the original publication of this method,[6] BW313 was incorrectly described as F′*lysA*. BW313 is actually an Hfr strain with the integrated F factor providing the pilus needed for phage attachment. Since there is no selective pressure that can be used to maintain the Hfr phenotype, we store BW313 frozen (−70°) in multiple aliquots containing 3 ml of a mid-log culture (2×10^8 to 2×10^9 cells/ml) mixed with 0.3 ml of DMSO. When needed, a vial is thawed and cells are streaked on a YT plate (or any rich medium plate) to obtain single colonies. This plate may be used as a source of colonies for over 2 months when stored at 4°. We have not encountered problems in achieving M13 infections using this procedure. However, since others have observed a loss of infectability with this strain, a second strain was produced by introducing a selectable F′ that confers resistance to chloramphenicol into a BW313 strain that had lost its competence for M13 infection. This strain (CJ236, produced by Catherine Joyce at Yale) stably retains susceptibility to M13 infection when selective pressure is maintained, and it has been used successfully by us to prepare uracil-containing DNA for several site-directed mutagenesis experiments.

Since BW313 does not contain an amber suppressor, a third *E. coli dut$^-$ ung$^-$* strain was constructed for use with cloning vectors which contain amber mutations in essential genes (e.g., M13 mp8). Strain JM101 (*supE44*) was transduced to tetracycline resistance using P1 grown on strain SK2255 which carries the *tet*^{r} marker on a transposable element, Zbd-279: : Tn*10*. A tetracycline-resistant derivative of JM101 was used to grow P1 which were then used to transduce strain BW313 to tetracycline resistance. The BW313 Tet^r transductants were tested for their ability to support growth of phage M13mp8 (which contains two amber mutations) and for the ability to produce M13mp2 phage containing uracil in their DNA (due to the host *dut1* and *ung1* mutations). The resulting Hfr strain (RZ1032) fulfills these criteria and templates prepared from this strain perform well in subsequent *in vitro* mutagenesis experiments. The *supE44* is maintained with tetracycline selection. This strain grows somewhat more slowly than BW313, and characteristically produces smaller M13 plaques and lower phage yields in liquid cultures.

As with BW313, RZ1032 loses its Hfr phenotype at a low but bothersome frequency and a single-colony isolate may not support M13 infection. (At present, a selectable F′ has not been placed into RZ1032.) To overcome this problem with either BW313 or RZ1032, several individual colonies should be picked from a plate and liquid cultures (YT medium), each from a single colony, should be screened by plaque assays on plates to test for M13 infectability. Once a competent culture is identified, aliquots can be stored indefinitely at −70° in 10% DMSO. We have not encountered instability of the *dut*⁻ and *ung*⁻ markers in BW313, CJ236, or RZ1032 when they are grown in rich medium. Likewise, beyond the usual slight increase in mutation frequency associated with *dut*⁻ *ung*⁻ strains (see below), which is negligible in site-directed mutagenesis protocols, we have observed neither high mutation frequencies nor spurious mutations on growing vectors in these strains.

The other strains listed in Table I have been useful in establishing the utility of uracil-containing DNA, but are not required for most mutagenesis protocols. NR8052 (*ung*⁻) and its wild-type (*ung*⁺) parent, NR8051, can be used to measure the relative survival of uracil-containing DNA upon transfection. Similarly, the newly constructed derivatives of these strains, prepared by Kenneth Tindall of the National Institute of Environmental Health Sciences and designated KT8052 and KT8051, respectively, can be used for a similar analysis of intact phage since they contain an F′. Survival data can be obtained by comparing phage titers on BW313 (or CJ236 or RZ1032, all *ung*⁻ strains) with titers on any wild-type (*ung*⁺) strain. However, NR8052 and NR8051 have another advantage in that they can also be used to determine spontaneous mutation rates in the phage-borne *lacZα* gene by following the loss of α-complementation as previously described.[6,17] Such experiments were carried out to establish that growth in the *dut*⁻ *ung*⁻ host is not mutagenic.[6]

Escherichia coli CSH50 and NR9099 are *ung*⁺ and are routinely used for α-complementation experiments in our laboratory. These, as well as other *ung*⁺ *E. coli* strains (for example, the JM series of α-complementation strains), are acceptable hosts for transfection of the products of the *in vitro* mutagenesis reactions with uracil-containing templates.

Growth of Phage

Uracil-containing viral DNA template is isolated from intact M13 phage grown on an *E. coli dut*⁻ *ung*⁻ strain. Phage can be produced as previously described[6] or by a simpler method which we present here. Using a sterile pipet tip, remove one plaque [usually 10^9–10^{10} plaque-forming units (pfu)/plaque for M13] from a plate and place it in 1 ml of sterile YT medium in a 1.5-ml Eppendorf tube. Incubate the tube for 5 min at 60° to kill cells, vortex vigorously to release the phage from the agar, then pellet cells and agar with a 2-min spin in a microcentrifuge. Place 100 μl of the resulting supernatant (containing 10^8–10^9 pfu) into a 1-liter flask containing 100 ml of YT medium supplemented with 0.25 μg/ml uridine; we have found that neither the thymidine nor the adenosine supplementation originally described[6] is necessary, since omitting these effected neither phage yield nor mutation frequency. Add 5 ml of a mid-log culture of the appropriate *E. coli dut*$^-$ *ung*$^-$ strain. These proportions result in a multiplicity of infection of ≤1. Most or all phage infect cells and are thus "passaged" through the *dut*$^-$ *ung*$^-$ strain. Since few uracil-lacking phage remain, a single cycle of growth results in a sufficient survival difference (as measured by titers on *ung*$^+$ and *ung*$^-$ hosts) to make the DNA suitable for the *in vitro* mutagenesis protocol.

The flask is incubated with vigorous shaking at 37°. We have prepared phage from cultures incubated for as short as 6 hr or as long as 24 hr. Shorter times are recommended for vectors that contain unstable inserts, since this will help to avoid the growth advantage of phages which have deleted the insert. (A *recA*$^-$ derivative of a *dut*$^-$ *ung*$^-$ strain might be useful to stabilize otherwise unstable DNA sequences, but at present we do not have such a strain.)

After incubation at 37°, the culture is centrifuged at 5000 *g* for 30 min. The clear supernatant contains the phage at about 10^{11} pfu/ml. (This yield may vary depending on the vector and strain used; our experience with RZ1032 suggests that phage titers of 2–5 × 10^{10} pfu/ml are not unusual.) Before preparing viral template DNA, the phage titers should be compared on *ung*$^-$ and *ung*$^+$ hosts. Phage which contain uracil in the DNA have normal biological activity in the *ung*$^-$ host but greater than 100,000-fold lower survival in the *ung*$^+$ host. Phage produced in the *dut*$^-$ *ung*$^-$ host show only a slight (~2-fold) increase in mutation frequency when compared to phage produced in wild-type *E. coli*. Loss of α-complementation occurs in about 0.1% of the uracil-containing phage,[6] a negligible background compared to frequencies of 50–90% in site-directed mutagenesis experiments.

Preparation of Template DNA

Phage are precipitated from the clear supernatant by adding 1 volume of 5× PEG/NaCl to 4 volumes of supernatant, mixing, and incubating the phage at 0° for 1 hr. The precipitate is collected by centrifugation at 5000 *g* for 15 min, and the well-drained pellet is resuspended in 5 ml of PEB in a 15-ml Corex tube. After vigorous vortexing, the resuspended phage solution is placed on ice for 60 min and then centrifuged as above to remove residual debris. (This step has proved useful in reducing the level of endogenous priming in subsequent *in vitro* DNA polymerase reactions.) The supernatant containing the intact phage is extracted twice with phenol (previously equilibrated with PEB) and twice with chloroform : isoamyl alcohol (24 : 1). The DNA is precipitated [by adding 0.1 volume of 3 *M* sodium acetate (pH 5.0) and 2 volumes of ethanol and chilling the mixture to −20°], collected by centrifugation and resuspended in TE buffer. The DNA concentration is determined spectrophotometrically at 260 nm using an extinction coefficient of 27.8 ml/mg cm (i.e., 1 OD_{260} = 36 μg/ml) for single-stranded DNA. The purity of the DNA is examined by agarose gel electrophoresis as described below, overloading at least one lane to visualize trace contaminants.

We have found no need to further purify the DNA in order to achieve high efficiencies of *in vitro* mutagenesis. If problems related to template purity are encountered, or if mutant production approaching 100% is needed, the DNA can be subjected to any standard purification procedure, since the substitution of a small percentage of thymine residues by uracil should not affect the physical properties of the DNA.

In principle, any cloning vector that can be passaged through an *E. coli dut*$^-$ *ung*$^-$ strain can be used with the uracil selection technique. Once the uracil-containing DNA is prepared, it can be used as would be any standard template, in a variety of *in vitro* methodologies for altering DNA sequences.[1] We present below a typical oligonucleotide-directed mutagenesis experiment to make several points of interest. Some of these have previously appeared in the literature (see Smith[1] and references therein, as well as several other chapters in this volume[2–4]) and are well known to investigators acquainted with this field. However, these notes may be useful to those less familiar with *in vitro* mutagenesis.

Example of the Method

For reasons to be published elsewhere, we required a mutant of M13mp2 containing an extra T residue in the viral-strand run of four consecutive Ts at positions 70 through 73 (where position 1 represents the first transcribed base) in the coding sequence for the α peptide of β-galactosidase.[21]

Description and Phosphorylation of the Oligonucleotide

A 22-base oligonucleotide, complementary to positions 62–82 and containing an extra (i.e., fifth) A residue, was purchased from the DNA

[21] T. A. Kunkel, *J. Biol. Chem.* **260,** 5787 (1985).

Synthesis Service, Dept. of Chemistry, Univ. of Pennsylvania. (The ability of this or any oligonucleotide to prime *in vitro* DNA synthesis at the appropriate position should be examined.[1]) The 5′ OH of the 22-mer was phosphorylated (for subsequent ligation) in a 20-μl reaction containing 50 mM Tris–HCl (pH 7.5), 10 mM $MgCl_2$, 5 mM dithiothreitol, 1 mM ATP, 2 units of T4 polynucleotide kinase, and 9.0 ng of the oligonucleotide. The reaction was incubated at 37° for 1 hr and terminated by adding 3 μl of 100 mM EDTA and heating at 65° for 10 min.

Hybridization of the Oligonucleotide to the Uracil-Containing Template

To the phosphorylated oligonucleotide was added 1 μg (in 0.6 μl) of single-stranded, uracil-containing, circular wild-type M13mp2 DNA and 1.2 μl of 20× SSC. After mixing and spinning the sample briefly (5 sec) in a microcentrifuge, the tube was placed in a 500-ml beaker of water at 70° and allowed to cool to room temperature. After another 5-sec centrifugation to spin down condensation, the tube was placed on ice.

We typically perform hybridization at a primer : template ratio between 2 : 1 and 10 : 1 since higher ratios do not yield more of the desired product and in some cases inhibit ligation. Hybridization conditions should be chosen to optimize heteroduplex formation with the particular oligonucleotide and template being used, and these conditions are expected to vary widely depending on the resulting heteroduplex (see Smith[1] and references therein for more details).

In Vitro DNA Synthesis and Product Analysis

The sequence contained within the oligonucleotide is converted to a biologically active, covalently closed circular (CCC) DNA molecule by DNA synthesis and ligation. These reactions are performed in a volume of 100 μl containing 20 μl of the above hybridization mixture; 20 mM HEPES (pH 7.8); 2 mM dithiothreitol; 10 mM $MgCl_2$; 500 μM each of dATP, TTP, dGTP, and dCTP; 1 mM ATP; 2.5 units (as defined by the supplier) of T4 DNA polymerase; and 2 units of T4 DNA ligase. All components are mixed at 0° (enzymes being added last), and the reaction is incubated at 0° for 5 min. The tube is placed at room temperature for 5 min, then at 37° for 2 hr. The rationale for this pattern of incubation is as follows.

The reaction is begun at lower temperatures (0°, then room temperature) to polymerize a small number of bases onto the 3′ end of the oligonucleotide, thus stabilizing the initial duplex between the template phage DNA and the mutagenic oligonucleotide primer. However, since T4 DNA polymerase does not utilize long stretches of single-stranded DNA template well at low temperature, synthesis is then completed at 37°. We have not encountered significant pausing by T4 DNA polymerase under these conditions. The high concentration of dNTPs (500 *mM*) serves to optimize DNA synthesis and to reduce the 3′-exonuclease activity of the T4 DNA polymerase.

The reaction is terminated by addition of EDTA to 15 m*M* (3 μl of a 500 m*M* stock). The products of the reaction are then examined by subjecting 20 μl (to which is added 2.5 μl of SDS dye mix) to electrophoresis in a 0.8% agarose gel (in 1× TAE buffer containing 0.5 μg/ml ethidium bromide). For comparison, an adjacent lane should contain the appropriate standards: single-stranded, circular viral DNA, and double-stranded replicative form I (supercoiled CCC) and form II (nicked circular) DNAs.

The product of the *in vitro* DNA synthesis reaction should migrate at the same rate as the RF I standard, indicating that the DNA has been converted from primed circles to RF IV (duplex, CCC relaxed DNA) by the combined action of DNA polymerase and ligase. (Note that, in gels containing ethidium bromide, the dye will bind to the CCC relaxed DNA, generating positive supercoils and causing the RF IV to migrate like RF I.) Our experiments typically yield 80% conversion to primarily double-stranded DNA, but only 10–50% of this material is ligated to form covalently closed circles. Lower quality enzymes, unusual inserts, contaminants, or less than optimum reaction conditions may reduce the yield. (For more comments on the DNA synthesis step, see Troubleshooting below.)

Transfection and Plating

After incorporation of the oligonucleotide into double-stranded DNA *in vitro,* the DNA can be used to transfect (or transform) any competent *E. coli* strain (prepared with either $CaCl_2$[22] or by the method of Hanahan[23]). Provided the strain is *ung*$^+$, one can take advantage of the selection against the uracil-containing template strand. Unless high biological activity is required, the products of the *in vitro* DNA synthesis reaction can be used directly without further manipulation to remove reaction components.

For the reaction described above, 10 μl of the DNA synthesis mixture (containing approximately 80 ng of the double-stranded DNA) was added to 1 ml of competent CSH50 cells (prepared according to Hanahan[23]) in a sterile glass tube. A second transfection with a known amount of RF

[22] A. Taketo, *J. Biochem.* **72,** 973 (1972).

[23] D. Hanahan, *J. Mol. Biol.* **166,** 557 (1983).

DNA was performed to determine the transfection efficiency. The cells were gently mixed and incubated on ice for 30 min, heat shocked at 42° for 2 min, then returned to ice. Small volumes of this mixture (1, 5, 10, 50, or 100 μl) were added to tubes containing 2.5 ml of soft agar (at 50°), 2.5 mg Xgal (previously mixed well with the media to disperse the DMF), 0.24 mg IPTG, and 0.2 ml of mid-log culture of CSH50 cells. The mixtures were poured onto minimal plates and allowed to solidify. Plates were incubated at 37° overnight to allow blue color to develop as a measure of α-complementation. Nonconfluent plates (in this case the 1- and 5-μl plates) were scored for total plaques. Transfection using 10% (~80 ng in 10 μl) of the *in vitro* DNA synthesis reaction produced 193,000 plaques, an efficiency of about 2400 pfu/ng. In this particular instance, the desired mutants are expected to be colorless due to the addition of an extra base in the *lacZα* coding region. Colorless plaques comprised 70% of the total.

To confirm that the colorless mutants contained the expected sequence alteration, DNA from several phage was prepared for sequencing. First, 10 colorless plaques were harvested into 0.9% NaCl, serially diluted, and replated (as above) to obtain plaques derived from single phage particles. This genetic purification eliminates the possibility that the DNA to be sequenced comes from a plaque which contained two genotypes (as is possible in the original plaques) due either to transfection by a heteroduplex DNA molecule or to plaque overlap. This step becomes unnecessary if, following heat shock of the transfection mixture, one adds rich medium and continues incubation at 37° to allow production of progeny phage, which can then be diluted and plated. We have not used this extra incubation in experiments described here; thus, quantitation of the efficiency of mutant production does not require a correction for differences in growth rates of mutant versus wild-type phages. In either instance, once a purely mutant shock is obtained, the DNA can be sequenced by standard techniques. In our example, all 10 colorless plaques contained the expected change, an extra T residue in the viral strand.

Troubleshooting

Unsuccessful experiments are usually characterized by one of two outcomes. The first is low biological activity on transfection. To eliminate the obvious possibility that the *E. coli* cells were not competent, we always perform a parallel transfection with a known quantity of normal DNA to determine the transfection efficiency. Since biological activity depends on a complete complementary strand, inefficient DNA synthesis may also lead to low numbers of plaques. When this problem appears, a careful product analysis is warranted. In our experience, the presence of completely double-stranded DNA in the transfection mix has always been a harbinger of good biological activity. However, even when the yield of fully double-stranded DNA is poor, the strong selection against incompletely copied or uncopied uracil-containing templates which occurs on transfection permits the desired mutant to be recovered. In fact, we have obtained the desired sequence alterations from transfection of reactions that contained no double-stranded DNA as judged by agarose gel electrophoresis. These reactions yielded very few plaques, but 40% of the surviving DNA carried the mutant sequence.

In two instances (out of ~50 separate experiments), however, we observed no biological activity from mixtures which exhibited a band of DNA at the position of RF I on agarose gel electrophoresis in the presence of ethidium bromide. In these two cases, a further analysis was performed to determine the nature of the reaction products. If the product band were indeed the desired species, that is covalently closed circles of RF IV, this DNA would migrate like RF II DNA in the absence of the intercalating dye. Another possibility is that the product was the result of incomplete synthesis of the complementary strand and that its migration at the position of RF I in the initial agarose gel was fortuitous. If this were true, the migration of this DNA would not be affected by ethidium bromide. In both cases of low biological activity, the synthesis products migrated at the position of RF I in both the presence and absence of ethidium bromide, indicating that DNA synthesis was incomplete and explaining the lack of biological activity. (In both cases, the incomplete synthesis was due to the use of an old T4 DNA polymerase preparation that was no longer fully active.)

Incomplete synthesis can result from several factors, including inefficient hybridization of the oligonucleotide primer, inactive (or excess) DNA polymerase, contaminants in the DNA, the polymerase, or the reagents, or a DNA template which contains structures (e.g., hairpin loops) that block polymerization. (For example, with several T4 DNA polymerase preparations we have found as much as 10 units of enzyme may be required to achieve complete synthesis.) Such problems must be dealt with on an individual basis, e.g., by varying hybridization conditions, by ensuring the activity of the DNA polymerase, by repurifying the DNA, or by using alternative incubation temperatures or single-stranded DNA binding protein to assist the polymerase in synthesis on unusually difficult templates.

Low biological activity could also result from dNTP contamination by dUTP (e.g., by deamination of dCTP), which, when incorporated *in vitro*, provides targets for the production of lethal AP sites.[24] For this reason,

[24] P. D. Baas, H. A. A. M. van Teeffelen, W. R. Teerstra, H. S. Jansz, G. H. Veeneman, G. A. van der Marel, and J. H. van Boom, *FEBS Lett.* **110,** 15 (1980).

high quality dNTP substrates should be used to eliminate the need for dUTPase treatment of the deoxynucleoside triphosphates.[6,14] (dUTPase is not commercially available.)

The second undesirable outcome is a low percentage of mutants among the progeny. To confirm that there is indeed strong selection for the mutant strand over the template strand due to uracil residues in the latter, a titer of the phage from which the template is purified should be performed on *ung*$^-$ and *ung*$^+$ hosts. The difference in titers should be greater than 100,000-fold, although phage that show a smaller difference will still yield mutants at high efficiencies.[6] If a large difference in titer is not obtained, there is insufficient uracil in the viral DNA and new templates should be prepared. (Another but less likely possibility is that the putative *ung*$^+$ host is genetically and/or phenotypically *ung*$^-$.)

A low percentage of mutants can also result from impurities in the template DNA which provide endogenous primers for complementary strand synthesis. The amount of endogenous priming can be determined by performing a DNA synthesis reaction without added oligonucleotide and examining the products by gel electrophoresis. When template is prepared as described above, the amount of RF IV DNA produced *in vitro* should be negligible in the absence of oligonucleotide. With impure template preparations, essentially all the single-stranded circular DNA can be converted to double-stranded product without added primer, in which case biological activity but no mutants will be obtained. The impurities can be removed by standard techniques (e.g., alkaline gradients) or the template can be prepared anew.

Another source of low mutant yield is displacement of the oligonucleotide during *in vitro* DNA synthesis of the strand which carries the mutation (see line 4, Table 2, in Ref. 6). This possibility is not a concern when one uses T4 DNA polymerase, since this enzyme does not perform strand displacement synthesis under the conditions given above. However, such problems may arise with some primers and/or templates that have more, or stronger, polymerase pause sites than do our templates. In these situations, the polymerase can be assisted by its homologous single-stranded DNA binding protein (the T4 gene *32* protein). Alternatively, the Klenow fragment of *E. coli* DNA polymerase I, which efficiently synthesizes complete complementary strands without significant pausing, can be used at low temperature (to prevent displacement of the oligonucleotide). Even this enzyme has difficulty with some templates at reduced temperatures; in these instances *E. coli* single-stranded DNA binding protein may be helpful.

One pathway by which the mutant, complementary strand of DNA may be selected against is the methyl-directed mismatch correction system present in most *E. coli* strains.[25] Repair synthesis which uses the uracil-containing strand as a template before it is destroyed will eliminate the artificially produced mutation from the complementary strand. Although this process is not a major concern (since we routinely achieve high efficiency, 50–70%), two means of reducing mismatch repair and improving mutant production can be employed in those situations where one desires the highest efficiency possible. The first is to transfect the DNA into competent cells made from *E. coli* mutator strains (*mutH*, *mutL*, or *mutS*) which are deficient in mismatch correction.[25] [In a typical experiment, the efficiency of mutant production was improved from 51% (wt) to 57% (*mutH*), 67% (*mutL*), and 59% (*mutS*).] A second strategy is to treat the product DNA with uracil *N*-glycosylase as described[6] to form AP sites in the template strand, followed by alkali treatment to hydrolyze the phosphodiester bonds at AP sites and to disrupt hydrogen bonding. This protocol eliminates the transforming activity of the parental strand and leaves only the covalently closed, single-stranded DNA which contains the desired sequence alteration as a source of transforming DNA. In practice, this treatment improved efficiency from 51 to 89%.[6] We do not routinely utilize either mutator strains or glycosylase, since, for our purposes, 50–70% efficiency is more than sufficient and since uracil *N*-glycosylase is not commercially available.

We originally observed good biological activity but low efficiency of mutagenesis when ligase was omitted from the *in vitro* reaction (see line 4, Table 2, in Ref. 6). This observation was made with DNA products synthesized by the DNA pol I Klenow fragment under conditions that allow strand displacement (37°). More recently, using T4 DNA polymerase as described above, we attempted oligonucleotide-directed mutagenesis of the gene for the *E. coli* cyclic AMP receptor protein. We used a 36-mer which primed DNA synthesis efficiently but could not be ligated to form CCC DNA (i.e., only RF II but no RF IV DNA was produced), perhaps because of abberant chemistry during oligonucleotide synthesis or impurities in the oligonucleotide preparation. Despite the fact that the product DNA was not covalently closed, the mutant yield, with no selection, was 67% (six of the nine clones sequenced having the appropriate mutation). We conclude that, with T4 DNA polymerase (i.e., in the absence of strand displacement), ligation *in vitro* is not required for highly efficient mutant production. Presumably, on transfection, the cell performs the necessary processing at the termini before the template DNA is destroyed.

Lest these comments on troubleshooting discourage use of the technique, we reiterate that, after the preparation of the uracil-containing

[25] B. W. Glickman and M. Radman, *Proc. Natl. Acad. Sci. U.S.A.* **77**, 1063 (1980).

template, oligonucleotide-directed mutagenesis at or above 50% efficiency is a simple procedure, consisting of a polymerization reaction and a transfection.

Variations

We have presented here a simple oligonucleotide-directed mutagenesis protocol to demonstrate the utility of the uracil selection technique for generating mutants with high efficiency (≥50%). Uracil-containing DNA can be prepared for any vector that can be passaged through an *E. coli* *dut*$^-$ *ung*$^-$ strain. Such templates can be used in conjunction with the wide variety of established procedures (gapped duplexes, double priming, etc.) and vectors (single-stranded phage, pBR derivatives, shuttle vectors, etc.). This procedure may also prove to be useful for investigating the mutagenic potential of specific DNA adducts located at defined positions in genes and for studies of mutational specificity of *in vitro* DNA synthesis.[21] The applications of these techniques for engineering DNA sequences and the proteins for which they code are limited only by the need and the imagination of the investigator.

[20] Improved Oligonucleotide-Directed Mutagenesis Using M13 Vectors

By PAUL CARTER

Introduction

Precisely defined mutations may be constructed in DNA fragments cloned into M13 using synthetic oligonucleotides.[1] An oligonucleotide is synthesized which is complementary to part of the DNA template but contains an internal mismatch to direct the required change (point mutation, multiple mutation, insertion, or deletion).[2] The simplest approach to mutagenesis in M13 is "single priming" (Fig. 1), where the mutagenic primer is annealed with the single-stranded M13 template and extended

[1] C. A. Hutchison III, S. Phillips, M. H. Edgell, S. Gillam, P. Jahnke, and M. Smith, *J. Biol. Chem.* **253,** 6551 (1978).

[2] M. Smith, *Annu. Rev. Genet.* **19,** 423 (1985).

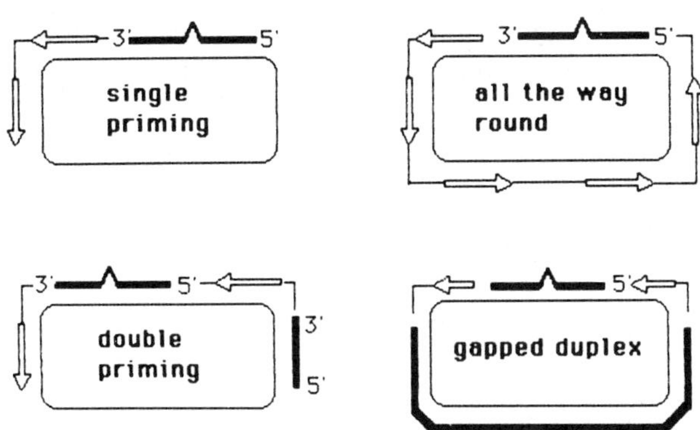

FIG. 1. Different strategies for site-directed mutagenesis in M13 (Carter *et al.*[27]).

with DNA polymerase I (Klenow fragment) using deoxynucleoside triphosphates (dNTPs). The resulting partial heteroduplex is used to transfect an *Escherichia coli* host, and, after DNA replication, mutant and wild-type daughter molecules result.

The most general method for screening for mutants is by hybridization to the 5′-^{32}P-labeled mutagenic oligonucleotide.[3] Under nonstringent conditions (room temperature wash) the probe hybridizes both to the mutant DNA to which it is perfectly matched and also to wild-type DNA to which it is mismatched. By increasing the stringency of washing (by elevating the temperature) the mutagenic oligonucleotide can be selectively dissociated from wild-type DNA, leaving it bound to mutant DNA.[3]

Having identified a good hybridization-positive clone, it is necessary to plaque purify the mutant since plaques may contain mixed populations of phage; in fact most plaques (>80%) contain predominantly (>95%) one kind of phage.[4] Single-stranded DNA is then prepared from individual plaques and sequenced by the dideoxy method[5,6] to verify the mutation. In constructing point mutations, often only the required mutation is obtained. However in constructing deletions, usually only a fraction of the hybridization positive clones carry the required mutation.[7,8] The whole of

[3] R. B. Wallace, P. F. Johnson, S. Tanaka, M. Schold, K. Itakura, and J. Abelson, *Science* **209,** 1396 (1980).

[4] V. Enea, G. F. Vovis, and N. D. Zinder, *J. Mol. Biol.* **96,** 495 (1975).

[5] F. Sanger, S. Nicklen, and A. R. Coulson, *Proc. Natl. Acad. Sci. U.S.A.* **74,** 5463 (1977).

[6] M. D. Biggin, T. J. Gibson, and G. F. Hong, *Proc. Natl. Acad. Sci. U.S.A.* **80,** 3963 (1983).

[7] K. A. Osinga, A. M. Van der Bliek, G. Van der Horst, M. J. A. Groot Koerkamp, H. F. Tabak, G. H. Veeneman, and J. H. Van Boom, *Nucleic Acids Res.* **11,** 8595 (1983).

[8] V.-L. Chan and M. Smith, *Nucleic Acids Res.* **12,** 2407 (1984).

the recombinant fragment should ideally be sequenced to check that no other mutations have inadvertently been introduced into the clone.[9,10]

Design, Synthesis, and Purification of Mutagenic Oligonucleotides

Several factors should be considered in designing mutagenic primers[11]:

1. The mismatch(es) should be located at least four nucleotides away from the 3′ end of the primer to avoid being removed by the 3′ → 5′ exonuclease activity of Klenow.[12] By positioning the mismatch in the middle of the oligonucleotide maximum destabilization of the heteroduplex with the template is obtained, allowing good discrimination in hybridization screening.[13]

2. The choice between a single or double mismatch between template and mutagenic primer is unimportant since good discrimination in oligonucleotide hybridization screening may be obtained in either case (with the possible exception of a single G/T mismatch which is relatively stable). It may be convenient to remove or generate a unique restriction site along with the required mutation to allow enrichment for mutants[14] or screening for mutants.[14–16]

3. Likely competing priming sites may be avoided by comparing the proposed sequence of the oligonucleotide with that of the vector and cloned insert using a computer program such as ANALYSEQ.[17]

4. The length of a mutagenic oligonucleotide synthesized is a compromise between the specificity of priming and ease of hybridization screening. For single or double point mutations 16 to 20-mer oligonucleotides are frequently used with the mismatch(es) located at the middle. Many simultaneous nucleotides changes may be constructed using long oligonu-

[9] J. E. Villafranca, E. E. Howell, D. H. Voet, M. S. Strobel, R. C. Ogden, J. N. Abelson, and J. Kraut, *Science* **222,** 782 (1983).

[10] A. J. Wilkinson, A. R. Fersht, D. M. Blow, P. Carter, and G. Winter, *Nature (London)* **307,** 187 (1984).

[11] M. J. Zoller and M. Smith, this series, Vol. 100, p. 468.

[12] S. Gillam and M. Smith, *Gene* **8,** 81 (1979).

[13] J. W. Szostak, J. I. Stiles, B.-K. Tye, P. Chiu, F. Sherman, and R. Wu, this series, Vol. 68, p. 419.

[14] J. Corden, B. Wasylyk, A. Buchwalder, P. Sassone-Corsi, C. Kedinger, and P. Chambon, *Science* **209,** 1406 (1980).

[15] S. Gillam, C. R. Astell, P. Jahnke, C. A. Hutchison III, and M. Smith, *J. Virol.* **52,** 892 (1984).

[16] S. Gillam, T. Atkinson, A. Markham, and M. Smith, *J. Virol.* **53,** 708 (1985).

[17] R. Staden, *Nucleic Acids Res.* **12,** 521 (1984).

cleotides with multiple mismatches to the template with several (~6–10) nucleotides at each end which are perfectly matched to the template.[18] For deletions and insertions an overlap of at least 10–12 perfectly matched nucleotides at each end should be used.

The last few years have seen rapid advances in the synthesis of oligonucleotides,[19,20] which has led to their routine use in molecular biology. Detailed protocols are available for the synthesis of oligonucleotides, using either phosphotriester or phosphite triester chemistries and using either manual, semiautomated, or fully automated methods.[21–24]

Synthetic oligonucleotides may be purified by ion-exchange or reverse-phase HPLC[25] or more conveniently by polyacrylamide gel electrophoresis.[26] Oligonucleotides are routinely detected in gels by UV shadowing, but ethidium bromide staining is a convenient alternative.[18,27] Oligonucleotides may be sequenced by the "wandering spot" technique,[28,29] by the chemical sequencing procedure[30] (modified for oligonucleotides[31]), or by fast atom bombardment (FAB) mass spectrometry.[32] However, synthetic procedures are now so reliable that oligonucleotides are only routinely checked for purity (and length) on denaturing polyacrylamide gels.

[18] P. Carter, H. Bedouelle, and G. Winter, *Nucleic Acids Res.* **13,** 4431 (1985).

[19] M. J. Gait, *in* "Oligonucleotide Synthesis: A Practical Approach" (M. J. Gait, ed.), p. 1. IRL Press, Oxford, England, 1984.

[20] M. H. Caruthers, *Science* **230,** 281 (1985).

[21] H. W. D. Mathes, A. Staub, and P. Chambon, this volume [14].

[22] M. H. Caruthers, A. D. Barone, S. L. Beaucage, D. R. Dodds, E. F. Fisher, L. J. McBride, M. Matteucci, Z. Stabinsky, and J.-Y. Tang, this volume [15].

[23] B. S. Sproat and M. J. Gait, *in* "Oligonucleotide Synthesis: A Practical Approach" (M. J. Gait, ed.), p. 83. IRL Press, Oxford England, 1984.

[24] T. Atkinson and M. Smith, *in* "Oligonucleotide Synthesis: A Practical Approach" (M. J. Gait, ed.), p. 35. IRL Press, Oxford, England, 1984.

[25] L. W. McLaughlin and N. Piel, *in* "Oligonucleotide Synthesis: A Practical Approach" (M. J. Gait, ed.), p. 117. IRL Press, Oxford, England, 1984.

[26] R. Wu, N.-H. Wu, Z. Hanna, F. Georges, and S. Narang, *in* "Oligonucleotide Synthesis: A Practical Approach" (M. J. Gait, ed.), p. 135. IRL Press, Oxford, England, 1984.

[27] P. Carter, H. Bedouelle, M. M. Y. Waye, and G. Winter, *in* "Oligonucleotide Site-Directed Mutagenesis in M13." Anglian Biotechnology, Colchester, England, 1985.

[28] G. G. Brownlee and F. Sanger, *Eur. J. Biochem.* **11,** 395 (1969).

[29] R. Frank and H. Blöcker, *in* "Chemical and Enzymatic Synthesis of Gene Fragments" (H. G. Gassen and A. Lang, ed.), p. 225. Verlag-Chemie Weinheim, Federal Republic of Germany, 1982.

[30] A. M. Maxam and W. Gilbert, *Proc. Natl. Acad. Sci. U.S.A.* **74,** 560 (1977).

[31] A. M. Banaszuk, K. V. Deugau, J. Sherwood, M. Michalak, and B. R. Glick, *Anal. Biochem.* **128,** 281 (1983).

[32] L. Grotjahn, R. Frank, and H. Blöcker, *Nucleic Acids Res.* **10,** 4671 (1982).

Factors Reducing Mutant Yield

In "single priming" mutagenesis (Fig. 1) the 5′ end of the oligonucleotide is exposed and the mismatch may be edited out by 5′ → 3′ exonucleases *in vivo*. Several approaches have been used to protect the 5′ end of the mutagenic oligonucleotide by ligation (Fig. 1). After extending "all the way round" from the mutagenic primer and ligating with T4 DNA ligase, covalent closed circular DNA may be separated from partially extended DNA on an alkaline sucrose gradient[33,34] or by agarose gel electrophoresis in the presence of ethidium bromide.[35] In the "double priming" technique a partial extension is made from a mutagenic primer and a second primer located 5′ to the mutagenic primer to protect the mutation.[36,37] Finally in the "gapped duplex" technique,[38] mutagenesis is conducted in a short single-stranded "window" made by annealing the M13 template to be mutated with a restriction fragment from the M13 vector. After extension and ligation, the 5′ end of the mutagenic primer is protected.

The main problem with these techniques is that the frequency of mutants obtained is often low and very variable (commonly 0.1–30%), so mutants are routinely detected by hybridization with the mutagenic primer. If, however, the frequency of mutations were routinely over 50% then mutants could be obtained by simply sequencing a few clones. Several factors in the *in vitro* construction of heteroduplex DNA may act to reduce the frequency of mutants obtained:

1. After extending "all the way round" the mutagenic primer may be displaced by Klenow. By titration of the ratio of Klenow and ligase used, ligation of the ends of the nascent strand may be favored over the competing strand displacement rection.[1,12]
2. Spurious priming by the mutagenic primer at sites other than the target site may be discouraged by careful computer-aided design of the mutagenic primer to minimize complementarity to sites other than the target site.
3. Any contaminating DNA polymerase I activity in the Klenow may remove the mismatch by nick translation after extending "all the way

[33] I. Kudo, M. Leineweber, and U. L. RajBhandary, *Proc. Natl. Acad. Sci. U.S.A.* **78,** 4753 (1981).

[34] M. J. Zoller and M. Smith, *Nucleic Acids Res.* **10,** 6487 (1982).

[35] G. F. M. Simons, G. H. Veeneman, R. N. H. Konings, J. H. Van Boom, and J. G. G. Schoenmakers, *Nucleic Acids Res.* **10,** 821 (1982).

[36] K. Norris, F. Norris, L. Christiansen, and N. Fiil, *Nucleic Acids Res.* **11,** 5103 (1983).

[37] M. J. Zoller and M. Smith, *DNA* **3,** 479 (1984).

[38] W. Kramer, K. Schughart, and H.-J. Fritz, *Nucleic Acids Res.* **10,** 6475 (1982).

round." With improvements in the quality of commercially available Klenow in the last couple of years this problem has largely disappeared.

4. Small RNA molecules present in the template preparation may give rise to spurious priming, but this is readily overcome by RNase treatment of the template.[18,27]

5. Trace amounts of dUTP in the dNTPs used in the *in vitro* extension (dUTP results from deamination of dCTP) produces a DNA strand containing deoxyuridine residues. These initiate strand breakage and nick translation *in vivo* which reduces mutant yield.[39] This problem may be overcome by HPLC purification or dUTPase treatment of dNTPs.

In vivo there are two main effects which act to reduce the yield of mutants:

1. In *E. coli* there is a methylation-dependent mismatch repair system which is directed by methylation of GATC sites by the *dam* methylase.[40,41] In DNA replication the daughter strand is transiently undermethylated, and mismatch repair is directed toward the methylated parent strand.[42–45] After transfection of *E. coli* with M13 heteroduplex DNA, the *dam* system will direct mismatch repair toward the methylated template strand to reduce the yield of mutants.[44–47]

2. Progeny phage may be derived from either plus or minus strands of M13 although there is a 2 : 1 bias in favor of the minus strand as a result of the asymmetric nature of M13 replication.[4,44]

"Coupled Priming" Mutagenesis

The improved mutagenesis procedures described in this chapter aim to overcome the *in vivo* problems reducing mutant yield (see above) by eliminating point mismatch repair and selecting genetically against prog-

[39] T. A. Kunkel, *Proc. Natl. Acad. Sci. U.S.A.* **82,** 488 (1985).

[40] M. G. Marinus and N. R. Morris, *Mutat. Res.* **28,** 15 (1975).

[41] S. Hattman, J. E. Brooks, and M. Masurekar, *J. Mol. Biol.* **126,** 367 (1978).

[42] M. Radman, R. E. Wagner, Jr., B. W. Glickman, and M. Meselson, *in* "Progress in Environmental Mutagenesis" (M. Alacevic, ed.), p. 121. Elsevier/North-Holland, Amsterdam, 1980.

[43] B. W. Glickman, *in* "Molecular and Cellular Mechanisms of Mutagenesis" (J. F. Leman and W. M. Generoso, eds.), p. 65. Plenum New York, 1982.

[44] B. Kramer, W. Kramer, and H.-J. Fritz, *Cell* **38,** 879 (1984).

[45] C. Dohet, R. Wagner, and M. Radman, *Proc. Natl. Acad. Sci. U.S.A.* **82,** 503 (1985).

[46] W. Kramer, V. Drutsa, H.-W. Jansen, B. Kramer, M. Pflugfelder, and H.-J. Fritz, *Nucleic Acids Res.* **12,** 9441 (1984).

[47] A. Marmenout, E. Remaut, J. Van Boom, and W. Fiers, *Mol. Gen. Genet.* **195,** 126 (1984).

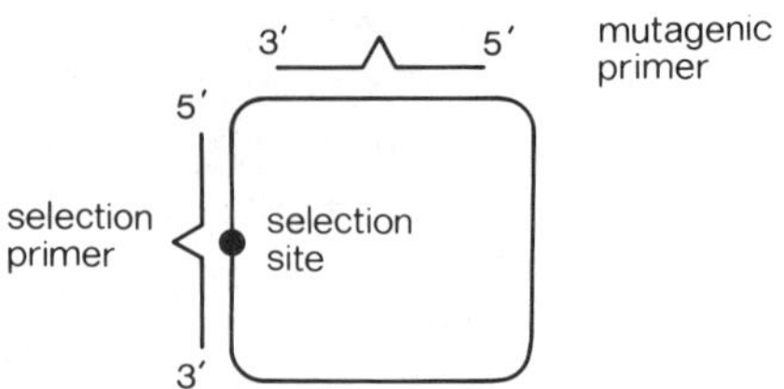

FIG. 2. "Coupled priming" mutagenesis (Carter *et al.*[18]).

eny phage from the template strand of M13. A "coupled priming" technique was devised, using one oligonucleotide to construct the "silent" mutation of interest and a second primer to remove a selectable marker in the template (Fig. 2). After extending from these two primers, the partial heteroduplex material is used to transfect an *E. coli* host strain which is deficient in point mismatch repair and selects genetically against the template. Thus only progeny phage where the selectable marker has been removed will be viable.

Use of Amber Selection Vectors

A selectable marker was obtained by constructing an amber mutation in gene IV of the vectors M13mp18 and M13mp19[48] to give M13mp18amIV and M13mp19amIV, respectively.[18] The amber mutation at position 5327 was generated by mutating the glutamine codon (CAG) to the stop codon (TAG).

Template is prepared after growing these amber phage in a suppressor ($Su2^+$) host (such as TG1) which suppresses the amber mutation. [These amber IV phage will not grow in a nonsuppressor ($Su2^-$) strain.] For mutagenesis, the template is primed with a mutagenic oligonucleotide and the selection primer, SEL1 (5′-AAGAGTCTGTCCATCAC-3′) which restores the glutamine codon in gene IV. The heteroduplex DNA is then transfected into competent HB2154 cells ($Repair^-$, $Su2^-$) using a lawn of HB2151 cells ($Repair^+$, $Su2^-$) to select against the starting template. [$Repair^-$ (*mut*) strains have a higher intrinsic mutation rate than normal strains. It is therefore advisable to transfect into the $Repair^-$ strains and use a lawn of the $Repair^+$ strain to minimize the exposure of the phage to the mutator phenotype of $Repair^-$ strains. However, the occurrence of spontaneous mutations does not appear to be a significant problem using *mut* strains.]

[48] J. Norrander, T. Kempe, and J. Messing, *Gene* **26,** 101 (1983).

EcoK 5' AACNNNNNNGTGC 3'

✳

EcoB 5' TGANNNNNNNNTGCT 3'

FIG. 3. Recognition sequence of *Eco*K and *Eco*B restriction and modification systems (Carter[61]).

"Cyclic Selection" Mutagenesis

A drawback of using amber mutations for strand selection is that after one round of mutagenesis the amber mutation has been removed, so further mutations cannot be constructed using selection. This problem may be overcome by using a pair of reciprocating markers in which a second selectable mutation is introduced at the same time that the first one is removed. Several features of the *Eco*K and *Eco*B restriction–modification systems[49,50] have enabled them to be used as genetic markers for site-directed mutagenesis.

If double-stranded DNA containing an *Eco*K site is introduced into a K strain ($r_K^+m_K^+$) then the DNA will be either modified if one strand is already modified or it will be restricted if neither strand is modified. This similarly applies to double-stranded DNA containing an *Eco*B site in B strains ($r_B^+m_B^+$). However, hybrid sites containing the *Eco*K or *Eco*B recognition sequence on one strand but a mutationally altered site on the other strand are not recognized by the corresponding restriction or modification enzymes. The recognition sequence of the *Eco*K and *Eco*B systems are very similar, allowing interconversion by site-directed mutagenesis using an oligonucleotide with a single mismatch (Fig. 3).

A scheme for "cyclic selection" using the *Eco*K and *Eco*B markers is shown in Fig. 4. In the first round of mutagenesis, one primer is used to remove the *Eco*K site and one primer is used to make a "silent" mutation of interest. After transfection into a repair-deficient K strain, only phage with the *Eco*K site removed will be viable. In removing the *Eco*K site an *Eco*B site may be created, providing a basis for selection in the next round of mutagenesis.

In constructing a second mutation, one primer is used to convert the *Eco*B site to an *Eco*K site and one primer is used to make a "silent" mutation. By transfecting into a repair-deficient B strain only phage where the *Eco*B site has been removed—converted to an *Eco*K site—will

[49] P. Modrich, *Q. Rev. Biophys.* **12,** 315 (1979).

[50] R. Yuan, *Annu. Rev. Biochem.* **50,** 285 (1981).

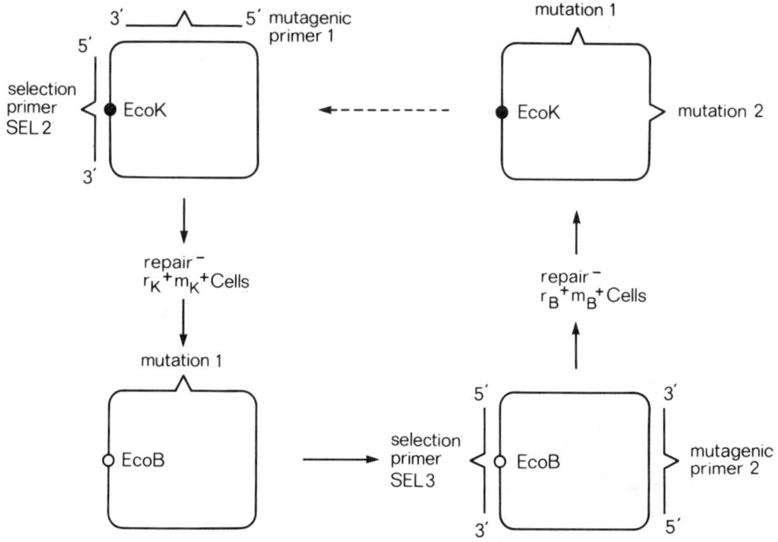

FIG. 4. Scheme for "cyclic selection" mutagenesis (Carter *et al.*[18]).

be viable. Hence additional mutations may be made by cycling between the two selectable markers.

Use of *Eco*K and *Eco*B Selection Vectors

For "coupled priming" mutagenesis using *Eco*K and *Eco*B selection, novel M13 vectors were constructed by cloning a 30-bp selection cassette into the *Hin*cII site of M13mp18 and M13mp19.[48] These new vectors M13K18 and M13K19 contain an *Eco*K site and additional unique sites for *Stu*I, *Nru*I and *Xho*I (Fig. 5).[18] In addition an *Eco*B site in gene II of M13 (present in all Messing's M13mp vectors) was removed by site-directed mutagenesis. The *Eco*K selections site in M13K18 and M13K19 was mu-

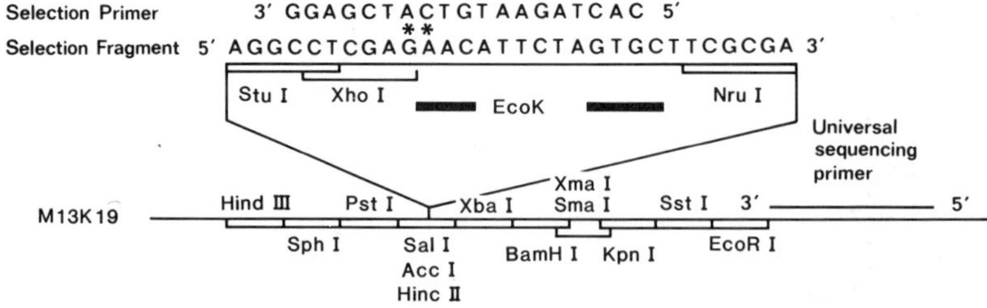

FIG. 5. Multiple cloning site of the selection mutagenesis vector M13K19 (Carter *et al.*[18]).

tated to an *Eco*B site to obtain the vectors M13B18 and M13B19, respectively.[18] A DNA fragment to be mutated should be cloned such that the *Eco*K/*Eco*B site lies downstream of the mutagenic primer (on the 3′ side of the mutagenic primer). This is to avoid possible displacement of the mutagenic primer by Klenow extending from the selection primer.[18]

Template is prepared for the *Eco*K vectors in a nonrestrictive ($r_K^-m_K^-$) host such as TG1 (they should not be grown in K-restrictive host strains). For mutagenesis the template is primed with a mutagenic oligonucleotide and the selection primer, SEL2 (5′-CACTAGAATGT-CATCGAGG-3′), which removes the *Eco*K site and substitutes an *Eco*B site. The heteroduplex DNA is transfected into competent HB2154 cells (Repair$^-$, $r_K^+m_K^+$) using a lawn of HB2151 cells (Repair$^+$, $r_K^+m_K^+$) to select against the starting template.

Template is prepared for the *Eco*B vectors in a nonrestrictive ($r_B^-m_B^-$) host such as TG1 (they should not be grown in B-restrictive host strains). For mutagenesis the template is primed with a mutagenic oligonucleotide and the selection primer, SEL3 (5′-CACTAGAATGTTATCGAGG-3′), which removes the *Eco*B site and substitutes an *Eco*K site. The heteroduplex DNA is transfected into competent HB2155 cells (Repair$^-$, $r_B^+m_B^+$), using a lawn of AC2522 cells (Repair$^+$, $r_B^+m_B^+$) to select against the template.

In constructing mutants using the *Eco*K/*Eco*B reciprocating markers, it is critical that the required "silent" mutant is tightly coupled to the new selection marker. However, a small proportion (~2–5%) of phage escape restriction by modification (methylation) rather than mutation of the selection marker.[18] After the plaque purification procedure using TG1 ($r_K^-m_K^-$, $r_B^-m_B^-$), selection sites which escaped restriction by being modified will now be unmodified and hence susceptible to restriction. It is very easy to check whether phage carry the new marker by comparing the plaque yield on lawns of nonrestrictive hosts (TG1) and restrictive hosts (BMH 71-18 or AC2522) in parallel. This comparison is most conveniently done using the plaque purification procedure, where the differences in plaque yield are readily seen by casual inspection.

Evaluation of the "Coupled Priming" Technique

The "coupled priming" technique was first evaluated by comparing the frequency of a model point mutation using a variety of strategies for mutagenesis. A 6- to 8-fold improvement in the frequency of the model mutation was obtained by removing methylation-dependent mismatch repair. A further 2- to 3-fold improvement was obtained using amber, *Eco*K, or *Eco*B selection against the progeny phage derived from the template strand to a total of about 70% yield.[18] "Coupled priming" mutagenesis using amber selection was used in constructing more than 30 mutants in the tyrosyl-tRNA synthetase of *Bacillus stearothermophilus* with yields of 2–68%.[18,51] This variability in the frequency presumably reflects some of the problems in constructing heteroduplex DNA discussed above. Amber selection using repair-deficient host strains has also been used very successfully with the "gapped-duplex" technique.[46]

Blue → White Model Mutation

A very convenient way of comparing mutagenic strategies and of checking the efficiency of mutagenesis is by using a model mutation which changes the phenotype of M13 plaques. Messing's M13mp vectors[52] contain the *lacZ′* gene which is induced by IPTG to express the *lacZ* α-peptide. The α-peptide complements a deletion in several widely used *E. coli* host strains (such as JM101, TG1, and BMH 71-18) to give a functional β-galactosidase. The substrate analog, BCIG, is hybrolyzed by the β-galactosidase to a blue dye, and hence blue plaques result.[53] The mutagenic oligonucleotide, B/W (5′-GGTTTTCCTAGTCACGA-3′), introduces an amber mutation in the *lacZ′* fragment (Trp16 → Amber: TGG → UAG, not suppressible in Su2$^+$ hosts) which prevents complementation of the host β-galactosidase and hence results in white plaques.[46]

The results from a typical blue → white model mutagenesis experiment are shown in Table I. A 2- to 5-fold improvement in the frequency of this mutation was obtained by removing methylation-dependent mismatch repair. A further 1.5- to 2-fold improvement was obtained using amber or *Eco*K selection against the progeny phage derived from the template strand to a total of approximately 70% yield. (*Eco*B selection cannot be checked in this way using the B strains HB2155 and AC2522 since they have a wild-type *lacZ* gene and hence give rise to blue lawns in the presence of IPTG and BCIG.) The following section provides detailed protocols for "coupled priming" mutagenesis using amber, *Eco*K, or *Eco*B selection.

Materials and Methods

T4 polynucleotide kinase, *E. coli* DNA polymerase I Klenow fragment, and T4 DNA ligase were obtained from Anglian Biotechnology

[51] H. Bedouelle, P. Carter, and G. Winter, *Philos. Trans. R. Soc. London, A* **317,** 433 (1986).

[52] J. Messing, this series, Vol. 101, p. 20.

[53] J. Messing, B. Gronenborn, B. Müller-Hill, and P. H. Hofschneider, *Proc. Natl. Acad. Sci. U.S.A.* **74,** 3642 (1977).

TABLE I
BLUE → WHITE MODEL MUTAGENESIS EXPERIMENT[a]

		% White plaques in host strains			
Template	Primers	TG1 (Repair$^+$ Su2$^+$ $r_K^- m_K^-$)	HB2151 (Repair$^+$ Su2$^-$ $r_K^+ m_K^+$)	BMH 71-18 *mutL* (Repair$^-$ Su2$^+$ $r_K^+ m_K^+$)	HB2154 (Repair$^-$ Su2$^-$ $r_K^+ m_K^+$)
M13mp19	B/W, SEL1	10.8	17.7	29.8	26.5
M13mp19amIV	B/W, SEL1	9.3	47.8	47.5	68.0
M13K19	B/W, SEL2	6.5	34.2	62.1	66.0

[a] The mutagenic primer for the blue → white model mutation (B/W) and a selection primer (SEL1 or SEL2) were annealed to a M13 template (M13mp19, M13mp19amIV, or M13K19) extended and ligated for 4 hr at 12°. The heteroduplex DNA was then used directly to transfect various *E. coli* host strains (TG1, HB2151, BMH 71-18 *mutL*, and HB2154) in the presence of BCIG and IPTG. The percentage of white plaques obtained is shown. Data were obtained by participants on the EMBO site-directed mutagenesis course held at the European Molecular Biology Laboratory, Heidelberg, FRG, in November 1985, organized by Prof. R. Cortese and Dr. G. Winter. See Carter *et al.*[18,27]

Limited. [γ-^{32}P]ATP (3000 Ci/mmol) was obtained from Amersham International plc. HPLC-grade dNTPs were obtained from Pharmacia PL Biochemicals. Nitrocellulose filters (82 mm diameter, BA85) were obtained from Schleicher and Schuell. Other reagents were obtained as analytical grade.

Solutions

10× TM buffer: 100 m*M* Tris–HCl (pH 8.0), 100 m*M* $MgCl_2$
10× Kinase buffer: 500 m*M* Tris–HCl (pH 8.0), 100 m*M* $MgCl_2$
10× TBE buffer: 900 m*M* Tris base, 890 m*M* boric acid (pH 8.3), 25 m*M* EDTA
TE buffer (10 : 0.1): 10 m*M* Tris–HCl (pH 8.0), 0.1 m*M* EDTA
TE buffer (10 : 10): 10 m*M* Tris–HCl (pH 8.0), 10 m*M* EDTA
100× Denhardt's solution[54]: 2% (w/v) bovine serum albumin, 2% (w/v) Ficoll 400, 2% (w/v) poly(vinylpyrrolidone)
10× SSC: 1.5 *M* NaCl, 150 m*M* sodium citrate, 5 m*M* EDTA, pH adjusted to 7.2 with HCl

[54] D. T. Denhardt, *Biochem. Biophys. Res. Commun.* **23,** 641 (1966).

Formamide dyes: 0.1% (w/v) xylene cyanol, 0.1% (w/v) bromophenol blue, 10 m*M* EDTA in formamide

Media

2× TY: 16 g Bactotryptone, 10 g yeast extract, 5 g NaCl, add water to 1 liter and adjust pH to 7.4 with NaOH
TYE plates: 15 g agar, 10 g Bactotryptone, 10 g yeast extract, 8 g NaCl, add water to 1 liter
H Top agar: 8 g Bactoagar, 10 g Bactotryptone, 8 g NaCl, add water to 1 liter
10× M9 salts: 70 g $Na_2HPO_4 \cdot 2H_2O$, 30 g KH_2PO_4, 5 g NaCl, 10 g NH_4Cl, add water to 1 liter
Minimal glucose plates: 15 g Bactoagar, 888 ml water; after autoclaving add 100 ml 10× M9 salts, 10 ml 20% (w/v) glucose solution, 1 ml 1 *M* $MgSO_4$ and 1 ml thiamin (2 mg/ml)

Bacterial Strains

The following *E. coli* strains were used:

TG1: K12, Δ(*lac–pro*), *supE*, *thi*, *hsdD5*/F′ *traD36*, *proA*$^+$*B*$^+$, *lacI*^{q}, *lacZ*ΔM15 (Gibson[55])
BMH 71-18: K12, Δ(*lac–pro*), *supE*, *thi*/F′ *proA*$^+$*B*$^+$, *lacI*^{q}, *lacZ*ΔM15 (Gronenborn[56])
BMH 71-18 *mutL*: Same as BMH 71-18 but *mutL* : : Tn*10* (Kramer *et al.*[44])
HB2151: K12, *ara*, Δ(*lac–pro*) *thi*/F′ *proA*$^+$*B*$^+$, *lacI*^{q}, lacZΔM15 (Carter *et al.*[18])
HB2154: Same as HB2151 but *mutL* : : Tn*10* (Carter *et al.*[18])
AC2522: B/r, Hfr, *sulI* (Boyer[57])
HB2155: Same as AC2522 but *mutL* : : Tn*10* (Carter *et al.*[18])

The strains HB2151, HB2154, BMH 71-18, BMH 71-18 *mutL*, and TG1 should be restreaked on minimal glucose plates (once a month) to keep the episome, whereas AC2522 and HB2155 may be grown either on complete medium or on minimal glucose plates. The *mutL* ($Repair^-$) strains mutate at a higher frequency than $Repair^+$ strains and should therefore be stored at −20° or −70° in glycerol.

[55] T. J. Gibson, Ph.D. thesis. University of Cambridge, Cambridge, England, 1984.
[56] B. Gronenborn, *Mol. Gen. Genet.* **148,** 243 (1976).
[57] H. Boyer, *J. Bacteriol.* **91,** 1767 (1966).

Typical Timetable for a Mutagenesis Experiment

Day 1
  Kinase oligonucleotides for mutagenesis
  Set up extension and ligation reaction
  Prepare competent cells
Day 2
  Transfect heteroduplex DNA
  Grid out plaques resulting from transfection
Day 3
  Colony blot hybridization screening using mutagenic oligonucleotide
  Plaque purify suspected mutants
Day 4
  Prepare template from individual plaques
Day 5
  Dideoxy sequencing of suspected mutants

Kinasing Oligonucleotides for Site-Directed Mutagenesis

Kinase enough mutagenic and selection oligonucleotides for several experiments.

1. To a 1.5-ml Eppendorf tube add:
 5 μl oligonucleotide (10 pmol/μl)
 2 μl 10× kinase buffer
 1 μl 100 m*M* DTT
 1 μl 20 m*M* rATP

Add water to a total volume of 20 μl.

2. Add 5 units T4 polynucleotide kinase then incubate for 30 min at 37° followed by 10 min at 70° (to stop the reaction).
3. Use immediately or store at −20° if required.

Preparation of Template for Mutagenesis

Template is prepared as for dideoxy sequencing, except on a 5-ml scale, and a RNase step is included to remove small RNA primers (not essential). TG1 is a suitable host strain for preparing template from all M13 strains.

1. Inoculate 5 ml 2× TY medium with 50 μl of a saturated fresh overnight culture of TG1 cells.
2. Toothpick a single plaque (transfers $\sim 10^7$ phage) into the medium and incubate at 37° with shaking at around 200 rpm for 4.5–5.5 hr.

3. Pour the culture into five 1.5-ml Eppendorf tubes.

4. Centrifuge for 5 min in a microfuge to pellet the cells.

5. Pour the supernatant into five tubes, each containing 200 μl 20% (w/v) PEG 6000, 2.5 *M* NaCl. Do not try to transfer the supernatant completely, as it is important to avoid carrying across any cells.

6. Leave at room temperature for 15 min and then centrifuge for 5 min in a microfuge.

7. Remove the supernatant by aspiration using a drawn-out pasteur pipet attached to a water pump (pellet should be visible). Respin for a few seconds to bring the liquid off the walls of the tube and then aspirate off traces of PEG.

8. To each tube add 100 μl TE buffer (10 : 0.1) and vortex for about 30 sec to resuspend the pellet. Add 5 μl RNase A (10 mg/ml) and incubate at 37° for 30 min. (RNase A is prepared by boiling in 300 m*M* sodium acetate pH 5.5, for 15 min to denature any DNase present, and is stored frozen at −20°.)

9. Add 50 μl neutralized phenol (prepared as Maniatis *et al.*[58]), vortex for 30 sec and spin for 1 min in a microfuge to separate the phases. Remove the aqueous (upper) phase and transfer to another Eppendorf tube.

10. Remove traces of phenol by one extraction with 0.5 ml diethylether, and add 10 μl 3 *M* sodium acetate (pH 5.5) and 250 μl ethanol. Vortex and quick freeze in dry ice (5–10 min).

11. Spin for 5 min in a microfuge to pellet the DNA and resuspend all five tubes in a total of 25 μl TE buffer (10 : 0.1). Measure the absorbance at 260 nm after adding a 1-μl aliquot to 1 ml water in a 1-ml quartz cuvette. [One OD unit of single-stranded DNA corresponds to ~40 μg/ml. The total yield of single-stranded DNA is normally 3–5 μg/ml culture, although up to 10 μg/ml may be obtained from large-scale (50 ml) cultures under conditions of good aeration.]

12. Adjust the concentration of the template to 0.5 μg/μl with TE buffer (10 : 0.1).

In Vitro DNA Synthesis: "Coupled Priming"

A. Annealing Reaction

1. Anneal the mutagenic primer and the selection primer with the M13 template in a 1.5-ml Eppendorf tube:

2 μl kinased mutagenic primer (2.5 pmol/μl)

[58] T. Maniatis, E. F. Fritsch, and J. Sambrook, *in* "Molecular Cloning: A Laboratory Manual." Cold Spring Harbor Lab., Cold Spring Harbor, New York, 1982.

2 μl kinased selection primer (2.5 pmol/μl)
2 μl template (0.5 μg/μl, ~15-fold molar excess of primer)
1 μl 10× TM buffer
Add water to a total volume of 10 μl.

2. Place the tube containing the sample in a small beaker of water at 80° and then allow to cool to room temperature (~30 min).

3. Spin briefly in a microfuge to recover any condensation on the lid of the tube.

B. Extension and Ligation Reactions

1. Place the annealed mixture (10 μl) on ice and add the following:
1 μl 10× TM buffer
1 μl 5 m*M* rATP
1 μl 5 m*M* dNTPs
1 μl 100 m*M* DTT
Add water to a total volume of 20 μl.

2. Add 10 units T4 DNA ligase and 1 unit Klenow fragment of DNA polymerase I and incubate at 12–15° for 4–20 hr.

3. Dilute the extension/ligation mixture with 30 μl TE buffer (10 : 10).

In Vitro DNA Synthesis: "Single Priming"

For mutagenesis without selection, the procedure above for "coupled priming" is modified by leaving out the selection primer from the annealing reaction:

1. Anneal the mutagenic primer with M13 template in a 1.5-ml Eppendorf tube:
2 μl kinased mutagenic primer (2.5 pmol/μl)
2 μl template (0.5 μg/μl, ~15-fold molar excess of primer)
1 μl 10× TM buffer
Add water to a total volume of 10 μl.

The annealing reaction and then the extension and ligation reactions are carried out according to the procedure above.

Transfection of Heteroduplex DNA

A. Preparation of Competent Cells[59]

1. Take a colony of cells from a fresh plate and inoculate 5 ml 2× TY medium. Grow overnight at 37° to saturation.

[59] S. N. Cohen, A. C. Y. Chang, and L. Hsu, *Proc. Natl. Acad. Sci. U.S.A.* **69,** 2110 (1972).

2. Inoculate 20 ml 2× TY medium with 200 μl of saturated cells and grow until an OD_{550} of about 0.3. Cool on ice for 5 min and then spin the cells down gently (15 min, 2,500 rpm, 4°).

3. Resuspend the cells in 10 ml ice-cold 100 m*M* $CaCl_2$ and leave on ice for at least 30 min.

4. Spin the cells down gently (10 min, 2,500 rpm, 4°).

5. Resuspend the cells in 2 ml ice-cold 100 m*M* $CaCl_2$.

The competent cells may be used directly or stored overnight on ice before use. The transfection efficiency increases overnight and then decreases to an unacceptably low level after a few days.

B. Transfection of Competent Cells

1. Dispense 0.2-ml aliquots of competent cells into sterile glass culture tubes (prechilled) and keep on ice.

2. Add 0.1-, 1-, and 10-μl aliquots of the diluted heteroduplex DNA to the competent cells and leave on ice for at least 30 min.

3. Dispense 3-ml aliquots of H top agar into sterile glass culture tubes and allow to cool to 45°.

For the blue → white model mutagenesis (*only*) add 25 μl IPTG (25 mg/ml) and 25 μl BCIG (25 mg/ml in DMF) to the H top agar.

4. Heat shock the transfected cells for 2 min at 42°. Then add 0.2 ml of lawn cells (conveniently provided by an overnight culture diluted 1 : 10 with 2× TY).

5. Add the H top agar to the cells and plate out on TYE plates.

6. Allow the plates to cool for about 5 min, then invert them and incubate at 37° for 8–20 hr. Plaques may start to become visible after 5–6 hr, but for the blue → white model mutation the blue color takes at least 9 hr to develop.

The dNTPs present with the heteroduplex DNA may inhibit transfection to a small extent. However they are readily removed by a PEG precipitation (not essential):

7. To the extension/ligation mixture (20 μl), add 80 μl TE buffer (10 : 10) add 100 μl 13% PEG 6000, 1.6 *M* NaCl. Leave on ice for 15 min, then spin in a microfuge for 10 min.

8. Remove the supernatant completely by aspirating using a drawn-out pasteur pipet and then dissolve the pellet in 50 μl water.

Colony Blot Hybridization Screening

1. Grid out the plaques using toothpicks onto TYE plates (or L plates) to give about 200 per plate. One plate per mutant is invariably sufficient.

The use of an asymmetric grid will help orientate the grid later on. Grow at 37° as colonies of infected bacteria (8–20 hr).

2. To make the probe, dry down 30 μl [γ-^{32}P]ATP (3000 Ci/mmol, 1 mCi/ml) in a 1.5-ml Eppendorf tube. [Where the radiaoactive ATP is obtained as an aqueous solution rather than in the presence of 50% (v/v) ethanol (as above) it may be used directly without drying down.] Then add:

1.5 μl mutagenic oligonucleotide (10 pmol/μl)
3 μl 10× kinase buffer
1 μl 100 m*M* DTT

Add water to a total volume of 30 μl.

2. Add 2 units T4 polynucleotide kinase, incubate at 37° for 30 min and then dilute with 4.0 ml 6× SSC in disposable plastic petri dish.

3. Store at −20° until required.

The primer is in 1.5-fold molar excess over ATP and it is not usually necessary to remove excess label. If, however, a high background of ATP spots is obtained then this may be overcome by filtering the probe through a 0.2-μm Millex filter. Alternatively unincorporated label may be removed by DE-52, Sephadex, G-50, C_{18} Sep-Pak chromatography as described by Zoller and Smith.[37] The probe may be reused several times over a few days (stored at −20° between uses), but it is advisable to filter the probe through a Millex filter between uses to avoid an increased background.

4. Label a nitrocellulose filter with a soft pencil and lay the filter on the colonies. Leave at room temperature for at least 1 min (until the filter is completely wetted). The filter may be stabbed with a pin to help orientate the filter with respect to the colonies, but this is not necessary if an asymmetric grid is used.

5. Gently peel off the filter and transfer it to a sheet of Whatman 3MM paper soaked in 500 m*M* NaOH with the colonies *faceup* for 3 min. Excess liquid should first be poured off from the Whatman 3MM and any bubbles smoothed out. Large plastic petri dishes (225 × 225 mm) are convenient for this step. Store the colony plates at 4° for later plaque purification of mutants.

6. Neutralize the filters by transferring them onto sheets of Whatman 3MM soaked in 1 *M* Tris–HCl (pH 7.4) twice for 1 min and finally to 500 m*M* Tris–HCl (pH 7.4), 1.5 *M* NaCl (or 2× SSC) for 5 min.

7. Allow the filters to dry (colonies *faceup*) for about 20 min at room temperature or 5–10 min at 80°. Then bake the filters for at least 10 min at 80° in a vacuum oven, sandwiched between sheets of dry Whatman 3MM. Prolonged baking is unnecessary and may lead to the filters becoming

very brittle and so readily damaged.[60] (Alternatively incubate for at least 30 min at 80° at atmospheric pressure.)

8. Prewet the filters in 6× SSC (~5 min) and prehybridize at 67° for at least 5 min in 10× Denhardt's solution,[54] 6× SSC, and 0.2% SDS. There is no harm in prehybridizing for several hours, but this is unnecessary.[61] (Do not shake the filters vigorously or the colonies tend to drop off! This will result in a reduced and very uneven hybridization signal.) Then rinse the filters in 100 ml 6× SSC.

9. Place the filter in the probe (colonies *facedown*), being careful not to trap bubbles under the filter, and swirl the probe gently in the dish. Use only one filter per petri dish to be sure of even hybridization. Hybridize at room temperature for about 1 hr or at 37° for around 30 min. Then wash at room temperature 3 times in 100 ml 6× SSC (~1 min in fresh 6× SSC per wash).

10. Gently shake off the excess 6× SSC, cover the filter with SaranWrap, and autoradiograph using preflashed film and an intensifying screen at −70°. Do not let the filters dry out as they then easily tear and also the colonies may stick to the SaranWrap.

11. The nonstringent wash should give the background grid after a 15–60 min exposure. Allow the filters to warm to room temperature before removing the SaranWrap, otherwise the colonies may stick to it. Now wash at 5° below the calculated dissociation temperature (T_d). The T_d for mutant DNA hybridized to the mutagenic oligonucleotide may be estimated from the number of GC and AT base pairs in the duplex[62]:

$$T_d\ (°) = 4 \times \text{GC pairs} + 2 \times \text{AT pairs}$$

This empirical relationship is a good approximation to the behavior of 11- to 20-mer oligonucleotides in 6× SSC. The T_d of the mismatched oligonucleotide is reduced by destabilization of the double helix as well as by removal of one or more G/C or A/T base pairs. For oligonucleotides longer than about 20-mers this rule no longer applies, so a series of washes at increasing temperatures is used to determine the ideal washing conditions.

12. Wash for 1 min and check the strength of the hybridization signal before and after washing with a minimonitor. If there is a large reduction in the signal, autoradiograph again (this may take several hours). Other-

[60] J. Meinkoth and G. Wahl, *Anal. Biochem.* **138,** 267 (1984).

[61] P. J. Carter, Ph.D. thesis. University of Cambridge, Cambridge, England, 1985.

[62] S. V. Suggs, T. Hirose, T. Miyake, E. H. Kawashima, M. J. Johnson, K. Itakura, and R. B. Wallace, *in* "Developmental Biology Using Purified Genes" (D. Brown, ed.). Academic Press, New York, 1981.

wise repeat the wash up to 4 times, checking after each 1-min wash that the counts have not fallen dramatically, and then autoradiograph.

13. Wash again at the T_d (as above) and autoradiograph again. Sometimes it is necessary to exceed the calculated T_d by a few degrees to obtain good discrimination between mutants (perfectly matched to oligonucleotide) and wild type (one or more mismatches).

Plaque Purification

1. Toothpick from the center of the original colony into 1 ml 2× TY medium (transfers 10^7–10^8 plaque-forming units).
2. Streak the diluted phage over a TYE plate with a sterile wire loop.
3. Add 0.2 ml TG1 cells (overnight culture diluted 1 : 10 in 2× TY) to 3 ml H top agar, and pour on to the plate (as for the transfection procedure).
4. Invert the plates and incubate at 37° for 8–20 hr. Take phage from well-isolated plaques for sequencing (three plaques should be enough).

Diagnostic Procedures

If mutagenesis does not give rise to strongly hybridizng clones on screening 200 plaques, then it is advisable to carry out some of the following diagnostic tests.

Specificity of Priming of Mutagenic Oligonucleotide

The specificity of priming of the mutagenic oligonucleotide may be checked by using the oligonucleotide in a dideoxy sequencing reaction or in a primer-directed extension reaction followed by restriction endonuclease digestion as described by Zoller and Smith.[11] These assays should be done over a range of primer concentrations since spurious priming at sites other than the target site may sometimes be overcome by reducing the molar excess of primer used in the annealing reaction.

Extent of Phosphorylation of Mutagenic Oligonucleotide

In order to be ligated, the 5′-OH of the mutagenic and selection oligonucleotides must be phosphorylated. For a new batch of kinase, quantitative phosphorylation of the oligonucleotide may be checked by running the oligonucleotide on a native 20% polyacrylamide gel prestained with ethidium bromide as described by Carter *et al.*[18,27] Take 40 pmol of oligonucleotide and load on a narrow slot (2–3 mm) with unkinased oligonucleotide in an adjacent slot. Each sample must be loaded in *identical salt*, as this will affect the mobility. The kinased oligonucleotide runs ahead of the unkinased material by virtue of its extra negative charge, and the extent of phosphorylation of the primer (usually about 100%) can be assessed.

Purity of Oligonucleotide

The purity of the mutagenic oligonucleotide will affect the mutant yield obtained. The purity may be checked by running an aliquot of the oligonucleotide on a narrow slot (3 mm) of an ethidium bromide-stained gel[18,27] or by kinasing with [γ-^{32}P]ATP and running on a 20% polyacrylamide urea gel. Mix together

1 μl [γ-^{32}P]ATP (3000 Ci/mmol, 1 mCi/ml, dried down)
1 μl oligonucleotide (10 pmol/μl)
1 μl 10$\times$ kinase buffer
1 μl 100 m*M* DTT
7 μl water

Add 0.5 units T4 polynucleotide kinase and incubate for 15 min at 37°. Boil for 3 min in formamide dyes and load on a 1-cm slot of a 20% polyacrylamide–urea gel. (Preelectrophorese the gel for 30 min at 37 W to remove the salt front which would otherwise interfere with electrophoresis of the oligonucleotides.) Run the gel until the bromophenol blue reaches the bottom (corresponds to ~10 nucleotides). Cover with SaranWrap (do not attempt to fix as the oligonucleotide will diffuse out) and autoradiograph for 30 min (or less, as appropriate).

Formation of Covalent Closed Circular DNA

Many failures in obtaining mutants have probably been due to unsuitable batches of Klenow. Enzyme which works well for dideoxy sequencing will not always be suitable for mutagenesis. The levels of Klenow and ligase used should be titrated to promote ligation of the nascent strand (after extending "all the way around") over the competing strand displacement reaction. The efficiency of the extension and ligation steps in forming covalent closed circular DNA may be followed by alkaline sucrose gradient centrifugation as described by Zoller and Smith[11] or by agarose gel electrophoresis in the presence of ethidium bromide.[35]

In a "coupled priming" experiment, M13 heteroduplex DNA from an overnight extension and ligation reaction was either used directly to transfect an *E. coli* host or first enriched for covalent closed circular material on an alkaline sucrose gradient. The high yield (~70%) of mutants obtained from material transfected directly was not improved using the covalent closed circular material.[18]

Efficiency of Mutagenesis

The blue → white model mutation provides a convenient way of checking the overall efficiency of mutagenesis and hence of checking new batches of Klenow and ligase. Any mutation which prevents expression of the *lacZ* α-peptide will give rise to white plaques. Hence a more rigorous model system would be a white → blue mutation. However the blue → white mutation is a very convenient test mutation for use with all of Messing's standard M13mp vectors.

Choice of Mutagenesis Strategy

The *Eco*K/*Eco*B selection system provides a convenient way of constructing a series of mutations at high yield.[18] However "single priming" without strand selection using a Repair$^-$ host strain generally gives very acceptable results (P. Carter, unpublished result). Thus this very simple and reliable approach is recommended if only a few mutations are being constructed. *Eco*K selection has been extended to the construction of large deletions in high yield and to blunt-end cloning.[63]

Acknowledgments

The methods described here were developed at the MRC Laboratory of Molecular Biology, Hills Road, Cambridge CB2 2QH, England, with invaluable help from Drs. Greg Winter, Hugues Bedouelle, Mary Waye, and Kiyoshi Nagai. P.C. is supported by a Research Fellowship from Gonville and Caius College, Cambridge, England.

[63] M. M. Y. Waye, M. E. Verhoeyen, P. T. Jones, and G. Winter, *Nucleic Acids Res.* **13,** 8561 (1985).

[21] Site-Specific Mutagenesis to Modify the Human Tumor Necrosis Factor Gene

By DAVID F. MARK, ALICE WANG, and COREY LEVENSON

The development of oligonucleotide-directed site-specific mutagenesis on single-stranded phage template[1] has provided a powerful new technique in the study of structure–function relationships of proteins.[2] More recently, the development of the M13mp series of M13 phage-derived

[1] M. Smith and S. Gillam, *in* "Genetic Engineering. Principles and Methods" (J. K. Setlow and A. Hollaender, eds.), Vol. 3, p. 1. Plenum, London, 1981.

[2] M. J. Zoller and M. Smith, this series, Vol. 100, p. 468.

vectors which provided a large number of unique restriction enzyme cloning sites and a simple phenotypic screen for DNA inserts,[3,4] has led to the widespread use of this technique in a variety of biological applications.[5] These include the modification of genes for expression in heterologous systems,[6–8] the structure–function relationships of cloned proteins,[9–11] the alteration of enzyme substrate specificity,[12] and the improvement of enzyme thermostability.[13]

This chapter describes a simplified procedure for oligonucleotide-directed site-specific mutagenesis on M13 single-stranded phage DNA templates. The procedure described here improves on that previously described by Zoller and Smith[2] by elimination of a number of steps, such as the phosphorylation of the oligonucleotide primer by T4 kinase, which logically led to the omission of the T4 ligase in the reaction mixture, and by shortening of the time needed to anneal the oligonucleotide to the template as well as the time for the extension of the primer by *Escherichia coli* DNA polymerase I (Klenow fragment). The rationale behind this approach is to make use of the DNA repair mechanism present in the *E. coli* host to repair the gapped M13 phage DNA into the covalently closed circular form before replication of the DNA molecules. Although this faster method is less efficient in the conversion of parent DNA to mutant DNA molecules, the ability to use the same oligonucleotide primer as probe to identify the mutant phage plaques effectively compensates for this inefficiency. In addition, if a new restriction enzyme site can be created by the oligonucleotide used as a mutagenesis agent, this then provides another quick method to confirm the presence of mutated DNA molecules and aid in the screening for the correct mutant clones.

Experimental Rationale

The cDNA for the human tumor necrosis factor (TNF) has been cloned from the human promyelocytic leukemia cell line, HL60.[6,14] The

[3] J. Messing, this series, Vol. 101, p. 20.

[4] C. Yanisch-Perron, J. Vieira, and J. Messing, *Gene* **33,** 103 (1985).

[5] M. Smith, *Annu. Rev. Genet.* **19,** 423 (1981).

[6] A. M. Wang, A. A. Creasey, M. B. Ladner, L. S. Lin, J. Strickler, J. N. Van Arsdell, R. Yamamoto, and D. F. Mark, *Science* **228,** 149 (1985).

[7] J. P. Adelman, J. Hayflick, M. Vasser, and P. Seeburg, *DNA* **2,** 183 (1983).

[8] A. J. Brake, J. Merryweather, D. Coit, U. Herberlein, F. Masiarz, G. Mullenbach, M. Urdea, P. Valenzuela, and P. Barr, *Proc. Natl. Acad. Sci. U.S.A.* **81,** 4642 (1984).

[9] D. F. Mark, S. D. Lu, A. A. Creasey, R. Yamamoto, and L. S. Lin, *Proc. Natl. Acad. Sci. U.S.A.* **81,** 5662 (1984).

[10] T. Shiroishi, G. A. Evans, E. Appella, and K. Osato, *Proc. Natl. Acad. Sci. U.S.A.* **81,** 7544 (1984).

[11] M. Mishina, T. Tobimatsu, K. Imoto, K. Tanaka, Y. Fujita, K. Fukuda, M. Kurasaki, H. Takahashi, Y. Morimoto, T. Hirose, S. Inayama, T. Takahashi, M. Kuno, and S. Numa, *Nature (London)* **313,** 364 (1985).

cDNA sequence predicts the presence of a 76-amino acid signal sequence, which has to be removed before the mature TNF protein can be expressed in *E. coli* (Fig. 1). This was accomplished by the insertion of a restriction enzyme recognition site and an ATG initiation codon between the end of the signal peptide coding sequence and the beginning of the mature protein coding region, using site-directed oligonucleotide mutagenesis.[6]

A functionally related protein, human lymphotoxin (LT), has also been cloned recently,[15] and comparison of the amino acid sequence of the two cytotoxic proteins reveal that although they have regions of homology throughout the two protein sequences there are two distinct regions of divergence.[6,14,15] The first ten residues of the mature TNF protein has very little homology with the N-terminal sequence of LT. In addition, TNF has two cysteine residues at positions 69 and 101 (Fig. 1), which are not found in LT, and the amino acid sequence between these two cysteines also has little homology when compared to LT. Based on these comparisons of amino acid sequence homology, one would predict that the N-terminal ten amino acids and the cysteine residues of TNF may not be necessary for biological activity. To test this hypothesis, we have constructed mutants of TNF with deletions of up to residues 9 and 10 of the mature TNF protein sequence, and we have also substituted the two cysteine residues at positions 69 and 101 with serine residues, using oligonucleotide-directed site-specific mutagenesis.

Design and Synthesis of Oligonucleotides

The oligonucleotides used in the mutagenesis of the TNF cDNA are listed in Table I. Oligonucleotides used to generate insertion or deletion mutants were designed to have 10–12 matched base pairs flanking the mutation and an approximate melting temperature (T_m) of 71–75° for the correctly base-paired oligonucleotide. The approximate T_m of an oligonucleotide was determined with the formula[16]

$$\text{Approximate } T_m = 81.5° + 16.6 \log M + 0.41(\% \text{ G} + \text{C}) - 500/n - 0.61(\% \text{ formamide})$$

where M is the ionic strength (mol/liter) and n the length of the duplex.

[12] C. S. Craik, C. Largman, T. Fletcher, S. Roczniak, P. Barr, R. Fletterick, and W. Rutter, *Science* **228,** 291 (1985).

[13] L. Jeanne-Perry and R. Wetzel, *Science* **226,** 556 (1984).

[14] D. Pennica, G. Nedwin, J. Hayflick, P. Seeburg, R. Derynck, M. Palladino, W. Kohr, B. Aggarwal, and D. Goeddel, *Nature (London)* **312,** 724 (1984).

[15] P. W. Gray, B. Aggarwal, C. Benton, T. Bringman, W. Henzel, J. Jarrett, D. Leung, B. Moffat, P. Ng, L. Svedersky, M. Palladino, and G. Medwin, *Nature (London)* **312,** 721 (1984).

[16] J. Meinkoth and G. Wahl, *Anal. Biochem.* **138,** 269 (1984).

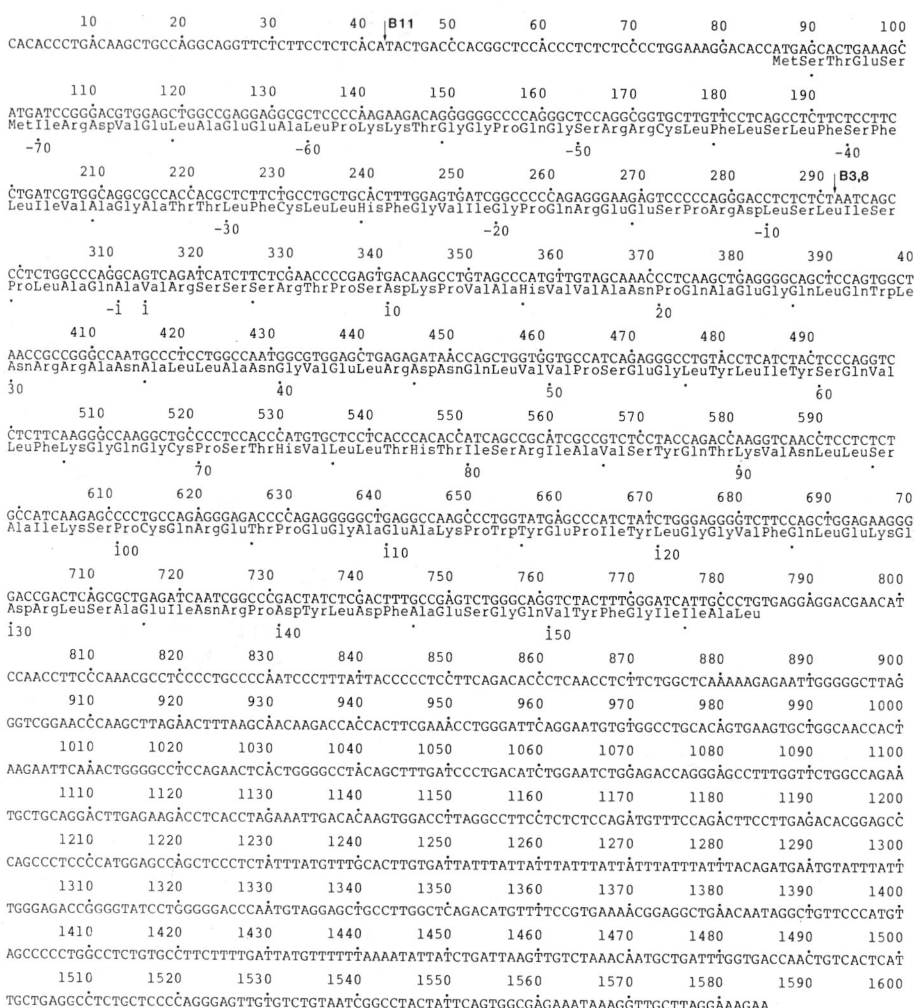

FIG. 1. Nucleotide sequence and predicted protein sequence of human tumor necrosis factor, clone pE4. The N-terminal amino acid (valine) of mature TNF is labeled as amino acid residue 1. The arrows point to the 5′ end of the shorter cDNA clones (B3 and B8) and of another full-length clone, B11. (From Wang *et al.*[6] Copyright 1985 by the AAAS.)

Oligonucleotides used for generating amino acid substitutions usually have 8–10 matched base pairs flanking the mutations and an approximate T_m of 60–68°. Also indicated in Table I are the restriction enzyme recognition sites created by the oligonucleotide-directed mutagenesis. These oligonucleotides were synthesized via phosphoramidite chemistry on an automated DNA Synthesizer manufactured by Biosearch, San Rafael, CA. Synthesis and deprotection of oligonucleotides followed protocols pro-

TABLE I
OLIGONUCLEOTIDES USED IN THE MUTAGENESIS OF THE TNF cDNA

	Oligonucleotides[a]	Length	T_m (°)	Restriction site
TNF1	GAAGATGATCTGAC**CAT**<u>**AAGCTT**</u>TGCCTGGGCC	33	71	*Hin*dIII
TNF2	GGGCTACAGGCTTGTC∧CATAAGCTTTGCCTGGGCC	35	75	—
TNF3	CATGGGCTACAGGCTT∧CATAAGCTTTGCCTGGGCC	35	74	—
TNF4	CATGGGTG**CTC**GGGCT GCCTT	21	68	*Ava*II
TNF5	GAGGGAG<u>ACT**CT**C</u>CCCGAGAAC	22	68	*Dde*I

[a] The bases in bold print indicate mismatches which will generate a substitution mutation. The ∧ indicates the position where deletion mutations will be generated. The underlined bases indicate the newly generated restriction enzyme sites.

vided by Biosearch. The oligonucleotides were purified by polyacrylamide gel electrophoresis followed by reverse-phase high-pressure liquid chromatography and were characterized by base composition analysis following enzymatic digestion to nucleosides.

Preparation of M13 Phage Template

The replicative form (RF) of the M13mp18 DNA was prepared as described by Messing and co-workers.[4] The *Pst*I fragment of pE4,[6] containing the entire coding region of human TNF, was isolated by agarose gel electrophoresis and cloned into the *Pst*I site of M13mp18 RF-DNA. The ligated phage DNA was transfected into competent *E. coli* K12 strain DG98 (*thyl endA1 hsdR17 supE44 lacZ*–M15 *proC*;TN*10*/F′ *lacI*^{q} *lacZ*–M15 *proC*$^+$) (gift from D. Gelfand) and cultured by plating on media containing 5×10^{-4} *M* IPTG and 40 μg/ml X-Gal. Non-α-complementing white plaques were picked, and minicultures were screened for the presence of the correct sized insert (1.1 kb). One recombinant phage with the correct insert was designed as clone 4.1.

Procedure for Oligonucleotide-Directed Mutagenesis

Step 1. Insertion of ATG Initiation Codon for Expression of Mature TNF Protein

The chemically synthesized 33-mer oligonucleotide, TNF1 (Table I), was used to introduce a *Hin*dIII restriction enzyme site and an ATG initiation codon before the GTC codon coding for the first amino acid (valine) of the mature TNF protein (Figs. 1 and 2).

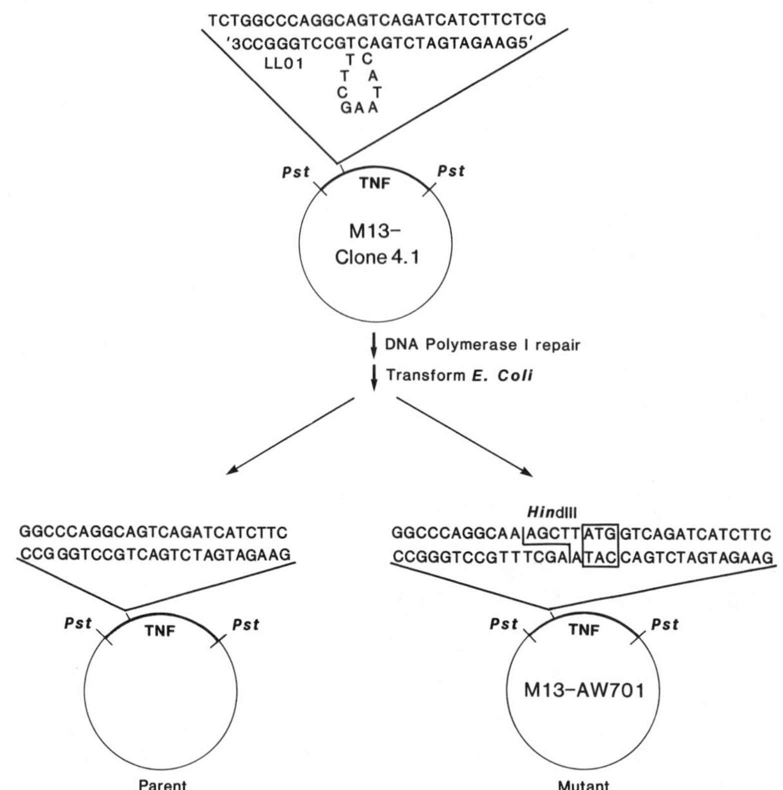

FIG. 2. Scheme for the insertional mutagenesis of TNF clone 4.1. The location of the *Hin*dIII restriction site and the initiation codon are as indicated in the mutant M13-AW701 clone.

Ten picomoles of the oligonucleotide, TNF1, were hybridized to 2.6 μg of ss clone 4.1 DNA in 15 μl of a mixture containing 100 mM NaCl, 20 mM Tris–HCl (pH 7.9), 20 mM $MgCl_2$, and 20 mM 2-mercaptoethanol, by heating at 67° for 5 min and 42° for 30 min. The annealed mixtures were chilled on ice and then adjusted to a final volume of 30 μl of a reaction mixture containing 0.5 mM of each deoxynucleoside triphosphate (dNTP), 17 mM Tris–HCl (pH 7.9), 17 mM $MgCl_2$, 83 mM NaCl, 17 mM 2-mercaptoethanol, 5 units of DNA polymerase I (Klenow fragment), incubated at 0° for 5 min and 37° for 20 min. The reaction was terminated by heating to 70° for 5 min, and the reaction mixture used to transform competent DG98 cells, plated onto agar plates, and incubated overnight to obtain phage plaques.

Plates containing mutagenized clone 4.1 plaques as well as two plates containing unmutagenized clone 4.1 plaques (controls), were chilled to 4°, and the phage DNA from each plate was transferred onto two nitrocellulose filter circles by layering a dry filter on the agar for 5 min for the first filter and 15 min for the second filter. The filters were then placed on thick filter papers soaked in 0.2 *N* NaOH, 1.5 *M* NaCl, and 0.2% Triton X-100 for 5 min and neutralized by layering onto filter papers soaked in 500 m*M* Tris–HCl (pH 7.5) and 1.5 *M* NaCl for another 5 min. The nitrocellulose filters were washed in a similar fashion on filters soaked in 2× SSC, dried, and then baked in a vacuum oven at 80° for 1 hr. The duplicate filters were prehybridized at 60° for 1 hr with 5 ml per filter of DNA hybridization buffer, 5× SSC (pH 7.0), 4× Denhardt's solution [poly(vinylpyrrolidine), Ficoll, and bovine serum albumin; 1× = 0.02% of each], 0.1% SDS, 50 m*M* sodium phosphate buffer (pH 7.0), and 100 μg/ml of denatured salmon sperm DNA. The primer, TNF1, was labeled with [γ-^{32}P]ATP by T4 kinase and used (at 5×10^6 cpm/ml) to hybridize with the filters in 1–5 ml of hybridization buffer per filter, at 64°, for 1 hr. The filters were washed once at room temperature for 10 min in 0.1% SS, 20 m*M* sodium phosphate, and 6× SSC and then at 60° for 20 min in 20 m*M* sodium phosphate and 1× SSC, at which time the control filters were monitored with a Geiger counter to ensure that no radioactivity could be detected. The filters were air dried and autoradiographed at −70° for 4 hr.

Since the oligonucleotide primer, TNF1, was designed to create a new *Hin*dIII restriction site (Fig. 2, Table I), RF-DNA from a number of clones which hybridized with the primer was characterized by digestion with *Hin*dIII. One clone, M13-AW701, which was cleaved by *Hin*dIII, was further characterized by nucleotide sequence analysis using the method described by Sanger *et al.*[17] to confirm the presence of the TNF1 nucleotide sequence.

Step 2. Expression of Mature TNF Protein in E. coli

The coding sequence for the mature TNF protein was subcloned from M13-AW701 into plasmid pAW711 containing the bacteriophage λ P_L promoter and gene *N* ribosome binding site and transformed into *E. coli* strain DG95 ($\lambda N_7 N_{53} cI_{857} sus P_{80}$) as described by Wang *et al.*[6] (Fig. 3).

Step 3. Construction of TNF N-Terminal Deletion Mutants

The (−9)TNF and (−10)TNF mutants were constructed using oligonucleotides TNF2 and TNF3, respectively (Table I). Since both of these oligonucleotides have T_ms very similar to TNF1, we were able to use the

[17] F. Sanger, S. Nicklen, and A. R. Coulson, *Proc. Natl. Acad. Sci. U.S.A.* **74,** 5463 (1977).

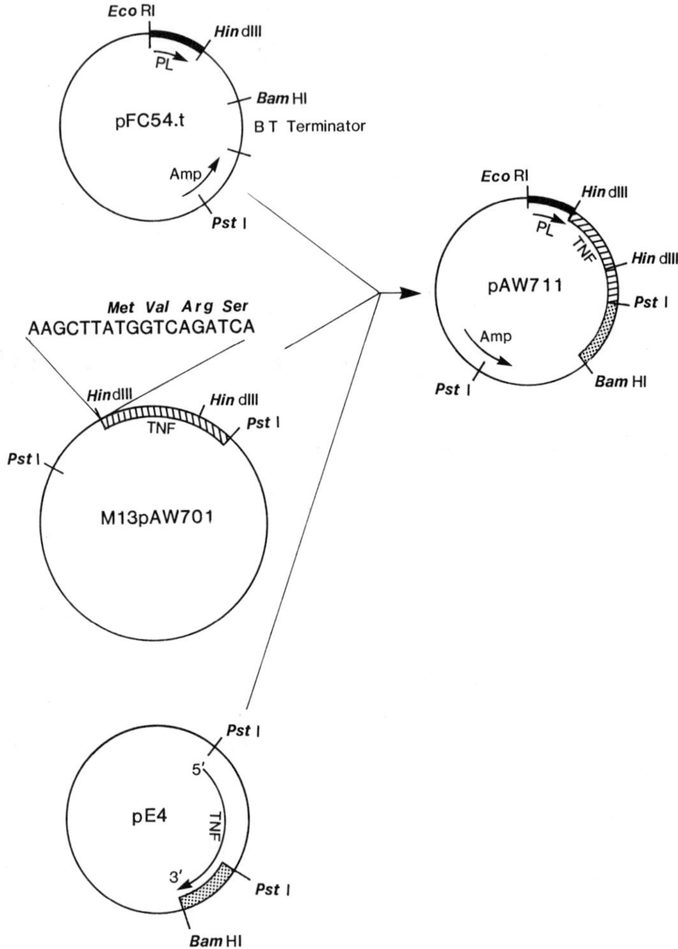

FIG. 3. Expression of TNF in *E. coli*. The clone M13-AW701 was digested with *Pst*I and then digested partially with *Hin*dIII to obtain the *Hin*dIII–*Pst*I TNF coding sequence. The *Pst*I–*Bam*HI fragment containing the 3′ noncoding sequence of the TNF gene was purified from pE4 after digestion of the DNA with *Pst*I and *Bam*HI. The temperature-sensitive Cop$^-$ plasmid, pFC54.t,[6] was digested with *Hin*dIII and *Bam*HI, and the vector fragment was purified on an agarose gel. The isolated fragment was then ligated with the above *Hin*dIII–*Pst*I and *Pst*I–*Bam*HI segments in a three-fragment ligation, and the mixture was used to transform *E. coli* K12 strain DG95 to the ampicillin-resistant phenotype, resulting in the plasmid pAW711. (From Wang *et al.*[6] Copyright 1985 by the AAAS.)

same reaction conditions as described above (Step 1), with the exception that ssDNA from M13-AW701 was used as the parental template. Following the identification of the mutant phage plaques with the respective ^{32}P-labeled probes, the mutant clones were further characterized by restriction enzyme analysis, using the enzymes *Hin*dIII and *Pvu*II. With two enzymes, the parental clone, M13-AW701, shows a characteristic restriction fragment of 146 bp, while the (−9)TNF and (−10)TNF mutants, M13-AW741 and M13-AW742, respectively, have smaller restriction fragments. Both clones were confirmed by nucleotide sequence analysis.[16]

Mutant expression plasmids, pAW741 and pAW742, were constructed by switching the respective *Hin*dIII restriction fragments, containing the entire TNF coding region, with the *Hin*dIII fragment of pAW711 (Fig. 3).

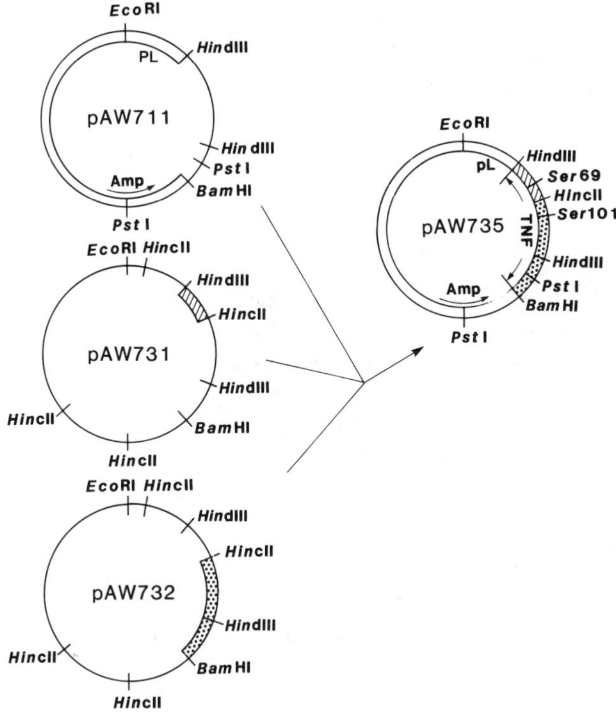

FIG. 4. Construction of the plasmid pAW735 encoding the $Ser_{69}Ser_{101}$TNF mutant. The *Hin*dIII–*Hin*cII fragment containing the Ser_{69} mutation and the *Hin*cII–*Bam*HI fragment containing the Ser_{101} mutation were isolated from plasmids pAW731 and pAW732, respectively. The large *Hin*dIII–*Bam*HI fragment from pAW711, containing the P_L promoter, was ligated with the TNF mutant DNA fragments to form the plasmid pAW735, containing the $Ser_{69}Ser_{101}$TNF mutant coding sequence.

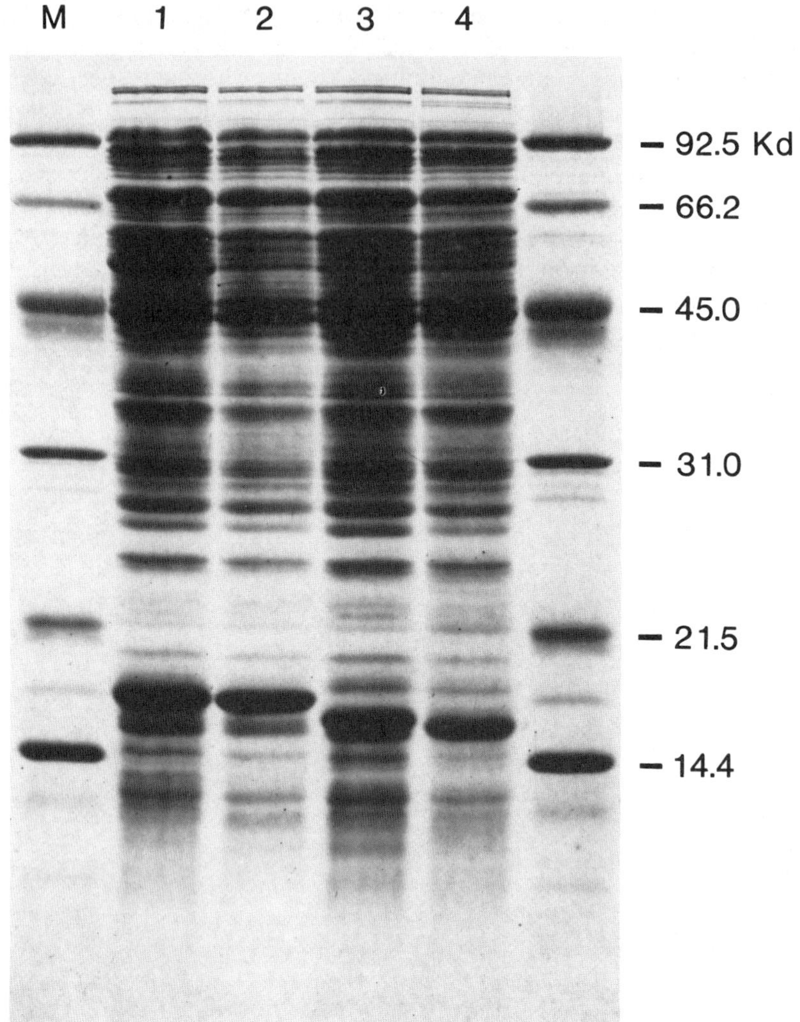

FIG. 5. SDS–polyacrylamide gels of bacterial extracts. Bacterial extracts were prepared as previously described.[6] Lane 1, pAW711 (parental TNF); lane 2, pAW735 ($Ser_{69}Ser_{101}$TNF); lane 3, pAW741 [(−9)TNF]; lane 4, pAW742 [(−10)TNF]. Protein bands were visualized by staining with Coomassie blue.

Step 4. Construction of Serine Substitution Mutants

The oligonucleotides TNF4 and TNF5 were used to construct Ser_{69}TNF and Ser_{101}TNF, respectively (Table I). The template used for TNF4 was ssDNA from M13-AW701 as described above (Step 1), and the same reaction conditions were used with the exception that the hybridization temperature for screening the mutant plaques with ^{32}P-labeled TNF4 was 60°. The template for TNF5 is ss clone 4.5, which contains the non-sense strand of TNF nucleotide sequence. The conditions employed were identical to that of TNF4 described above.

Since both TNF4 and TNF5 were designed to introduce a new restriction enzyme recognition site, *Ava*I and *Dde*I, respectively, the resultant mutant clones were characterized for the presence of the *Ava*I site (M13-AW731) and the *Dde*I site (M13-AW732), followed by nucleotide sequence analysis. The Ser_{69}TNF coding sequence was recloned into an expression plasmid (pAW731) by switching the *Hin*dIII restriction fragment as described above. Similarly, the Ser_{101}TNF was subcloned into expression plasmid, pAW732 (Fig. 4).

The $Ser_{69}Ser_{101}$TNF mutant was constructed from the Ser_{69}TNF mutant, pAW731, and the Ser_{101}TNF mutant, pAW732, by the scheme shown in Fig. 4. Briefly, the *Hin*dIII–*Hin*cII fragment from pAW731, containing the Ser_{69} mutation, and the *Hin*cII–*Bam*HI fragment from pAW732, containing the Ser_{101} mutation, were cloned into the *Hin*dIII and *Bam*HI sites of plasmid pFC54.5,[6] resulting in a plasmid, pAW735, encoding $Ser_{69}Ser_{101}$TNF.

Results

The results are summarized in Table II. Cell-free extracts of *E. coli* containing pAW711 (mature TNF), pAW735 ($Ser_{69}Ser_{101}$TNF), pAW741 [(−9)TNF], and pAW742 [(−10)TNF] expressed similar levels of TNF biological activity between 3×10^5 and 1×10^6 U/ml. Since the level of

TABLE II
TNF BIOLOGICAL ACTIVITY IN BACTERIAL EXTRACTS[a]

Plasmid	Mutation	Activity (U/ml)[b]
pAW711	$Cys_{69}Cys_{101}$	1×10^6
pAW735	$Ser_{69}Ser_{101}$	4×10^5
pAW741	(−9)TNF	1×10^6
pAW742	(−10)TNF	3×10^5

[a] Bacterial extracts were prepared as previously described.[6]

[b] TNF biological activity was determined using the murine L929 cell line as previously described.[6]

expression of these modified TNF proteins was about the same as the mature TNF, as judged by the intensity of the TNF protein band on SDS–polyacrylamide gels (Fig. 5), it is likely that the specific activity of these TNF proteins are also very similar. This has subsequently been confirmed by purification of these proteins to homogeneity (data not shown). Therefore, we can conclude that the cysteines and the 10 N-terminal acids of TNF are not necessary for the maintenance of a biologically active conformation, and they can be modified or deleted without affecting the ability of TNF to kill its target cell.

Comments

The mutagenesis procedure described above provides a rapid method for the modification of cloned genes. It takes advantage of the ability of *E. coli* to efficiently repair gapped circular DNA molecules, to shorten the overall time of the mutagenesis, from initiation of the mutagenesis reaction to the identification of the mutant phage plaques, down to just 2 days instead of 5–6 days as previously described.[2]

The improvements described here did sacrifice efficiency for speed, such that only about 1% of the phage plaques contain the desired mutation. In most instances, this lowered efficiency should not be a concern since only one correct site-specific mutant plaque is required. Besides, the simplification of the plaque hybridization procedure and the introduction of a new restriction site during mutagenesis makes it easier to screen a larger number of plaques for the correct mutant. Furthermore, although not mentioned in the precedure, the use of tetramethylammonium chloride[18] in the washing procedure may further enhance the specificity of the oligonucleotide for the mutant plaque.

[18] W. I. Wood, J. Gitschier, L. A. Lasky, and R. M. Lawn, *Proc. Natl. Acad. Sci. U.S.A.* **82,** 1585 (1985).

[22] An Improved Method to Obtain a Large Number of Mutants in a Defined Region of DNA

By RICHARD PINE and P. C. HUANG

Introduction

In recent years production of mutations *in vitro* has supplemented the ability to learn from naturally occurring mutations about specific genes and gene products. Two widely used methods of mutagenesis include oligonucleotide-directed mutagenesis of specific nucleotides and treatment of target regions with base-specific chemical mutagens.[1-36] These

[1] D. Botstein and D. Shortle, *Science* **229,** 1193 (1985).
[2] P. J. Carter, G. Winter, A. J. Wilkinson, and A. R. Fersht, *Cell* **38,** 835 (1984).
[3] P. Carter, H. Bedouelle, and G. Winter, *Nucleic Acids Res.* **13,** 4431 (1985).
[4] P. Carter, this volume [20].
[5] D. J. Der, T. Finkel, and G. M. Cooper, *Cell* **44,** 167 (1986).
[6] R. D. Everett and P. Chambon, *EMBO J.* **1,** 433 (1982).
[7] O. Fasano, T. Aldrich, F. Tamanoi, E. Taparowsky, M. Furth, and M. Wigler, *Proc. Natl. Acad. Sci. U.S.A.* **81,** 4008 (1984).
[8] W. Kramer and H.-J. Fritz, this volume [18].
[9] S. S. Ghosh, S. C. Bock, S. E. Rokita, and E. T. Kaiser, *Science* **231,** 145 (1986).
[10] S. Gutteridge, I. Sigal, B. Thomas, R. Arentzen, A. Cordova, and G. Lorimer, *EMBO J.* **3,** 2737 (1984).
[11] M. Hannink and D. J. Donoghue, *Proc. Natl. Acad. Sci. U.S.A.* **82,** 7894 (1985).
[12] K. Itakura, J. J. Rossi, and R. B. Wallace, *Annu. Rev. Biochem.* **53,** 323 (1984).
[13] D. K. Jemiolo, C. Zwieb, and A. E. Dahlberg, *Nucleic Acids Res.* **13,** 8631 (1985).
[14] J. T. Kadonaga and J. R. Knowles, *Nucleic Acids Res.* **13,** 1733 (1985).
[15] T. A. Kunkel, J. D. Roberts, and R. A. Zakour, this volume [19].
[16] R. J. Leatherbarrow, A. R. Fersht, and G. Winter, *Proc. Natl. Acad. Sci. U.S.A.* **82,** 7840 (1985).
[17] S.-M. Liang, D. R. Thatcher, C.-M. Liang, and B. Allet, *J. Biol. Chem.* **261,** 334 (1986).
[18] D. F. Mark, A. Wang, and C. Levenson, this volume [21].
[19] R. Pine, M. Cismowski, S. W. Liu, and P. C. Huang, *DNA* **4,** 115 (1985).
[20] R. Rohan and G. Ketner, *J. Biol. Chem.* **258,** 11576 (1983).
[21] M. Schold, A. Colombero, A. A. Reyes, and R. B. Wallace, *DNA* **3,** 469 (1984).
[22] D. Shortle and D. Nathans, *Proc. Natl. Acad. Sci. U.S.A.* **75,** 2170 (1978).
[23] D. Shortle, D. DiMaio, and D. Nathans, *Annu. Rev. Genet.* **15,** 265 (1981).
[24] D. Shortle and D. Botstein, this series, Vol. 100, p. 457.
[25] M. Smith, *Annu. Rev. Genet.* **19,** 423. (1985).
[26] J. W. Taylor, J. Ott, and F. Eckstein, *Nucleic Acids Res.* **13,** 8765 (1985).
[27] P. G. Thomas, A. J. Russell, and A. R. Fersht, *Nature (London)* **318,** 375 (1985).
[28] J. A. Tobian, L. Drinkard, and M. Zasloff, *Cell* **43,** 415 (1985).
[29] P. V. Viitanen, D. R. Menick, H. K. Sarkar, W. R. Trumble, and H. R. Kaback, *Biochemistry* **24,** 7628 (1985).

methods embody two rationales for obtaining mutant genes. In the use of random target-directed methods, the genotype must be correlated with the phenotype after the mutagenesis. In contrast, each mutation is exactly specified in advance by oligonucleotide-directed mutagenesis. The desired mutants should be examined by DNA sequencing to determine or confirm the mutations produced by either method.

With oligonucleotide-directed protocols, specific individual mutations can be obtained by targeting relatively few residues of known interest without reliance on the phenotype. This approach has been useful especially for structural gene modifications which have been used to confirm and elaborate X-ray crystallographic data or specific biochemical evidence on the precise amino acid involvement in a catalytic or folding interaction.[2,5,9–11,16–18,27,29–33] Creating various replacements of a single key residue can reveal how the native amino acid functions in the protein.[2,5,11] Although in theory every base can be altered by oligonucleotide-directed mutagenesis, a significant limit is the fact that the entire procedure must be undertaken to obtain a given mutant. Thus this method is best suited to production of a small number of mutant molecules.

Mutations produced randomly at relatively low frequency (0.5–1.6% of mutable sites) can be screened or selected for specific molecular characteristics or desired biological phenotypes. The fraction of target molecules having mutations will be determined by the number of bases and the proportion that are mutable in the target sequence. Systems in which the DNA of interest can be mapped to a particular gene segment but for which more specific information is not available have been used in applications of base specific mutagenesis.

Several examples of this approach have used sodium bisulfite as the mutagen. This chemical can react specifically with cytosine in the single-stranded regions of DNA, causing deamination of the cytosine to produce uracil. On copying of the mutated single-strand DNA by DNA polymerase, adenosine pairs with the uracil, and finally thymidine replaces the original cytosine. The end result is replacement of a CG base pair with a TA base pair; thus each strand in the target has undergone a transition

[30] J. E. Villafranca, E. E. Howell, D. H. Voer, M. S. Strobel, R. C. Ogden, J. N. Abelson, and J. Kraut, *Science* **222,** 782 (1983).

[31] G. Weinmaster, M. J. Zoller, M. Smith, E. Hinze, and T. Pawson, *Cell* **37,** 559 (1984).

[32] A. J. Wilkinson, A. R. Fersht, D. M. Blow, and G. Winter, *Biochemistry* **22,** 3581 (1983).

[33] G. Winter, A. R. Fersht, A. J. Wilkinson, M. Zoller, and M. Smith, *Nature* (*London*) **299,** 756 (1982).

[34] M. J. Zoller and M. Smith, this series, Vol. 100, p. 468.

[35] M. J. Zoller and M. Smith, *DNA* **3,** 479 (1984).

[36] M. J. Zoller, this volume [17].

mutation. Various experiments have identified alterations in properties of viral growth[22] and gene regulation[6,20] or *ras* protein transforming potential.[7] Structure–function studies of the 3′ minor domain of *E. coli* 16 S rRNA also have been based on *in vitro* mutagenesis with sodium bisulfite.[13] Analogous procedures which produce CT to GA transitions with hydroxylamine[14] or methoxylamine[28] have been used in studies of protein or RNA processing, respectively.

A significant advantage of random mutagenesis is the ability to obtain mutations at various sites and in different combinations with a single treatment. In addition, conditions can be controlled to alter the extent of mutagenesis and thus the distribution of mutations from one experiment to another.

An approach in which large numbers of altered molecules are produced with random mutations in single or multiple occurrence is required for certain targets. Such a method can be designed by adaptation of low frequency mutagenesis with sodium bisulfite in combination with appropriate cloning and screening. In addition to the criteria on which other uses of random mutagenesis are based, these targets should be sequences with many putatively important sites. Mutants containing varied but specific base alterations throughout a gene would provide a sufficient choice of replacements to be compared and studied. This method is well suited to investigation of conserved proteins or the conserved sequences of gene families in general. The degenerate triplet code ensures that certain conserved codons always will be subject to coding changes that are not found naturally, while some may be mutated both in a degenerate and nondegenerate manner. Others though will never be changed by a given base-specific mutagen. High frequency mutagenesis and production of large numbers of mutants allow a great likelihood of obtaining all possible alterations of each conserved site. Such an approach is especially desirable when many sites of interest occur within the target and when the effects of single, double, or multiple mutations may reveal important interactions between the bases of nucleic acids or the amino acids in proteins.

The metallothionein (MT) system is an excellent model for the above approach to mutagenesis *in vitro*. MT is a ubiquitous, highly conserved protein which binds 7 g atoms of transition class IIB (or IB) metals with tetrahedral coordination via mercaptide bonds.[37–46] Typical MTs have no

[37] J. H. R. Kagi and M. Nordberg, *in* "Metallothionein" (J. H. R. Kagi and M. Nordberg, eds.), p. 41. Birkhauser Verlag, Basel, Switzerland, 1979.

[38] Y. Boulanger, I. M. Armitage, K.-A. Miklossy, and D. R. Winge, *J. Biol. Chem.* **257,** 13717 (1982).

[39] Y. Boulanger, C. M. Goodman, C. P. Forte, S. W. Fesik, and I. M. Armitage, *Proc. Natl. Acad. Sci. U.S.A.* **80,** 1501 (1983).

aromatic amino acids or histidine, and 20 of 61 amino acids are cysteines.[46–65] The protein is comprised of two metal binding domains.[38–41,66,67] It is suggested that at least the 20 cysteines[37,39,40,43–45,68] and probably many other conserved amino acids such as lysines[69] act together in the structure and function of MT. Thus this system is a candidate for production of a large number of point mutations which may provide more

[40] R. W. Briggs and I. M. Armitage, *J. Biol. Chem.* **257,** 1259 (1982).

[41] K. B. Nielson and D. R. Winge, *J. Biol. Chem.* **258,** 13063 (1983).

[42] K. B. Nielson, C. L. Atkin, and D. R. Winge, *J. Biol. Chem.* **260,** 5342 (1985).

[43] M. Vasak and J. H. R. Kagi, *Proc. Natl. Acad. Sci. U.S.A.* **78,** 6709 (1981).

[44] M. Vasak and J. H. R. Kagi, *in* "Metal Ions in Biological Systems" (H. Sigel, ed.), Vol. 15, p. 213. Dekker, New York, 1983.

[45] M. Vasak, G. E. Hawkes, J. K. Nicholson, and P. J. Sadler, *Biochemistry* **24,** 740 (1985).

[46] P. D. Whanger, S. M. Oh, and J. T. Deagan, *J. Nutr.* **111,** 1207 (1981).

[47] R. D. Andersen, B. W. Birren, T. Ganz, J. E. Piletz, and H. R. Herschman, *DNA* **2,** 15(1983).

[48] R. D. Andersen, B. W. Birren, S. J. Taplitz, and H. R. Herschman, *Mol. Cell. Biol.* **6,** 302 (1986).

[49] D. M. Durnam, F. Perrin, F. Gannon, and R. D. Palmiter, *Proc. Natl. Acad. Sci. U.S.A.* **77,** 6511 (1980).

[50] N. Glanville, D. M. Durnam, and R. D. Palmiter, *Nature (London)* **292,** 267 (1981).

[51] B. B. Griffith, R. A. Walters, M. D. Enger, C. E. Hildebrand, and J. K. Griffith, *Nucleic Acids Res.* **11,** 901 (1983).

[52] I.-Y. Huang, A. Yoshida, H. Tsunoo, and H. Nakajima, *J. Biol. Chem.* **252,** 8217 (1977).

[53] I.-Y. Huang, M. Kimura, A. Hata, H. Tsunoo, and A. Yoshida, *J. Biochem.* **89,** 1839 (1981).

[54] M. Karin and R. I. Richards, *Nature (London)* **299,** 797 (1982).

[55] M. M. Kissling and J. H. R. Kagi, *in* "Metallothionein" (J. H. R. Kagi and M. Nordberg, eds.), p. 145. Birkhauser Verlag, Basel, Switzerland, 1979.

[56] S. Koizumi, N. Otaki, and M. Kimura, *J. Biol. Chem.* **260,** 3672 (1985).

[57] Y. Kojima, C. Berger, B. L. Vallee, and J. H. R. Kagi, *Proc. Natl. Acad. Sci. U.S.A.* **73,** 3413 (1976).

[58] Y. Kojima, C. Berger, and J. H. R. Kagi, *in* "Metallothionein" (J. H. R. Kagi and M. Nordberg, eds.), p. 153. Birkhauser Verlag, Basel, Switzerland, 1979.

[59] K. Munger, U. A. Germann, M. Beltramini, D. Niedermann, G. Baitella-Eberle, J. H. R. Kagi, and K. Lerch, *J. Biol. Chem.* **260,** 10032 (1985).

[60] R. I. Richards, A. Heguy, and M. Karin, *Cell* **37,** 263 (1984).

[61] C. J. Schmidt and D. H. Hamer, *Gene* **24,** 137 (1983).

[62] C. J. Schmidt, M. F. Jubier, and D. H. Hamer, *J. Biol. Chem.* **260,** 7731 (1985).

[63] P. F. Searle, B. L. Davison, G. W. Stuart, T. M. Wilkie, G. Norstedt, and R. D. Palmiter, *Mol. Cell. Biol.* **4,** 1221 (1984).

[64] U. Varshney, N. Jahroudi, R. Foster, and L. Gedamu, *Mol. Cell. Biol.* **6,** 26 (1986).

[65] D. R. Winge, K. B. Nielson, R. D. Zeikus, and W. R. Gray, *J. Biol. Chem.* **259,** 11419 (1984).

[66] K. B. Nielson and D. R. Winge, *J. Biol. Chem.* **260,** 8698 (1985).

[67] D. R. Winge and K.-A. Miklossy, *J. Biol. Chem.* **257,** 3471 (1982).

[68] J. D. Otvos and I. M. Armitage, *Proc. Natl. Acad. Sci. U.S.A.* **77,** 7094 (1980).

[69] J. Pande, M. Vasak, and J. H. R. Kagi, *Biochemistry* **24,** 6717 (1985).

detailed information concerning the secondary structural folds and tertiary structural domains of MT, its mechanism of metal binding, as well as the regulation of MT genes and the physiological function of MT proteins.

Figure 1 shows the alterations of the Chinese hamster ovary cell (CHO) MT II coding sequence that can be induced by sodium bisulfite mutagenesis of either its sense or antisense strand. Naturally occurring differences in MT sequences which have evolved in human, monkey, horse, cow, mouse and rat, as well as CHO MT I, are also included for comparison. This representation of natural differences is culled from published reports of protein[46,52,53,55–59,65] and/or nucleic acid sequences.[47–51,54,60–64] As can be seen, the sequence of the target for these studies, CHO MT II cDNA, ensures that mutagenesis with sodium bisulfite can be used on its sense or antisense strand to allow significant alteration of conserved codons. At the same time, mutagenesis of the complementary strand subjects a given codon to only degenerate mutations. Also, the codons of many nonconserved amino acids may be mutated to encode residues which have not been detected at the respective position

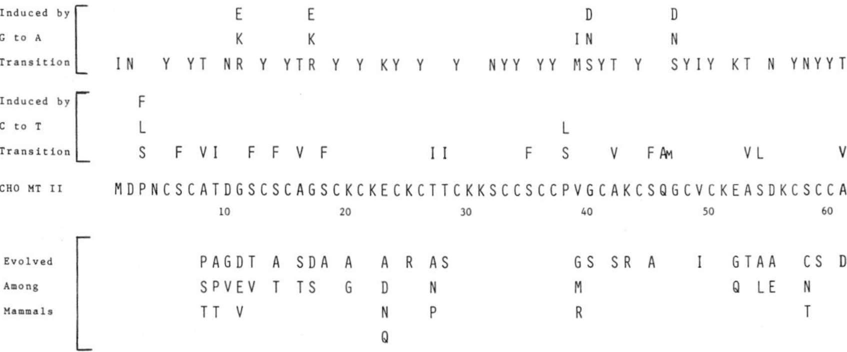

FIG. 1. Comparison of CHO MT II amino acid sequence to mutations that can be induced by sodium bisulfite and to differences evolved among other mammalian MTs. The single letter abbreviations for the amino acids are used; an amber nonsense codon is indicated by Am. Amino acid changes encoded due to induced G to A or C to T transition mutations are shown above the amino acid sequence of CHO MT II. Amino acid differences that have evolved among other mammalian MTs, as determined from protein analysis or deduced from DNA sequences, are shown below the respective positions in the CHO MT II sequence. It should be noted that the occurrence of glutamine at position 52 is reported only for mouse MT II,[53] while glutamate is deduced for that position from DNA sequence data.[63] Also, cysteine at position 58 and serine at position 59 are only reported in the amino acid sequences of MT I and MT II from horse[58] and human.[55] However, analysis of four apparently functional human MT genes has shown serine and cysteine to be encoded at positions 58 and 59, respectively.[54,60,62,64] At this time, human genes known to be functional and which encode valine at positions 10 or 12, threonine at position 53, or aspartate at position 55 have not been cloned.

in any naturally occurring MT. Only an asparagine and the lysine codons cannot be mutated with sodium bisulfite, while the conserved glutamine codon at position 46 can be mutated only to an amber nonsense codon. Thus, determination of the importance and role of conserved residues as well as the leeway for changes at nonconserved positions in MT is made accessible if large numbers of mutants are produced.

Principles

Conditions for mutagenesis of nucleic acids with sodium bisulfite have been developed based on the well-characterized reactions of this reagent with the bases of nucleosides, nucleotides, and nucleic acids.[70] Sulfonation at the 6 position of cytosine, uracil, and thymine occurs readily at 37° between pH 3 and 8. The reaction with uracil and thymine is fully reversible, while the adduct of cytosine can undergo hydrolytic deamination to produce 5,6-dihydrouracil-6-sulfonate. The 5,6-dihydrouracil-6-sulfonate adduct yields uracil and bisulfite irreversibly at pH 9. Thus bisulfite can be used to specifically convert cytosine to uracil. Although the rate for deamination of cytosine is maximal at pH 5, protection of acid-labile components of nucleic acids is effected by the buffering capacity of bisulfite when the reaction is carried out at pH 6–7. Oxidation of bisulfite can produce free radicals which react in various ways with DNA. However, this process can be prevented by using a relatively high concentration of bisulfite (≥ 1 *M*) and a free radical scavenger such as hydroquinone. In addition, these relatively high concentrations are necessary for treatment of nucleic acids since the susceptible residues in polynucleotides react manyfold more slowly than the corresponding constituents free in solution. Importantly, the reaction of bisulfite with bases that are hydrogen bonded in Watson–Crick pairs is negligible. Thus for all practical purposes, only single-stranded portions of nucleic acids are reactive.

These considerations underlie the protocol for reaction with sodium bisulfite to mutagenize DNA.[22,24] The initial reaction is carried out at pH 6–7 and 37° using 1–3 *M* bisulfite. During the incubation 5,6-dihydrocytosine-6-sulfonate adduct formation and deamination occur within the single-strand portion of a target molecule. Dialysis at pH 6.8 and 0° stops the reaction as the bisulfite is removed. Further dialysis at pH 9 desulfonates any remaining adducts and thus completes the conversion of cytosine to uracil.

In the method described here, M13 phage have been used as vectors

[70] H. Hayatsu, *Prog. Nucleic Acid Res. Mol. Biol.* **16,** 75 (1976).

for the target DNA. Several advantages make this an attractive choice. Phage particles are very stable at −20 or 4°. Single-stranded genomic DNA can be easily isolated from these particles. To obtain material for recombinant DNA manipulations, closed circular replicative form (RF) DNA can be extracted from bacterial cells infected by the phage. A library can be propagated initially and individual elements recovered from the phage first secreted after chimeric M13 DNA is transformed into *Escherichia coli.* However, M13 is not ideal for the amplification of a library of cloned DNA since deletions are more likely than in some other vectors. Strains of M13 have been constructed to allow rapid cloning and sequencing of DNA fragments.[70] The design of these strains includes a multiple cloning site which provides numerous restriction sites to facilitate insertion of DNA into the vector. Additionally, the cloned DNA can later be subcloned via fragments isolated with enzymes other than those used in the first place. Pairs of M13 strains have a given multiple cloning site inserted in opposite orientations. A target sequence can be cloned into each of the pair of strains so that the recombinant fragment will have the same flanking sequence in the double-stranded chimeric DNA derived from either strain. Each strand of the cloned DNA will be part of the single-stranded phage DNA in one strain or the other.

Single-stranded recombinant phage DNA is a ready target for reaction with sodium bisulfite, but the phage genes must be protected from mutagenesis since they are all essential to the phage life cycle. Double-stranded vector DNA, linearized by restriction digestion at the same site or sites used for construction of the chimeric DNA, is annealed to the respective recombinant single-stranded phage DNA to form a gapped duplex molecule. Thus, each strand of the cloned sequence remains available as a target while the viability of the M13 vector is protected.

The final step needed to obtain fixed mutations with this method is propagation of the treated DNA. Since these molecules contain deoxyuridine in a single-strand region they are susceptible to attack by uracil-DNA glycosylase and degradation by further nucleolytic action in wild-type *E. coli* cells. However, cells which are *ung*$^-$ lack uracil-DNA glycosylase and will not carry out the first step of this process. Thus gapped duplex DNA containing uracil in the single-stranded region can be directly transformed into such *E. coli* strains. Adenine will pair with uracil and then thymine will pair with adenine; thus a cytosine to thymine transition is fixed during replication of the phage. Phage initially secreted from these cells constitute a mutant library which can be stored safely. Individual elements are obtained by plating the phage with a normal host strain to give single plaques. They can be exactly and efficiently characterized by extracting and sequencing the DNA of phage amplified from the plaques. Subsequently, the RF DNA of chosen mutants can be prepared to provide material for whatever manipulations are needed for further studies. Thus the combination of sodium bisulfite mutagenesis, M13 vectors, and an *ung*$^-$ strain of *E. coli* provides a powerful yet simple and direct way to produce and handle large numbers of mutants having the desired distribution of random mutations.

Materials

Bacteria, Phage, and Plasmids

The following bacterial strains were used in this study: *Escherichia coli* BD1528,[71] which lacks uracil-DNA glycosylase, as a recipient for DNA mutagenized by sodium bisulfite and *E. coli* JM101[70] as a host for M13 phage. Phage M13mp8 and M13mp9[71] were used as vectors for DNA to be mutagenized and for DNA sequencing. *Escherichia coli* cultures carrying chimeric plasmids containing CHO MT cDNA clones were obtained from the Genetics group, Los Alamos National Laboratory, courtesy of Drs. Barbara and Jack Griffith.[51] Subcloning of MT sequences into M13mp9 and M13mp8, which produced M13MT2S and M13MT2A, respectively, has been described.[19] These contain the sense and antisense strands of the MT coding sequence in the single-strand genomic phage DNA.

Reagents

Chemicals. Acrylamide and *N,N'*-methylenebisacrylamide were purchased from Bio-Rad. Deoxynucleotide triphosphates and dideoxynucleotide triphosphates were from P-L Biochemicals. Pentadecamer sequencing primer was from New England Biolabs. Other specialty reagents were obtained from Bethesda Research Laboratories. [α-^{32}P]dATP (>3000 Ci/mmol) was purchased from Amersham. Hydroquinone (99+ %), from Aldrich, was stored under N_2. All other chemicals were reagent grade from various standard suppliers.

Enzymes. T4 DNA ligase was purchased from either Boehringher Mannheim Biochemicals or Bethesda Research Laboratories. All other enzymes were purchased from Bethesda Research Laboratories.

[71] J. Messing, this series, Vol. 101, p. 20.

Methods and Results

Microbiology

Escherichia coli BD1528 was maintained on NZYDT plates[72] or grown in NZYDT broth at 37°. M13 phage, recombinants therein, and the host, *E. coli* JM101, were kept and grown as described,[71] using 2× YT broth or B media for agar plates and top agar. Transformation of DNA into bacterial cells was by a calcium chloride procedure essentially as described,[73] except that cells were 10-fold concentrated from the log-phase culture.

Preparation of DNA

Phage from a single plaque were first amplified. An overnight culture of JM101 was diluted 1 : 100 into 1.5- and 10-ml aliquots of 2× YT broth. The small aliquot was inoculated with phage from a single plaque and both aliquots were grown at 37° with vigorous aeration. After 2.5 hr the small culture was added to the large culture and growth was continued for 5 hr. The culture was centrifuged at 13,000 *g* for 10 min, and the supernatant, containing phage at 10^{12} pfu/ml, was decanted and stored at −20°. M13 RF DNA was obtained as follows. Mid-log cultures of JM101 (400 ml at 2×10^8 cells/ml) were infected with 4 ml of phage at 10^{12} pfu/ml and incubation was continued for 4 hr at 37° with aeration. Chloramphenicol at 25 μg/ml was added for the last 30 min, then DNA was extracted according to Birnboim,[74] omitting RNase treatment and adding CsCl banding[72] for 16 hr at 190,000 *g* (45,000 rpm) in a Beckman VTi65 rotor. M13 genomic DNA was prepared as described.[74]

Sodium Bisulfite Mutagenesis of MT Coding Sequences

A gapped heteroduplex was formed by annealing the chimeric M13 plus strand with the denatured double-strand M13 vector DNA, which had been linearized by restriction digestion at the multilinker site originally used for subcloning the MT sequence.[19] The appropriate vector DNA was mixed with recombinant DNA, M13MT2S (sense strand insert in M13qmp9) or M13MT2A (antisense strand insert in M13mp8), at 2.5 : 1.0 (w/w) in 3 m*M* sodium citrate, 30 m*M* sodium chloride (pH 7.0),

[72] T. Maniatis, E. F. Fritsch, and J. Sambrook, "Molecular Cloning: A Laboratory Manual." Cold Spring Harbor Lab., Cold Spring Harbor, New York, 1982.

[73] R. W. Davis, D. Botstein, and J. R. Roth, "Advanced Bacterial Genetics," p. 140. Cold Spring Harbor Lab., Cold Spring Harbor, New York, 1980.

[74] H. L. Birnboim, this series, Vol. 100, p. 243.

10 mM magnesium chloride to give a final total DNA concentration of 15 μg/ml. The DNA was heated to 90° and held for 2.5 min, cooled quickly to 80° and held for 3 min, cooled quickly to 70° and held for 10 min, cooled quickly to 65° and held for 50 min, and then slowly cooled to 25°. The extent of annealing can be assessed by electrophoresis of a small aliquot on a 1% agarose gel in Tris–acetate buffer,[72] which resolves linear duplex, gapped duplex circular, and single-stranded circular molecules. However, samples should not be heated before being loaded, and the gel should be run slowly, at or below 2 V/cm, until the bromophenol blue marker dye is at the end of the gel. Gapped duplex circular molecules have the same mobility as fully duplex relaxed circular molecules. Conversion of single-stranded circular DNA to gapped circular duplex DNA should be observed. This assay has been used to determine that the annealed DNA can be stored for several days at 4°.[75] For routine use, the annealing reaction is so reliable that this analysis is generally not necessary.

Fresh stocks of 50 mM hydroquinone and 4 M sodium bisulfite (pH 6.0) (1.56 g $NaHSO_3$, 0.64 g Na_2SO_3, 4.3 ml H_2O) were used to produce uracil in the single-strand MT portion of the heteroduplex essentially as described.[22,24] The annealed DNA was mixed with bisulfite stock and water in 1.5-ml polypropylene tubes to yield varying final concentrations of mutagen. Addition of 0.04 volumes of hydroquinone completed the reaction mixture, which was then overlaid with paraffin oil and incubated at 37° in the dark. Reactions with 1 M sodium bisulfite were in a final volume of 190 μl or 380 μl with a final DNA concentration of 7.5 μg/ml while reactions with 3 M sodium bisulfite were in 760 μl with a final DNA concentration of 3.75 μg/ml. Thus reactions in 190 μl contained 0.2 μg of chimeric DNA and 0.5 μg of vector DNA while the others contained 0.4 μg of chimeric DNA and 1.0 μg of vector DNA. The final concentration of bisulfite and the length of reaction prior to dialysis were varied to meet the goal of producing a large number of singly or multiply mutated molecules.

Following the incubation samples were recovered from below the paraffin oil with a pasteur pipet and transferred to $\frac{1}{4}$-inch dialysis tubing. Dialysis was performed at 0° versus 5 mM potassium phosphate (pH 6.8), 0.5 mM hydroquinone, twice for 2 hr each time, then once again at 0° for 2 hr versus 5 mM potassium phosphate (pH 6.8). Each change was greater than about 200 volumes relative to the sum of the reaction mixture volumes. Further dialysis against 200 mM Tris (pH 9.2), 50 mM NaCl, 2 mM EDTA was performed for 16–20 hr at room temperature, followed by dialysis against 10 mM Tris, 1 mM EDTA (pH 8.0) for 4 hr at 4°. These dialyses used at least 400 volumes relative to the sum of the reaction

[75] R. Pine, unpublished observations (1985).

mixture volumes. Finally mutagenized DNA was recovered and propagated. The use of dialysis to complete the conversion of cytosine to uracil is especially appropriate for handling many samples at one time, as may be required to produce a range of mutation frequencies for a given target sequence.

The target in this example, CHO MT II cDNA, consists of 62 codons, inclusive of termination,[51] as expected of that protein in mammalian species. It has 55 G and 55 C residues in its coding sequence. The choice of M13 as a vector allows use of sodium bisulfite to obtain exclusively G to A transitions in the codons (GA clones) as the indirect result, via base pairing, of C to T transitions in the original antisense single-stranded DNA. Codons of the sense strand directly reflect C to T mutations (CT clones) made in that single-stranded target. As described for the general case in the section Principles, each strand of the MT coding sequence was obtained with identical vector flanking sequence by cloning the sense and antisense strands of the CHO MT II sequence into the complementary M13 strains mp9 and mp8, respectively. These clones, M13MT2S and M13MT2A, served as a target for *in vitro* mutagenesis with sodium bisulfite. However, different proportions of degenerate and nondegenerate mutations will result from C to T transitions compared to G to A transitions in the coding sequence. Thus it was necessary to use different reaction times or concentrations of sodium bisulfite for the mutagenesis of the sense and antisense single strands to achieve comparable numbers of coding changes due to C to T or G to A mutations in the codons. The conditions were chosen based on earlier reports of mutagenesis with sodium bisulfite.[7,20,22,24] As the data will show, this empirical approach produced the desired results, which now provide an additional guide for other investigators.

Propagation of Mutated DNA and Analysis of the Mutant Library

The library of mutant MT sequences is composed of collections obtained from reactions with 1 *M* sodium bisulfite carried out for 3, 5, or 7 hr, and from incubations of 20 or 40 min with 3 *M* sodium bisulfite. *Escherichia coli* BD1528 (*ung*$^-$)[76] was transformed with 10-μl aliquots of the mutagenized DNA. Cells were incubated with 1 ml 2× YT broth for 2.5 hr at 37°. DNA replication produced C to T transitions, thus fixing the C to U changes generated *in vitro*. At this time, the beginning of phage secretion, cells were removed by centrifugation. To allow sequencing of individual mutants, single plaques at a density of 50–150 per plate were

[76] B. K. Duncan and J. A. Chambers, *Gene* **28,** 211 (1984).

obtained by infecting JM101 with 0.1-ml portions of the secreted phage. The remainder of the phage in the supernatants, constituting the mutant library, was stored at −20°.

Dideoxy sequencing of single-stranded recombinant phage DNA was carried out as described[77] except that annealing of primer to template was done in 500-μl polypropylene tubes, and sequencing reactions were done at 30° in 400-μl polypropylene tubes. Reaction products were analyzed on 50 cm long by 0.4 mm thick 6% (w/v) polyacrylamide gels (40 : 1 acrylamide : *N*,*N*′-methylenebisacrylamide) containing 7 *M* urea, 100 m*M* Tris–borate (pH 8.3), 2 m*M* EDTA. Aliquots of samples cloned in M13mp8 were electrophoresed until the xylene cyanol marker dye had migrated 31 cm, while two aliquots of samples cloned in M13mp9 were electrophoresed so that either the bromophenol blue or the xylene cyanol FF had migrated 47 cm. The gels were dried and autoradiographed for 16–24 hr at room temperature.

For the library as a whole,[19] mutations were obtained at every G residue. Thus, all the possible coding changes caused by G to A transitions indicated in Fig. 1 were produced, either alone or in various combinations. Only 24 of 34 possible degenerate C to T transition mutations were found, but 19 of 21 possible nondegenerate C to T transitions encoding all the amino acid changes shown in Fig. 1 except Pro_{38} to Ser and Ala_{42} to Val were obtained. Overall, only 56 of the 183 GA clones examined were found to have no alterations, and an additional 14 clones had only degenerate mutations. Of the 81 CT mutants sequenced only one had no changes, while 23 more had only degenerate mutations. Extrapolating from the data collected, 45% of the GA elements and 34% of the CT elements are unique within their collections. Since only 264 plaques of a possible 10^5 have been examined to date, the phage stocks are expected to contain still more unique mutated sequences.

The sequence analysis of individual mutated DNA molecules showed that the collections within the library had different desired characteristics. Treatment of the antisense strand with 3 *M* sodium bisulfite for 20 and 40 min or with 1 *M* sodium bisulfite for 3 or 5 hr produced G to A mutations in the codons (GA clones). The sense strand was treated for 5 or 7 hr with 1 *M* sodium bisulfite to produce C to T changes in the codons (CT clones).

Another view of the data obtained from sequence analysis of this library is shown in Table I as the percent of cytosine residues which reacted with bisulfite for the GA clones. The results validate the empirical

[77] F. Sanger, A. R. Coulson, B. G. Barrell, A. J. H. Smith, and B. A. Roe, *J. Mol. Biol.* **143**, 161 (1980).

TABLE I
EXTENT OF MUTAGENESIS FROM VARIOUS REACTION CONDITIONS[a]

GA collection	Reaction time (min)	Sodium bisulfite concentration (*M*)	Number of clones sequenced	Number of target sites	Target sites mutagenized (%)
A	20	3	42	2,310	5.3
B	20	3	37	2,035	10.5
C	180	1	56	3,080	1.6
D	300	1	48	2,640	2.2

[a] GA collections were obtained by treatment of the CHO MT II cDNA antisense strand, which contains 55 dC residues. The number of target sites is calculated from the number of sequenced clones multiplied by 55.

choice of conditions based on earlier reports, and can serve as a guide for future experiments. These clones and other collections in the library essentially show a Poisson distribution of mutation frequency which varies as expected in accord with the average percent of mutations. These distributions reveal important characteristics when various collections are compared. Figure 2A shows that treatment with 1 *M* sodium bisulfite for 5 hr allowed more total changes per template for the sense strand than was seen from 3 hr of treatment with 1 *M* sodium bisulfite for the antisense strand, as expected. As intended, the extent of coding changes caused by these treatments was comparable for both collections. Among the mutated sequences, the modal number of coding changes from these reactions is one per template (Fig. 2B). Obtaining additional and more frequent changes desired in the G to A collection required a greater degree of reaction with sodium bisulfite. The occurrence of more multiple mutations and codon changes from 20 and 40 min reactions with 3 *M* sodium bisulfite is shown in Figs. 3A and 3B, respectively. The former collection again has a modal number of one coding change per template, but includes elements with up to four mutations. The latter collection has a modal number of four to five coding changes per template, and contains elements having up to ten coding changes per template.

Mutants with single coding changes are the most obviously useful. However, sets of mutants with one common and one or more other coding changes may help reveal interactions between amino acids. Since many proteins, including MT, have more than one structural domain, multiple mutations can be very useful if they fall in different domains. Given the appropriate distribution of mutations, highly mutated molecules may eliminate the function of one domain while only slightly altering that of

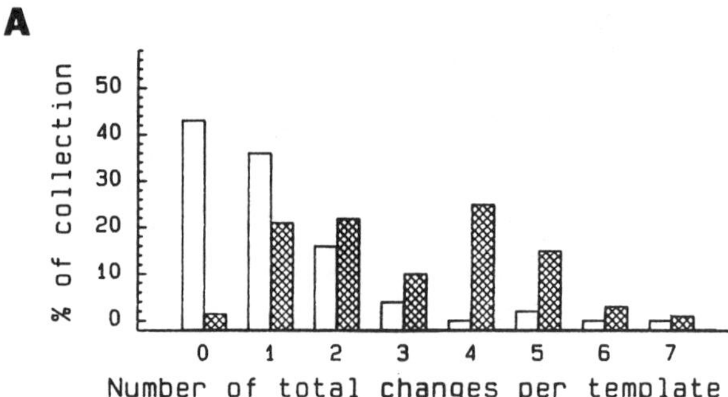

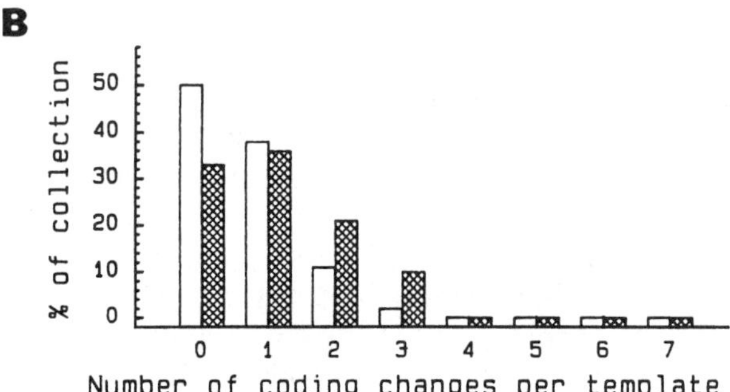

FIG. 2. Transition mutations per template for GA and CT collections in the MT coding sequence point mutant library. The data are shown as histograms (GA clones, open bars; CT clones, crosshatched bars). (A) Total changes in GA clones (n = 56) produced by treatment of CHO MT II cDNA antisense strand for 3 hr with 1 M sodium bisulfite; and in CT clones (n = 67) produced by treatment of CHO MT II cDNA sense strand for 5 hr with 1 M sodium bisulfite. (B) Coding changes in GA clones analyzed in (A) and in CT clones analyzed in (A).

another. Examples of such DNA sequences have been found in the MT point mutant library.[19]

The differences between the induced and evolved amino acid changes in MT shown by Fig. 1 presumably reflect evolutionary constraints on MT. It is perhaps coincidental, but of the 31 codons that could mutate by a single base change to encode an aromatic amino acid, only 10 are found in CHO MT II cDNA. Five of those encode cysteine or serine. In fact, a nonfunctional human MT I pseudogene encodes Tyr in place of Cys_5 and Phe in place of Ser_{35}, as well as a nonsense codon at position 40.[62]

Although this library is quite extensive, it is also limited. To resolve this paradox, other techniques for point mutagenesis can be employed.[1,4,8,15,18,23,25,36] Alternative libraries can be made with other target-directed protocols, as described in this volume and elsewhere.[1,14,23,25]

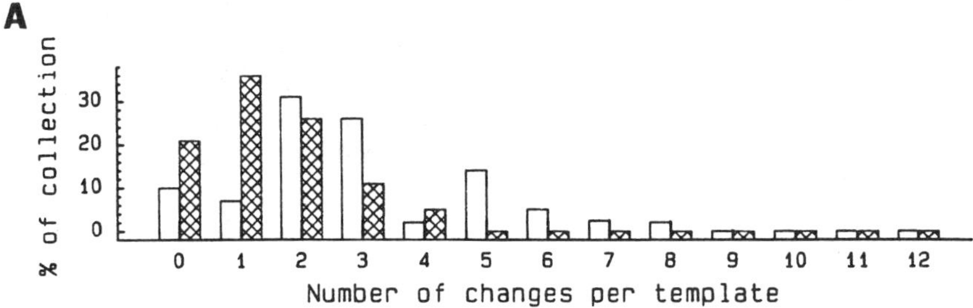

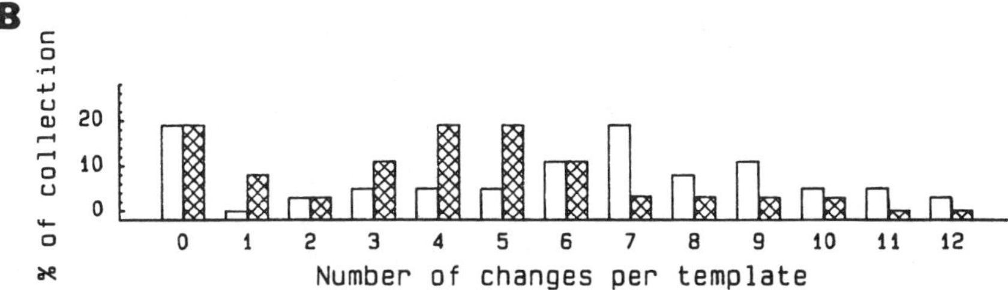

FIG. 3. High frequency collections of GA transition mutants in the CHO MT II coding sequence point mutant library. The data are shown as histograms (total changes, open bars; coding changes, crosshatched bars). (A) Total and coding changes in GA clones ($n = 42$) produced by treatment of the CHO MT II cDNA antisense strand for 20 min with 3 *M* sodium bisulfite. (B) Total and coding changes in GA clones ($n = 37$) produced by treatment of the CHO MT II cDNA antisense strand for 40 min with 3 *M* sodium bisulfite.

However, specific site-directed mutations are likely to be required in the future. While the CT mutants have the advantage of not altering any cysteine codon, and the GA clones include mutations of each cysteine codon (Fig. 1), oligonucleotides can be designed to allow introduction of particular, desired mutations which are rare or unavailable in the bisulfite-generated library.

It is necessary to actually obtain mutant proteins to reap the benefits of available mutated structural gene sequences. Some of these sequences[78] as well as MT deletion mutants[79] have been expressed as hybrid proteins in *E. coli*. This has allowed the use of a common scheme to obtain substantial purification of those proteins. Their structure and metal binding properties are under investigation.

The MT coding sequence provides a broadly applicable model of how the reaction with sodium bisulfite can be controlled and analyzed. MT also is an example of the type of system that initially requires a large

[78] M. Cismowski, R. Pine, S. W. Liu, and P. C. Huang, manuscript in preparation (1986).
[79] R. Pine and P. C. Huang, manuscript in preparation (1986).

number of mutants and varying frequency of mutations to help resolve the role of the many structurally and functionally important amino acids. In this study, several aspects of the available technology were chosen and combined to constitute an improved method to obtain a large number of mutant sequences in a defined region of DNA.

Acknowledgment

Work was supported in part by the National Institutes of Health, Grant R01 GM 32606. R.P. is an NIA postdoctoral trainee.

[23] Molecular Mechanics and Dynamics in Protein Design

By PETER KOLLMAN and W. F. VAN GUNSTEREN

Introduction

Given the capabilities of genetic engineering methods, which often allow one to make specific mutants of a given protein, one asks: Can one use theoretical methods to *predict* the physicochemical consequence of this site-specific mutation? It is the thesis of this chapter that recently developed free energy perturbation methods could allow successful predictions to be made in a large number of interesting cases.

What are these physicochemical consequences one would like to predict? First, when one changes a single amino acid residue of the protein, one might like to know how this alters the three-dimensional structure of the protein, not only of the residue changed but also surrounding residues. Second, one might like to know whether the altered protein is more or less stable to heat denaturation. Third, if the protein is a catalyst, how are its ligand binding strengths, catalytic efficiencies, and ligand selectivities altered when the site-specific mutation is carried out? There may be other physicochemical consequences of mutation, e.g., other measurable structural or dynamical properties, and one would like to have at one's disposal theoretical methods that can predict them.

Are such theoretical methods available? In principle, yes. But these theoretical methods work best in the "small perturbation" limit, so that, if one has accurately characterized the properties (structure and stability) of a protein, then one has some hope that the methods can predict something useful about a protein which differs in only a single amino acid from the reference protein.

The methods we describe are *not* aimed at describing protein folding to a native state from an arbitrary unfolded structure. To attempt to "fold a protein" from first principles is beyond the range of current methodologies, as we shall discuss more fully below.

Application of the methods requires that one knows, to a reasonable accuracy, the X-ray crystal structure, or perhaps a well-determined 2-D NMR structure, for the native protein as a starting structure to predict the consequences of changing the chemical nature of one or a few amino acids in the protein. In fact X-ray crystallographic studies of both the native structure and mutants of T4 lyzozyme by Matthews[1] and subtilisin suggest that the structural perturbations due to single amino acid substitutions are indeed localized and should be "predictable." Nonetheless, one should be aware that perhaps the crystallographic studies have "selected out" those mutants which involve small perturbations. Thus other single site mutations could cause large-scale perturbations and/or protein denaturation, and this consequence would be very difficult for current theoretical methods to predict. Below we describe some of the basic principles behind molecular mechanical and molecular dynamical methods, some applications of these methods to proteins, and protein design.

Molecular Mechanics

What is molecular mechanics? The term is used here to refer to simulations on complex systems in which the potential energy of the system is represented by a simple analytical equation such as Eq. (1). This is dis-

$$V = \sum_{\text{bonds}} k_b(b - b_0)^2 + \sum_{\text{angles}} k_\theta(\theta - \theta_0)^2 + \sum_{\substack{\text{dihedrals,}\\ \text{periodicities,}\\ \text{improper torsions}}} \frac{V_n}{2}[1 + \cos(n\theta + \delta)] + \sum_{\substack{\text{nonbonded}\\ i<j}} \frac{A_{ij}}{R_{ij}^{12}} - \frac{B_{ij}}{R_{ij}^{6}} + \frac{q_i q_j}{\varepsilon R_{ij}} + \sum_{\text{H bonds}} \frac{C_{ij}}{R_{ij}^{12}} - \frac{D_{ij}}{R_{ij}^{10}} \quad (1)$$

tinct from quantum mechanics, in which one solves for the electronic structure of the system within the Born–Oppenheimer approximation and, in this fashion, determines a potential surface for the different positions of the nuclei. Molecular mechanics has the advantage that Eq. (1) and its analytical first and second derivatives can be evaluated very rapidly for complex, aperiodic systems with thousands of atoms, something that is impossible with the more time-consuming quantum mechanical approaches. The disadvantage of molecular mechanics is that its application is limited to systems where the nature of the chemical *bonding* is well

[1] T. Albers and B. W. Matthews, this volume [27].

defined and can be specified at the beginning of a given simulation and does not change during the simulation. On the other hand, quantum mechanical calculations on the active site residues of a protein can be combined with molecular mechanics on the remaining part of the protein to study enzyme catalysis with a combined quantum/molecular mechanical approach.[2]

Given that the energy of the system is represented by Eq. (1) and its first and second derivatives are evaluated analytically, how can one use this information? *Energy minimization* involves using the first derivatives and, often, the second derivatives to move the system to a local energy minimum. When this minimum has been located to suitable accuracy, *normal mode analysis*, i.e., diagonalization of the mass-weighted second derivatives of the system, allows one to calculate the normal modes of the system (in the particular local minimum) suggested by the system. Finally, *molecular dynamics* involves using the energy gradients of Eq. (1), together with atomic masses to create a trajectory of the molecule at a given temperature. Even though the nature of such a trajectory is strongly influenced by the starting structure, this method, unlike energy minimization, has the ability to climb over energy barriers and not remain in the initial energy minimum. A related approach is the *Monte Carlo* method, which has been used extensively with small molecules in simple solvents, and which uses Eq. (1) to sample configurational space subject to a Bolzmann distribution; in this way, Monte Carlo methods can also surmount energy barriers.

In order to apply Eq. (1) to the molecular system of interest, one must determine the appropriate parameters (k_b, b_0, k_θ, θ_0, V_n, δ, A_{ij}, B_{ij}, q_i, C_{ij}, D_{ij}) to be used. The particular form of Eq. (1) and the set of parameters used for a particular collection of molecules is called a *force field,* and there are many of these in the literature. We describe some of the approaches to force field development and refinement below.

Historical Overview

A brief historical overview of the application expressions such as Eq. (1) to chemical systems is in order. In the pre-computer era, Westheimer and Mayer[3] used an analysis of the intramolecular energy [first three terms in Eq. (1)] to analyze the *strain energy* of cyclic systems compared to their acyclic analogs. When digital computers began to become available in the late 1950s and early 1960s, the application of molecular me-

[2] S. J. Weiner, G. L. Seibel, and P. A. Kollman, *Proc. Natl. Acad. Sci. U.S.A.* **83,** 649 (1986).

[3] F. H. Westheimer and J. E. Mayer, *J. Chem. Phys.* **14,** 733 (1946).

chanics to calculate the structure and strain of a wide variety of complex organic molecules (mainly hydrocarbons or relatively nonpolar species) was carried out by a number of groups. They used equations such as Eq. (1), but, particularly for a more quantitative treatment of highly strained molecules, more complex expressions than those shown there for the intramolecular energies were required. The most completely developed and refined of these force fields derived by the various research groups is the MM2 force field of Allinger and co-workers.[4]

In a parallel development, the use of simple analytical models to study the properties of water clusters, liquid and ices using equations such as Eq. (1) (usually keeping the intramolecular geometry of water fixed and using only the sum over nonbonded interactions) led in the middle 1960s to molecular dynamics simulations of liquid H_2O by Rahman and Stillinger.[5] There have been many useful molecular dynamics and Monte Carlo simulations on water and organic liquids and solutions since that time.

In the middle 1960s the group of Scheraga and co-workers began developing a force field for peptides and proteins,[6] using a fixed bond length/angle model and varying only dihedral angles and intermolecular variables. Also, one of the groups that had begun working on organic molecules, the Lifson group at the Weizmann Institute, "spun off" into molecules of biophysical interest. There were a number of important developments from this group, for example, the realization that full normal mode analysis and optimization of large molecules using Eq. (1) might proceed more effectively by using Cartesian rather than internal coordinates,[7] and the macromolecular energy program developed in that group by Levitt and Lifson[8] became the seed for the software currently used at Harvard,[9] Groningen,[10] Pasadena,[11] UCSF,[12] and La Jolla.[13]

[4] N. L. Allinger, *J. Am. Chem. Soc.* **99,** 8127 (1977); U. Burkert and N. L. Allinger, "Molecular Mechanics." Am. Chem. Soc., Washington, D.C., 1982.

[5] A. Rahman and F. Stillinger, *J. Chem. Phys.* **55,** 3336 (1971).

[6] H. A. Scheraga, *Chem. Rev.* **71,** 195 (1971).

[7] S. Lifson and A. Warshel, *J. Chem. Phys.* **49,** 5116 (1968).

[8] M. Levitt and S. Lifson, *J. Mol. Biol.* **46,** 269 (1969).

[9] B. R. Brooks, R. E. Bruccoleri, and B. D. Olafson, D. J. States, S. Swaminathan, and M. Karplus, *J. Comp. Chem.* **4,** 187 (1983).

[10] GROMOS, Groningen Molecular Simulation System. Observations by W. F. van Gunsteren (1984, 1986) are described in some detail in Ref. 54.

[11] BIOGRAPH, by B. Olafson *et al.,* is a set of graphics and molecular mechanics/dynamics programs developed at California Institute of Technology, Pasadena, California.

[12] P. Weiner and P. Kollman, *J. Comp. Chem.* **2,** 287 (1981); U. C. Singh, P. Weiner, J. Caldwell, and P. Kollman, Amber UCSF 2.0 (1985) and AMBER 30 (1986).

[13] BIOSYM represents graphics and molecular mechanics/dynamics programs developed by A. Hagler and co-workers.

Molecular Force Fields

In all of the developments described above—organic molecules, simple liquids, peptides—it was critical to develop force fields which gave a reasonable representation of experimental reality. The development of force fields has some element of subjectivity in it, in that the force fields that come out depend on the choice of which sets of experimental data should be fit by the model. In other words, the force field developer must choose a specific functional form for Eq. (1), a set of molecules, and a set of experimental data on these molecules to derive the force field, which can then be tested by its ability to reproduce either other types of experimental data or the same type of experimental data on another set of molecules.

With MM2,[4] the most widely used force field for organic molecules, the model was calibrated mainly on gas-phase structures, conformer energies, and heats of formation of a collection of test molecules. It is clear that the MM2 force field works excellently for a wide variety of relatively nonpolar hydrocarbons, but less well for more polar organic molecules, and, indeed, how best to handle the nonbonded electrostatic interactions (and whether to create a more complex model including polarization effects) is a matter of current concern and research.[14]

A wide variety of models for the *nonbonded* part of Eq. (1) have been developed for water–water interactions, almost all assuming pair-wise addivity. This assumption is, of course, quite correct for water dimer interactions. However, to reproduce the enthalpy of vaporization of water liquids, one must use an effective dipole moment for the water model of 2.3–2.5 D, considerably larger than the gas-phase value of 1.85 D; this fact can be attributed to the mutual polarization and induced dipole moments of the water molecules by the electric fields of their neighbors. The most "successful" liquid water models, in so far as their ability to reproduce more accurately the various properties of liquid water, such as heat of vaporization, density, heat capacity, density of liquid as a function of temperature, and X-ray radial distribution function, have been those that have been *empirically* derived by Monte Carlo (TIPS3P and TIPS4P)[15] or molecular dynamics (SPC)[16] simulations to force the *density* and *vaporization enthalpy* to reproduce the experimental values. Those models then give other properties that are in reasonable agreement with experiment.

[14] L. Dosen-Micovic, D. Jeremic, and N. L. Allinger, *J. Am. Chem. Soc.* **105,** 1716 and 1723 (1983).

[15] W. Jorgensen, J. Chandresekhar, and J. Madura, *J. Chem. Phys.* **79,** 926 (1983).

[16] H. J. C. Berendsen, J. P. M. Postma, W. F. van Gunsteren, and J. Hermans, *in* "Intermolecular Forces" (B. Pullman, ed.), p. 331, Reidel, Dordrecht, The Netherlands, 1981.

However, if one wishes to carry out simulations of aqueous solutions, a correct density and enthalpy of vaporization for the water model is essential.

An alternative *ab initio* approach is to use quantum mechanical calculations and then fit these to a simple model such as the nonbonded term in Eq. (1). It is clear from the results of such an approach that such models (e.g., MCY water)[17] give a rather poor representation of the vaporization enthalpy and density of liquid water. The reasons for this are two: the model is inherently two-body, and thus leads to a heat of vaporization that is considerably too small; and the accurate *ab initio* quantum mechanical calculation of dispersion attraction is very difficult. The improvements of such *ab initio* models is an active area of research by Clementi and co-workers,[18] but there is still considerable work to be done if they are to give an accurate representation of all the properties of liquid water and still be computationally efficient.

A recent exciting development is the careful optimization of the nonbonded parameters in Eq. (1) for organic liquids such as hydrocarbons, ethers, and amides in order to optimally fit the density and enthalpy of vaporization of such liquids.[19] The use of such nonbonded parameters in peptide and protein simulations in solution appears promising, particularly since they have been derived in an analogous fashion to the water parameters and thus water–water, protein–protein, and protein–water interactions will be well balanced. This last is, we feel, a critical feature of simulations, and the use of water potentials from one source and peptide parameters from another without insuring that the hydrogen bond strengths are balanced is certain to be less than satisfactory.

Other sources of empirical parameters for Eq. (1) are the vibrational frequencies of the molecules (mainly intramolecular parameters)[7] and crystal packing properties (nonbonded parameters). The reproduction of crystal properties (unit cell parameters) and sublimation energies has played an important role in the development of the peptide force field by the Scheraga and Lifson schools. A classic study in this area is the study of amide crystals by Hagler *et al.*[20] The study by Scheraga and co-workers[21] of hydrocarbon crystals, using a united atom force field (CH, CH_2 and CH_3 groups treated as single "extended" atoms), was the first clear demonstration that one had to use larger than van der Waals radii for such

[17] O. Matsuoka, E. Clementi, and M. Yoshimine, *J. Chem. Phys.* **64,** 1351 (1976).
[18] E. Clementi, *J. Phys. Chem.* **89,** 4426 (1985).
[19] W. Jorgensen, J. Madura, and C. Swenson, *J. Am. Chem. Soc.* **106,** 6638 (1984).
[20] A. Hagler, E. Euler, and S. Lifson, *J. Am. Chem. Soc.* **94,** 5319 (1974).
[21] L. G. Dunfield, A. W. Burgess, and H. A. Scheraga, *J. Phys. Chem.* **82,** 2609 (1978).

groups than had been the case previously. This was confirmed by subsequent simulations on liquid hydrocarbons by Jorgensen.[22]

The construction of a protein force field with all the requisite parameters for Eq. (1) is a large undertaking and, to our knowledge, only Momany *et al.*[23] and Weiner *et al.*[24] have written sufficiently detailed accounts of the development of their protein force fields to allow a critical evaluation of what they have done. Proteins are relatively unstrained molecules, and thus the most critical parameters in a protein force field are the nonbonded: van der Waals and electrostatic. The unique aspect of the Weiner *et al.* study, even though there was some precedent in the literature, was its use of quantum mechanically calculated electrostatic potentials[25] to derive the partial charges q_i of the protein atoms. These charges were used in model H bonding calculations using Eq. (1) to determine hydrogen bond structures and geometries to ensure they were compatible with experiments or highly accurate quantum mechanical calculations. As we have suggested above, one of the critical subject decisions to make in protein simulations is the choice of dielectric constant ε. In an accurate molecular representation, there would be no need for a dielectric constant, since the molecules would undergo electronic polarization and dipolar reorientation and the "dielectric constant" would be a consequence of these molecular factors. However, given that Eq. (1) is a two-body additive potential (no electronic polarization effects) and that the computational efficiency dictates that solvent water is often not included explicitly in the calculations, one needs to ensure that long-range electrostatic effects are not too large. Scheraga and co-workers use $\varepsilon = 4$,[23] the dielectric constant of amide crystals, and Weiner *et al.*[24] use $\varepsilon \propto R_{ij}$ (dielectric constant proportional to the distance between the atoms) when using their force field without explicit solvent and $\varepsilon = 1$ if solvent is included explicitly. Although one can give arguments for/against various dielectric models, none of them is very aesthetically satisfying, and the best situation is to be able to include waters explicitly in the simulation.

Hall and Pavitt[26] have recently tested various force fields in the litera-

[22] W. Jorgensen, *J. Am. Chem. Soc.* **103,** 335 (1981).

[23] F. Momany, R. F. McGuire, A. W. Burgess, and H. A. Scheraga, *J. Phys. Chem.* **79,** 2361 (1975).

[24] S. J. Weiner, P. Kollman, D. Case, U. C. Singh, C. Shio, G. Alagona, S. Profeta, and P. Weiner, *J. Am. Chem. Soc.* **106,** 765 (1984).

[25] E. Scrocco and J. Tomasi, *Adv. Quant. Chem.* **11,** 115 (1978); F. Momany, *J. Phys. Chem.* **82,** 592 (1918); S. R. Cox and D. E. Williams, *J. Comp. Chem.* **2,** 304 (1981); U. C. Singh and P. Kollman, *J. Comp. Chem.* **5,** 129 (1984).

[26] D. Hall and N. Pavitt, *J. Comp. Chem.* **5,** 441(1984).

ture by carrying out energy minimizations on three peptide crystals and seeing how well they "preserved" the crystal structure. Although it is clear that molecular dynamics is indeed a more critical test of force fields, it is not unlikely that many of the relative rankings of force fields would be preserved in a molecular dynamics study. The force field proposed by Weiner *et al.*[24] performed best in that study,[26] notwithstanding that it was of a "united atom" (CH, CH_2, CH_3 treated as extended atoms) variety and despite the fact that it was not "derived" to fit crystal structures. The use of $\varepsilon = R_{ij}$ or $\varepsilon = 1$ gave similar results with that force field as did the use of either the smaller or larger CH, CH_2, CH_3 van der Waals radii. Nonetheless, as stated by Weiner *et al.*[24] and Tilton *et al.*,[27] the use of the larger van der Waals radii gave smaller "protein compaction" than the smaller and a subsequent 100 psec molecular dynamics study by Tilton *et al.*[28] confirms that such larger radii do a very good job of preserving the interior and total volume of the protein. A molecular dynamics study by van Gunsteren and Berendsen[29] on a cytidine analog crystal and the many Monte Carlo simulations by Jorgensen and co-workers also[19,22] make it clear that the use of the larger van der Waals radii is essential for simulations to lead to the correct density in liquids and solid systems.

The three sets of potentials mentioned above (that by Weiner *et al.*,[24] van Gunsteren *et al.*,[29,30] and Jorgensen *et al.*[19,22]) differ in the way they treat hydrogen bonds; Weiner *et al.*[24] use a 10–12 H bond parameter with a small well depth [see Eq. (1)] between the H atom and the H bond acceptor atom to ensure correct H bond distances, and van Gunsteren *et al.*[10,30] and Jorgensen use larger van der Waals radii between the two heteroatoms in the H bond with no van der Waals repulsion from the hydrogen. Jorgensen *et al.*[19] use the geometric mean rule for van der Waals interactions, namely, $A_{ij} = \sqrt{A_{ii}A_{jj}}$, whereas van Gunsteren *et al.* use two sets of van der Waals repulsive parameters for H bonding heteroatoms, a larger one when interacting with other H bonding atoms and a smaller one when interacting with non-H bonding atoms. All of these models lead to reasonable H bond structure and energies, but the Jorgensen and van Gunsteren models are computationally more efficient.

[27] R. F. Tilton, S. J. Weiner, U. C. Singh, I. D. Kuntz, P. Kollman, N. Max, and D. Case, *J. Mol. Biol.* **192,** 443 (1986).

[28] R. F. Tilton, U. C. Singh, G. Petsko, I. D. Kuntz, and P. A. Kollman, *J. Mol. Biol.,* in press (1987).

[29] W. F. van Gunsteren and H. J. C. Berendsen, *in* "Molecular Dynamics and Protein Structure" (J. Hermans, ed.), p. 5. Polycrystal Press, Western Springs, Illinois, 1985.

[30] J. Hermans, H. J. C. Berendsen, W. F. van Gunsteren, and J. P. M. Postma, *Biopolymers* **23,** 1513 (1984).

Recent studies by Wodak *et al.*[31] suggest that, in some cases, "united atom" models may be inadequate for some studies, and an all-atom protein force field has been reported by Weiner *et al.*[32] as well. In this context, it should be noted that Jorgensen and Swenson[33] have found that the Hagler *et al.* "all-atom" parameters[21] for amides, when employed in Monte Carlo simulations on liquid formamide, gave errors in the density of 5% and enthalpy of vaporization of 13%. Thus, it is of concern that parameters fit to crystal properties give such relatively large errors when used on liquids. In summary, it is clear that the use of such a simple nonbonded potential functional form as in Eq. (1) is surprisingly accurate and effective, that recent force fields for peptides are an improvement over older ones. The use of such a function at an all-atom level should allow a more accurate representation of protein properties. Going beyond the simple functional form in Eq. (1) to include other nonadditive effects is critical to accurately representing the energies for highly charged systems and is an area of active research for the next generation of force fields.[34] But it is clear that even the current generation can give an accurate, although far from perfect, description of the structural properties and a "reasonable" description of the energetics of complex peptides and proteins.

One of the difficulties in the evaluation of force fields is the most critical that faces simulators of complex systems, the local minimum problem. A force field can "retain" a peptide crystal structure when starting from this configuration and give a good value for the enthalpy of sublimation, but in almost every protein X-ray structure, with the possible exception of crambin,[35] the water structure is not known accurately enough to evaluate whether a movement of the protein atoms from their X-ray values in the simulation is due to force field errors or inaccuracy in the initial water placement. In any case, the "radius of convergence" of molecular dynamics and, more so, energy minimization, is discouragingly small. With this cautionary note in mind, such simulations can still be of use, as we will attempt to demonstrate below.

How does one choose an initial geometry for the system, to which to apply an expression like Eq. (1)? For a protein, one usually starts with the

[31] S. J. Wodak, P. Alard, P. Delhaise, and C. Renseborg-Squilbin, *J. Mol. Biol.* **181,** 317 (1984).

[32] S. J. Weiner, P. A. Kollman, D. T. Nguyen, and D. A. Case, *J. Comp. Chem.* **7,** 230 (1986).

[33] W. L. Jorgensen and C. J. Swenson, *J. Am. Chem. Soc.* **107,** 569 (1985).

[34] P. Barnes, J. L. Finney, J. P. Nicholas, and J. E. Quinn, *Nature (London)* **282,** 459 (1979); T. Lybrand and P. Kollman, *J. Chem. Phys.* **83,** 2923 (1985).

[35] M. Teeter, personal communication.

protein X-ray structure. The availability of many atom–atom distances from 2-D NMR experiments can allow the construction of three-dimensional models using computer graphics model building as a starting geometry followed by restrained molecular dynamics "refinement."[36]

Energy Minimization

Given an initial geometry, the application of energy minimization to a large protein system involves calculating the energy [Eq. (1)] and the analytical derivatives ($\partial V/\partial x_i$, $\partial V/\partial y_i$, $\partial V/\partial z_i$) of each atom i and using a minimization algorithm such as conjugate gradients to move the system to a lower energy structure. The appropriate convergence criterion to use is not clear, but one needs to test that any conclusions drawn do not depend on it. A typical choice is when the root mean square energy gradient is less than some specified value, e.g., 0.1 kcal/mol Å. The rate-limiting factor in such minimizations (and in the molecular dynamics calculations described below) is the cutoff radius used for the nonbonded interactions. Typical values range from 8 to 20 Å, and there is no well-defined criterion on what to choose. The optimized structure is quite insensitive to this choice, but the energy, which will include "long-range" electrostatic effects, is not. More complete reviews of the treatment of "long-range" forces in simulations are given in Ref. 37. The nonbonded cutoff chosen can influence the convergence of the energy minimization because of errors introduced in the energy gradients when a charged or polar atom moves in and out of the cutoff radius.

Molecular Dynamics Simulations

In classical molecular dynamics, one solves Newton's equations of motion [Eq. (2)], where the force on an atom i is the gradient of the

$$\mathbf{F}_i = \boldsymbol{\nabla}_i V = m_i(d^2_{\mathbf{r}_i}/dt^2) \tag{2}$$

potential energy of Eq. (1). One starts with a Maxwellian distribution of velocities $\mathbf{v}_i$ and the starting coordinates $\mathbf{r}_i$ from, e.g., the X-ray structure and uses Eq. (2) to determine the acceleration, i.e., change in velocity, of each atom. The coupled differential Eqs. (2) must be solved numerically.

[36] R. Kaptein, E. R. P. Zuiderweg, R. M. Scheek, R. Boelens, and W. F. van Gunsteren, *J. Mol. Biol.* **182,** 179(1985).

[37] A. Warshel and S. T. Russell, *Q. Rev. Biophys.* **17,** 283 (1984); H. J. C. Berendsen, *in* "Molecular Dynamics and Protein Structure" (J. Hermans, ed.), p. 1. Polycrystal Press, Western Springs, Illinois, 1985.

Other questions which must be addressed when considering molecular dynamics (MD) simulations on protein systems include the following:

1. What are the appropriate boundary conditions to use?
2. What type of MD will be applied?
3. What algorithm will be used?

In assessing the appropriate boundary conditions to use, one is often guided by computational considerations. Thus, most protein simulations have been performed *in vacuo*, that is, without any wall or boundary. In this case the pressure is certainly correct (0 atm), but other properties, especially those of atoms near or at the protein surface will be distorted by the vacuum boundary condition.

The classic way to minimize edge effects is to use *periodic boundary conditions*. The atoms of the system that is to be simulated are put into a cubic (or any periodically space-filling shaped) box, which is surrounded by 26 identical translated images of itself. When calculating the forces in the central box, all interactions with atoms in the central box or images in the surrounding boxes that lie within the spherical cutoff (circle) are taken into account. Thus in fact a crystal is simulated. For a protein in solution the periodicity is an artifact of the computation, so the effects of periodicity on the forces on the atoms should not be significant. This means that an atom should not simultaneously interact with another atom and its image. Consequently, the box size R_{box} should exceed twice the cutoff radius R_c.

Possibly distorting effects of the periodic boundary condition may be traced by simulating systems of different size. When applying periodic boundary conditions, the protein is put into a large enough (rectangular) box, and the remaining empty space in the box is filled with solvent molecules. This is a very costly treatment. For example, BPTI contains about 500 atoms, but filling a rectangular box around it requires about 2600 water molecules (7800 atoms),[38] which enlarges the calculation by about a factor of 17.

For larger proteins application of periodic boundary conditions is far too expensive. In that case the number of atoms in the simulation can be limited by simulating only part of the molecule. Edge effects can be minimized by restraining the motion of the atoms in the outer shell of the system, which can be kept fixed[39] or harmonically restrained to stationary positions.[40] Such an approach has been used by Brünger *et al.*[41] in their study of the active site of ribonuclease.

[38] W. F. van Gunsteren and M. Karplus, *Biochemistry* **21,** 2359 (1982).
[39] M. Berkowitz and J. McCammon, *Chem. Phys. Lett.* **90,** 215 (1982).
[40] C. L. Brooks III, A. Brünger, and M. Karplus, *Biopolymers* **24,** 843 (1985).
[41] A. Brünger, C. Brooks, and M. Karplus, *Proc. Natl. Acad. Sci. U.S.A.* **82,** 8458 (1985).

Considerable reduction of the number of degrees of freedom can be obtained in this way. For example, the active site of LADH (liver alcohol dehydrogenase) is lined with atoms of two LADH monomers plus waters. The dimer contains about 7000 atoms. A sphere with radius of 1.6 nm around the center of the active site contains about 900 atoms. Application of an extended wall region boundary condition reduces the size of the system by roughly a factor 8. One must still test to what extent the properties of the atoms in the inner region are distorted by restraining the dynamics of the atoms in the wall region, but it is clear that this approach should be useful.

In answering the second question, one realizes that the initial formulation of molecular dynamics used constant volume and constant energy yielding a microcanonical ensemble. In some cases, this is not convenient and various approaches have appeared in the literature to yield a type of dynamics in which temperature and pressure are independent variables rather than derived properties. When MD is performed in nonequilibrium situations in order to study irreversible processes, catalytic events, or transport properties, the need to impress external constraints or restraints is apparent. In such cases the temperature should be controlled as well in order to absorb the dissipative heat produced by the irreversible process. But also in equilibrium cases the automatic control of temperature and pressure as independent variables is very convenient. Thus slow temperature drifts that are the unavoidable result of truncation errors are corrected, while rapid transitions to new desired conditions of temperature and pressure are also more easily accomplished.

Several methods have been proposed, ranging from *ad hoc* rescaling of velocities in order to adjust temperatures, to consistent formulation in terms of modified Lagrangian equations of motion that force the dynamics to follow the desired constraints. We shall not review these methods here but briefly describe a method that has been found to be very useful.

The basic idea is to modify the equations of motion in such a way that the net result on the system is a first-order relaxation of temperature T and pressure P toward given reference values T_0 and P_0:

$$dT/dt = (T_0 - T)/\tau_T \tag{3}$$

and

$$dP/dt = (P_0 - P)/\tau_P \tag{4}$$

The modification of the equations of motion is such that local disturbances are minimized while the global effects of Eqs. (3) and (4) are conserved. This is effected by scaling the velocities in every MD time step with a factor λ:

$$v_i \leftarrow \lambda v_i \tag{5}$$
$$\lambda = [1 + (T_0/T - 1)\Delta t/\tau_T]^{1/2} \tag{6}$$

where

$$T = E_{\text{kin}}/(3Nk/2) \tag{7}$$

$$E_{\text{kin}} = \sum_{i=1}^{N} \tfrac{1}{2} m_i v_i^2 \tag{8}$$

and scaling coordinates r_i and size R_{box} of the periodic box in every step with a factor μ:

$$r_i \leftarrow \mu r_i; \qquad R_{\text{box}} \leftarrow \mu R_{\text{box}} \tag{9}$$
$$\mu = [1 - \beta(P_0 - P)\Delta t/\tau_P]^{1/3} \tag{10}$$

where β denotes the isothermal compressibility of the system and the pressure P is computed from the relation

$$P = (2/3V)(E_{\text{kin}} - \Xi) \tag{11}$$

Here, V denotes the volume of the box, E_{kin} the kinetic energy, and the virial Ξ is defined as

$$\Xi = -\tfrac{1}{2} \sum_{i<j}^{N} \mathbf{r}_{ij}\mathbf{F}_{ij} \tag{12}$$

In the calculation of the pressure, forces within a molecule can be omitted, together with kinetic contributions of internal degrees of freedom. The equations can be modified to include nonisotropic systems.[42,43] The coupling time constants τ_T and τ_P can be arbitrarily chosen but should exceed 10 time steps to ensure stability of the algorithm. The value of the compressibility used in Eq. (10) is not critical, since it only influences the accuracy of the pressure coupling time constant τ_P.

What MD algorithm should be used? We need to solve Eqs. (1) and (2) using numerical integration. Various algorithms for integrating Eq. (2) have been proposed, of which one of the simplest is the so-called leapfrog scheme[44]:

$$v_i(t + \Delta t/2) = v_i(t - \Delta t/2) + m_i^{-1} F_i(t)\Delta t \tag{13}$$
$$r_i(t + \Delta t) = r_i(t) + v_i(t + \Delta t/2)\Delta t \tag{14}$$

[42] H. J. C. Berendsen, J. P. M. Postma, W. F. van Gunsteren, A. di Nola, and J. R. Haak, *J. Chem. Phys.* **81,** 3684 (1984).

[43] M. Ferrario and J. P. Ryckaert, *Mol. Phys.* **54,** 587 (1985).

[44] R. W. Hockney and J. W. Eastwood, "Computer Simulation Using Particles." McGraw-Hill, New York, 1981.

The leapfrog algorithm is exactly equivalent to that due to Verlet,[45] which is one of the most accurate, stable, and yet simple and efficient algorithms presently available for macromolecular systems.[46]

One can make an argument for the use of the leapfrog scheme [Eqs. (13) and (14)] for integrating the equations of motion [Eq. (2)] instead of the algorithms of Gear,[47] Verlet,[45] or Beeman,[48] which have also been applied in protein simulations, for the following reasons:

1. The leapfrog, Verlet, and Beeman algorithms generate exactly the same trajectories.
2. The application of a higher order, more accurate algorithm like that of Gear is of no use as long as the noise in the generated trajectories is mainly due to the cutoff applied to the long-range Coulomb forces when the protein atoms bear nonzero partial charges.
3. The advantage of the leapfrog and Beeman algorithm over that of Verlet is that their explicit treatment of the atomic velocities $v(t)$ allows for a coupling of the system to a heat bath, e.g., by Eq. (5).
4. The advantage of the leapfrog scheme over the Beeman algorithm is that its formulas are less complicated and require at least 25% less computer memory for storage, which can be very convenient when large macromolecules are simulated.

The complete procedure for simulating a macromolecular system including internal constraints using the leapfrog algorithm and coupling to a temperature bath and a pressure bath (isotropic case) has been given in Ref. 41.

One of the fundamental problems with MD is the fact that, because the force an atom experiences will depend on the positions of the surrounding atoms, and given the mass of an atom, the time step used in solving Eq. (2) numerically must be of the order of 10^{-15} sec. Thus, one must evaluate Eqs. (1) and (2) approximately 100,000 times to carry out a simulation of 100 psec and, in this way, molecular dynamics simulations become quite computer-time intensive.

The first molecular dynamics simulation on a protein was reported by McCammon *et al.*[49] in 1976, and this method has been applied to a wide

[45] L. Verlet, *Phys. Rev.* **159,** 98 (1967).

[46] W. F. van Gunsteren and H. J. C. Berendsen, *Mol. Phys.* **34,** 1311 (1977).

[47] C. W. Gear, "Numerical Initial Value Problems and Ordinary Differential Equations." Prentice-Hall, Englewood Cliffs, New Jersey, 1971.

[48] D. Beeman, *J. Comput. Phys.* **20,** 130 (1976).

[49] J. McCammon, B. Gelin, and M. Karplus, *Nature (London)* **267,** 585 (1976).

variety of proteins, e.g., trypsin inhibitor,[50] cytochromes,[51] myoglobin,[28,52] L7/L12 ribosomal protein,[53] ribonuclease,[41] and crambin,[54] mostly in the "gas phase," but in two cases using periodic boundary conditions in crystal and solution.[55,56] The drift from the native X-ray structure varies from about 1 to 2–3 Å[57] and depends on protein and "environment," but to this point the major contact with experiment has been on average structural and dynamic properties, such as X-ray B factors or NMR relaxation times. This field has been recently reviewed by McCammon and Karplus[58] and by McCammon and Harvey.[59] Nonetheless, 10^{-10} sec is a very short time to get an extensive sampling of the conformations of the system, and to do a much longer time simulation requires stochastic dynamics[60] and an abandonment of a complete atomic-level description of the system. However, many chemically and biochemically interesting processes do not take a long time; rather, they are rare events, so that if one knows what process one is looking for, one can use an approach to "force" the system to undergo the process by adding an extra potential term to Eq. (1) and, in this way, estimate the energetics and nature of the process.

Such an approach to add a constraining potential to the system and evaluate the *free energy* of the system corrected for the effect of the constraining potential has been applied to the flipping of a tyrosine ring in trypsin inhibitor,[61] the passage of an ion through the acetylcholine recep-

[50] M. Levitt, *J. Mol. Biol.* **168,** 595 and 621 (1983).

[51] S. H. Northrup, M. R. Pear, J. A. McCammon, M. Karplus, and T. Takano, *Nature (London)* **287,** 659 (1980).

[52] R. Levy, R. Sheridan, J. Keepers, G. Dubey, S. Swaminathan, and M. Karplus, *Biophys. J.* **48,** 509 (1985).

[53] J. Aquist, W. F. van Gunsteren, M. Leijonmarck, and O. Tapia, *J. Mol. Biol.* **183,** 461 (1985).

[54] M. Teeter, unpublished results on the molecular dynamics of crambin.

[55] W. F. van Gunsteren, H. J. C. Berendsen, J. Hermans, W. G. J. Hol, and J. P. M. Postma, *Proc. Natl. Acad. Sci. U.S.A.* **80,** 4315 (1983); W. F. van Gunsteren and H. J. C. Berendsen, *J. Mol. Biol.* **176,** 559 (1984).

[56] P. Krüger, W. Strassburger, A. Wollmer, and W. F. van Gunsteren, *Eur. Biophys. J.* **13,** 77 (1985).

[57] See Table 2 of W. F. van Gunsteren and H. J. C. Berendsen, *in* "Molecular Dynamics and Protein Structure" (J. Hermans, ed.), p. 5. Polycrystal Press, Western Springs, Illinois, 1985.

[58] J. A. McCammon and M. Karplus, *Annu. Rev. Phys. Chem.* **31,** 29 (1980).

[59] J. A. McCammon and S. Harvey, "Dynamics of Proteins and Nucleic Acids." Cambridge Univ. Press, London, 1987.

[60] W. F. van Gunsteren, H. J. C. Berendsen, and J. A. C. Rullman, *Mol. Phys.* **44,** 69 (1981); R. M. Levy, M. Karplus, and J. A. McCammon, *Chem. Phys. Lett.* **65,** 4 (1979).

[61] S. H. Northrup, M. R. Pear, C.-Y. Lee, J. A. McCammon, and M. Karplus, *Proc. Natl. Acad. Sci. U.S.A.* **79,** 4035 (1982).

tor channel,[62] and the rate of noncovalent dissociation of a diatomic ligand from myoglobin,[63] using molecular dynamics techniques, and to the solvent effect on the $Cl^- + CH_3Cl^-$ exchange reaction, using Monte Carlo methods.[64] Although such an approach is quite computer-time intensive and requires assumptions on the nature of the "reactive process," it is clear that it can be a powerful tool in leading to further understanding of events in complex molecules.

A related recent approach,[65] which, it is clear, should be very useful for protein design, is to evaluate free energy differences between related systems A and B using molecular dynamics or Monte Carlo methods by "gradually" changing the empirical parameters used in Eq. (15) from those characteristic of system A to those characteristic of system B.

Let us review the fundamentals of such an approach. If one has generated a canonical ensemble (which is true for the usual Monte Carlo simulations and for certain MD simulations carried out at constant temperature; for the microçanonical MD simulations the results are also valid), the fundamental formula for the Helmholtz free energy A is

$$A = -kT \ln Z \tag{15}$$

where Z is the partition function, determined by the Hamiltonian $H(p,q)$ that describes the total energy of the system in terms of momenta p and coordinates q:

$$Z = (h^{3N}N!)^{-1} \int dp\, dq \exp[-N(p,q)/kT] \tag{16}$$

Now let the Hamiltonian $H(p,q)$ be a function of a parameter λ: $H = H(\lambda,p,q)$. It now follows that the derivative of the free energy with respect to λ is equal to the ensemble average of the derivative of the Hamiltonian with respect to λ:

$$\frac{\partial A}{\partial \lambda} = \left\langle \frac{\partial H}{\partial \lambda} \right\rangle \tag{17}$$

Here the angular brackets denote an ensemble average. This equation follows simply from taking the derivative in Eq. (15), using Eq. (16):

$$\frac{\partial A}{\partial \lambda} = -\frac{kT}{Z}\frac{\partial Z}{\partial \lambda} = \frac{\int dp\, dq\, (\partial H/\partial \lambda) \exp(-H/kT)}{\int dp\, dq \exp(-H/kT)} \tag{18}$$

[62] P. Bash, Ph.D. thesis. University of California, Berkeley, California, 1986; P. Bash, U. C. Singh, R. Stroud, R. Langridge, and P. Kollman, manuscript in preparation.

[63] D. A. Case and J. A. McCammon, *Ann. N.Y. Acad. Sci.* **482,** 222 (1986).

[64] J. Chandresekhar, S. Smith, and W. Jorgensen, *J. Am. Chem. Soc.* **197,** 154 (1985).

[65] H. J. C. Berendsen, J. P. M. Postma, and W. F. van Gunsteren, *in* "Molecular Dynamics and Protein Structure" (J. Hermans, ed.), p. 43. Polycrystal Press, Western Springs, Illinois, 1985.

The latter is by definition equal to the ensemble average of $\partial H/\partial\lambda$. This quantity can easily be evaluated during the simulation and hence the free energy's relative to λ can be found.

Instead of the free energy's derivative relative to λ, the free energy change due to a small perturbation in the Hamiltonian can be obtained during a simulation. let a system with Hamiltonian H_0 be simulated; let its partition function be Z_0 and free energy A_0. Now consider a perturbed Hamiltonian $H = H_0 + \Delta H$, which yields a partition function Z and free energy $A = A_0 + \Delta A$. Now

$$\Delta A = -kT \ln(Z/Z_0)$$

$$= -kT \ln \frac{\int dp\, dq \exp(-\Delta H/kT)\exp(-H_0/kT)}{\int dp\, dq \exp(-H_0/kT)}$$

or

$$\Delta A = -kT \langle \exp(-\Delta H/kT)\rangle_0 \tag{19}$$

where $\langle\ \rangle_0$ means the average over an ensemble generated by a simulation using H^0.

In practice, a perturbation of only the potential energy will be considered, and it is sufficient to average over configurations in space only. By either method (derivative or perturbation) the behavior of the free energy is found in the vicinity of the parameters of the simulated system. By combining many small perturbations or by numerical integration over a path, large free energy differences can also be calculated.

For example, Jorgensen and Rachovin[66] have carried out such a simulation using Monte Carlo methods on the following thermodynamic cycle

$$\begin{array}{ccc} CH_3OH\ (g) & \xrightarrow{\Delta G_1} & CH_3CH_3\ (g) \\ \downarrow \Delta G_{sol}(CH_3OH) & & \downarrow \Delta G_{sol}(CH_3CH_3) \\ CH_3OH\ (aq.) & \xrightarrow{\Delta G_2} & CH_3CH_3\ (aq.) \end{array} \tag{20}$$

where the two vertical processes refer to the transfer of CH_3OH and CH_3CH_3 [$\Delta G_{sol}(CH_3OH)$ and $\Delta G_{sol}(CH_3CH_3)$] from the gas phase to aqueous solution and can be measured experimentally. Because this is a thermodyamic cycle, it is clear that

$$\begin{aligned} \Delta\Delta G = \Delta G_2 - \Delta G_1 &= \Delta G_{sol}(CH_3CH_3) - \Delta G_{sol}(CH_3OH) \\ &= 6.93 \quad \text{kcal/mol} \end{aligned} \tag{21}$$

Even though the horizontal processes are not "real", they have been simulated using Monte Carlo methods (using a single solute in a box with 125 H_2O molecules for the aqueous solution process), and a $\Delta\Delta G = 6.6 -$

[66] W. Jorgensen and C. Rachovin, *J. Chem. Phys.* **83,** 3015 (1985).

6.8 kcal/mol has been calculated, in very good agreement with experiment. Similar agreement with experiment has been achieved by evaluating the relative solvation free energy of Cl^- and Br^- [67] the relative binding affinity of a cryptand[68] $SC24/4H^+$ for these ions as well as the relative free energy of binding of a benzamidine and fluorobenzamidine ligand to trypsin.[69] Thus, this method should prove to be very useful in protein design, despite its computer-intensive nature and the fact that much care must be taken to ensure adequate sampling of the conformational space of the system during the simulation.

To our knowledge, there have been three molecular mechanics studies on single site-specific simulations, all employing energy minimization. Alagona *et al.*[70] studied the site-specific mutation in triose phosphate isomerase (TIM) of His 95 to Gln 95. They carried out a combination of quantum and molecular mechanical calculations on the catalytic proton abstraction from the substrate dihydroxyacetone phosphate (DHAP) to the base Gln 165 and showed the critical role that His 95 N_ε—H and Lys 13 NH_3^+ played in creating the positive electrostatic potential that made proton abstraction facile and the enzyme catalysis so efficient that product dissociation from the enzyme was the rate-limiting step. The original hypothesis[71] that led to the creation of the His 95 → Gln 95 mutant was the idea that the Gln NH_2 and C=O should be able to mimic the His N_ε—H and N_δ:, respectively, if only electrostatic/H bonding effects were critical in structure and catalysis, but that the Glu 95 mutant should be inactive if the pK_a and the ability of the residue at position 95 to become positively charged was critical.

The calculations brought out a third, unanticipated difference between the His 95 and Gln 95 mutants. Whereas in the TIM–DHAP complex native structure during energy minimization the His 95 remained in its original position (model built on the basis of X-ray difference densities), in the Gln 95 mutant, the amide side chain moved away from this location to form an H bond with Glu 165. This could reduce the rate of catalysis in two possible ways: first, by "holding onto" the abstracting base and preventing it from abstracting the substrate proton and, second, by not being in the *right* position to provide the critical electropositive field near the DHAP and facilitate proton abstraction. These possibilities were not anticipated prior to the calculations and show how theoretical calculations can sometimes lead to unanticipated new ideas about the consequences of site-specific mutagenesis. It appears, on the basis of subse-

[67] T. Lybrand and A. McCammon, *J. Am. Chem. Soc.* **107,** 7793 (1985).

[68] T. Lybrand, A. McCammon, and G. Wipff, *Proc. Natl. Acad. Sci. U.S.A.* **83,** 833 (1986).

[69] C. F. Wong and A. McCammon, *Isr. J. Chem.*, in press (1986).

[70] G. Alagona, P. Desmeules, C. Ghio, and P. Kollman, *J. Am. Chem. Soc.* **106,** 3623 (1984).

[71] R. Davenport and G. Petsko, personal communication.

quent experiments, that the His 95 → Gln 95 mutant has much reduced catalytic activity,[72] but the reason for this has not been established. It is of interest nonetheless that Fersht and co-workers have found another example where a His → Gln mutant[73] has much reduced catalytic activity and have interpreted this result in terms of the greater flexibility/different intrinsic geometry of the Gln residue.

Brady *et al.*[74] have studied a Gly → Asp mutant in an influenza hemagglutinin protein, and have used conformational energy minimization and conformational scans to suggest the orientation of the Asp residue in the mutant protein. The lowest energy position had the Asp side chain forming a salt bridge/H bond with a neighboring Arg residue, close to the subsequently determined X-ray position. Although one might have anticipated this result from model building, it is very encouraging that the actual calculations reproduce the experimental results well.

In the third study, by Kollman *et al.*,[75] a set of energy minimization calculations on native trypsin and mutants Gly 194 → Ala, Gly 204 → Ala, Gly 194, 204 → Ala, and Asp 102 → Asn was done subsequent to the experiments. But the results were qualitatively encouraging, in that the Gly → Ala mutants were calculated to have a positive $\Delta\Delta E$ (binding) and $\Delta\Delta E$ (transition state energy), relative to the native structure. The quantitative results paralleled more accurately experiment for the $\Delta\Delta E$ (transition state energy) and predicted that the Asp 102 → Asn 102, whose detailed kinetics have not been reported as yet, should bind substrate well, but be an even worse catalyst than Gly 194, 204 → Ala.

All the above studies calculate structure and internal energy, but, ultimately, $\Delta\Delta G$ or free energy differences are what is needed. So far, only small molecule calculations using the free energy perturbation method have been reported, but it is not hard to relate the CH_3OH → CH_3CH_3 perturbation[66] result to a calculation of the Thr → Val site-specific mutation. Seibel *et al.*[76] and Wilson[77] are currently applying the free energy perturbation method to trypsin mutants, and McCammon and

[72] R. Davenport and G. Petsko, unpublished results; see T. Alber, W. Gilbert, D. Ringe-Ponze, and G. Petsko, *Ciba Found. Symp.* **93,** 4 (1983).

[73] D. M. Lowe, A. R. Fersht, A. Wilkinson, P. Carter, and G. Winter, *Biochemistry* **24,** 5106 (1985).

[74] H. H.-L. Shih, J. Brady, and M. Karplus, *Proc. Natl. Acad. Sci. U.S.A.* **82,** 1697 (1985).

[75] P. Kollman, S. Weiner, G. Leibel, T. Lybrand, U. C. Singh, J. Caldwell, and S. Rao, *Ann. N.Y. Acad. Sci.* **482,** 234 (1986); see also P. Bash, U. Singh, F. Brown, R. Langridge, and P. Kollman, *Science* **235,** 574 (1987); P. Bash, U. Singh, R. Langridge, and P. Kollman, *Science* **236,** 564 (1987); U. Singh, F. Brown, P. Bash, and P. Kollman, *J. Am. Chem. Soc.* **109,** 1607 (1987).

[76] G. Seibel, U. C. Singh, and P. Kollman, unpublished results.

[77] K. Wilson (University of California—San Diego), personal communication.

co-workers[69] have evaluated the difference in free energy of binding of an analog of a benzamidine inhibitor to trypsin within experimental error.

A study of the excellent data bases being created for subtilisin mutants by the Genentech group and T4 lyzozyme mutants by the Oregon group[1] should be very interesting. In both these cases, one has high-resolution structural data on both native and mutant proteins as well as data on relative protein stability of native and mutant proteins. Thus, by carrying out free energy perturbation molecular dynamics calculations and gradually changing the native structure to the mutant, one can see both how well the calculated structural perturbation reproduces the observed, but also, by carrying out similar perturbations on the analogous mutation on an extended chain (as a model for the denatured protein), one can see how well one can predict protein stability. In the long run, we anticipate that it may be possible to design, with calculations: (1) more stable proteins and (2) more effective catalysts and, combined with quantum mechanical calculations, (3) catalyst with altered function/mechanism. We should not lose sight of the fact, as noted earlier, that such calculations will work best in the "small perturbation" limit, and, if a mutation causes a large conformational change, it is unlikely that calculations can "anticipate" them. Nonetheless, as the experiments show, there are many examples where mutations cause small structural perturbations and the theory is in a good position to understand the free energy consequences of such changes.

A final point must be made, given the goal of the chapters in this volume to give enough technical detail so the uninitiated can learn the technique in question. Obviously, it is a very difficult task for the person with little training in the mathematical aspect of physical chemistry to pick up requisite aspects of the molecular simulations touched on in this chapter. The learning process requires an understanding of many of the articles referred to, or, preferably, some time spent learning the ins and outs of the computer codes[10–14] that are available for little or no cost to those at nonprofit institutions. A more detailed analysis of the manipulation of the molecular mechanics energy functions such as Eq. (1) is given in Refs. 4, 78, and 79.

Acknowledgments

P.A.K. would like to thank the National Institutes of Health (GM-29072) for research support. We are also grateful to H. J. C. Berendsen for stimulating discussions.

[78] S. R. Niketic and K. Rasmussen, "The Consistent Force Field: A Documentation." Springer-Verlag, Berlin, Federal Republic of Germany, 1979.

[79] O. Ermer, "Aspekte von Krafteldrechnungen." Wolfgang Bauer Verlag, Munich, Federal Republic of Germany, 1981.

[24] The Use of Structural Profiles and Parametric Sequence Comparison in the Rational Design of Polypeptides

By SÁNDOR PONGOR

Introduction

"Protein engineering," i.e., the design and construction of novel or mutated proteins, may involve different levels of structural information. In the case of well-studied proteins the design can rely on a wealth of X-ray, NMR, and chemical modification data. On the other hand, mutations often have to be designed in cases when the amino acid sequence is the only structural information available. For example, some proteins are not readily available in quantities sufficient for structural analysis, and some short peptides have no characteristic structures in solution. Mutated proteins, on the other hand, can now be obtained more and more readily due to advances in recombinant DNA techniques, which make it possible to test structural hypotheses in practice. If the goal is the exchange of a single amino acid residue then the problem can be attacked by replacing the residue in question with all the 19 remaining amino acids in parallel mutagenesis experiments. On the other hand, this approach would be clearly too expensive and time consuming if several residues of a region have to be changed. The number of possibilities can be narrowed down if a potentially important structural pattern, such as an amphiphilic α-helix, can be recognized in the given region. In this case the mutations can be rationally designed based on the hypothesis that those amino acid replacements leaving the original structural pattern intact are likely to be accepted. There are two generally used methods available for the recognition of local structures in amino acid sequences: (1) the use of symbolic diagrams such as the Schiffer–Edmundson[1] "helical wheels" or the "helical nets,"[2] and (2) the use of structural profiles.

This chapter briefly describes how the structural profiles are constructed and used. A particular goal of this chapter is to review the numeric and graphic methods developed for the comparative evaluation of structural profiles.[3,4] The advantage of comparing structural profiles rather than primary sequences originates from the simple fact that pro-

[1] M. Schiffer and A. B. Edmundson, *Biophys. J.* **7,** 121 (1967).

[2] P. Dunnill, *Biophys. J.* **8,** 865 (1968).

[3] S. Pongor and A. A. Szalay, *Proc. Natl. Acad. Sci. U.S.A.* **82,** 366 (1985).

[4] S. Pongor, M. J. Guttieri, L. M. Cohen, and A. A. Szalay, *DNA* **4,** 319 (1985).

files, unlike primary sequences, can be subjected to arithmetic operations (averaging, subtraction, etc.) and their similarities can be characterized in quantitative terms such as correlation coefficients and standard deviation. This is essentially a *parametric approach of sequence comparison* which makes it possible, e.g., to compare groups of sequences, to carry out a semiquantitative comparison (ranking) of sequences in structural terms, etc. These operations which are not feasible with primary sequences can be easily programed on laboratory microcomputers and allow quantitative characterization of the mutations planned. The molecular biologist can use these techniques in two closely related areas: (1) comparison of mutated or "designed" sequences to their wild-type counterparts; (2) comparison of newly determined amino acid sequences to their known homologs.

Principle and Scope

Structural profiles make it possible to compare wild-type and designed sequences in quantitative terms. Average profiles and standard deviation profiles calculated from homologous sequences allow identification of conserved structural features that can help to formulate a structural hypothesis for the design of engineered analogs. The present chapter contains a compilation of the most important parameter sets used to construct structural profiles from primary sequence data. Numeric and graphic methods designed for the comparative evaluation of structural profiles are described. The methods can be used for the simple quantitative evaluation (ranking) of sequences containing multiple amino acid replacements.

Definitions

The *structural profile* is one of the simplest representations of a protein structure. It is constructed by plotting a structural parameter P_i against the sequential position i. The P_i structural parameters are obtained either statistically, such as the Chou-Fasman parameters,[5] or by physical measurement, such as the various hydrophobicity parameters. (Profile representations of three-dimensional structures are not discussed here).

In the computational sense it is useful to divide the structural profiles into two groups, single residue and multiple residue methods. In the *single residue methods* each of the 20 amino acids is characterized by a constant structural parameter which is independent from the sequence. In this case the profile is constructed simply by plotting these constant val-

[5] P. Y. Chou and G. M. Fasman, *Adv. Enzymol.* **47,** 45 (1978).

ues against the sequential position. In the *multiple residue methods* the P_i parameters of the ith amino acid depends both on the ith amino acid and on the sequential neighbors, and is calculated separately for every position in the sequence. For example, the algorithm of Garnier *et al.*[6] uses contributions of the eight preceding and the eight subsequent sequential neighbors for the calculation of the secondary structure propensity parameter of the ith amino acid [Eq. (3)].

For the presentation of the data, the plotted P_i parameters have to be processed by *curve smoothing* that visualizes trends within the profiles by removing dispersion. For our purposes basically any smoothing technique is acceptable, and commercially available softwares usually contain such options. In these procedures each P_i parameter is replaced by a new P'_i value which is an average calculated within a *window,* i.e., a sequential environment of the ith residue. Usually, the *window width* is an optional odd number $(2k + 1)$ and the resulting average value is assigned to the central position of the window. In the simplest case, P'_i can be an arithmetic average [Eq. (1a)] or a weighted arithmetic average [e.g., Eqs. (1b) and (c)]:

$$P'_i = [1/(2k + 1)] \sum_{j=i-k}^{j=i+k} P_j \tag{1a}$$

$$P'_i = [17P_i + 12(P_{i+1} + P_{i-1}) - 3(P_{i+2} + P_{i-2})]/35 \tag{1b}$$

$$P'_i = \{7P_i + 3[2(P_{i-1} + P_{i+1}) + P_{i-2} + P_{i+2}] - 2(P_{i-3} + P_{i+3})\}/21 \tag{1c}$$

More advanced methods can produce continuously differentiable curves.[7-9] While the latter procedures more efficiently remove dispersion than arithmetic averaging [Eq. (1a)], they also distort the values numerically and may give rise to chain-end artifacts at the first and last residues. Repeated use of the same algorithm also improves smoothing efficiency. We note that (1) if the profiles are to be used for prediction, then the optimized procedures recommended by the original papers have to be employed (see below). (2) The unprocessed P_i parameters should be used for numeric and graphic comparison of profiles as well as for the calculation of average and standard deviation profiles.

The different structural parameters (Table I) are usually scaled so as to meet some statistical or other requirement. For better comparison and

[6] F. Garnier, D. J. Osguthorpe, and B. Robson, *J. Mol. Biol.* **120,** 97 (1978).

[7] P. R. Bevington, "Data Reduction and Error Analysis for the Physical Sciences." McGraw-Hill, New York, 1969.

[8] A. Savitzky and M. Golay, *Anal. Chem.* **36,** 1627 (1974).

[9] G. D. Rose, *Nature (London)* **272,** 586 (1978).

TABLE I
STRUCTURAL PARAMETERS (SINGLE RESIDUE METHODS)[a]

Amino acid residue	a	b	c	d	e	f	g	h	i	j	k	l	m
Gly	0.57	0.75	1.56	−0.40	0.48	0.62	1.00	0.00	0.00	0.49	3.40	0.00	0
Ala	1.42	0.83	0.74	1.80	0.62	1.56	2.20	−0.50	−4.20	1.81	11.50	0.00	0
Val	1.06	1.70	0.59	4.20	1.08	1.14	2.90	−1.50	−8.40	1.08	21.57	0.13	0
Leu	1.21	1.30	0.50	3.80	1.06	2.93	5.00	−1.80	−10.10	3.23	21.40	0.13	0
Ile	1.08	1.60	0.47	4.50	1.38	1.67	3.30	−1.80	−10.50	1.45	21.40	0.13	0
Ser	0.77	0.75	1.43	−0.80	−0.18	0.81	−0.87	0.30	6.30	0.97	9.47	1.67	0
Thr	0.83	1.19	0.98	−0.70	−0.05	0.91	0.04	−0.40	3.80	0.84	15.77	1.66	0
Asp	1.01	0.52	1.52	−3.50	−0.90	0.14	−2.10	3.00	31.00	0.05	11.68	49.70	−1
Glu	1.51	0.37	0.95	−3.50	0.74	0.23	−2.30	3.00	24.70	0.11	13.57	49.90	−1
Asn	0.67	0.89	1.46	−3.50	0.78	0.27	−2.40	0.20	12.20	0.23	12.82	3.38	0
Gln	1.11	1.10	0.95	−3.50	−0.85	0.51	−2.40	0.20	10.10	0.72	14.45	3.53	0
Lys	1.16	0.74	1.19	−3.90	−1.50	0.15	−2.40	3.00	17.60	0.06	15.71	49.50	+1
His	1.00	0.87	0.97	−3.20	−0.40	0.29	−2.80	−0.50	14.30	0.31	13.69	51.60	0
Arg	0.98	0.93	1.01	−4.50	−2.53	0.45	−2.40	3.00	47.30	0.20	14.28	52.00	+1
Phe	1.13	1.38	0.66	2.80	1.19	2.03	1.00	−2.50	−14.20	1.96	19.80	0.35	0
Tyr	0.69	1.47	1.14	−1.30	0.26	0.68	−0.79	−2.30	4.70	0.39	18.03	1.61	0
Trp	1.08	1.37	0.60	−0.90	0.81	1.08	−1.10	−3.40	−8.40	0.77	21.67	2.10	0
Cys	0.70	1.19	0.96	2.50	0.29	1.23	−2.00	−1.00	−6.30	1.89	13.46	1.48	0
Met	1.45	1.05	0.60	1.90	0.64	2.96	−0.21	−1.30	−11.30	2.67	16.25	1.43	0
Pro	0.57	0.55	1.56	−1.60	0.12	0.76	−1.60	0.00	13.90	0.76	17.43	1.58	0

[a] Key to column heads: (a) Chou–Fasman helical propensity[5]; (b) Chou–Fasman β-sheet propensity[5]; (c) Chou–Fasman turn propensity[5]; (d) hydrophobicity according to Kyte and Doolittle[16]; (e) consensus hydrophobicity scale according to Eisenberg *et al.*[22]; (f) membrane-buried helical potential according to Argos and Palau[18]; (g) membrane propensity scale according to Kuhn and Leigh[19]; (h) hydrophilicity scale according to Hopp and Woods[22]; (i) signal sequence helical potential scale according to Von Heinje[24] (kJ/mol); (j) signal sequence propensity scale according to Argos and Palau[18]; (k) side-chain bulkiness; (l) polarity; (m) charge.

uniformity, we have normalized the single residue amino acid parameter sets to a mean 0 and a standard deviation 1 (Table II).

Description of Structural Profiles

Secondary Structure Propensity Profiles. The most widely used parameter set, that of Chou and Fasman,[5] is statistically derived from protein structures determined by X-ray crystallography. The parameters represent the propensity of each amino acid to form an α-helix, β-sheet, or turn (Table I). For example, the α-helix propensity parameter of an amino acid a, $P_{a,\alpha}$ is calculated by dividing the frequency of the amino acid a in α-helix, $f_{a,\alpha}$ by the average frequency of residues in the helix conformation:

$$P_{a,\alpha} = f_{a,\alpha}/\langle f_{\alpha}\rangle \tag{2}$$

In their original paper Chou and Fasman used structural data on 15 proteins for the calculation of the conformational parameters. The Chou–Fasman parameters in Table I were calculated from a database containing 29 proteins. In the original method the profiles are smoothed with a four-residue window, and the average value is assigned to the first position within the window. Levitt[10] has published a Chou–Fasman type parameter set derived from a database of 60 protein structures.

The method of Chou and Fasman is a single residue method, in which neighbor interactions are not included. Robson and co-workers developed the so-called directional information algorithm,[6] a statistically based multiple residue method in which contributions from eight preceding and eight subsequent sequential neighbors are included. For example, the propensity of the ith amino acid to be in an α-helix, $P_{i,\alpha}$ is calculated by the following equation:

$$P_{i,\alpha} = \sum_{i-8}^{i+8} I_{j,\alpha} - D_{\alpha} \tag{3}$$

where I_j values are the contributions of the neighbors as well as of the ith residue to the propensity value, taken from Table III for α-helix. There are 20×17 P values for each of the four conformational states (α-helix, β-sheet, β-turn, and coil, Tables III to VI). D_{α} is the so-called decision constant which is selected from Table VII if the secondary structure content of the protein is known from independent measurement. For comparative purposes, unbiased profiles ($D_{\alpha} = 0$, etc.) can be used. Strongly predicted segments have P_i values of several hundred centinats.

[10] M. Levitt, *Biochemistry* **18,** 4277 (1978).

TABLE II
NORMALIZED[a] STRUCTURAL PARAMETERS (SINGLE RESIDUE METHODS)[b]

Amino acid residue	a	b	c	d	e	f	g	h	i	j	k	l
Gly	−1.53	−0.73	1.55	1.37	0.48	−0.48	0.61	0.08	−0.35	−0.56	−2.58	−0.62
Ala	1.49	−0.54	−0.68	1.28	0.62	0.64	1.14	−0.19	−0.61	0.91	−0.84	−0.62
Val	0.21	1.82	−1.09	1.28	1.08	0.14	1.45	−0.71	−0.87	0.09	1.34	−0.61
Leu	0.75	0.74	−1.13	1.36	1.06	2.27	2.38	−0.87	−0.97	2.48	1.30	−0.61
Ile	−0.28	1.55	−1.55	1.32	1.38	0.77	1.63	−0.87	−1.00	0.51	1.30	−0.61
Ser	−0.82	−0.76	1.19	−0.10	−0.18	−0.25	−0.21	0.24	0.04	−0.03	−1.27	−0.54
Thr	−0.60	−0.44	−0.03	−0.08	−0.05	−0.13	0.19	−0.13	−0.11	−0.17	0.09	−0.54
Asp	0.04	−1.33	1.44	−1.27	−0.90	−1.05	−0.75	1.67	1.60	−1.05	−0.80	1.65
Glu	1.81	−1.79	−0.11	−1.12	−0.74	−0.94	−0.84	1.67	1.18	−0.98	−0.39	1.66
Asn	1.17	−0.38	1.28	−1.03	−0.77	−0.89	−0.89	0.19	0.41	−0.85	−0.55	−0.47
Gln	0.39	0.19	−0.09	−0.96	−0.85	−0.61	−0.89	0.19	0.28	−0.31	−0.20	−0.46
Lys	0.57	−0.78	0.54	−0.99	−1.50	−1.04	−0.89	1.67	0.74	−1.04	0.07	1.64
His	0.00	−0.43	−0.11	−1.13	−0.40	−0.87	−1.06	−0.19	0.54	−0.76	−0.36	1.73
Arg	−0.07	−0.27	0.05	−1.28	−2.53	−0.68	−0.89	1.67	2.58	−0.88	−0.24	1.76
Phe	0.46	0.95	−0.90	0.74	1.19	1.20	0.61	−1.24	−1.23	1.07	0.96	−0.60
Tyr	−1.10	1.20	0.40	−0.32	0.26	−0.41	−0.18	−1.14	−0.06	−0.67	0.57	−0.55
Trp	0.28	0.93	−1.07	−0.27	0.81	0.07	−0.31	−1.72	−0.87	−0.25	1.36	−0.52
Cys	−1.07	0.44	−0.09	0.65	0.29	0.25	−0.71	−0.45	−0.74	0.99	−0.41	−0.55
Met	1.60	0.06	−1.07	0.06	0.64	2.31	0.08	−0.61	−0.05	1.86	0.19	−0.56
Pro	−1.53	−1.30	1.55	−0.09	0.12	−0.31	−0.53	0.08	0.51	−0.26	0.44	−0.55

[a] Structural parameters normalized to a mean 0 and a standard deviation 1.

[b] Key to column heads: (a) Chou–Fasman helical propensity[5]; (b) Chou–Fasman β-sheet propensity[5]; (c) Chou–Fasman turn propensity[5]; (d) hydrophobicity according to Kyte and Doolittle[16]; (e) consensus hydrophobicity scale according to Eisenberg *et al.*[22]; (f) membrane-buried helical potential according to Argos and Palau[18]; (g) membrane propensity scale according to Kuhn and Leigh[19]; (h) hydrophilicity scale according to Hopp and Woods[22]; (i) signal sequence helical potential scale according to Von Heinje[24]; (j) signal sequence propensity scale according to Argos and Palau[18]; (k) side-chain bulkiness; (l) polarity.

Both the Chou–Fasman and the Robson profiles were originally designed for secondary structure prediction; for the details of prediction methodology the reader is referred to the original papers. In addition, both methods can be used for the predictive *recognition* of supersecondary structures as suggested by Taylor and Thornton.[11,12] Super–secondary structures are sequentially linked pieces of secondary structures that are in contact in three dimensions. The simple supersecondary structures include the $\beta\alpha\beta$ unit (found in the β/α protein family), the β-hairpin

[11] W. L. Taylor and J. M. Thornton, *Nature (London)* **301,** 540 (1983).

[12] W. L. Taylor and J. M. Thornton, *J. Mol. Biol.* **173,** 487 (1984).

TABLE III

STRUCTURAL PARAMETERS (DIRECTIONAL INFORMATION MEASURES) FOR THE α-HELICAL CONFORMATION[a]

Amino acid residue	Residue position																
	$j-8$		$j-6$		$j-4$		$j-2$		j		$j+2$		$j+4$		$j+6$		$j+8$
Gly	−5	−10	−15	−20	−30	−40	−50	−60	−86	−60	−50	−40	−30	−20	−15	−10	−5
Ala	5	10	15	20	30	40	50	60	65	60	50	40	30	20	15	10	5
Val	0	0	0	0	0	0	5	10	14	10	5	0	0	0	0	0	0
Leu	0	5	10	15	20	25	28	30	32	30	28	25	20	15	10	5	0
Ile	5	10	15	20	25	20	15	10	6	0	−10	−15	−20	−25	−20	−10	−5
Ser	0	−5	−10	−15	−20	−25	−30	−35	−39	−35	−30	−25	−20	−15	−10	−5	0
Thr	0	0	0	−5	−10	−15	−20	−25	−26	−25	−20	−15	−10	−5	0	0	0
Asp	0	−5	−10	−15	−20	−15	−10	0	5	10	15	20	20	20	15	10	5
Glu	0	0	0	0	10	20	60	70	78	78	78	78	78	70	60	40	20
Asn	0	0	0	0	−10	−20	−30	−40	−51	−40	−30	−20	−10	0	0	0	0
Gln	0	0	0	0	5	10	20	20	10	−10	−20	−20	−10	−5	0	0	0
Lys	20	40	50	55	60	60	50	30	23	10	5	0	0	0	0	0	0
His	10	20	30	40	50	50	50	30	12	−20	−10	0	0	0	0	0	0
Arg	0	0	0	0	0	0	0	0	−9	−15	−20	−30	−40	−50	−50	−30	−10
Phe	0	0	0	0	0	5	10	15	16	15	10	5	0	0	0	0	0
Tyr	−5	−10	−15	−20	−25	−30	−35	−40	−45	−40	−35	−30	−25	−20	−15	−10	−5
Trp	−10	−20	−40	−50	−50	−10	0	10	12	10	0	−10	−50	−50	−40	−20	−10
Cys	0	0	0	0	0	0	−5	−10	−13	−10	−5	0	0	0	0	0	0
Met	10	20	25	30	35	40	45	50	53	50	45	40	35	30	25	20	10
Pro	−10	−20	−40	−60	−80	−100	−120	−140	−77	−60	−30	−20	−10	0	0	0	0

[a] According to Garnier *et al.*[6]

TABLE IV

STRUCTURAL PARAMETERS (DIRECTIONAL INFORMATION MEASURES) FOR THE EXTENDED CONFORMATION[a]

Amino acid residue	Residue position																
	$j-8$		$j-6$		$j-4$		$j-2$		j		$j+2$		$j+4$		$j+6$		$j+8$
Gly	10	20	30	40	40	20	0	−20	−42	−20	0	20	40	40	30	20	−10
Ala	0	0	0	0	−5	−10	−15	−20	−23	−20	−15	−10	−5	0	0	0	0
Val	0	0	−10	−20	0	20	40	60	68	60	40	20	0	−20	−10	0	0
Leu	0	0	0	0	0	5	10	20	23	20	10	5	0	0	0	0	0
Ile	0	−10	−20	−10	0	20	40	60	67	60	40	20	0	−10	−20	−10	0
Ser	0	10	20	10	0	−5	−10	−15	−17	−15	−10	−5	0	10	20	10	0
Thr	5	10	15	20	15	15	10	10	13	10	10	15	15	20	15	10	5
Asp	0	5	10	15	20	0	−20	−30	−44	−30	−20	0	0	0	0	0	0
Glu	−10	−15	−20	−25	−30	−35	−40	−45	−50	−55	−60	−60	−50	−40	−30	−20	−10
Asn	10	30	50	30	20	0	−15	−30	−41	−30	−15	0	20	30	50	30	10
Gln	0	0	0	0	0	−5	−10	0	12	20	30	40	50	50	40	30	15
Lys	−5	−10	−15	−20	−30	−40	−50	−40	−33	−20	−10	0	10	10	0	0	0
His	−10	−20	−40	−20	−10	0	−10	−20	−25	−35	−30	−25	−20	−15	−10	−5	0
Arg	0	0	0	0	0	0	0	0	4	0	0	0	0	0	0	0	0
Phe	0	0	0	0	0	5	10	20	26	10	−10	−30	−60	−65	−60	−40	−20
Tyr	0	5	10	15	20	25	30	35	40	35	30	25	20	15	10	5	0
Trp	0	0	0	0	0	−10	−10	−10	−10	−10	−10	−15	−20	−25	−30	−20	−10
Cys	0	0	0	0	0	10	20	30	44	30	20	10	0	0	0	0	0
Met	−10	−20	−30	−40	−40	−30	0	10	23	10	0	−30	−40	−40	−30	−20	−10
Pro	10	20	30	30	20	10	0	−10	−18	−20	−10	10	30	40	30	20	10

[a] According to Garnier *et al.*[6]

TABLE V
STRUCTURAL PARAMETERS (DIRECTIONAL INFORMATION MEASURES) FOR THE TURN CONFORMATION[a]

Amino acid residue	Residue position																
	j − 8		*j* − 6		*j* − 4		*j* − 2		*j*		*j* + 2		*j* + 4		*j* + 6		*j* + 8
Gly	0	0	0	0	10	30	55	55	57	40	0	0	0	0	0	0	0
Ala	0	0	0	−10	−20	−30	−40	−50	−50	−40	−30	−20	−10	0	0	0	0
Val	0	0	0	0	−10	−20	−30	−40	−60	−40	−30	−20	−10	0	0	0	0
Leu	0	0	0	−10	−20	−30	−40	−50	−56	−20	−10	0	0	0	0	0	0
Ile	0	0	0	0	0	−10	−20	−30	−46	−40	−10	0	0	20	30	20	10
Ser	0	−10	−20	−20	10	15	20	25	26	25	20	15	10	0	0	0	0
Thr	0	10	20	20	20	15	18	5	3	5	10	15	20	20	20	10	0
Asp	0	0	0	0	0	0	5	10	31	10	5	0	0	0	0	0	0
Glu	0	−5	−10	−15	−20	−30	−40	−45	−47	−20	0	10	5	0	0	0	0
Asn	0	0	0	10	20	30	35	40	42	40	35	30	20	10	5	0	0
Gln	10	20	30	25	20	15	10	5	4	20	30	40	50	60	50	40	20
Lys	−10	−20	−30	−40	−25	−10	0	10	10	10	0	−20	−30	−20	−10	5	0
His	0	0	0	0	0	0	0	0	−3	0	10	20	30	20	10	0	0
Arg	0	0	0	0	0	0	0	10	21	30	40	30	20	10	0	0	0
Phe	0	0	0	0	0	−5	−10	−15	−18	−15	0	15	30	25	20	10	0
Tyr	0	0	0	5	15	15	20	25	29	25	20	15	15	5	0	0	0
Trp	0	0	0	10	20	30	40	80	36	−30	30	40	50	60	70	40	20
Cys	20	40	50	60	60	55	50	45	44	40	35	30	25	20	15	10	5
Met	−5	−15	−20	−25	−30	−35	−40	−45	−48	−45	−40	−35	−30	−25	−20	−15	−5
Pro	10	20	30	40	50	70	10	−90	36	90	10	0	0	0	0	0	0

[a] According to Garnier *et al.*[6]

TABLE VI

STRUCTURAL PARAMETERS (DIRECTIONAL INFORMATION MEASURES) FOR THE COIL CONFORMATION[a]

Amino acid residue	Residue position																
	$j-8$		$j-6$		$j-4$		$j-2$		j		$j+2$		$j+4$		$j+6$		$j+8$
Gly	0	0	0	0	10	30	40	45	49	45	40	30	10	0	0	0	0
Ala	0	0	0	0	−5	−10	−20	−25	−25	−25	−20	−15	−10	−5	0	0	0
Val	0	0	0	0	−10	−20	−25	−30	−35	−30	−25	−20	−10	0	0	0	0
Leu	0	0	0	−10	−20	−30	−40	−30	−20	−20	−10	0	0	0	0	0	0
Ile	0	0	0	0	0	−10	−20	−30	−33	−30	−10	0	10	20	30	20	0
Ser	0	−10	−20	−20	10	15	20	25	50	25	20	15	10	0	0	0	0
Thr	0	10	20	30	20	15	10	15	17	15	10	15	20	30	20	10	0
Asp	0	0	0	0	0	0	0	0	0	0	0	0	0	0	0	0	0
Glu	0	0	10	20	40	20	0	−10	−44	−40	−20	−10	0	0	0	0	0
Asn	0	0	0	10	20	30	35	40	46	40	35	30	20	10	0	0	0
Gln	10	20	30	25	20	15	10	0	−5	20	30	40	50	60	50	40	20
Lys	−10	−20	−30	−40	−25	−20	−10	−8	−8	0	0	−20	−30	−20	−10	−5	0
His	0	0	0	0	0	0	0	10	16	15	10	10	10	10	5	0	0
Arg	0	0	0	0	0	0	0	0	−12	0	20	30	20	10	0	0	0
Phe	0	0	0	0	0	−5	−10	−20	−41	−20	0	15	30	25	20	10	0
Tyr	0	0	0	0	0	0	0	0	−6	0	0	0	0	0	0	0	0
Trp	0	0	0	10	20	30	40	20	12	20	30	40	50	60	70	40	20
Cys	0	0	0	0	0	0	−10	−30	−47	−30	−10	0	0	0	0	0	0
Met	0	−5	−10	−15	−20	−25	−30	−40	−41	−40	−30	−25	−20	−15	−10	−5	0
Pro	0	0	10	20	30	40	50	55	58	50	10	0	0	0	0	0	0

[a] According to Garnier *et al.*[6]

TABLE VII
DECISION CONSTANTS RELATED TO THE SECONDARY STRUCTURE CONTENT OF THE PROTEIN[a]

% Secondary structure (α-helix or extended)	Decision constants[b]	
	α-Helix, D_α	Extended, D_β
I. Less than 20%	158	50
II. Between 20 and 50%	−75	−87.5
III. Over 50%	−100	−87.5

[a] According to Garnier *et al.*[6]

[b] These values have to be substituted into Eq. (3). Decision constants for the turn and coil conformations are equal to 0.

(β-turn-β or $\beta \times \beta$, frequently found in the all-β family), and the $\alpha \times \alpha$ hairpin. An examination of 31 proteins showed that two-thirds of the identified pieces of secondary structures were parts of such units.[13] The principle is illustrated in Fig. 1. For example, the $\beta\alpha\beta$ structure can be recognized from a peak in the α-helix propensity profile flanked on both sides by peaks in the β-sheet profile. Naturally, not all supersecondary structures may be clearly detectable from structural profiles; their importance lies in the fact that they represent interpretation of predictive data (structural profiles) at a higher level of complexity (supersecondary structures).

Membrane Preference Profiles. Transmembrane segments in proteins are characterized by a large number of nonpolar residues. Based on the experimental data of Nazaki and Tanford[14] and Wolfenden *et al.*,[15] Kyte and Doolittle[16] developed a hydropathy scale that characterizes the hydrophobicity of the amino acid side chains. Nonpolar side chains have large positive hydropathy parameters (Ile = 4.50), and polar ones are assigned negative values (Arg = −4.50). Hydropathy profiles constructed with these parameters and smoothed with a 19-residue window [Eq. (1), $k = 9$], will display large positive peaks at membrane-bound segments. Proteins located within the membranes are frequently characterized by a series of transmembrane regions that appear in the graphs as positive

[13] M. Levitt and C. Chotia, *Nature* (*London*) **261,** 552 (1976).

[14] T. Nozaki and C. Tanford, *J. Biol. Chem.* **246,** 2211 (1971).

[15] R. Wolfenden, L. Anderson, P. M. Cullis, and C. C. B. Southgate, *Biochemistry* **20,** 849.

[16] J. Kyte and R. F. Doolittle, *J. Mol. Biol.* **157,** 105 (1982).

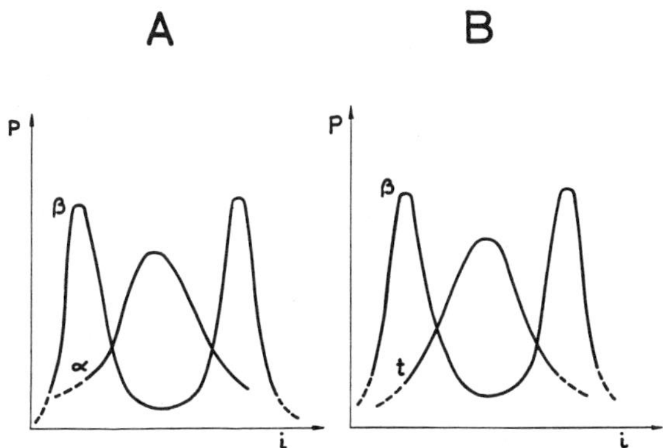

FIG. 1. Schematic structural profiles representing two simple supersecondary structures, the $\beta\alpha\beta$ (A) and the β-turn-β units (B). A conserved β-turn-β unit can be tentatively identified in Fig. 2 at the C-terminus of the ribulose 1,5-bisphosphate carboxylase small subunit polypeptide.

peaks interspersed by negative valleys corresponding to regions located outside the membrane.

Several other hydrophobicity scales were developed. Tables I and II include that of Eisenberg *et al.*,[17] the "consensus hydrophobicity scale," which incorporates features of several other scales. We note that in our applications differences between the various scales are negligible and that the scale of Kyte and Doolittle[16] seems to have the widest acceptance by molecular biologists.

Transmembrane segments are frequently helical. Argos and Palau developed a membrane-buried helix propensity scale, using statistical data on 1125 residues.[18] Kuhn and Leigh developed another statistical membrane propensity scale, using structural data on 24 known transmembrane segments.[19] The parameters range from −2.8 (His) to 5.0 (Leu) and are scaled so that a protein of average composition will have an average membrane propensity of 0. The profiles are smoothed with a running average algorithm [Eq. (1a)], using a window width of 7. Eisenberg *et al.* used a combination of hydrophobic moment profiles and hydrophobicity profiles for the identification of transmembrane helices[20–23] (see below).

[17] D. Eisenberg, R. M. Weiss, T. C. Terwilliger, and W. Wilcox, *Faraday Symp. Chem. Soc.* **17,** 109.

[18] P. Argos and J. Palau, *Int. J. Pept. Protein Res.* **19,** 380 (1982).

[19] L. Kuhn and J. S. Leigh, *Biochim. Biophys. Acta* **828,** 351 (1985).

[20] D. Eisenberg, E. Schwarz, M. Komaromy, and R. Wall, *J. Mol. Biol.* **179,** 125 (1984).

[21] D. Eisenberg, *Annu. Rev. Biochem.* **53,** 595 (1984).

[22] D. Eisenberg, R. M. Weiss, and T. C. Terwilliger, *Nature (London)* **299,** 371 (1982).

[23] D. Eisenberg, R. M. Weiss, and T. C. Terwilliger, *Proc. Natl. Acad. Sci. U.S.A.* **81,** 140 (1984).

Signal Sequence Helical Potential Profiles. Signal sequences are characterized by a stretch of nonpolar residues of high helix-forming propensity, and their amino acid composition significantly differs from the average protein composition. Von Heinje developed a helix hydrophobicity scale based on the free energy required to transfer a membrane-bound helix into solution.[24] The parameters range from −14.42 kJ/mol for Phe to 47.3 kJ/mol for Arg. In the profiles the helical part of the signal sequence displays a negative trough. Argos *et al.* developed a signal sequence helical propensity parameter set from a 705-residue database compiled from signal sequences using Eq. (2).[25]

The Hydrophilicity Profile: Prediction of Antigenic Determinants. Antigenic determinants of proteins are located on the surface of the molecule and are frequently found in regions highly exposed to the solvent. This, together with the fact that charged, hydrophilic amino acids are common features of antigenic determinants led Hopp and Woods to use a hydrophilicity profile to locate antigenic determinants in a number of proteins.[26] Their hydrophilicity parameters are derived from the solvent parameters of Levitt.[27] These parameters are based on the same experimental data as the hydrophobicity scales and indeed contain the same information. In this scale hydrophilic amino acids have large positive parameters (Arg = 3.0) whereas hydrophobic residues have negative ones (Trp = −3.4) (Table III). The hydrophilicity profiles are smoothed by simple arithmetic averaging using a window width of 6 residues, and the tentative antigenic regions are identified as those displaying the highest positive peaks. As peptide chain turns are often hydrophilic,[8] this technique is likely to select turns as potential epitopes.

The Hydrophobic Moment Profile. Amphiphilic structures in proteins and peptides are those in which hydrophobic and hydrophilic side chains are separated on opposite sides. Amphiphilic structures are "surface seeking"; i.e., they bind to membrane–solution surface boundaries. Amphiphilic segments of soluble proteins are frequently located on the protein surface with the hydrophilic side chains pointing toward the solution. Amphiphilic structures in peptides can be easily recognized by diagramatic representations such as the Schiffer–Edmundson wheel diagrams[1] (see Chap. [25], this volume). Graphic analysis of long sequences would be cumbersome, however, so Eisenberg and co-workers developed a numeric method.[20–23] The method is based on the concept of *hydrophobic*

[24] G. Von Heinje, *Eur. J. Biochem.* **116,** 419 (1981).

[25] P. Argos, J. K. Mohana Rao, and P. A. Hargrave, *Eur. J. Biochem.* **128,** 565 (1982).

[26] T. Hopp and K. R. Woods, *Proc. Natl. Acad. Sci. U.S.A.* **78,** 3824.

[27] M. Levitt, *J. Mol. Biol.* **104,** 59 (1976).

moment, a numeric index of amphiphilicity defined by the following equation:

$$\mu_H = \left(\left[\sum_{j=1}^{n} H_j \sin(\delta j) \right]^2 + \left[\sum_{j=1}^{n} H_j \cos(\delta j) \right]^2 \right)^{1/2} \tag{4}$$

H_j is the hydrophobicity (Table I, column e) of the jth amino acid, δ the periodicity parameter (100° for α-helix and 180° for β-sheet, expressed in radians), and n is the total number of residues. μ_H is a Fourier transform type expression that has large values if residues of adverse hydrophobicity are clustered on opposite sides of the helix. Membrane-seeking amphiphilic helices can be located in protein sequences according to their hydrophobic moment μ_H and average hydrophobicity, $\langle H \rangle$. μ_H of amphiphilic helices is greater than 0.47 and $\langle H \rangle$ is in the range of -0.22 to 0.34. Transmembrane regions on the other hand have greater hydrophobicity values ($\langle H \rangle \geq 0.68$ for a 21-residue segment or ≥ 1.10 for two adjacent segments) and lower hydrophobic moments ($\mu_H < 0.3$).[21]

For the construction of the hydrophobic moment profile, the value of μ_H is calculated for a window of 21 residues, and its value is plotted at the central position of the window:

$$\mu_{H,i} = \left(\left[\sum_{j=i-10}^{i+10} H_j \sin(\delta j) \right]^2 + \left[\sum_{j=i-10}^{i+10} H_j \cos(\delta j) \right]^2 \right)^{1/2} \tag{5}$$

The profile calculated with $\delta = 100°$ gives maxima at amphiphilic helices; amphiphilic β-sheets give peaks at $\delta = 180°$. A simple analysis of amphiphilicity can be carried out by plotting hydrophobicity (smoothed by arithmetic averaging over a 21-residue window) and the hydrophobic moments for α-helix ($\delta = 100°$) and β-sheet ($\delta = 180°$) in the same coordinate system. Finer-Moore and Stroud introduced a two-dimensional hydrophobic moment plot in which the sequential position i and δ are the two dimensions.[28] In this representation the amphiphilic segments appear as "hills" at their respective δ values.

Physical Parameters. Basically any physical parameter of the amino acids could be used to construct a structural profile. The available softwares usually contain three of these, *electric charge, side-chain bulkiness,* and the *polarity index* (Tables I and II). The bulkiness parameter is defined as the ratio of side-chain volume to length (in Å^2).[29] The polarity index is proportional to the electric force (originating from charge and dipole effects) due to an amino acid side chain acting on its immediate

[28] J. Finer-Moore and R. M. Stroud, *Proc. Natl. Acad. Sci. U.S.A.* **81,** 155 (1984).

[29] J. M. Zimmerman, N. Eliezer, and R. Simha, *J. Theor. Biol.* **21,** 170 (1968).

surroundings.[29] These parameter sets can be useful if a newly derived sequence is compared to its known homologs.

Average Profiles and Standard Deviation Profiles: Identification of Structurally Conserved and Variable Regions[4]

Average profiles can be used to identify structurally conserved regions in homologous sequences such as members of a gene family. This feature is especially useful when the degree of primary sequence homology is low between the proteins in question. For the calculation of an average profile, the homologous sequences have to be aligned for maximum homology. The average profile of n aligned sequences is defined as

$$\langle P_i \rangle = \left(\sum_{j=1}^{n} P_{i,j} \right) \Big/ n \tag{6}$$

The structural variability that exists among the sequences at a given sequential position can be characterized by standard deviation of the mean $\langle P_i \rangle$:

$$\sigma_i = \left[\sum P_{i,j}^2 - \left(\sum P_{i,j} \right)^2 \Big/ n \right]^{1/2} / (n - 1) \tag{7}$$

The number n is not constant throughout the sequence; it includes only those sequences in which the P_i value is not missing because of an alignment gap. The $\langle P_i \rangle$ and the σ_i profiles, defined by Eqs. (6) and (7), respectively, can be shown in the same coordinate system by one of the following methods: (1) The σ_i profile is drawn symmetric to 0 as a shaded area ($\pm\sigma_{i\,\text{profile}}$). This makes it easy to identify regions where the $\langle P_i \rangle$ profile is significantly different from 0 (i.e., it is outside the shaded area, see Fig. 2). (2) The σ_i profile can be added symmetrically to the $\langle P_i \rangle$ profile as a shaded area ($\langle P_i \rangle \pm \sigma_{i\,\text{profile}}$). This makes it possible to compare an additional sequence to those included in the average. For example, a structural profile of a newly determined sequence can be visually compared to a group of homologous sequences (presented as an average ± standard deviation profile) and segments identified where the new sequence is expected to be different from those already known (i.e., the profile is outside the shaded area). Since the secondary structures are stabilized by short-range interactions, regions displaying no conserved structural propensity ($P_i < \sigma_i$) may be those stabilized by long-range interactions. These regions are likely to "accept" modifications without affecting biological activity of the protein.

The use of average profiles is illustrated on the ribulose 1,5-bisphosphate carboxylase small subunit (Rubisco SSU) gene family.[4] Figure 2 shows averaged α-helix, β-sheet, and turn propensity profiles calculated

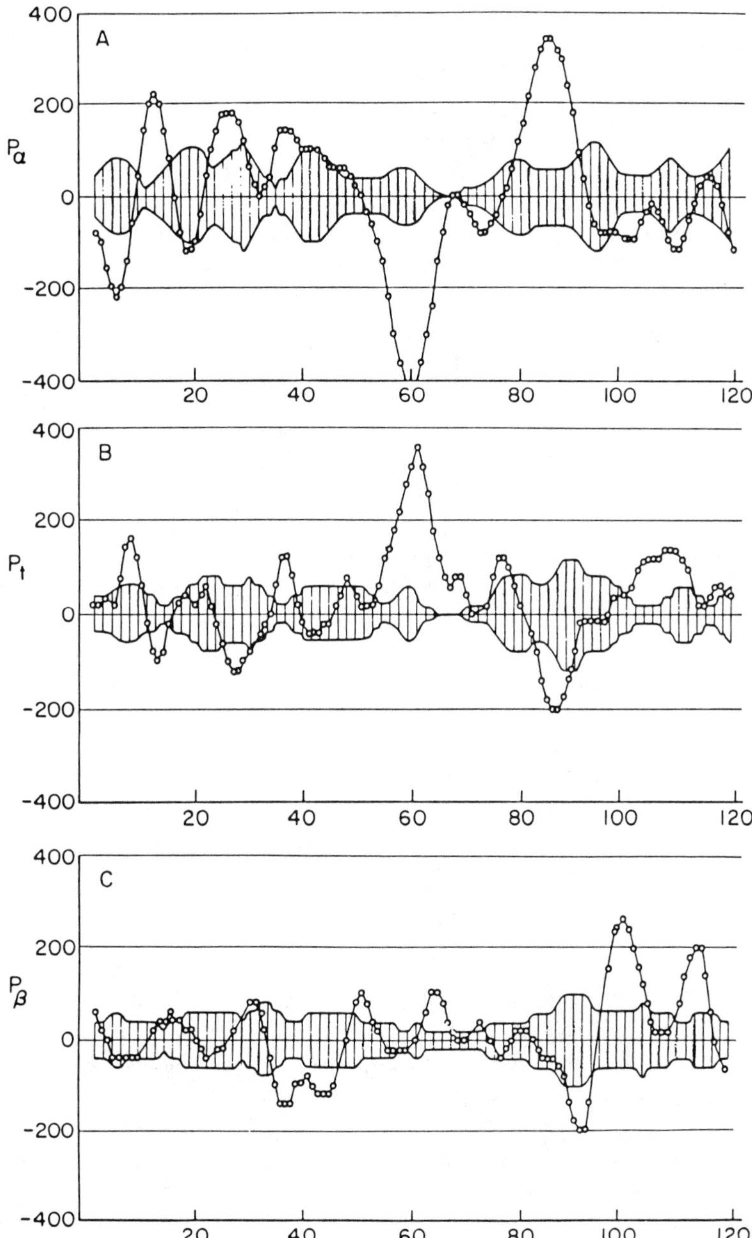

FIG. 2. Averaged α-helix (A), turn (B), and β-structure (C) propensity profiles calculated from ribulose 1,5-bisphosphate carboxylase small subunit sequences of five higher plants (tobacco, spinach, petunia, soybean, and pea).[4] Shaded areas indicate the standard deviation ($\pm\sigma_i$) of the respective structural parameter.

for six ribulose 1,5-bisphosphate carboxylase small subunit proteins from different higher plants. It follows from the probabilistic definition of the P_i values that only those regions for which $P_i > \sigma_i$ (i.e., the structure forming propensity is substantially different from 0) can be considered conserved in the structural sense. It appears that there are only a few regions in which this condition is fulfilled; the rest of the protein has no characteristic structure forming potential. There are two plausible explanations for this finding. First, the structurally determined regions, serving as folding nuclei, may be sufficient for the formation of the native conformation. For example, a conserved ($P_i > \sigma_i$) β-turn-β supersecondary structure can be tentatively identified in the C-terminal region. Second, external factors, such as the large subunit of the enzyme, or long-range interactions may contribute to the formation of the native conformation, so there may be no rigid selectional constraints on the secondary structure forming propensity of the small subunit. Region 54–66 displays a pronounced β-turn forming propensity which is conserved in all the six sequences. However, this region is entirely missing in two homologous proteins isolated from prokaryotic organisms (the cyanobacteria *Synechococcus* R2 and *Anabaena variabilis*). In the higher plant genes this region is located near an intron–exon junction. A comparison of the three-dimensional structures of homologous eukaryotic and prokaryotic proteins has shown that intron–exon junctions in eukaryotic genes frequently coincide with loop structures present on the surface of these proteins.[30] These "variable surface loop" structures were implied to account for functional differences among prokaryotic and eukaryotic members of a gene family. The strong turn forming potential of region 54–66 is in fact consistent with a surface loop structure, which may explain functional differences between higher plant and cyanobacterial Rubisco SSUs.

Comparison of Structural Profiles[3,4]

Homology of proteins has been studied at various levels of structural organization. Comparison of gene sequences at the DNA or amino acid level provides quantitative measures of homology which is essential for establishing phylogenetic relationships. Comparison of three-dimensional structures makes it possible to reveal similarities between proteins that are not demonstrably homologous in their primary structure. Comparison of structural profiles is essentially a parametric approach to the analysis of sequence homology that has several advantages over the comparison of

[30] C. S. Craik, W. J. Rutter, and R. Fletterick, *Science* **220,** 1125 (1983).
[31] R. M. Sweet and D. Eisenberg, *J. Mol. Biol.* **171,** 479 (1983).

primary sequences. First, it makes it possible to describe sequence differences in terms of protein structure. For example one can rank homologous sequences according to a structural parameter (average helicity, hydrophobicity, hydrophobic moment, etc.) and then correlate biological activity with the structural parameter in question. Second, the use of average and standard deviation profiles makes it possible to compare groups of sequences, which is not possible through primary sequence comparison. For example, one can answer questions whether a newly determined protein sequence is within the range of the known homologous sequences in terms of secondary structure, hydrophobicity, etc. Also, the expected structural differences can be located and characterized. Third, the numeric indices described below allow optimization of sequence alignment in equivocal cases.

Before structural profiles are compared, the two sequences have to be aligned for maximum homology by introducing gaps at appropriate positions. Similarly, gaps should be introduced into the calculated structural profiles at the same positions so as to maintain homology. As a rule, positions where one of the residues is missing due to an alignment gap are omitted from the further calculations.

Graphic Comparison. If $P_{i,1}$ and $P_{i,2}$ denote structural parameters of the ith residue in sequences 1 and 2, respectively, then the difference profile can be written as

$$\Delta P_i = P_{i,2} - P_{i,1} \tag{8}$$

and plotted against the sequential position i. Difference plots give a zero baseline for completely homologous regions. Regions that are different in the two sequences will give peaks or valleys, depending on the sign of the difference. The difference profiles are usually scattered graphs that need to be smoothed for data presentation. Quantitative evaluation of ΔP_i is meaningful only on a comparative basis. Such an evaluation is possible if a profile ($P_{i,2}$) is compared to an average profile ($P_{i,1}$) calculated from known homologs with a standard deviation σ_i. In this case σ_i can be considered as a measure of "acceptable" variability and ΔP_i should substantially exceed σ_i for structural differences to be predicted in a given region.

The difference plots shown in Fig. 3 were calculated between α-helix and β-sheet propensity profiles of Rubisco SSU from the prokaryote *Synechococcus* R2 ($P_{i,1}$), with the average of six higher plant Rubisco SSU proteins ($P_{i,2}$).[3] The figure shows the N-terminal part of the ΔP_i profile and the $\pm\sigma_i$ profile which characterizes the variability of the plant sequences. In this region, the *Synechococcus* R2 protein seems to have a higher α-helix forming potential ($\Delta P_{i,\alpha} > 0$) whereas the higher plant proteins

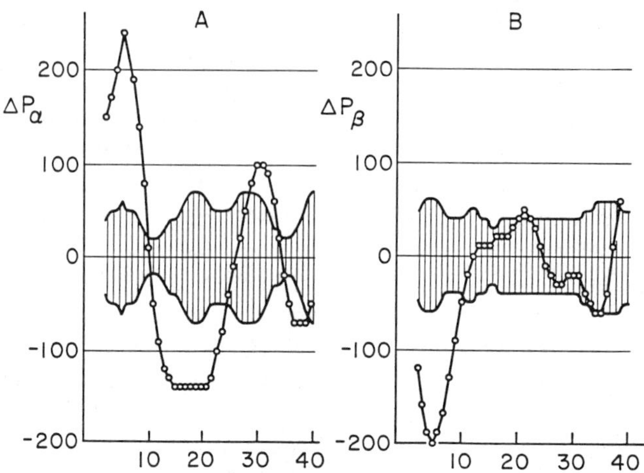

FIG. 3. α-Helix and β-structure propensity difference profiles comparing the N-terminal region of the *Synechococcus* R2 ribulose 1,5-bisphosphate carboxylase small subunit [$P_{i,2}$ in Eq. (8)] to the average profile [$P_{i,1}$ in Eq. (8)] of five membrane-translocated ribulose 1,5-bisphosphate carboxylase small subunit sequences (from tobacco, spinach, petunia, soybean, and pea).

seem to have a greater tendency to form β-structures at the N-terminus ($\Delta P_{i,\beta} < 0$). Rubisco SSU is a membrane-translocated protein in higher plants that is synthesized with a transit peptide whereas its prokaryotic counterparts are cytoplasmic proteins. The finding is thus in accordance with the known propensity of membrane-translocated proteins to form β-structures at their N-termini.

Numeric Comparison. The concept of using *correlation coefficients* to compare structural profiles was developed overall similarities in hydrophobicity[30] and secondary structure.[3,4] With the terminology introduced at Eq. (8) the correlation coefficient calculated between two structural profiles can be written as follows:

$$R = \frac{\sum P_{i,1}P_{i,2}}{\left(\sum P_{i,1}{}^2 P_{i,2}{}^2\right)^{1/2}} \tag{9}$$

R is equal to 1 for two identical sequences, R approximates for two unrelated sequences, and R approaches -1 if the structural profiles are anticorrelated. The z-test of Fisher can be used as a test of significance,[32] i.e., to establish if R is significantly different from 0. To meet the criteria of the z-test, the structural parameters were normalized to mean 0 and

[32] C. M. Goulden, "Methods in Statistical Analysis," 2nd Ed., p. 122. Wiley, New York, 1952.

standard deviation 1. The rescaled structural parameters are summarized in Table I. If the original (not normalized) structural parameters are used to calculate correlation coefficients, the results can be erroneous. For example, the average of the Chou–Fasman parameters is 1 (i.e., >0), so sequences of similar composition are likely to give high correlation coefficients in the absence of any structural similarity if these parameters are used for the calculation of Eq. (9) without normalization. Several authors use the following formula for calculating correlation coefficients from nonnormalized parameters:

$$R = \frac{\sum ([P_{i,1} - \langle P_{i,1}\rangle][P_{i,2} - \langle P_{i,2}\rangle])}{\left(\sum [P_{i,1} - \langle P_{i,1}\rangle]^2[P_{i,2} - \langle P_{i,2}\rangle]^2\right)^{1/2}} \tag{10}$$

where $\langle P_{i,1}\rangle$ and $\langle P_{i,2}\rangle$ are mean values calculated in the sequential region included in the summation.

The correlation coefficient can be used as an indicator of structural similarity. The procedure is illustrated on synthetic peptide analogs that have known differences in secondary structure as compared to their natural counterparts.[33] The synthetic peptides were designed so as to have the following characteristics relative to the corresponding natural peptides: (1) equal or increased helix content; (2) conserved charged residues and hydrophobic/hydrophilic balance; (3) low (20–50%) primary sequence homology.

Melittin is a 26-residue hemolytic peptide with a segment that has the potential to form an amphiphilic helix. When the sequence of the amphiphilic segment was rearranged according to the above criteria, a biologically active analog was obtained that was higher in α-helix content (35%) than the native melittin (18%) as determined by circular dichroism.[34] Structural correlation values listed in Table VIII reflect the differences in α-helix formation: The value of the α-helix correlation R_α is the lowest among all values compared. The correlation coefficient can also be used to rank sequences according to their expected structural similarity. For example, analogs of β-endorphin representing a range of α-helix content were correctly ranked using R_α alone.[35] The correlation coefficient is an index of variability and gives no information about the direction of the difference (this can be established from a difference plot). Finally we mention that the structural correlation coefficient can also be used to describe evolutionary changes in protein secondary structure.[3]

There are two other numeric indices that allow quick evaluation of

[33] E. T. Kaiser and F. J. Kézdy, *Proc. Natl. Acad. Sci. U.S.A.* **80,** 1137.

[34] W. F. DeGrado, G. F. Musso, M. Lieber, and E. T. Kaiser, *Biophys. J.* **37,** 329.

[35] J. W. Taylor, R. J. Miller, and E. T. Kaiser, *J. Mol. Pharmacol.* **22,** 657.

TABLE VIII
STRUCTURAL CORRELATION CALCULATED BETWEEN MELITTIN AND ITS SYNTHETIC ANALOG[a]

Structural profile	R^b
α-Helix propensity (Garnier *et al.*[6])	0.11
Extended structure (Garnier *et al.*[6])	0.89
β-Turn (Garnier *et al.*[6])	0.81
Coil (Garnier *et al.*[6])	0.54
α-Helix (Chou–Fasman[5]	0.18
β-Sheet (Chou–Fasman[5])	0.79
β-Turn (Chou–Fasman[5])	0.53
Hydrophobicity (Kyte–Doolittle[16])	0.94
Hydrophilicity (Hopp–Woods[22])	0.90
Charges	0.89

[a] Sequences[34]: Melittin, G I G A V L K V L T T G L P A L I S W I K R K R Q Q: Synthetic analog, L L Q S L L S L L Q S L L S L L L Q W L K R K R Q Q.

[b] *R* values given were calculated by using Eq. (9) (P_1, melittin; P_2, synthetic analog). *R* values above 0.40 are significant at the 99.99% level.

multiple amino acid replacements in shorter segments: (1) $\langle \Delta P_i \rangle$ is the average difference (sum of the ΔP_i values calculated between the two sequences divided by the number of residues in the segment and (2) $\langle |\Delta P_i| \rangle$, the average absolute difference $\langle \Delta P_i \rangle$ is the average value of the difference profile, i.e., it contains the same information. $\langle |\Delta P_i| \rangle$ is an indicator of variability. The use of these indices is illustrated in Ref. 4.

Computer Programs

Sequence management programs used in most molecular biology laboratories (for reviews, see Ref. 36 and F. Lewitter and W. Rindone, Vol. 155 [36]) are generally capable of drawing structural profiles from primary sequence data, even though they may not contain all parameter sets summarized in this chapter. Averaging of profiles is feasible in some computer programs (e.g., Ref. 37), but the features crucial for engineering applications (standard deviation profiles and the numeric indices of structural

[36] L. J. Korn and C. Queen, *DNA* **3,** 421 (1984).

[37] J. Novotny and C. Auffray, *Nucleic Acids Res.* **12,** 243 (1984).

comparison) are not standard options in the generally used molecular biology softwares. These capabilities can be either incorporated into separate programs (such as the protein sequence evaluation package developed by the author[4]), or integrated into available sequence management programs. The design of the programs should make it possible to carry out any combination of the numeric and graphic procedures outlined above. It is also recommended that segments of sequences could be separately analyzed. Curve smoothing and graphic output can be performed by commercially available microcomputer software.

Comments: Evaluation of the Results

Since engineering of a protein segment can have a variety of goals (e.g., reinforcement or disruption of structural elements, removal of sites for enzymatic or chemical attack), there are no generally applicable rules for the evaluation of the results. In most cases the problem itself defines the solution, and the experimenter can simply select the most promising replacement alternatives, using the numeric and graphic methods outlined here. General guidelines can be given, however, for two experimental situations: (1) The sequence to be redesigned has no known homologs. In this case it is recommended to carry out secondary structure prediction and to draw helical wheel and helical net diagrams for the segment of interest. On this basis one might be able to recognize a structural motif (e.g., Fig. 1) which makes it possible to formulate a structural hypothesis for amino acid replacements. Then the sequences containing the proposed alterations can be ranked in computer experiments using the methods presented here. (2) If the sequence to be redesigned has several known homologs (with differences in their primary sequence) then average profiles and standard deviation profiles can be constructed with the parameters thought to be important in protein folding, i.e., secondary structure propensities, hydrophobicity, polarity, and side-chain bulkiness. Through visual comparison of the profiles it can then be established which of these properties is the most conserved within the group. (For the comparison between the parameters, the normalized parameter sets summarized in Table II have to be used.)

Additional information can be obtained from the predicted secondary structure of the individual sequences as well as from averaged predictions.[6,36] As an example, Fig. 4A shows the primary sequence of transit peptides that mediate passage into the chloroplasts of proteins synthesized in the cytoplasm. Although chloroplasts of higher plants reciprocally recognize each other's transit peptides, the primary sequences show little homology. On the other hand, the predicted secondary structures

A

```
                 -50       -40       -30       -20       -10       -1 1
Chl. AB80                     MAASSSSSMALSSPTLAGKOLKLNPSSQELCAARFTMRK SATTK
Soybean SS    MASSMISSPAVTTVNRAGAGMVAPFTGLKSMAGFPTRKTNNDITSIASNGGRVQC MQVWP
Pea SS3.6   MASMISSSAVTTVSRASRGQSAAVAPFGGLKSMTGFPVKKVNTDITSITSNGCRVKC MQVWP
Wheat SSW9            MAPAVMASSATTVAPFQGLKSIACLPISCRSGSTGLSSVSNGGRIRC MQVWP
```

B

```
                 -50       -40       -30       -20       -10       -1 1
Chl. AB80                     hhhhcccEEEEccccEEEEEhhtccchhhhhhhhhhhh hhhhh
Soybean SS    hhhhecccEEEEEEEcccchEEEEEccctcEEEEEcctctttEEEEEcctttttEE EEEtt
Pea SS3.6   hhhhhcccEEEEEEEcctchhhhEEEcctctcEEEEEEEtcttEEEEEEcctttttEE Ettcc
Wheat SSW9            hhhhhhhhhhccEEEEctttcccttEEEEtttttttEEEEcctttttEE EEhtt
```

FIG. 4. Amino acid sequence (A) and predicted secondary structure (B) of transit peptides that mediate the posttranslational transport of precursor proteins into the chloroplast.[4] Chl AB80, chlorophyll *a*/*b* binding polypeptide; SS, ribulose 1,5-bisphosphate carboxylase small subunit transit peptide. Method was by Garnier *et al.* (unbiased prediction).[6]

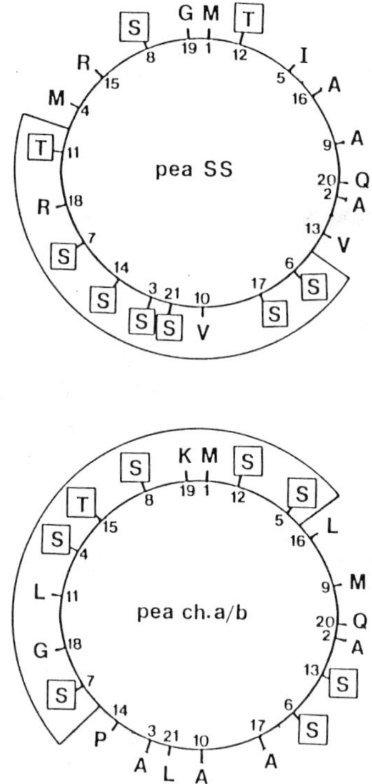

FIG. 5. Helical wheel representation of the N-termini of two chloroplast transit peptides. The hydroxyamino acids are boxed (abbreviations are as in Fig. 4).

show a distinct similarity: they all contain an α-helix at the N-terminus followed by a repetitive pattern of β-strands (Fig. 4B). In addition, the helical wheel diagrams shown in Fig. 5 show that the hydroxyamino acids, serine and threonine, seem to form asymmetric clusters on the N-terminal helices. Naturally, these and similar findings are only hypothetical and need experimental confirmation. On the other hand, they can provide a rational framework for experimental design and thereby reduce the number of directed mutagenesis experiments necessary for obtaining biologically active protein analogs.

[25] Structure–Function Analysis of Proteins through the Design, Synthesis, and Study of Peptide Models

By JOHN W. TAYLOR and E. T. KAISER

In recent years, a large number of intermediate-sized biologically active peptides have been identified and their amino acid sequences determined. Unless they contain multiple disulfide linkages, these peptides usually do not form very stable secondary or tertiary structures in aqueous solution. However, they often have the potential to form segments of amphiphilic secondary structures that might be induced on binding to a matching amphiphilic environment provided by the biological interfaces at which their activities are expressed.[1,2] Understanding structure–function relationships in such peptides requires the identification of these potential structures and rigorous testing of their importance.

A variety of peptides, including serum apolipoproteins,[3,4] toxins,[5–7] chemotactic factors,[8] and peptide hormones,[9–12] contain segments of their

[1] E. T. Kaiser and F. J. Kezdy, *Proc. Natl. Acad. Sci. U.S.A.* **80,** 1137 (1983).

[2] E. T. Kaiser and F. J. Kezdy, *Science* **223,** 249 (1984).

[3] W. M. Fitch, *Genetics* **86,** 623 (1977).

[4] A. D. McLachlan, *Nature (London)* **267,** 465 (1977).

[5] W. F. DeGrado, F. J. Kezdy, and E. T. Kaiser, *J. Am. Chem. Soc.* **103,** 679 (1981).

[6] A. Argiolas and J. J. Pisano, *J. Biol. Chem.* **260,** 1437 (1985).

[7] D. Andreu, R. B. Merrifield, H. Steiner, and H. G. Boman, *Biochemistry* **24,** 1683 (1985).

[8] D. G. Osterman, G. F. Griffin, R. M. Senior, E. T. Kaiser, and T. F. Deuel, *Biochem. Biophys. Res. Commun.* **107,** 130 (1982).

[9] J. W. Taylor, D. G. Osterman, R. J. Miller, and E. T. Kaiser, *J. Am. Chem. Soc.* **103,** 6965 (1981).

[10] G. R. Moe, R. J. Miller, and E. T. Kaiser, *J. Am. Chem. Soc.* **105,** 4100 (1983).

[11] S. H. Lau, J. Rivier, W. Vale, E. T. Kaiser, and R. J. Kezdy, *Proc. Natl. Acad. Sci. U.S.A.* **80,** 7070 (1983).

METHODS IN ENZYMOLOGY, VOL. 154

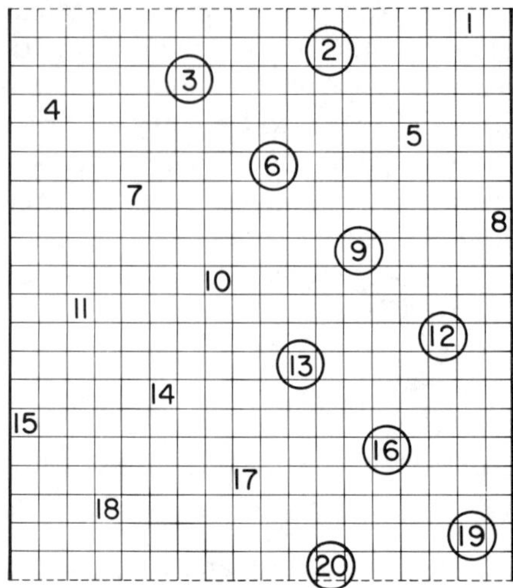

FIG. 1. Construction of a helical net diagram. The relative positions of the amino acid side chains on the surface of an α helix (3.6 residues/turn) are marked by the numbers of the residues in a 20-residue peptide. If, for example, the amino acid residues that are circled are the only hydrophobic residues in the sequence, the entire peptide will be amphiphilic in an α-helical conformation.

linear amino acid sequences that have the potential to form amphiphilic α-helical structures, where the hydrophobic residues are clustered together in a single domain on the helix surface. A diagramatic representation that is suitable for identifying such structures rapidly is the α-helical net.[13] This is a two-dimensional representation of the relative positions of the amino acid side chains on a helix surface that is based on the regular α-helical structure of 3.6 residues per turn, so that 18 residues form exactly five complete turns of the helix.[14] When drawn on squared paper, the helix surface is represented by a rectangle 18 squares wide and 1 square long for each residue to be included. Successive residues in the linear sequence are then connected to one another as shown in Fig. 1 by moving 5 squares leftward and 1 square downward per residue, and by remembering that the perpendicular edges of the diagram are connected behind the plane of the paper when the cylindrical surface of the α helix is

[12] R. M. Epand, R. F. Epand, S. W. Hui, N. B. He, and M. Rosenblatt, *Int. J. Pept. Protein Res.* **25,** 594 (1985).

[13] P. Dunnill, *Biophys. J.* **8,** 865 (1968).

[14] R. E. Dickerson and I. Geis, *in* "The Structure and Action of Proteins," p. 28. Benjamin, London, 1969.

reconstructed. Any areas of this diagram having only contiguous hydrophobic residues represent regions of potential amphiphilic α-helical structure in the amino acid sequence. In the particular case when a hydrophobic domain lying parallel to the axis of the helix is revealed, the end-on view of a helical wheel diagram is a suitable alternative representation that illustrates the amphiphilic nature of the structure.[2]

It is necessary, of course, to define what is meant by a hydrophobic or hydrophilic residue. Based on the parameters of Levitt,[15] it is clear that Phe, Tyr, Trp, Leu, Ile, Val, Pro, and Met may be classified as hydrophobic, and that the charged residues Lys, Arg, Asp, and Glu are hydrophilic. In addition, we have generally categorized Ser, Thr, Asn, Gln, and Gly as moderately hydrophilic and Ala, His, and Cys as moderately hydrophobic. However, the occurrence of a single residue from these latter groups in the "wrong" domain may not represent a serious disruption of any extensive amphiphilic structure, depending on the function of that structure. Thus, for example, any such apparently misplaced residues may be tolerated in interactions with phospholipid surfaces, which are expected to be less specific than protein–protein interactions, and may be specifically accommodated by the protein receptor of a peptide hormone.

Furthermore, in other studies, the amino acids have been classified somewhat differently so that, in particular, Tyr, His, and Trp may be less hydrophobic than is suggested above.[16] It is also worthwhile considering that certain residues—particularly Thr and Lys—have side chains which are themselves amphiphilic and could extend a region of hydrophobicity if they are located on its border, and that His is hydrophilic when it is protonated at lower pH. Finally, Pro and Gly are considered to be particularly disruptive of α-helical structure on the basis of their infrequent occurrence in the helices of known protein structures,[17] although Pro might be expected to form part of a hydrophobic domain if it were located in the first position at the amino-terminal end of a helix, where its inability to participate in the regular α-helical structure would not be important.

In order to test the significance of a potential amphiphilic α-helical structure in determining the properties of a biologically active peptide, we have adopted the approach of synthesizing and testing peptide models of the proposed structure.[1,2] In these peptide models, the helical segment is replaced by a sequence of amino acid residues designed to have minimal homology to the natural sequence, but which retains the general features

[15] M. Levitt, *J. Mol. Biol.* **104,** 59 (1976).

[16] R. Wolfenden, L. Andersson, P. M. Cullis, and C. C. B. Southgate, *Biochemistry* **20,** 849 (1981).

[17] P. Y. Chou and G. D. Fasman, *Annu. Rev. Biochem.* **47,** 251 (1978).

of the amphiphilic structure in an idealized form. These general features would usually include the length of the helix, the size, shape, and relative orientation of its hydrophobic domain, and the distribution of positive and negative charge in the hydrophilic domain. They may be entirely reproduced using combinations of naturally occurring residues that are expected to have a high propensity for α-helix formation, such as the hydrophobic residue Leu and the neutral, basic, and acidic hydrophilic residues Gln, Lys, and Glu, respectively.[17] Understanding the functional significance of a potential amphiphilic α helix by this approach is then dependent on the ability of the model peptide to reproduce the properties of the natural peptide, and upon the extent to which the amino acid sequence has been changed. The subsequent demonstration that the properties tested have a definite requirement for the model helix, through the study of "negative" models where features of the helical structure are purposefully absent, is also an essential part of this approach.[18]

Analysis of the Structure of β-Endorphin

Human β-endorphin, a naturally occurring opioid peptide, consists of 31 amino acid residues and has no disulfide bridges (Fig. 2a).[19,20] The amino-terminal 5 residues are identical to the complete structure of [Met^5]enkephalin, which is itself a potent, naturally occurring opioid peptide,[21] and small changes in the structure of this segment usually result in a considerable or complete loss of opioid activities, indicating that its interactions with opiate receptors are highly specific. An initial glance at the remainder of the β-endorphin structure reveals that all of the other hydrophobic residues are located on the carboxy-terminal side of Pro^{13}. When this segment of the peptide is displayed on a helical net diagram (Fig. 2b), it is clear that a regular α-helical structure would be amphiphilic for four complete turns, possessing a hydrophobic domain covering about two-thirds of its surface from Pro^{13} to Tyr^{27} and lying along the length of the helix but not parallel to the helix axis—instead, twisting around the helix in a clockwise direction. The likely end points of this potential helix are the structure breakers Pro^{13} and Gly^{30}, with the former residue included in the helix to form part of the hydrophobic domain and the latter

[18] J. P. Blanc, J. W. Taylor, R. J. Miller, and E. T. Kaiser, *J. Biol. Chem.* **258,** 8277 (1983).

[19] A. F. Bradbury, D. G. Smyth, C. R. Snell, N. J. M. Birdsall, and E. C. Hulme, *Nature (London)* **260,** 793 (1976).

[20] B. M. Cox, A. Goldstein, and C. H. Li, *Proc. Natl. Acad. Sci. U.S.A.* **73,** 1821 (1976).

[21] A. Waterfield, R. W. J. Smokcum, J. Hughes, H. W. Kosterlitz, and G. Henderson, *Eur. J. Pharmacol.* **43,** 107 (1977).

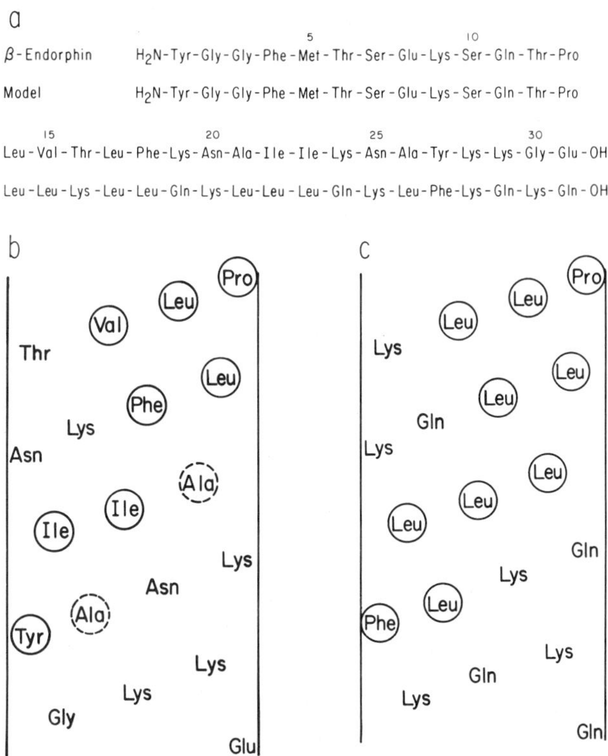

FIG. 2. (a) Amino acid sequences of β-endorphin and the model peptide. (b) α-Helical net diagram of β-endorphin residues 13–31. (c) α-Helical net diagram of the model peptide residues 13–31. Hydrophobic residues are circled.

residue possibly excluded. In this case, the hydrophilic domain would have a strongly basic character, the only charged residues being Lys^{19}, Lys^{24}, Lys^{28}, and Lys^{29}.

The functional significance of an amphiphilic α-helical structure in the carboxy-terminal region of β-endorphin is suggested by several studies. The decrease in the binding potency of carboxy-terminal deletion analogs to rat brain membrane receptors labeled with tritiated porcine β-endorphin correlates with decreased α-helical structure in that domain in 75% TFE.[22] Similarly, the successive removal of carboxy-terminal residues from β-endorphin leads to a sudden drop in potency on the types of opiate receptors mediating analgesic activity in the central nervous sys-

[22] R. G. Hammonds, Jr., A. S. Hammonds, N. Ling, and D. Puett, *J. Biol. Chem.* **257,** 2990 (1982).

tem[23,24] and the inhibition of electrically stimulated contractions in the rat vas deferens.[25] Furthermore, the resistance of β-endorphin toward the degradative actions of amino- and endo-peptidases that can rapidly hydrolyze [Met[5]]enkephalin has been demonstrated under a variety of conditions and appears to be related to the formation of a tertiary structure involving the carboxy terminus.[9,26–28] Thus, the potential amphiphilic α helix in residues 13 through 29 of β-endorphin may form an important binding site for certain types of opiate receptors, resulting in a range of activities that differ from those of the enkephalins, and also confer enough stability to the whole peptide so that it can act at receptors more distant from its point of release than can the enkephalins.

Design of a β-Endorphin Model Peptide

Several peptide models of β-endorphin have been designed and studied in order to investigate the important characteristics and functions of the potential amphiphilic α-helical segment in residues 13–29 of the natural peptide. These include (1) peptide models in which the shape of the hydrophobic domain was altered so that it lay parallel to the α-helix axis, in case β-endorphin formed such a structure by adopting a π-helical or discontinuous α-helical conformation at receptor surfaces[9,29]; (2) a peptide model in which the amphiphilic α-helical structure was reproduced in an idealized left-handed form by incorporation of only D-amino acid residues in that region[30]; and (3) a negative model in which the carboxy-terminal region of the peptide was carefully designed to minimize the possibility of any type of helix having significant amphiphilicity.[18]

Presently, however, we will describe the design and properties of our most conservative peptide model of β-endorphin,[31] as this illustrates the best initial approach to such an investigation. This model peptide (Fig. 2a) retained residues 1–12 of the β-endorphin sequence, which were connected to a linear sequence of Leu, Gln, and Lys residues chosen to

[23] J. F. W. Deakin, O. J. Dostrovsky, and D. G. Smyth, *Biochem J.* **189,** 501 (1980).

[24] P. Ferrara and C. H. Li, *Int. J. Pept. Protein Res.* **19,** 259 (1982).

[25] R. Schulz, M. Wuster, and A. Herz, *J. Pharmacol. Exp. Ther.* **216,** 604 (1981).

[26] L. Graf, G. Coeh, E. Barat, A. Z. Ronai, J. I. Szekely, A. Kennesey, and S. Bajusz, *Ann. N.Y. Acad. Sci.* **297,** 63 (1977).

[27] B. M. Austin and D. G. Smyth, *Biochem. Biophys. Res. Commun.* **76,** 477 (1977).

[28] C. H. Li and D. Chung, *Int. J. Pept. Protein Res.* **26,** 113 (1985).

[29] J. W. Taylor, R. J. Miller, and E. T. Kaiser, *Mol. Pharmacol.* **22,** 657 (1982).

[30] J. P. Blanc and E. T. Kaiser, *J. Biol. Chem.* **259,** 9549 (1983).

[31] J. W. Taylor, R. J. Miller, and E. T. Kaiser, *J. Biol. Chem.* **258,** 4464 (1983).

minimize homology to β-endorphin residues 13–31, but arranged in a sequence that allowed the formation of a similar amphiphilic α helix. Projected on a helical net diagram (Fig. 2c), these residues form a hydrophobic domain of the same length, shape, size, and orientation relative to the rest of the peptide as the corresponding structure in β-endorphin and a hydrophilic domain that consists of only neutral and basic residues. An exception to the rule of minimizing sequence homology to the natural peptide was made to include Pro^{13} in the model peptide for its unique properties as a helix breaker that would define the amino-terminal end of the helix when it forms,[17] and which might also limit the conformational flexibility of the rest of the peptide relative to the helix on a receptor surface. Similarly, an examination of a Corey–Pauling–Koltun model of β-endorphin indicated that the Tyr^{27} side chain was a very prominent feature on the surface of the potential amphiphilic α-helix which could not be reproduced by inclusion of a Leu residue at this point, and so Phe^{27} was incorporated into the model structure instead.

The decision to include these residue-specific features in this model peptide illustrates the additional considerations that need to be applied to the design of model peptide hormones. Earlier studies of peptide models of the serum apolipoproteins[32,33] and melittin,[5] proteins which interact with phospholipid surfaces, were successful when residue-specific features were completely excluded and only the correct general features of the natural amphiphilic structures were retained. An amphiphilic α helix in a peptide hormone may, however, have additional side-chain-specific interactions with, for example, the surface of a protein receptor. Depending on the importance of these interactions, the absence of such features in a peptide model may result in a substantial or complete loss of the related activities. Even in extreme cases, though, such specific interactions are likely to be limited to the hydrophobic domain of the amphiphilic structure, and then the modeling approach can be applied exclusively to the hydrophilic side of this structure and still give convincing results based on minimizing sequence homology. In the case of the present β-endorphin model, there are fourteen differences in residues 13–31 of the model peptide compared to the same region of β-endorphin, without including the conservative Tyr^{27} to Phe substitution. This corresponds to only 26% homology in that region.

[32] D. Fukushima, J. P. Kupferberg, S. Yokoyama, D. J. Kroon, E. T. Kaiser, and F. J. Kezdy, *J. Am. Chem. Soc.* **101,** 3703 (1979).

[33] S. Yokoyama, D. Fukushima, J. P. Kupferberg, F. J. Kezdy, and E. T. Kaiser, *J. Biol. Chem.* **255,** 7333 (1980).

Materials

Most of the reagents for peptide synthesis are available from several sources, including Bachem, U.S. Biochemical Corporation, Applied Biosystems, and Peninsula. These include the Merrifield Resin (chloromethylated, 1% cross-linked polystyrene–divinylbenzene copolymer, 200–400 mesh, having approximately 0.6 mmol Cl/g resin), benzhydrylamine (BHA) resin or *p*-methyl BHA resin (2% cross-linked, 200–400 mesh, having 0.2–0.6 amino groups/g resin), and the N^{α}-*tert*-butyloxycarbonyl (Boc) derivatives of the following side-chain protected L-amino acids: Ala; N^{ω}-tosyl-Arg; Asp, β-benzyl ester; Asn; *S*-(4-methylbenzyl)-Cys; Glu, γ-benzyl ester; Gln; Gly; N^{im}-tosyl-His; Ile; Leu; N^{ε}-(2-chlorobenzyloxycarbonyl)-Lys; Met; Phe; Pro; *O*-benzyl-Ser; *O*-benzyl-Thr; N^{i}-formyl-Trp; *O*-(2,6-dichlorobenzyl)-Tyr; Val. Amino acid derivatives should be checked for purity by elution on silica gel TLC plates with chloroform containing 5% acetic acid and 0–10% methanol, followed by visualization using a ninhydrin spray (1 mg/ml in *n*-butanol containing 10% acetic acid) and heating for 15 min at 100°.

The solvents employed in peptide synthesis are prepared as follows. Methylene chloride (CH_2Cl_2) is dried by refluxing with added P_2O_5 (30 g/liter) for 30 min and then distilling, or is used directly, and is stored over molecular sieves (3 Å). Diisopropylethylamine (DIEA) is dried by standing over KOH pellets for 2 days, then is decanted, refluxed with ninhydrin (30 g/liter) for 30 min, and distilled, collecting the middle fraction. Isopropanol is refluxed with added CaO (30 g/liter) for 30 min and then distilled and stored over molecular sieves (4 Å). Trifluoroacetic acid (TFA) is distilled slowly, collecting the first two-thirds of the distillate, and is always used within 2 days. Anhydrous dimethylformamide (DMF) and trifluoroethanol (TFE), both purchased from Aldrich (Gold Label grade), and ethanol (U.S.I. Absolute) are used directly.

Dicyclohexylcarbodiimide (DCC) is distilled under reduced pressure and is then used as a 500 m*M* stock solution in dry CH_2Cl_2. Gloves should always be worn when handling this reagent and solutions containing it, as it can cause severe allergic reactions. Picric acid is used as a 100 m*M* solution in CH_2Cl_2. Reagents for the Kaiser test for completion of coupling[34] are as follows: ninhydrin (250 mg) in ethanol (5 ml); phenol (80 g) in ethanol (20 ml); 1 m*M* aqueous KCN (2 ml) in pyridine (99 ml). Other required reagents include anhydrous HF, anisole, anhydrous KF, acetic anhydride, 1-hydroxybenzotriazole, polyethylene glycol (molecular

[34] E. Kaiser, R. L. Colescott, C. D. Bossinger, and P. I. Cook, *Anal. Biochem.* **34,** 595 (1970).

weight 15,000–20,000) or polyethylenimine, tris(hydroxymethyl)aminomethane (Aldrich, Gold Label grade), and Ultrapure KCl (Alfa Products).

Methods

A. Attachment of the Carboxy-Terminal Amino Acid to the Merrifield Resin Using Anhydrous KF[35]

1. Place the resin (4 g containing ~2.4 mmol Cl) in a sintered glass funnel (70–100 μm pore size). Wash with DMF (3 × 25 ml) and then transfer to a long-necked round-bottom flask.
2. Add the appropriate amino acid derivative (1.2 mmol), anhydrous KF (2.4 mmol), and DMF (20 ml). Stopper the flask securely and shake it in an oil bath at 50° for 24 hr, using a wrist-action shaker.
3. Transfer the resin back to the sintered glass funnel and wash with DMF (2 × 25 ml), 50% aqueous DMF (4 × 50 ml), 50% aqueous ethanol (4 × 50 ml), ethanol (2 × 25 ml), CH_2Cl_2 (50 ml), and ethanol (2 × 50 ml). Dry the resin overnight in a vacuum desiccator containing P_2O_5 pellets and then determine the substitution level by picric acid titration.

B. Attachment of the Carboxy-Terminal Amino Acid to the BHA Resin or the p-Methyl-BHA Resin

These resins are used to produce a carboxy-terminal amide. The *p*-methyl BHA resin is preferred only when the carboxy-terminal residue is aromatic, to ensure a high yield when the peptide is ultimately cleaved from the resin with anhydrous HF.[36]

1. Wash the resin (3 g containing 0.6–2.4 mmol amine) in a sintered glass funnel (70–100 μm pore size) with DMF (3 × 25 ml), 5% DIEA in CH_2Cl_2 (25 ml), and CH_2Cl_2 (3 × 25 ml), and then transfer to a long-necked round-bottom flask.
2. Prepare the symmetric anhydride of the appropriate amino acid derivative by adding 500 m*M* DCC in CH_2Cl_2 (1.8 ml) to an ice-cold solution of the amino acid derivative (1.8 mmol) in CH_2Cl_2 (15 ml). Stir at 4° for 15 min, during which a white precipitate of dicyclohexylurea (DCU) will appear.
3. Filter the symmetric anhydride solution through a sintered glass funnel (10–20 μm pore size) and wash the DCU precipitate with

[35] K. Horiki, K. Igano, and K. Inouye, *Chem. Lett.* **1978,** 165 (1978).

[36] G. R. Matsueda and J. M. Stewart, *Peptides* **2,** 45 (1981).

CH_2Cl_2 (5 ml). Collect the combined filtrate and immediately add this to the washed resin and shake the flask at room temperature for 30 min.

4. Repeat step 1, then add a mixture of acetic anhydride (1 ml) in CH_2Cl_2 (20 ml) to the washed resin, and shake this at room temperature for 30 min.
5. Wash the resin as in step 1, dry it over P_2O_5 pellets in a vacuum desiccator overnight, and then determine the substitution level by picric acid titration.

C. *Picric Acid Titration of Resins*

This titration is performed in triplicate, according to the method of Gisin.[37]

1. Place accurately weighed samples of the substituted resin (30–40 mg) in 3-ml sintered glass funnels (10–20 μm pore size) and wash with CH_2Cl_2 (2 × 2 ml) and 50% TFA in CH_2Cl_2 (2 ml).
2. Suspend the resin samples in 50% TFA in CH_2Cl_2 (2 ml) without filtration for 20 min. Then filter and wash the deprotected resin samples with CH_2Cl_2 (2 × 2 ml) and isopropanol (2 × 2 ml).
3. Change the filtering flasks and wash the resin samples with 5% DIEA in CH_2Cl_2 (2 × 2 ml), CH_2Cl_2 (5 × 2 ml), 100 m*M* picric acid solution in CH_2Cl_2 (2 × 2 ml), and then CH_2Cl_2 (5 × 2 ml).
4. Change the filtering flasks again and elute the yellow picric acid from the resin with 5% DIEA in CH_2Cl_2 until the eluate is clear, using less than 20 ml of the DIEA solution. Make this eluate up to 100 ml with EtOH and quantitate the picric acid (and hence the N^{α}-Boc-protected amino acid on the resin) by measuring the absorbance of this solution at 358 nm and using $\varepsilon = 14{,}500\ M^{-1}\ cm^{-1}$. If the amino acid substitution level is outside the range of 0.2–0.3 mmol/g resin, repeat the resin preparation using appropriately adjusted quantities of the reagents.

D. *Peptide Synthesis*[38,39]

Reactions should be performed at room temperature under dry nitrogen, using a programmable, automated instrument or a suitable, manually

[37] B. F. Gisin, *Anal. Chim. Acta* **58,** 248 (1972).

[38] G. Barany and R. B. Merrifield, *in* "The Peptides" (E. Gross and J. Meinenhofer, eds.), Vol. 2, pp. 3–254. Academic Press, London, 1979.

[39] J. M. Stewart and J. D. Young, *in* "Solid Phase Peptide Synthesis." Freeman, San Francisco, California, 1969.

TABLE I
CYCLIC PROCEDURE USED IN SOLID-PHASE PEPTIDE SYNTHESIS

Step	Procedure	Mix time (min)	Number of times performed
1	Add dry CH_2Cl_2, mix, drain	1	4
2	Add dry TFA/CH_2Cl_2 (1/1), mix, drain	1	1
3	Add dry TFA/CH_2Cl_2 (1/1), mix, drain	20	1
4	Add dry CH_2Cl_2, mix, drain	1	1
5	Add dry *i*-PrOH, mix, drain	1	2
6	Add CH_2Cl_2, mix, drain	1	3
7	Add 5% DIEA in CH_2Cl_2, mix, drain	2	1
8	Add CH_2Cl_2, mix, drain	1	2
9	Add 5% DIEA in CH_2Cl_2, mix, drain	2	1
10	Add CH_2Cl_2, mix, drain	1	5
11	Add symmetric anhydride, mix	20	1
12	Add DIEA/TFE,[a] mix, drain	10	1
13	Add CH_2Cl_2, mix, drain	1	3
14	Add EtOH/CH_2Cl_2 (1/2), mix, drain	1	3

[a] TFE should be 20% after addition to the coupling reaction and should contain 1.5 equivalents of DIEA, based on the number of amine groups on the resin.

operated reaction vessel. In either case, reaction volumes should be adjusted to ensure that all of the resin is subjected to all of the steps, and the stirring or shaking of the reaction mixtures must not cause mechanical breakdown of the resin.

1. Add the derivatized resin to the reaction vessel (0.25–0.5 mmol amine groups by picric acid titration is usually a convenient scale for synthesis) and subject it to the sequence of reactions and washings described in Table I.[40] Volumes should be adjusted to about 20–40 ml for each step, depending on the quantity and type of resin employed. BHA resins tend to swell more than the Merrifield resin, particularly during the deprotection steps using TFA, and require larger reaction volumes. Prepare the symmetric anhydride of the appropriate amino acid derivative immediately prior to its use. Use 6 equivalents of the amino acid derivative and 3 equivalents of DCC, based on the number of amine groups on the resin, and follow Procedure B, step 2, using appropriately larger volumes of CH_2Cl_2 as required. It is necessary to add up to 50% DMF to some symmetric anhydride preparations in order to solubilize the amino

[40] D. Yamashiro and C. H. Li, *J. Am. Chem. Soc.* **100,** 5174 (1978).

acid derivative or the symmetric anhydride. Provided that N^{im}-tosyl-His has not already been incorporated into the peptide chain, N^{α}-Boc-Asn and N^{α}-Boc-Gln are preferentially coupled using 3 equivalents of these compounds and 3 equivalents of 1-hydroxybenzotriazole in place of the 6 equivalents of the amino acid derivative, and extending the reaction time in step 11 (Table I) to 120 min.[38]

2. At the end of step 14, transfer a small quantity of the resin from the reaction vessel into a glass test tube and add 1 drop of each of the Kaiser test reagents.[34] Heat the tube in boiling water for at least 5 min. If the beads or the solution show any blue coloration, repeat the test to confirm the result, and then repeat steps 9–14 of the reaction sequence in Table I. In this recoupling procedure, the quantity of symmetric anhydride prepared may be reduced 3-fold. Alternatively, it may be entirely replaced by acetic anhydride (0.5 ml) in the appropriate volume of CH_2Cl_2, if chain termination at the unreacted sites is preferred.
3. When a negative Kaiser test is obtained, the reaction sequence in Table I is repeated using the appropriate derivatives of the next amino acid in the peptide sequence, so that the peptide chain is built onto the resin in the direction proceeding from the carboxy terminus toward the amino terminus.
4. After completion of the last reaction cycle coupling the amino-terminal amino acid derivative to the resin, the resin is subjected to steps 1 through 9, and then step 14 of the sequence in Table I, then washed with ethanol (2 × 25 ml) and dried over P_2O_5 in a vacuum desiccator overnight, prior to HF cleavage.
5. Using a purpose-built HF line, react the resin with a mixture of anisole (1 ml) and anhydrous HF (9 ml) at 4° for 60 min.[41] Distill the HF from the reaction flask to end the reaction and then triturate the resin–peptide mixture with ether (3 × 25 ml), recovering the solid material each time by centrifugation or filtration through a sintered glass funnel. In the same way, extract the peptide from the solid residue with 10% aqueous acetic acid (5 × 10 ml).
6. Evaporate the peptide extract down to a suitable volume by evaporation or lyophilization, and then desalt the peptide by gel permeation chromatography on, for example, Sephadex G-15 eluted with 1% aqueous acetic acid. Tryptophan-containing peptides must subsequently be deformylated by treatment with ice-cold piperidine,[42]

[41] J. Scotchler, R. Lozier, and A. B. Robinson, *J. Org. Chem.* **35,** 3151 (1970).

[42] M. Ohno, S. Tsukamoto, S. Makisumi, and N. Izumiza, *Bull. Chem. Soc. Jpn.* **45,** 2852 (1972).

and peptides containing two cysteine residues must be oxidized at a low concentration if an intramolecular disulfide bridge is to be formed.[43]

E. CD Spectra

CD spectra are recorded on a spectropolarimeter in buffered salt solutions that approximate physiological conditions, such as 160 m*M* KCl in 20 m*M* sodium phosphate (pH 7.4). In order to obtain accurate readings at low concentrations, it is essential that the sample cells are thoroughly cleaned with detergent and acid washed with cold dilute Nochromix (Taylor Chemical Company). It is often helpful to soak the cells in a 10% solution of polyethylene glycol (molecular weight 15,000–20,000) or polyethylenimine for 1 hr and then rinse thoroughly with water, to inhibit adsorption of peptides to the glass walls.

F. Surface Monolayer Isotherms

Monolayers are studied on a Langmuir trough apparatus, such as the Lauda Film Balance, connected to a calibrated chart recorder. A suitable buffer, such as 160 m*M* Ultrapure KCl (Alfa Products) in 1 m*M* Tris(hydroxymethyl)aminomethane–HCl (Aldrich, Gold Label grade) (pH 7.4), is prepared in acid-washed glassware using glass-distilled water, and freed from traces of surfactants by foaming for 10 min and aspirating the surface buffer. The surface of the film balance must be rigorously cleaned with a detergent paste and then completely freed of surfactants by washing with glass-distilled water. As a final precaution, the surface of the buffer is aspirated again after it has been poured onto the trough. A stock solution of the peptide (1 mg/ml) is then slowly applied directly to the buffer surface from a clean glass syringe or pipet that will not result in adsorptive losses. About 20 μg peptide is usually required for an initial surface area of 600 cm^2, but this will vary according to the peptide. The monolayer is allowed to stabilize for 30 min, and is then compressed at a rate of less than 0.5 cm^2/sec. The portion of the curve consisting of surface pressures below 1 dyn/cm is recorded separately on a suitably expanded scale for greatest accuracy. When the collapse pressure is reached (usually below 40 dyn/cm), as indicated by a leveling of the force–area curve, compression is stopped and the peptide thoroughly aspirated from the surface. Note that kinks in force–area isotherms are sometimes observed well below the collapse pressure, due to changes in the molecular conformation at the surface.[44] Isotherms are usually measured 3 times, as a check

[43] G. R. Moe and E. T. Kaiser, *Biochemistry* **24,** 1971 (1985).

[44] M. T. A. Evans, J. Mitchell, P. R. Mussell-White, and L. Iron, *in* "Surface Chemistry of Biological Systems" (M. Blank, ed.), pp. 1–22. Plenum, New York, 1970.

for consistency, and the temperature during each compression must be accurately recorded.

Peptide Synthesis

The synthesis of relatively short peptides of up to about 40 residues in length is most conveniently performed by standard solid-phase methods.[38,39] The solid support is a 1% cross-linked, styrene–divinyl-benzene copolymer which has been functionalized with chloromethyl or benzhydrylamine (BHA) groups to allow covalent attachment of the carboxy-terminal amino acid derivative. The former resin type is employed in the synthesis of peptides ending with a free carboxylic acid and the latter resin results in peptides having a carboxy-terminal amide. The substitution level of the first amino acid on the resin should be low enough (generally 0.2–0.3 mmol/g resin) to avoid steric crowding of the peptide chains as they are elongated during the synthesis. Elongation then proceeds through the stepwise addition of the appropriate N^{α}-*tert*-butyloxycarbonyl (Boc) derivatives of the side-chain-protected amino acids via repetition of the reaction sequence described in Scheme 1. This reaction cycle is carried out by means of the sequence of solvent washes and reagent

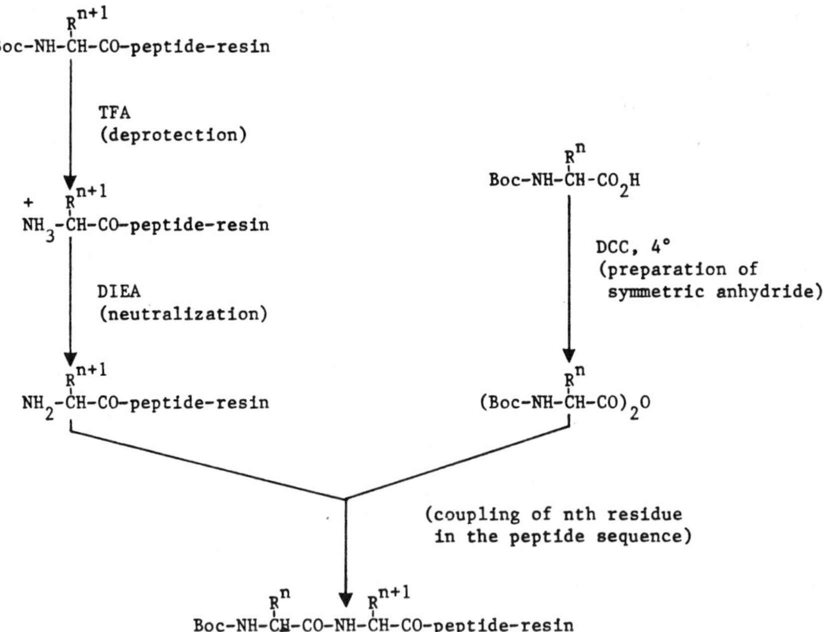

SCHEME 1. Reaction sequence for one coupling cycle in peptide synthesis.

additions described in Table I.[40] The repetitive nature of this process lends itself to programed automation, and there are currently several instruments available for this purpose. Nevertheless, a manual approach, using a suitably designed reaction vessel,[45] is equally effective, if more time and energy consuming.

When the complete, fully protected peptide has been assembled on the resin, the amino-terminal Boc group is removed and the amine neutralized; after washing and drying the resin the peptide is completely deprotected and cleaved from the solid support by reaction with liquid HF in the presence of anisole.[41] This reaction must be performed in apparatus suited to handling HF by competent personnel aware of the dangers of this powerful reagent. For many laboratories where peptides might be synthesized relatively infrequently, custom HF cleavage services are probably preferable. The resulting crude peptide is extracted and desalted before purification and characterization.

Peptide Purification

The purification and characterization of biologically active peptides that have been synthesized by the solid-phase method must be thorough. Many unwanted side reactions may occur during the synthesis and, particularly, the deprotection of the peptide, so that impurities may constitute a large part of the crude peptide material obtained from longer syntheses. The desired peptide should always be the major component, but the possibility that any measured biological activity arises from minor impurities that are orders of magnitude more potent than the major product must be minimized. The use of reverse-phase, high-performance liquid chromatography (RP-HPLC) is almost essential to this task, and we usually prefer to combine this with anion- or cation-exchange chromatography to provide a contrasting basis for separations. Gel permeation chromatography is generally not suited to this task, because of the nature of likely impurities and also because of the tendency of many peptides that have amphiphilic character to self-associate and produce broad or multiple peaks in elution profiles. Partition chromatography, however, is also often very useful as a powerful and contrasting purification technique.[46]

Typically, the ion-exchange chromatography is performed first on a relatively large scale (100 mg of crude peptide or more), and then the resultant material is desalted and subjected to RP-HPLC on a semi-pre-

[45] G. Barany and R. B. Merrifield, *in* "The Peptides" (E. Gross and J. Meienhofer, eds.), Vol. 2, p. 39. Academic Press, London, 1979.

[46] D. Yamashiro, *Int. J. Pept. Protein Res.* **13,** 5 (1979).

parative size column, using smaller quantities (usually 5 mg or less) so as not to lose resolution of the major peak from background impurities. The rapidity of the HPLC procedure allows enough material to be prepared for most physicochemical and biological studies (5–10 mg) from a few repeated chromatographic elutions. A suitable and convenient solvent system is 200 m*M* phosphoric acid–NaOH (pH 2.5)/acetonitrile. An initial elution of a small quantity of the peptide material on an analytical column, using a gradient of 0–60% acetonitrile over 90 min at 1.5 ml/min should identify the conditions required for the semi-preparative elutions. These should be performed using a shallow gradient of only a few percent change in acetonitrile concentration over about 20–30 min at 3–4 ml/min, under conditions where the desired peptide elutes in the middle of the gradient. Occasionally, we have found that the amphiphilic properties of our peptide models result in an unusually high affinity for reverse-phase columns, so that the binding cannot be reversed by this buffer system. In these cases, systems containing sodium perchlorate, as described elsewhere,[47] are often suitable. Multiple peaks arising from the self-association of amphiphilic β-strands have also been observed.

After RP-HPLC purification and a final desalting chromatography, the identity and purity of the peptide must be determined by amino acid analysis, sequence analysis by Edman degradation, and analytical RP-HPLC under isocratic conditions, preferably using more than one buffer system. At this point, quantitative amino acid analysis of a stock solution of the peptide will also allow an extinction coefficient at, for example, 230 or 280 nm to be determined under standard conditions, allowing subsequent stock solutions to be readily quantified by absorbance. This eliminates the difficulties of weighing small quantities of peptides, as well as the errors involved due to their water content, which may be high.

Physicochemical Studies

Solution Conformation

The solution conformation of a peptide is most readily investigated by examining the CD spectrum arising from amide bond absorption of UV light in the wavelength range of 180–250 nm.[48] By comparing this spectrum to those of polypeptide standards which are known to adopt particular types of secondary structure, a qualitative estimate of structure con-

[47] J. L. Meek, *Proc. Natl. Acad. Sci. U.S.A.* **77,** 1632 (1980).

[48] A. J. Adler, N. J., Greenfield, and G. D. Fasman, this series, Vol. 27, p. 675.

tent is possible.[49,50] For example, the fractional contributions of α helix, β sheet, β turn, and random coil to the complete peptide structure may be considered simultaneously by computing the best combination of the standard spectra for these conformations that fits the peptide spectrum. Alternatively, if α-helical structure appears to be dominant as indicated by minima at 208 and 222 nm, the helix content may be more simply estimated from the mean molar ellipticity per residue, θ, at 222 nm by using Eq. (1).[51] This equation assumes limiting values, derived from the standard spectra,[49] of $\theta_{222} = 3000$ deg cm^2/dmol for 0% α helix (i.e., random

$$\% \ \alpha \text{ helix} = [(\theta_{222} - 3000)/-39{,}000] \times 100 \qquad (1)$$

coil) and $\theta_{222} = -36000$ deg cm^2/dmol for 100% α helix. Similarly, Eq. (2) may be employed to estimate the amount of β sheet if a distinct minimum in the CD spectrum at 217 nm indicates that this structure is dominant.[52]

$$\% \ \beta \text{ sheet} = [(\theta_{217} - 4600)/-23{,}000] \times 100 \qquad (2)$$

All of these methods assume the validity of the standard spectra, and are certainly subject to considerable error. Their value lies in the use of the results for general, comparative purposes.

Peptides which have the potential to form amphiphilic α-helical or β-strand structures often aggregate in aqueous solution. The hydrophobic domains of amphiphilic helices may interact in an aggregate to form a micellelike structure with a hydrophobic core,[32] and amphiphilic β-strands may aggregate to form extended β-sheets with one face hydrophobic and one face hydrophilic which could then associate in pairs to bury the hydrophobic faces.[52] The concentration dependency of this self-association may readily be followed by means of the CD spectrum, since the secondary structure involved is normally more stable in the aggregated form. Amphiphilic α helices aggregate in small numbers and in a highly cooperative manner. A hydrophobic domain lying parallel to the helix axis is apparently ideal for self-association,[31] and model peptides that can form relatively hydrophobic helices of this type just 20 residues long begin to self-associate at concentrations as low as 200 nM.[5,29,43] This behavior illustrates the strength with which an amphiphilic α helix might bind to a protein receptor if it provides a suitable complementary amphiphilic envi-

[49] N. Greenfield and G. D. Fasman, *Biochemistry* **8,** 4108 (1969).

[50] A. Wollmer, W. Strassburger, and U. Glatter, *in* "Modern Methods in Protein Chemistry" (H. Tschesche, ed.), pp. 361–384. de Gruyter, New York, 1983.

[51] J. D. Morrissett, J. S. K. David, H. J. Pownall, and A. M. Gotto, Jr., *Biochemistry* **12,** 1290 (1973).

[52] D. G. Osterman and E. T. Kaiser, *J. Cell. Biochem.* **29,** 57 (1985).

ronment, even in the absence of any side-chain specificity. The potential additional importance of individual side chains on the hydrophobic side of the helix is illustrated by the observations in model peptides that a bulky Trp side chain will reduce the ability of an amphiphilic helix to self-associate[29] and that a charged Arg side chain will abolish self-association altogether.[53]

Cooperative self-association of a monomeric peptide, M, to form oligomers, L, comprising n monomer units is described by the equilibrium

$$nM \rightleftharpoons L$$

which has a dissociation constant, K_D, given by Eq. (3):

$$K_D = \frac{[M]^n}{[L]} \tag{3}$$

The experimentally determined mean molar ellipticity per residue at a particular wavelength, θ_{ex}, is related to the ellipticities of the peptide in its monomeric form, θ_m, and its oligomeric form, θ_l, by Eq. (4), where [P], the total peptide concentration, is given by Eq. (5).

$$\theta_{ex} = \frac{\theta_m[M] + n\theta_l[L]}{[P]} \tag{4}$$

$$[P] = [M] + n[L] \tag{5}$$

By combining Eqs. (3), (4), and (5) to eliminate [M] and [L], one obtains Eq. (6).

$$\left(\frac{\theta_{ex} - \theta_m}{[P]^{n-1}}\right)^{1/n} = \left(\frac{n}{K_D(\theta_l - \theta_m)^{n-1}}\right)^{1/n} (\theta_l - \theta_{ex}) \tag{6}$$

Experimentally, θ_{ex} may be determined as a function of [P] over a wide range of concentrations (10^{-7}–10^{-3} M) by using a 5-cm pathlength cell at low concentrations and 1- and 0.1-cm pathlength cells at successively higher concentrations. The most appropriate wavelength to study is 222 nm for α-helical aggregates and 217 nm for β-sheet-forming aggregates, and constant wavelength readings should be averaged over several minutes at each concentration for greatest accuracy. It is usually possible to obtain an accurate value for θ_m by determining the constant value of θ_{ex} at low concentrations where no aggregation occurs. In contrast, attempting to obtain θ_{ex} at high concentrations will require considerable quantities of peptide and may result in greater inaccuracies due to equilibria with

[53] D. Fukushima, S. Yokoyama, F. J. Kezdy, and E. T. Kaiser, *Proc. Natl. Acad. Sci. U.S.A.* **78,** 2732 (1981).

higher aggregation states, precipitation, or the solution having too high an absorbance. Once θ_m is determined, Eq. (6) shows that a graph of θ_{ex} (x axis) versus $[(\theta_{ex} - \theta_m)/[P]^{n-1}]^{1/n}$ (y axis) will give a straight line, provided that the correct value for n is employed. Thus, n may be determined by trying various integer values to find the best straight line fit of the CD data according to Eq. (6). Alternatively, the apparent molecular weight of the peptide in its aggregated state at high concentrations may be measured by an independent method such as analytical ultracentrifugation,[54] in order to obtain n. Once the correct graph has been drawn, θ_l is obtained from $\theta_l = -(y\text{-axis intercept/slope})$. The K_D is then calculated from either the y-axis intercept or the slope.

The CD spectrum of the β-endorphin model peptide in aqueous salt solutions buffered at pH 7.4 was of the type expected for mixtures of random coil and α-helical structure,[49] with a strong minimum at 208 nm and a weaker minimum at 222 nm. The ellipticity in this region of the spectrum became increasingly negative at peptide concentrations above 10 μM, indicating that self-association was occurring with a concomitant stabilization of α-helical structure. The concentration dependency of θ_{222}, shown in Fig. 3, is typical for the cooperative self-association of peptides containing regions of potential amphiphilic α-helical structure. Analysis of this curve, as described above using an experimentally determined value of $\theta_m = -6{,}450$ deg cm^2/dmol, gave a straight line fit according to Eq. (6) for $n = 4$ (Fig. 3, inset), but not for $n = 3$ or $n = 5$, indicating that the aggregated form of the model peptide is tetrameric. From this graph, $\theta_l = -10{,}650$ deg cm^2/dmol and $K_D = 2.63 \times 10^{-13}$ M^3 were calculated. These results show that, as expected, this β-endorphin model peptide self-associates less readily and with formation of less α-helical structure than several similar model peptides[5,26,43] because the shape of the hydrophobic domain of the amphiphilic helix formed is not ideally suited to aggregate formation. Its behavior in this respect is, therefore, closer to that of β-endorphin itself, which is monomeric at concentrations up to 40 μM at least and has little α-helical structure in the same buffered salt solutions.[9] When Eq. 1 is employed to calculate helical contents, the model peptide is found to be 24% α-helical in the monomeric form and 35% α-helical as the tetramer, whereas monomeric β-endorphin is only 12% α-helical. These differences are probably a consequence of the modeling strategy employed, in which the residues that have been incorporated into the model structure are expected to favor α helix formation strongly and will form a more hydrophobic face in that conformation than the corresponding natural sequence does.

[54] R. J. Pollet, B. A. Haase, and M. L. Standaert, *J. Biol. Chem.* **254,** 30 (1979).

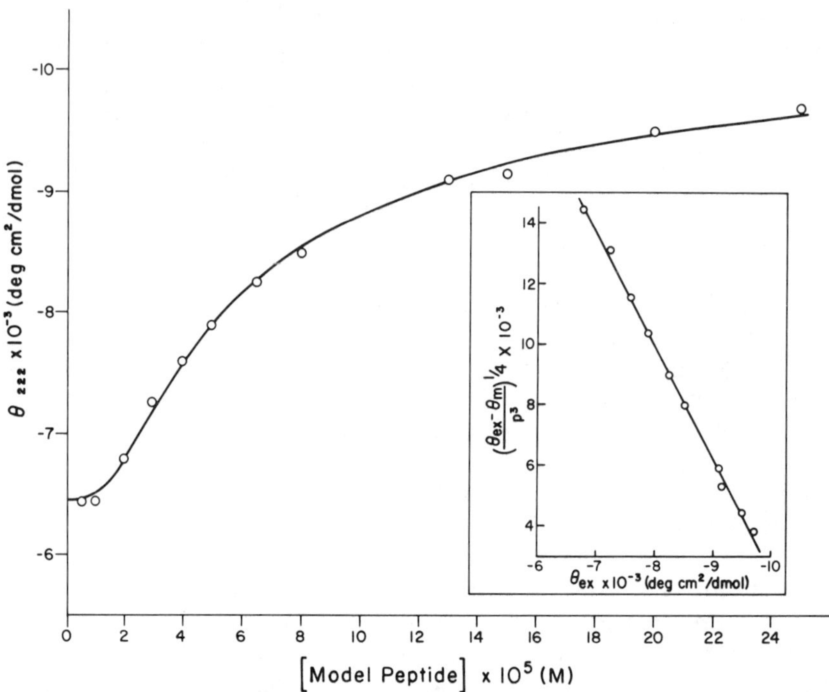

FIG. 3. Concentration dependency of θ_{222} for the β-endorphin model peptide in 160 mM KCl and 20 mM sodium phosphate (pH 7.4). Data points and a theoretical curve derived for cooperative tetramerization (see text) are shown. *Inset:* Linear regression analysis of the data points fitted to the equation describing cooperative tetramerization.

Conformation at the Air–Water Interface

The air–water interface represents an approximate model for the amphiphilic environment of biological interfaces, such as the surfaces of cells or lipoproteins. The abilities of biologically active peptides which normally act at these interfaces to form stable monolayers at the air–water interface are therefore of interest. Furthermore, a semi-empirical analysis of the force–area isotherms generated by compression of these monolayers yields quantitative data related to the conformational properties of the peptides in this environment.[32,55]

The experiment is conducted by slowly and gently applying a small quantity of peptide from a concentrated stock solution to the aqueous surface of a suitable Langmuir trough that has been rigorously freed of

[55] A. W. Adamson, *in* "Physical Chemistry of Surfaces," 3rd Ed., p. 94. Wiley, New York, 1976.

contaminating surfactants. The peptide monolayer is allowed to stabilize, and is then slowly compressed by means of a horizontal beam lying across the water surface. During compression, the increasing force exerted by the monolayer on one side of a second floating beam is measured as a function of the decreasing surface area. At low surface pressures (usually less than 1 dyn/cm), the force–area isotherms generally behave according to Eq. (7), where π is the surface pressure, A the total surface area, A_0 the

$$\pi(A - A_0) = nRT \tag{7}$$

limiting surface area of the monolayer (the area occupied by the peptide molecules alone), and n the apparent number of moles of peptide at the surface. This is a two-dimensional form of the ideal gas law. Since errors in the zero force calibration will significantly affect the apparent compliance with Eq. (7) at low surface pressures, the data are analyzed according to Eq. (8), where Y is the experimentally measured surface pressure

$$(Y - e)(A - A_0) = nRT \tag{8}$$

and e the constant error in the value of Y compared to the true surface pressure, π. Equation (8) is expanded to give the trivariant Eq. (9), to which the data are fitted by (trivariant) linear regression analysis. The results yield values for A_0, n, and e.

$$YA = nRT - eA_0 + YA_0 + eA \tag{9}$$

At higher surface pressure, the peptides occupy a significant fraction of the available surface area and become more highly organized. The error, e, is no longer significant so that $\pi = Y$, and regions of the isotherm will often conform to Eq. 10, where A_{00} is the limiting surface area occu-

$$\pi[A - A_{00}(1 - \kappa\pi)] = nRT \tag{10}$$

pied by the peptide molecules at zero surface pressure and κ the compressibility constant for the peptide. According to Eq. (10), pairs of data points (π_1, A_1) and (π_2, A_2) will be related as described by Eq. (11), where

$$\pi_1 A_1 - \pi_2 A_2 = 2\Delta A_{00}\pi_1 - \Delta\pi A_{00} + \Delta\pi^2\kappa A_{00} \tag{11}$$

$\Delta\pi = \pi_2 - \pi_1$. The experimental curves are initially analyzed for their conformation to Eq. (10) by choosing pairs of data points separated by a constant $\Delta\pi$ ($\Delta\pi = 0.5$ dyn/cm is usually convenient) and plotting a graph of $\pi_1 A_1 - \pi_2 A_2$ versus π_1. According to Eq. (11), this graph should give a straight line for regions of the force–area isotherm which obey Eq. (10). In these regions, the experimental data are fitted by trivariant linear regression analysis to Eq. (12), which is an expanded version of Eq. (10).

$$\pi A = nRT + \pi A_{00} - \pi^2\kappa A_{00} \tag{12}$$

From this analysis, A_{00}, n, and κ are obtained. Of course, data may also be fitted directly to either the low pressure Eq. (8) or the high pressure Eq. (10) by nonlinear methods.

The force–area isotherms obtained for β-endorphin and the β-endorphin model peptide spread on buffered salt solutions are compared in Fig. 4. Both peptides were able to form monolayers at the air–water interface, but that of the model peptide collapsed at a much higher surface pressure (about 21 dyn/cm) than that of β-endorphin (7 dyn/cm). At surface pressures below 0.9 dyn/cm, thc isotherms obeyed Eq. (7), and the corresponding analysis gave values of A_0 = 13.0 and 20.7 Å^2/residue for β-endorphin and the model peptide, respectively. At higher surface pressures, β-endorphin obeyed Eq. (10) in the range π = 2.5–6.5 dyn/cm, and the corresponding analysis yielded A_{00} = 13.1 Å^2/residue, suggesting that its conformation was essentially unchanged, and $\kappa = 3.90 \times 10^{-2}$ cm/dyn. In contrast, the model peptide obeyed Eq. (10) at higher surface pressures (π = 10–18 dyn/cm), where values of A_{00} = 16.2 Å^2/residue and $\kappa = 1.69 \times 10^{-2}$ cm/dyn were calculated. Thus, at these pressures the model peptide appears to have rearranged to a more compact structure which is less compressible than the β-endorphin structure.

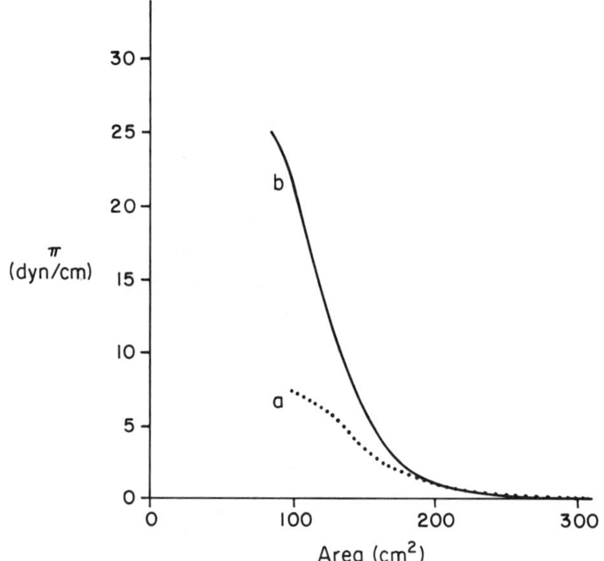

FIG. 4. Force–area isotherms obtained by compression of the insoluble monolayers formed by (a) β-endorphin (30 μl of a 2.0×10^{-4} M solution) and (b) the model peptide (25 μl of a 2.0×10^{-4} M solution) on a subphase consisting of 160 mM KCl and 1 mM Tris–Cl (pH 7.4).

The calculated values for n also showed that the model peptide appears to behave as a dimer at the air–water interface at low pressures and a monomer at high pressures, whereas β-endorphin is monomeric according to both analyses. The values for A_{00} obtained for both peptides are lower than those of model peptides consisting almost entirely of amphiphilic α-helical structure,[5,32] indicating that parts of the peptide chain—probably including the hydrophilic segments consisting of residues 6–12—are fully hydrated and oriented away from the air–water interface. Therefore, the formation of stable monolayers most likely depends on amphiphilic α-helical structures in the carboxy-terminal regions of these peptides. The larger limiting area of the monolayer formed by the model peptide suggests that its amphiphilic structure is more extensive. Its greater stability at the amphiphilic air–water interface compared to β-endorpin is very likely to be related to its ability to tetramerize in aqueous solution, since a regular tetrameric helical structure would provide a similar environment, and is also likely, therefore, to be a consequence of the modeling strategy adopted, as discussed above.

Biological Properties

The biological assays of a synthetic model peptide must be chosen carefully in order to evaluate the function and importance of the incorporated model structure. This has been particularly true of our studies of peptide models of β-endorphin since several naturally occurring opioid peptides have been characterized which share a variety of activities that are mediated by different types of cell surface receptors.[56,57] Furthermore, one of these opioid peptides, [Met^5]-enkephalin has a structure identical to that of the amino-terminal five residues of β-endorphin, so that the importance of the rest of the β-endorphin structure in determining each activity must be rigorously questioned. We have therefore concentrated on the abilities of our peptide models to reproduce (1) the selectivity of β-endorphin for different opiate receptor types in binding assays, which differs from that of other opioid peptides,[24] (2) its actions on the opiate receptors in the rat vas deferens, which are the most specific opiate receptor type for β-endorphin so far characterized,[25] and (3) its resistance to proteolytic degradation in brain,[9,27] which allows longer lasting effects at receptors more distant from the point of release than is possible for other rapidly hydrolyzed opioid peptides such as [Met^5]-enkephalin. In addition, the analgesic actions of the model peptides upon intracerebral injec-

[56] R. S. Zukin and S. R. Zukin, *Trends Neurosci.* **7,** 160 (1984).

[57] A. Goldstein and I. F. James, *Trends Pharmacol. Sci.* **5,** 503 (1984).

tion of mice[58] have also been studied as a test of their abilities to diffuse to receptor sites *in vivo* and reproduce the potent actions of β-endorphin at these sites.

The receptor binding selectivity of the model peptide described here was compared to that of human β-endorphin by studying the relative abilities of these peptides to displace the specific binding of radiolabeled ligands selective for δ- and μ-opiate receptors from guinea pig brain membrane preparations. The model peptide displaced each type of radiolabel at about 10-fold lower concentrations than β-endorphin, indicating that it has a higher affinity for both types of receptor and a similar δ/μ selectivity. In contrast, the model peptide was somewhat less potent than human β-endorphin at inhibiting electrically stimulated contractions of the rat vas deferens, its half-maximal effect being reached at concentrations of about 270 mM, compared to 40 nM for β-endorphin. However, this activity is much greater than that observed for all other opioid peptides and similar to the values reported for camel endorphin ([His^{27}, Gln^{31}]-human β-endorphin) and ovine endorphin ([Val^{23}, His^{27}, Gln^{31}]-human β-endorphin).[59]

The rate of degradation of the β-endorphin model peptide by proteolytic enzymes present in rat brain was examined by incubating the model peptide in diluted suspensions of a whole rat brain homogenate at 37° for various times. The suspensions were then heated in boiling water to destroy the proteolytic activities and centrifuged to remove the particulate matter. The remaining peptide was then quantitated by RP-HPLC of the supernatant. Under these conditions, [Met^5]-enkephalin was completely hydrolyzed in less than 5 min, and β-endorphin was much more slowly degraded over about 60 min. The model peptide was initially recovered in lower quantities than β-endorphin (about 30% recovery at $t = 0$ compared to 65% for β-endorphin), indicating a greater propensity for nonspecific binding to the brain membranes in the homogenate. However, degradation was even slower than for β-endorphin, and the amount of intact model peptide recovered decreased from this initial value only very slowly to about 15% after 90 min of incubation in the brain homogenate.

In hot plate assays of analgesic activity[60] after intracerebral injections of mice with the model peptide, a potent and long-lasting antinociceptive effect was observed. When equal doses of β-endorphin and the model peptide were compared (Fig. 5) the maximal effects of β-endorphin were

[58] T. J. Haley and W. G. McCormick, *Br. J. Pharmacol.* **12,** 12 (1957).

[59] J. P. Huidobro-Toro, E. M. Catury, N. Ling, N. M. Lee, H. H. Loh, and E. L. Way, *J. Pharmacol. Exp. Ther.* **222,** 262 (1982).

[60] S. I. Ankier, *Eur. J. Pharmacol.* **27,** 1 (1974).

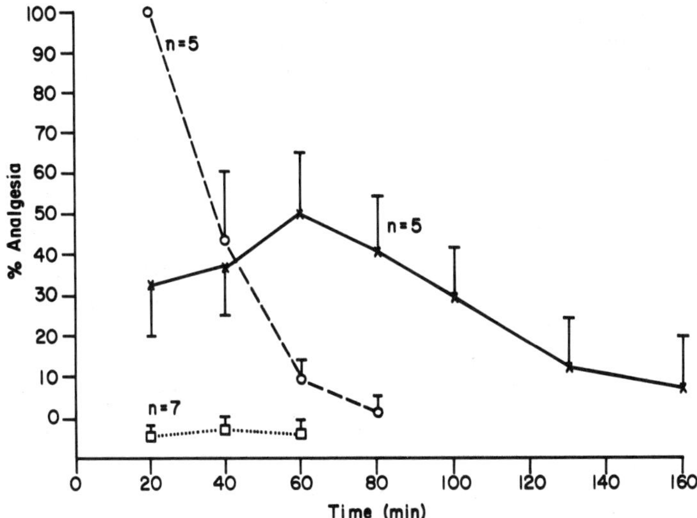

FIG. 5. Time courses of the analgesic effects on mice of equal doses (3 μg) of β-endorphin (○) and the β-endorphin model peptide (X). Error bars represent the standard error of the mean values from n experiments, and the effects of 0.9% saline injections in control experiments are also indicated (□).

observed 20 min after injection and thereafter rapidly diminished, whereas the model peptide diffused more slowly to the receptor sites and was more slowly cleared. These differences are consistent with the differences observed in the proteolysis experiment described above. The potency of the model peptide was about 15–30% that of β-endorphin, if the maximal effects of each dose are compared. This difference is similar to that observed for β-endorphin analogs with very minor sequence changes in the carboxy-terminal region.[24,61]

In summary, the peptide model of β-endorphin is able to reproduce all of the activities of the natural peptide tested,[31] including potent actions on highly specific receptors, despite the presence of 14 nonhomologous residues in the carboxy-terminal half of the molecule. This provides strong evidence that the potential for formation of an amphiphilic α helix in this region is an important functional aspect of the structure of β-endorphin. These results have encouraged the study of subsequent model peptides designed to determine the necessity for amphiphilicity in this region of β-endorphin (the "negative" model[18]), and to explore the possibility that a left-handed amphiphilic α helix consisting of D-amino acid residues

[61] J. Blake, L.-F. Tseng, and C. H. Li, *Int. J. Pept. Protein Res.* **15,** 167 (1980).

could also reproduce the properties of the natural sequence.[30] In this way, the role of the amphiphilic structure has been more thoroughly defined, and evidence suggesting that this structure is the regular α helix and not a π helix or other less regular structure has been obtained.

Differences observed in the physicochemical properties of the model peptide described here compared to those of β-endorphin provide a rationale for some of the observed differences in its behavior in the biological assays. Thus, tighter binding to δ- and μ-opiate receptors, which may not have much specificity for the carboxy-terminal portion of β-endorphin, as well as its strong nonspecific binding to brain membranes and greater resistance to the degradative actions of proteolytic enzymes are probably consequences of its enhanced stability at amphiphilic interfaces relative to β-endorphin. Just as these differences are a direct result of the model peptide design, as discussed previously, so too are the lower potencies of the model peptide in the most specific pharmacological assays employed.

The opiate receptors of the rat vas deferens and the opiate receptors mediating the analgesic action of β-endorphin in the central nervous system clearly have different additional specificities for individual side chains on the surface of the amphiphilic helical structure which the simplified structure of the model peptide does not provide. The present results, therefore, allow a rational approach to identifying these residues to be adopted so that opioid peptides with a high specificity for different opiate receptors and potent long-lasting *in vivo* activities may be developed. The common occurrence of potential amphiphilic α helices in peptide hormones[2] indicate that this approach should be widely applicable.

[26] Effect of Point Mutations on the Folding of Globular Proteins

By C. ROBERT MATTHEWS

General Introduction

Kinetic studies of protein folding have proven to be a richer source of information on the mechanism of folding than their equilibrium counterparts. The detection of multiple unfolded forms[1] and transient intermedi-

[1] B. T. Nall, *Comments Mol. Cell. Biophys.* **3,** 123 (1985).

ates[2] has improved our understanding of the types of structures that play significant roles in this complex process. An important goal of current experiments is to elucidate the conformations of folding intermediates or, equivalently, the details of the conformational changes that link native, intermediate, and unfolded forms.

Available spectroscopic methods that are sufficiently sensitive to detect the intermediates, e.g., difference ultraviolet or fluorescence spectroscopy, cannot, in general, provide detailed structural information. Methods that could provide such information, including X-ray crystallography or Fourier transform NMR spectroscopy, do not have the sensitivity to elucidate the conformations of transient species whose lifetimes may be in the millisecond time range. As discussed in a recent contribution to this series,[3] hydrogen exchange methods have the potential to follow the formation of secondary structure in identifiable segments of protein. However, the application of these methods to large proteins is sufficiently complex that the development of alternative methods would be desirable.

An approach that we have been developing in our laboratory is to study the effect of single amino acid replacements on the stability and folding of globular proteins. By identifying amino acids that play key roles in folding and stability, we hope to elucidate the structural basis for rate-limiting steps in folding. The ability to specifically replace a given amino acid using recombinant DNA technology makes such an approach feasible.

Principle

The rationale for this approach is based on the now well-accepted hypothesis of Anfinsen that the amino acid sequence of a protein determines its tertiary structure.[4] This hypothesis can be extended to propose that the primary sequence also determines the folding pathway and that replacements of amino acids which play key roles in this process will have observable effects on the folding rate and perhaps the stability of the native conformation. Quantitative comparisons of the changes in the free energy of folding and the rate constants for unfolding and refolding can reveal whether a given replacement alters the stability or a rate-limiting step in folding or both. This information can be used to map the residues involved in the rate-limiting steps and to eliminate potential structural

[2] P. S. Kim and R. L. Baldwin, *Annu. Rev. Biochem.* **51,** 459 (1982).

[3] P. S. Kim, this series, Vol. 131, p. 136.

[4] C. B. Anfinsen, *Science* **181,** 223 (1973).

models for conformational changes. Consideration of these results in terms of an X-ray crystal structure should eventually lead to a better understanding of the conformational changes involved in folding.

Materials and Reagents

Ultra-pure or the equivalent grade urea and guanidine hydrochloride should be used in unfolding studies and are commercially available. In some systems, the choice of denaturant will be dictated by the limited solubility of folding intermediates which are easily salted out by ionic denaturants. Reversible unfolding for proteins with the free sulhydryl groups usually necessitates the addition of a reducing agent in the buffer. For difference ultraviolet studies, 2-mercaptoethanol is a better choice than dithioerythritol because the oxidized form of 2-mercaptoethanol does not absorb strongly in the range 280–300 nm where the protein difference spectrum arises. Oxidation of the free sulfhydryls can also be minimized by degassing all buffers and by maintaining a low concentration, e.g., 0.1 m*M*, of ethylenediaminetetraacetic acid in all buffers.

Purified protein can usually be stored at 4° in a closed vial as an ammonium sulfate precipitate for a period of several months before a measurable loss of enzymatic activity occurs. The precipitate is suspended in a minimum amount of appropriate buffer and is dialyzed extensively to remove residual ammonium sulfate. The protein is then concentrated prior to folding studies by one of several commercially available products.

Method

Although the emphasis in this chapter is on the effects of amino acid replacements on the kinetics of folding, it is important to measure the effects on the equilibrium unfolding transition as well. The latter information is required for identification of an appropriate folding model, for designing kinetic experiments and for the correct interpretation of the results of kinetic studies.

Equilibrium Studies

The equilibrium unfolding transition for the wild-type protein can be monitored by a variety of spectroscopic techniques. Those with sufficient sensitivity to require protein concentrations of approximately 1 mg ml^{-1} or less are advantageous because intermolecular effects such as aggregation are minimized and because protein usage is low. Difference ultraviolet, fluorescence, and circular dichroism spectroscopies all satisfy this requirement; however, the first two methods are preferable because stopped-flow kinetic studies often necessary for this analysis can be performed on commercially available instruments.

Folding studies in our laboratory have relied principally on difference ultraviolet spectroscopy because these dual beam instruments have the stability to monitor small changes in absorbance over a period of hours. The principle contributions to the difference spectrum in the near ultraviolet region, 275–300 nm, come from changes in the exposure to solvent of buried tyrosine and tryptophan residues that occur on unfolding.[5] Therefore, the technique is applicable to a wide variety of proteins.

The unfolding can be induced by chemical denaturants, increases or decreases in pH, or by increase in the temperature. Because reliable analysis of the equlibrium data requires that the unfolding reaction be reversible, chemical denaturants, e.g., urea or guanidine hydrochloride, or decreases in pH are preferred over increases in pH or temperature. At alkaline pH or elevated temperatures, protein folding is often irreversible due, in part, to chemical damage, e.g., amide hydrolysis. For proteins with isoelectric points below pH 7, acid-induced unfolding can lead to potential problems with aggregation that can occur as the isoelectric point is reached during the titration. Therefore, the most generally useful method involves chemical denaturants.

The procedures that are used to obtain difference ultraviolet spectra for denaturant-induced protein unfolding have been described,[5,6] as has the procedure for determining the free energy of unfolding in the absence of denaturant, i.e., the stability.[7] The only additional comment to be made concerns the form of the dependence of the apparent free energy of unfolding, ΔG_{app}, on the denaturant concentration. Of the three models presented by Pace,[8] we prefer the linear dependence of ΔG_{app} on denaturant concentration because a thermodynamic justification has been provided.[9] In this model,

$$\Delta G_{app} = \Delta G_U^{H_2O} + A[\text{denaturant}]$$

where $\Delta G_U^{H_2O}$ is the free energy of unfolding in water and A is an empirical parameter that describes the dependence of ΔG_{app} on denaturant concentration. Estimates of the stability in water are obtained by linear extrapolation of ΔG_{app} to zero molar denaturant.

Once a satisfactory fit of the equilibrium unfolding data has been ob-

[5] T. T. Herskovits, this series, Vol. 11, p. 748.

[6] J. W. Donovan, this series, Vol. 27, p. 497.

[7] C. N. Pace, this series, Vol. 131, p. 266.

[8] C. N. Pace, *CRC Crit. Rev. Biochem.* **3,** 1 (1975).

[9] J. A. Schellman, *Biopolymers* **17,** 1305 (1978).

tained, the effect of amino acid replacements on the stability can be determined. In principle, one would prefer to compare the stabilities of wild-type and mutant proteins in the absence of denaturants. In practice, the necessity of extrapolating data which can only be measured accurately in the transition zone over several molar units of denaturant may lead to errors which are comparable to the differences in stability between wild-type and mutant proteins. Also, Pace has tested the validity of a linear extrapolation and has found that deviations of up to 30% can occur near zero molar denaturant.[10]

An alternative approach is to compare the free energies of unfolding at the concentration of denaturant corresponding to the midpoint of the unfolding transition for the wild-type protein.[11] This has the advantage that the comparisons are made where the free energies can be most accurately measured; however, it is strictly empirical.

Kinetic Studies

To obtain useful results from studies of the kinetics of folding of mutant proteins, one must establish a mechanism of folding for the wild-type protein. Mechanisms for a number of proteins have been proposed and the procedures by which these mechanisms were defined have been described.[12] Of importance to the present discussion are steps which actually involve folding, i.e., the interconversion of native, intermediate, and unfolded forms. Reactions that involve interconversions of multiple unfolded forms have been discussed in this series[1] and are not of interest here.

The data essential for this analysis are the relaxation times for unfolding and refolding and the dependence of these relaxation times on the final denaturant concentration. For unfolding studies, the protein is initially maintained in a buffered solution at neutral pH where the protein is known to be in the native conformation. Unfolding is than initiated by volumetrically diluting this sample into a buffered solution containing a high denaturant concentration. The procedure for refolding is similar except that the protein is first equilibrated in a high denaturant concentration where the unfolded form is favored. The equilibrium data define the appropriate conditions. Dilution to varying final denaturant concentrations and constant protein concentration is achieved by volumetrically adding this sample to solutions containing the required amounts of dena-

[10] C. N. Pace and K. E. Vanderburg, *Biochemistry* **18,** 288 (1979).

[11] J. F. Cupo and C. N. Pace, *Biochemistry* **22,** 2654 (1983).

[12] H. Utiyama and R. L. Baldwin, this series, Vol. 131, p. 51.

turant. The choice of the initial denaturant concentration for refolding studies does not affect the observed relaxation times because they are determined by the final solution conditions.[13] The accuracy of the measurements is usually improved by starting with the fully unfolded protein so as to maximize the amplitudes of the folding phases.

For reactions with relaxation times longer than 10 sec, manual mixing is satisfactory; faster reactions require stopped-flow instrumentation. The signal is recorded, and the relaxation times τ_i (or equivalently, the apparent rate constants, k_i, where $k_i = 1/\tau_i$) are extracted from the data. The transient response of a first-order system that undergoes a displacement in the equilibrium distribution can be described by a sum of exponentials as follows:

$$A(t) = \sum_i A_i \exp(-t/\tau_i) + A(\infty)$$

where $A(t)$ is the absorbance at time t, A_i the amplitude of phase i with relaxation time τ_i, and $A(\infty)$ the absorbance at infinite time. In simple cases where only two relaxation times that differ by at least a factor of 3 and preferably a factor of 5 or more are involved, exponential stripping[14] can be used with reasonable confidence. In more complex cases, a nonlinear least squares computer-fit of the data is advisable.[15] Appropriate software is normally available at major computational facilities and can be purchased commercially.

For phases that correspond to actual folding reactions, the relaxation times have been observed to have a characteristic dependence on the denaturant concentration (Fig. 1). The logarithm of the relaxation time first increases linearly with denaturant concentration, reaches a maximum, and then decreases with further increases in denaturant concentration. This dependence can be understood in terms of the expected behavior for a simple two-state reaction, e.g.,

$$\mathrm{N} \underset{k_\mathrm{R}}{\overset{k_\mathrm{U}}{\rightleftharpoons}} \mathrm{U}$$

where N and U are the native and unfolded forms, respectively, and k_U and k_R are the rate constants for unfolding and refolding, respectively. For this system, the observed relaxation time is related to the rate constants by

$$\tau^{-1} = k_\mathrm{U} + k_\mathrm{R}$$

[13] C. Tanford, *Adv. Protein Chem.* **23,** 121 (1968).

[14] P. J. Hagerman and R. L. Baldwin, *Biochemistry* **15,** 1462 (1976).

[15] M. C. Johnson and S. G. Frasier, this series, Vol. 117, p. 301.

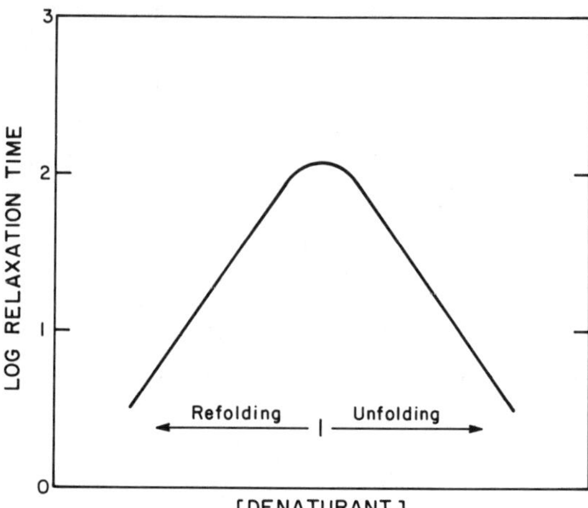

FIG. 1. Expected dependence of the logarithm of the relaxation time on the final denaturant concentration for a simple two-state folding reaction.

The equilibrium constant for unfolding, K_U, is defined in the usual way: $K_U = [U]/[N] = k_U/k_R$. For unfolding jumps where the unfolded form is favored, $K_U >> 1$ or equivalently $k_U >> k_R$. Under these conditions, $\tau^{-1} \cong k_U$, and the observed relaxation time is a measure of the unfolding rate constant. The progressive decrease in τ at high denaturant concentration reflects the shift of the equilibrium to favor the unfolded form. For refolding jumps where the native form is favored, $K_U << 1$ and $k_R >> k_U$. Under these conditions, $\tau^{-1} \cong k_R$, and the observed relaxation time is a measure of the refolding rate constant. The progressive decrease at low denaturant concentration reflects the shift of the equilibrium to favor the native form. At intermediate denaturant concentrations, the relaxation time is a composite of the two rate constants and proceeds through a maximum which corresponds closely to the midpoint of the appropriate equilibrium transition.

The relaxation times for the wild-type protein serve as a basis for comparison with the mutant proteins. The advantage of measuring the effects of amino acid replacements at a series of denaturant concentrations is that it improves the confidence in measurements of small differences in relaxation times. It also makes it possible to observe changes in the slopes of the plots of log τ versus denaturant concentration. Tanford has related this slope to the position of the transition state along the reaction coordinate, but no structural interpretations have been offered.[16] Studies on mutant proteins may shed light on this matter.

[16] C. Tanford, *Adv. Protein Chem.* **24,** 1 (1970).

Analysis

Comparison of the equilibrium and kinetic properties of folding for the wild-type and mutant proteins permits one to determine if the amino acid in question plays a key role in the rate-limiting step, alters the stability, or both. The scheme that we have developed for this analysis employs reaction coordinate diagrams and transition state theory. Such a diagram for a simple, two state N ↔ U folding reaction is shown in Fig. 2. The stability of the protein, in the absence of denaturant, is indicated by the difference in free energy between the native and unfolded forms. The relaxation time for unfolding or, equivalently, the rate constant for unfolding is related to the free energy differences between the native conformation and the transition state, and the corresponding quantities for refolding are related to the free energy difference between the unfolded conformation and the transition state. The quantitative relationships are

$$\Delta G_{\mathrm{U}} = -RT \ln K_{\mathrm{U}}$$
$$\Delta G_{\mathrm{U}}^{\ddagger} = -RT \ln[(hk_{\mathrm{U}}/k_{\mathrm{B}}T)]$$
$$\Delta G_{\mathrm{R}}^{\ddagger} = -RT \ln[(hk_{\mathrm{R}}/k_{\mathrm{B}}T)]$$

where ΔG_{U} is the free energy of unfolding, $\Delta G_{\mathrm{U}}^{\ddagger}$ and $\Delta G_{\mathrm{R}}^{\ddagger}$ the activation free energies for unfolding and refolding, respectively, R the gas constant, T the absolute temperature, and h and k_{B} the Planck and Boltzmann

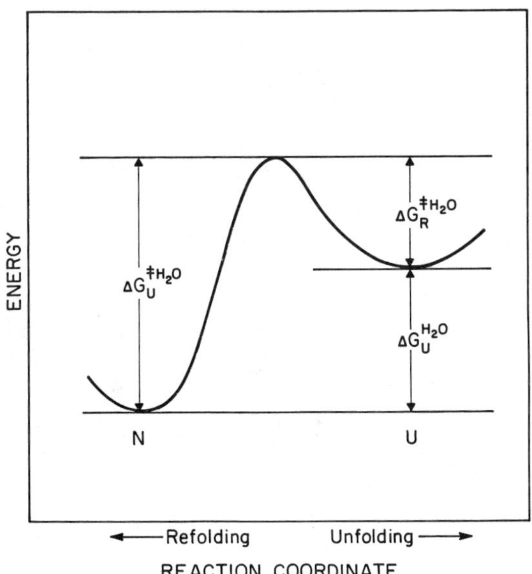

FIG. 2. Reaction coordinate diagram for a simple two-state folding reaction. Axes not drawn to scale.

constants, respectively. Therefore, measurements of the equilibrium and kinetic properties of folding permit one to measure the differences in free energy between the native, transition state, and unfolded forms.

Because K_U, k_U, and k_R and therefore ΔG_U, $\Delta G^{\ddagger}_{U}$, and $\Delta G^{\ddagger}_{R}$ all depend on the denaturant concentration, the question arises as to the appropriate concentration for comparing the wild-type and mutant proteins. The procedure described above for obtaining the free energy of folding in the absence of denaturant by extrapolating the linear dependence of free energy to zero molar denaturant can also be applied to the activation free energies to obtain the rate constants in the absence of denaturant. This assumption is based on the presumption from transition state theory that the transition state is in equilibrium with each of the stable conformations. Then, using Schellman's treatment,[9] the activation free energies should be a linear function of the denaturant concentration:

$$\Delta G^{\ddagger}_{U} = \Delta G^{\ddagger H_2O}_{U} + A_U[\text{denaturant}]$$
$$\Delta G^{\ddagger}_{R} = \Delta G^{\ddagger H_2O}_{R} + A_R[\text{denaturant}]$$

The observation that the log τ (or log k) is a linear function of the denaturant concentration for a number of proteins[13,17,18] supports this assumption. Therefore, the equilibrium and kinetic data for folding can be used to construct a reaction coordinate diagram for each particular protein in the absence of denaturant. In practice, the extrapolation of $\Delta G^{\ddagger}$ to zero molar denaturant has the same uncertainties as described above for the free energy difference. It is our opinion that comparisons of relaxation times for wild-type and mutant proteins are best made at urea concentrations where the measurements are made.

Figures 3A, 4A, and 5A show several possible effects of amino acid replacements on reaction coordinate diagrams for folding reactions. In Fig. 3A, the net effect of the replacement is to decrease the free energy of the native conformation with respect to the energies of the transition state and unfolded forms. The free energy of unfolding, i.e., the stability is increased (Fig. 3B), the relaxation time for unfolding is increased, and the relaxation time for refolding is unchanged (Fig. 3C). This behavior, where the free energy of one of the stable states is selectively altered, can be classified as an effect on the equilibrium properties and such mutant proteins termed equilibrium mutants. In this case, the amino acid involved plays a role in stabilizing the protein but is not involved in the rate-limiting step.

[17] R. R. Kelley, J. Wilson, C. Bryant, and E. Stellwagen, *Biochemistry* **25,** 728 (1986).
[18] M. M. Crisanti and C. R. Matthews, *Biochemistry* **20,** 2700 (1981).

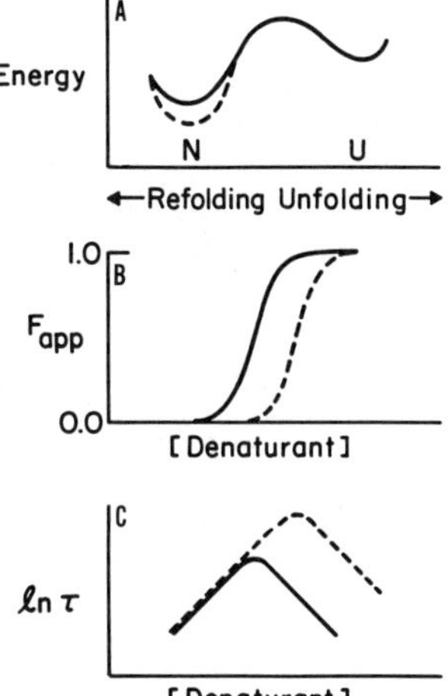

FIG. 3. Equilibrium and kinetic properties of the wild-type (—) and equilibrium mutant (---) proteins. (A) The reaction coordinate diagrams. (B) The equilibrium unfolding transition. F_{app} represents the fraction of unfolded protein. (C) The relaxation time as a function of the final denaturant concentration.

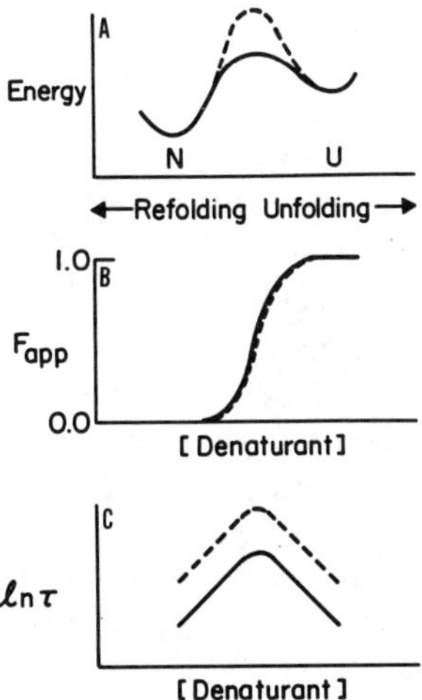

FIG. 4. Equilibrium and kinetic properties of the wild-type (—) and kinetic mutant (---) proteins. (A) The reaction coordinate diagrams. (B) The equilibrium unfolding transition. F_{app} represents the fraction of unfolded protein. (C) The relaxation time as a function of the final denaturant concentration.

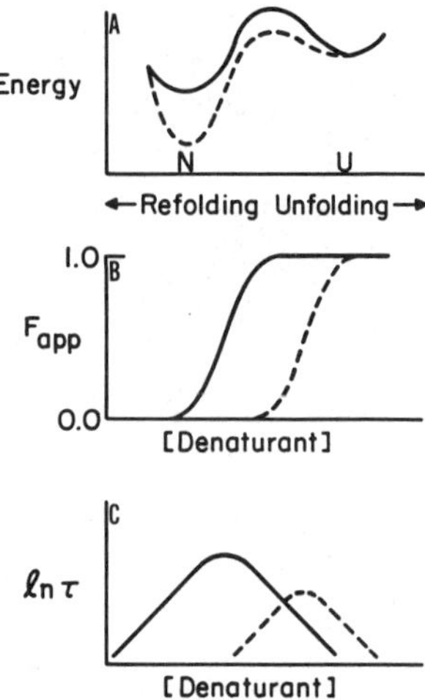

FIG. 5. Equilibrium and kinetic properties of the wild-type (—) and mixed equilibrium–kinetic mutant (---) proteins. (A) The reaction coordinate diagrams. (B) The equilibrium unfolding transition. F_{app} represents the fraction of unfolded protein. (C) The relaxation time as a function of the final denaturant concentration.

Another possible outcome for an amino acid replacement is shown in Fig. 4A. In this case, the free energy of the transition state is selectively altered. The stability of the protein is not changed (Fig. 4B); however, the relaxation times for both unfolding and refolding are increased. Such mutants can be classified as kinetic mutants and serve to identify amino acids that play key roles in the rate-limiting step in folding.

The third general class of mutants are those that affect both the equilibrium and kinetic properties, as shown in Fig. 5A–C. The free energies of the native conformation, transition state, and unfolded conformation are all altered with respect to each other. In the case shown, the stability is increased, the relaxation time for unfolding is increased, and the relaxation time for refolding is decreased. This behavior can be classified as a mixed equilibrium–kinetic mutant and, like the kinetic mutant described above, implicates the amino acid in the rate-limiting step in folding.

The reaction coordinate diagrams shown in Fig. 3–5 have been arbitrarily aligned by equating the free energies of the wild-type and mutant proteins in the unfolded state. The analysis of the equilibrium and kinetic data provides information on the *differences* in free energy between the various states. However, they do not provide information on the absolute free energy of any state. The critical factor in determining the role of a particular amino acid in folding is the *differential* effect of the replacement on the relative free energies of the two stable states and the transition state. Therefore, the alignment of the reaction coordinate diagrams is arbitrary.

Application

As an example of how amino acid replacements can be used to obtain structural information on folding reactions, a problem from the folding of the α subunit of tryptophan synthase from *Escherichia coli* will be described. The α subunit is a 29,000-dalton protein that has a larger, more stable amino domain, residues 1–188, and a small, less stable carboxyl domain, residues 189–268.[19] Studies on the mechanism of folding suggested that the final step in folding involves the conversion of a stable intermediate, which has a folded amino domain and an unfolded carboxyl domain, to the fully folded native conformation.[20] Left unanswered was the nature of the rate-limiting step in this process: Is it limited by the folding of the carboxyl domain or by the association of the two folded domains?

This problem was resolved by examining the effect of single amino acid replacements in both the amino and carboxyl domains on the unfolding and refolding relaxation times for the intermediate-to-native step. If the folding of the carboxyl domain is rate-limiting, only mutations in the carboxyl domain can have a real effect on the kinetics of this reaction. If domain association is rate limiting then mutations in either domain can act as kinetic or mixed equilibrium–kinetic mutants.

The effects of the Phe 22 → Leu and Gly 211 → Glu replacements on the relaxation times for unfolding and refolding are shown in Fig. 6.[21] The replacement in the amino domain, Phe 22 → Leu, is very close to a pure kinetic mutant; the unfolding relaxation time increases by 4-fold at 6 *M* urea while that for refolding increased by 2.5-fold at 2 *M* urea. The equilibrium data confirm the near absence of an effect on stability. The replacement in the carboxyl domain, Gly 211 → Glu, is clearly a mixed equilibrium–kinetic mutant; the relaxation time for unfolding increases 8-fold at 6 *M* urea and that for refolding increases 4-fold at 2 *M* urea. Consistent with these results, the equilibrium data show that this replacement causes a small increase in stability. Because replacements in both

[19] E. W. Miles, K. Yutani, and K. Ogaschara, *Biochemistry* **21,** 2586 (1982).

[20] C. R. Matthews, M. M. Crisanti, J. T. Manz, and G. L. Gepner, *Biochemistry* **22,** 1445 (1983).

[21] A. M. Beasty, M. R. Hurle, J. T. Manz, T. Stackhouse, J. J. Onuffer, and C. R. Matthews, *Biochemistry* **25,** 2965 (1986).

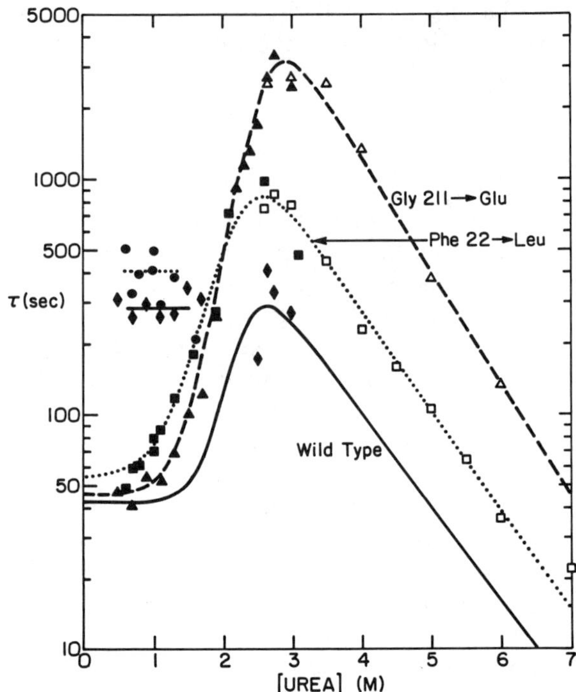

FIG. 6. Relaxation time at the indicated final urea concentration for the single phase in unfolding and two slow phases in refolding for the Phe 22 → Leu (□, ■, ●) and the Gly 211 → Glu (△, ▲, ◆) mutant α subunits from tryptophan synthase.[21] Open symbols represent unfolding data and filled symbols represent refolding data. For comparison, the relaxation times for the wild-type α subunit under the same conditions (25°, pH 7.8) are shown as the solid lines.

the amino and carboxyl domains affect the rate-limiting step, one can conclude that the reaction must correspond to domain association or, perhaps, to some other type of molecule-wide reaction.

The urea-independent relaxation times observed at urea concentrations below 1 *M* in Fig. 6 are thought to reflect Pro isomerization reactions which become rate limiting at low urea concentrations. A detailed account of their involvement in the folding mechanism has been presented elsewhere.[18]

Comments

The method presented for analyzing the effects of amino acid replacements on protein folding and stability is applicable to folding reactions in which both the unfolding and refolding rate constants or, equivalently, one of the rate constants and the associated equilibrium constant can be determined. Therefore, certain kinetic schemes may preclude its use. Also, it may be useful only for folding reactions that occur late in the folding pathway, near the native conformation, where the protein is highly constrained and is effectively moving from one particular conformation to another. Early folding reactions, nearer to the unfolded conformation, may proceed along a variety of pathways at nearly equal rates and preclude use of the method. Obviously, more experiments are required to test the extent of its applicability. The results described above, however, demonstrate its potential in elucidating the structural basis for reactions that occur during protein folding. It is our expectation that mutagenesis will dramatically improve our understanding of the mechanisms of protein folding and improve the possibility of predicting the three-dimensional structure of a protein from its amino acid sequence.

[27] Structure and Thermal Stability of Phage T4 Lysozyme

By TOM ALBER and BRIAN W. MATTHEWS

Introduction

Understanding the physical basis of the thermal stability of proteins is a major problem in molecular biology and a prerequisite for rational protein design. While the three-dimensional structure of a protein can be determined relatively accurately, the strategies for designing an amino acid sequence to stabilize that structure remain mysterious.

Reversible protein denaturation has proven to be an extremely complex reaction.[1–3] Dramatic changes in the solvation and flexibility of the polypeptide chain are accompanied by compensating changes in the enthalpy and entropy of the system. These thermodynamic state functions vary steeply with temperature. At high temperature, denaturation results in a large increase in entropy, presumably due to the added flexibility of the protein, and in a compensating increase in enthalpy, attributed to changes in interactions in the protein and solvent. Surprisingly, proteins can also denature at low temperature, and the system actually loses en-

[1] P. L. Privalov, *Adv. Protein Chem.* **33,** 198 (1979).
[2] J. A. Schellman, M. Lindorfer, R. Hawkes, and M. Grütter, *Biopolymers* **20,** 1989 (1981).
[3] P. L. Privalov, Y. V. Griko, S. Venyaminov, and V. P. Kutyshenko, *J. Mol. Biol.* **190,** 487 (1986).

tropy (becomes more ordered) and releases heat in the process.[3] The difference in free energy between the native and denatured states is also a complex function of temperature, characterized by a temperature of maximum stability.

The physical forces that underlie this complicated thermodynamic behavior—such as electrostatic forces, van der Waals forces, and hydrophobic interactions—are well known but have escaped quantitative enumeration.[4] The contributions of individual amino acids to the temperature-dependent stability of a protein are also poorly understood. Major barriers to further progress are presented by the complexities of protein structures, the lack of knowledge about the unfolded state, and the profound role of water in determining protein conformation. A fundamental practical difficulty is that experimental measurements of folding and denaturation provide information only about the sum of all interactions in the system. They do not directly identify the features of a complicated protein structure that are responsible for its stability.

A time-honored way to identify critical structural features is to select mutants with altered stability. Comparative studies of mutant and wild-type proteins offer the possibility of ascribing the observed differences to the small number of atoms that are changed. Modern methods of directed mutagenesis have increased the power of this approach. Despite these advances, comparison of mutant and wild-type proteins remains complicated by the inability to assess the absolute contribution of the mutation to the free energy of either the folded or the unfolded states.

A large number of genetic and biophysical techniques have been brought to bear on the lysozyme of bacteriophage T4, making this a powerful model system for studying protein thermal stability. Schellman and co-workers have developed methods for monitoring the reversible denaturation of the purified protein, and they have used these data to calculate ΔH, ΔS, ΔG, and ΔCp for the denaturation reaction.[2,5–8] Comparison of these values for mutant and wild-type lysozymes has provided an estimate of the magnitudes of the contributions of specific amino acids to thermal stability.

The X-ray crystal structures of wild-type and mutant lysozymes have been determined in order to search for structural patterns associated with

[4] N. Pace, *CRC Crit. Rev. Biochem.* **3,** 1 (1975).

[5] M. L. Elwell and J. A. Schellman, *Biochim. Biophys. Acta* **494,** 367 (1977).

[6] M. G. Grütter and R. B. Hawkes, *Naturwissenschaften* **70,** 434 (1983).

[7] R. Hawkes, M. G. Grütter, and J. Schellman, *J. Mol. Biol.* **175,** 195 (1984).

[8] W. J. Becktel and W. A. Baase, *Biopolymers* **26,** 619 (1987).

thermal stability.[9–12] Multidimensional heteronuclear NMR methods are being applied by Dahlquist and co-workers to investigate the relationship between stability and the motions of the molecule in solution.[13,14] Correlations between rapid motions and stability are being studied by Hudson *et al.*, using fluorescence spectroscopy.[15] Following the pioneering work of Streisinger *et al.*,[16] recent advances in genetic methods have been made by Muchmore *et al.*,[17] Perry and Wetzel,[18] and our laboratory.[12,19]

In this brief chapter we present an introduction to the phage T4 lysozyme system. Methods of rapidly generating mutants, characterizing their thermal stability, and accurately determining their X-ray crystal structures are discussed. Results and insights into the structural basis of protein thermal stability are summarized.

Methods

Mutagenesis

T4 lysozyme is produced late in the phage life cycle and is required for lysis of the host bacterium. T4 strains lacking a functional lysozyme gene must be grown in the presence of externally supplied lysozyme, usually from hen egg white. In addition to providing the basis for conditional

[9] S. J. Remington, W. F. Anderson, J. Owen, L. F. Ten Eyck, C. T. Grainger, and B. W. Matthews, *J. Mol. Biol.* **118,** 81 (1978).

[10] M. G. Grütter, R. B. Hawkes, and B. W. Matthews, *Nature (London)* **277,** 667 (1979).

[11] M. G. Grütter, L. H. Weaver, T. Gray, and B. W. Matthews, *in* "Bacteriophage T4" (C. K. Mathews, E. M. Kutter, G. Mosig, and P. M. Berget, eds.), p. 356. Am. Soc. Microbiol., Washington, D.C., 1983.

[12] T. Alber, M. G. Grutter, T. M. Gray, J. A. Wozniak, L. H. Weaver, B.-L. Chen, E. N. Baker, and B. W. Matthews, *UCLA Symp. Mol. Cell. Biol., New Ser.,* **39,** 307 (1986).

[13] R. H. Griffey, A. G. Redfield, R. E. Loomis, and F. W. Dahlquist, *Biochemistry* **24,** 817 (1985).

[14] F. W. Dahlquist, R. H. Griffey, L. P. McIntosh, D. C. Muchmore, T. G. Oas, and A. G. Redfield, *in* "Synthesis and Applications of Isotopically Labeled Compounds 1985" (R. R. Muccino, ed.), p. 533. Elsevier, Amsterdam, 1986.

[15] B. S. Hudson, D. L. Harris, R. D. Ludescher, A. Ruggiero, A. Cooney-Freed, and S. A. Cavalier, *in* "Applications of Fluorescence in the Biomedical Sciences" (D. L. Taylor, A. S. Waggoner, R. F. Murphy, F. Laani, and R. R. Birge, eds.), p. 159. Liss, New York, 1986.

[16] G. Streisinger, F. Mukai, W. J. Dreyer, B. Miller, and S. Horiuchi, *Cold Spring Harbor Symp. Quant. Biol.* **26,** 25 (1961).

[17] D. M. Muchmore, C. R. Russell, W. A. Baase, W. J. Becktel, and F. W. Dahlquist, *Gene,* manuscript in preparation.

[18] L. J. Perry and R. Wetzel, *Science* **226,** 555 (1984).

[19] T. Alber and J. Wozniak, *Proc. Natl. Acad. Sci. U.S.A.* **82,** 747 (1985).

phage growth, the presence of the enzyme can be assayed directly on petri plates.[16] This plate enzyme assay forms the basis of rapid screens for mutant phenotypes. After growing T4 phage on a bacterial lawn at the desired temperature, the (glass) petri plates are inverted and chloroform is added to the top. The combined action of the chloroform vapors and the lysozyme diffusing away from the phage plaque breaks the surrounding uninfected cells. The size of this zone of clearing (halo) around the plaque is a measure of the activity and the amount of enzyme produced during the infection.

This assay has been used to screen T4 stocks randomly mutagenized with chemicals[20] and base analogs[21] for a variety of phenotypes. Tight temperature-sensitive (ts) mutants form plaques at 30 but not at 42°C. Leaky ts mutants can form plaques at all growth temperatures, but they produce progressively less lysozyme activity at elevated growth temperatures (Fig. 1). Low activity mutants form smaller halos than wild type, but halo size is independent of growth temperature.

A critical advantage of the plate assay is that the screen for mutant phenotypes is carried out after phage growth. This permits screening for interesting mutants under conditions that kill the host. For example, after plating chemically mutagenized phage at 33°C, mutants resistant to heat inactivation (st), were identified by exposing the plates to a 55°C heat pulse prior to chloroform treatment.[19] A similar approach was used 30 years ago by Streisinger and Franklin[22] to distinguish wild type from mutants in the *h* gene of phage T2. In the 1920s and 1930s, resistance of phage lysozymes to a heat pulse was used by Sertic to distinguish and classify different phage types.[23–25] In principle, the plate assay could be used to screen lysozyme variants for sensitivity to a wide variety of conditions and chemicals including high pressure, cold, extremes of pH, metals, and organic solvents.

Genetic deletion mapping, peptide fingerprinting, and DNA cloning and sequencing have been used to identify the lesions in the mutant lysozyme genes.[26] Cloning T4 sequences is somewhat unusual, because the phage DNA contains monoglucosyl-5-hydroxymethylcytosine instead of cytosine.[27] This modification renders the DNA resistant to cleavage by

[20] E. Freese, E. Bautz, and E. B. Freese, *Proc. Natl. Acad. Sci. U.S.A.* **47,** 845 (1961).

[21] E. Freese, *J. Mol. Biol.* **1,** 87 (1959).

[22] G. Streisinger and N. C. Franklin, *Cold Spring Harbor Symp. Quant. Biol.* **21,** 103 (1956).

[23] V. Sertic, *Zentralbl. Bakteriol., Parasitenkd. Infektionskr.* **110,** 125 (1929).

[24] V. Sertic and N. Boulgakov, *C. R. Seances Soc. Biol. Ses Fil.* **108,** 948 (1931).

[25] V. Sertic and N. Boulgakov, *C. R. Seances Soc. Biol. Ses Fil.* **119,** 1270 (1935).

[26] T. M. Gray, Ph.D. thesis. University of Oregon, Eugene, Oregon, 1985.

[27] M. A. Jesaitis, *Nature (London)* **178,** 637 (1956).

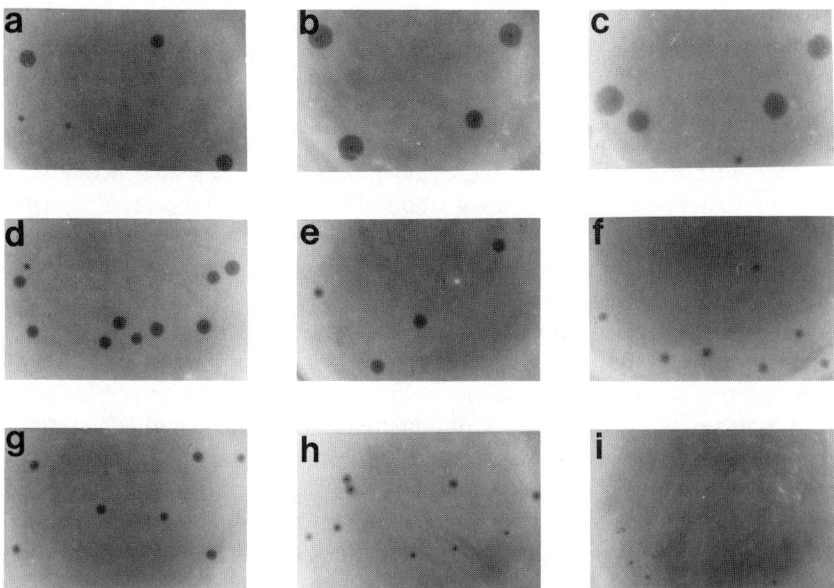

FIG. 1. Test for T4 phage producing a temperature-sensitive lysozyme. Bacteria mixed with a small number of T4 phage are incubated on petri plates at an elevated growth temperature for 6 hr and then exposed to chloroform vapors for 12–18 hr at room temperature. The size of the halo surrounding the dark central plaque is a measure of the amount, stability, and activity of the lysozyme present in the plaque. Panels (a), (b), and (c) show petri plates seeded with wild-type phage grown at 31, 37, and 43°C, respectively. The size of the halo increases with increasing growth temperature. The halos produced by ts lysozyme mutants Arg 96 → His (d–f) and Thr 157 → Ile (g–i) after growth at 31, 37, and 43°C are shown for comparison. While the ts mutants produce enough lysozyme to form plaques even at 43°C, the size of the halo decreases with increasing growth temperature. This correlates with reduced thermal stability of these phage lysozymes.

most restriction enzymes. To overcome this problem, the lysozyme mutation of interest is crossed into a strain lacking the enzymes that modify cytosine (*56*^{am}) and cleave unmodified DNA (*denA*, *denB*).[28] Such a recombinant T4 strain is propagated in the presence of externally supplied lysozyme in a host containing an amber suppressor tRNA. In infected cells lacking a suppressor (*su*$^-$), T4 sequences form the bulk of the DNA, but phage are not produced because the late genes are not expressed. DNA purified from infected *su*$^-$ cells[28] can be cut with restriction enzymes. The lysozyme gene is isolated on a unique 4-kilobase (kb) fragment by agarose gel electrophoresis of a *Xho*I digest of T4 DNA.[29] This

[28] J. E. Owen, D. W. Schultz, A. Taylor, and G. R. Smith, *J. Mol. Biol.* **165,** 229 (1983).
[29] P. H. O'Farrell, E. Kutter, and M. Nakanishi, *Mol. Gen. Genet.* **179,** 421 (1980).

4-kb fragment is extracted from the gel[30] and subcloned into m13 for sequencing[31] or into an expression plasmid[17] for production of protein.

While random mutagenesis has identified amino acids that are essential for protein stability, oligonucleotide-directed mutagenesis has been used to genetically saturate selected sites to quantitate the contributions of specific chemical groups. Directed mutagenesis has also provided a way to test specific hypotheses about the structural basis of thermal stability.

To make specific changes in the T4 lysozyme gene cloned in phage m13, we use the two-primer method of Zoller and Smith[32,33] with some variations. Many different mutations at a given site are rapidly constructed using a degenerate mixture of oligonucleotides, each of which differs from the wild-type sequence in at least one position. Since the wild-type sequence is missing from the mixture, mutants can be identified by differential hybridization to the ^{32}P-labeled oligonucleotides. Mutants are distinguished by DNA sequencing.

For example, a 21-mer mixture containing a triplet complementary to NNT (N = T, C, A, and G) was used to specify 16 different sequences at the codon for Pro 86 (CCG). This primer codes for 15 different amino acids, one of which is proline (CCT). Forty-three mutant plaques were sequenced to yield ten mutants—Leu, Ile, Ser, Thr, Ala, Gly, Cys, His, Arg, and Asp. The distribution of sequences is shown in Table I, Part a. This distribution favors single (6) and double (24) mutants over triple mutants (11). (The expected unbiased frequencies of these classes are 1/16, 6/16, and 9/16, respectively.) The observed bias may arise because excess mutagenic primer is used in the annealing step. There is also a clear bias against sequences containing only AT base pairs in the mutagenized codon (1/43). This can be explained by the lower melting temperature of these oligonucleotides. If any of the 16 sequences in the mixture (e.g., the single mismatch 5′—AGG—3′) bind to wild-type (CCG) phage as tightly as the AT sequences bind to perfectly matched mutants, hybridization will not distinguish these mutants from wild type. This difficulty can be overcome by avoiding the hybridization screen for mutants. This is accomplished by genetic selection for the mutant strand[34] or by engineering silent substitutions in linked restriction sites.

[30] S. A. Benson, *BioTechniques* **2,** 66 (1984).

[31] F. Sanger, S. Nickelsen, and A. R. Coulsen, *Proc. Natl. Acad. Sci. U.S.A.* **74,** 5463 (1977).

[32] M. J. Zoller and M. Smith, *DNA* **3,** 479 (1984).

[33] M. J. Zoller, this volume [17].

[34] T. A. Kunkel, *Proc. Natl. Acad. Sci. U.S.A.* **82,** 488 (1985).

TABLE I

FREQUENCY DISTRIBUTIONS OF SEQUENCES OBTAINED AFTER MUTAGENESIS WITH OLIGONUCLEOTIDE MIXTURES[a]

Part a. Oligonucleotide mixture was complementary to 5′-ATTAAAA$\underline{\text{NN}}$TGTTTATGATTC-3′; 21-mers were gel purified.[35]

First position	Second position[b]			
	T	C	A	G
T	Phe—0	Ser—2	Tyr—0	Cys—2
C	Leu—4	Pro—6	His—2	Arg—8
A	Ile—1	Thr—1	Asn—0	Ser—2
G	Val—0	Ala—7	Asp—1	Gly—5

Part b. Oligonucleotide mixture was complementary to 5′-AACTGGC$\underline{\text{GN}}$$^{\text{T}}_{\underline{\text{A}}}$TGGGACGC-3′; 18-mers were gel purified.[35]

First position	Second position				Third position
	T	C	A	G	
G	Val—0	Ala—11	Asp—7	Gly—4	T
G	Val—0	Ala—3	Glu—2	Gly—2	A

[a] Mismatches are underlined. Each entry gives the identity of the amino acid corresponding to the named codon, together with the number of times that the codon occurred.

[b] The third position is T. Two of 43 sequenced plaques were wild type (CCG).

When using mutagenic oligonucleotide mixtures, gel purification of the mixture should be avoided. The reason for this is shown in the experiment summarized in Table I, Part b. In this experiment an 18-mer containing a triplet complementary to the sequence GN$^{\text{T}}_{\text{A}}$ was used to mutagenize the ACT codon of Thr 157. As shown in Table I, Part b, mutants with T in the second position were not found among the 30 plaques sequenced. It was subsequently discovered that the oligonucleotide complementary to GTN runs anomalously fast on the 15% acrylamide 4 *M* urea gel[35] used for purification. As a result, this sequence had been removed from the mixture of 18-mers.

Mixtures of oligonucleotides have been used to make multiple mutants at over a dozen sites in T4 lysozyme. A typical reaction mix contains 0.1

[35] A. M. Maxam and W. Gilbert, this series, Vol. 65, p. 499.

pmol m13 template, 2pmol kinased mutagenic oligonucleotide, 0.5 pmol m13 sequencing primer, 0.5× ligase buffer, 0.5× Klenow buffer, 0.5 m*M* dNTPs, 0.5 m*M* rATP, 3 U of T4 DNA ligase, and 2 U of the large fragment of DNA polymerase. After incubation at 16°C for 8–16 hr, this mixture is used to transform *E. coli* JM101 in broth culture. This step separates heteroduplex heterozygotes. The number of independent transformants is estimated by plating aliquots of transformed cells. The frequency of mutagenesis is 2–20%. Sequencing of mutants has revealed a number of unexpected events including unplanned base substitutions and frameshift mutations under the primer. A 32-bp insertion of inverted lysozyme sequences and a deletion of approximately 100-bp have also been observed.

Expression and Protein Purification

Two expression systems have been developed to produce and purify phage T4 lysozyme.[13,17,36] On the pT4lystacII plasmid of Perry *et al.*,[36] the *tacII* promoter drives expression of the lysozyme gene. The system of Muchmore *et al.* (pHSe5)[17] contains the lysozyme gene flanked by tandem *lacUV5* and *tac* promoters and the *trp* terminator. Tight control of expression is provided by the presence of the *lacI*^{q} gene on the plasmid. The *trp* terminator eliminates selection against cells harboring the expression plasmid.

The plate assay for phage lysozyme described above has been adapted to measure enzyme activity produced by bacterial colonies.[17] A lawn of *E. coli* RR1/pBR322 is first grown on LB-H plates containing 100 μg/ml ampicillin. The cells in the lawn are killed by exposure to UV light and chloroform. Lysozyme-producing colonies subsequently grown on the lawn produce halos. This assay greatly simplifies identification of colonies harboring the expression plasmid and allows rapid screening for mutant phenotypes of cloned lysozyme genes.

T4 lysozyme is purified from lysates of induced plasmid-bearing cells by column chromatography on CM-Sepharose.[13,17] Cell lysis is aided by the fact that lysozyme-producing bacteria are hypersensitive to agents that disrupt the cell membrane (SDS, EDTA, chloroform, m13 infection, *tet*^{r} overexpression). Treatment with chloroform has been found to irreversibly alter the properties of the purified protein, so lysis is promoted with EDTA.[17] The enzyme is conveniently assayed on lysoplates[37] or in solution.[38] Typical yields using pHSe5 are 50–70 mg of lysozyme per liter

[36] L. J. Perry, H. L. Heyneker, and R. Wetzel, *Gene* **38,** 259 (1985).

[37] W. J. Becktel and W. A. Baase, *Anal. Biochem.* **150,** 258 (1985).

[38] A. Tsugita and M. Inouye, *J. Mol. Biol.* **37,** 201 (1968).

of induced cell culture. This massive overproduction of the protein has reduced the purification time (now 3 days) by an order of magnitude and has greatly facilitated crystallographic and magnetic resonance experiments that require large amounts of material.

Thermodynamic Studies

Thermal denaturation of T4 lysozyme is monitored by measuring the optical properties of a solution of the purified protein as a function of temperature. Fluorescence emission and circular dichroism are the most sensitive methods.[5] When the protein unfolds, the optical spectrum (e.g., the molar ellipticity at 223 nm, which is sensitive to protein secondary structure) undergoes a cooperative change. Assuming that the spectrum at any temperature is a linear combination of the spectra of only the native and denaturated states of the protein, the fraction of the material that is (un)folded can be calculated at each temperature. Although evidence for multistate denaturation has been reported,[39,40] the equilibrium measurements of Schellman and co-workers on wild-type lysozyme and a large number of mutants are consistent with the two-state assumption.[2,5–8] The ratio of the fraction of the protein in the unfolded and folded conformations is the equilibrium constant (K) for the denaturation reaction.

Thermodynamic values are derived from these data in several steps.[2] (1) The van't Hoff equation, $\partial \ln K/\partial(1/T) = -\Delta H/R$, gives the enthalpy change for the reaction over the temperature range where K can be measured accurately. (2) At the midpoint of the denaturation transition (T_m), half the material is folded and half is unfolded ($K = 1$). Since $\Delta G = -RT(\ln K)$, $\Delta G = 0$ at T_m. (3) Since $\Delta G = \Delta H - \mathrm{T}\Delta S$, ΔS can be evaluated at T_m. (4) T_ms of wild-type and mutant lysozymes vary as the pH is changed. This allows ΔH and ΔS to be evaluated as a function of temperature. These data fit the expression $\Delta H = \Delta H_0 + \Delta Cp(T - T_0)$ and $\Delta S = \Delta S_0 + \Delta Cp \ln(T/T_0)$, where ΔH_0 and ΔS_0 are the enthalpy and entropy changes at a selected reference temperature (T_0) and ΔCp is the difference in the heat capacity of the folded in unfolded forms. (5) Once ΔH_0, ΔS_0, and ΔCp are evaluated for the wild-type protein, the relations in steps (4) and (3) can be used to calculate the change in the free energy of stabilization due to mutation ($\Delta\Delta G$) at the T_m of the mutant protein.

In practice, great care is taken to ensure that the experimental measurements are reversible and reproducible.[7,8] Fresh protein samples purified in the absence of chloroform are extensively dialyzed against oxygen-free buffers and reducing agent. Ionic strength is kept above 0.15 with

[39] M. Desmadril and J. M. Yon, *Biochem. Biophys. Res. Commun.* **101,** 563 (1981).

[40] M. Desmadril and J. M. Yon, *Biochemistry* **23,** 11 (1984).

KCl or NaCl, and pH is adjusted with HCl (pH 2–3), 10 mM acetate buffer (pH 4–5), or 10 mM phosphate buffer (pH 5.5–7). Protein concentration is kept below 30 μg/ml to avoid irreversible aggregation at high temperature.

Circular dichroism is monitored with a Jasco J-500C instrument equipped with a Hewlett-Packard 89100A thermoionic controller. The temperature of the sample is changed at a constant rate, typically 1°K per minute, under the control of a Hewlett-Packard 87 XM computer. The temperature and optical signal are digitized for subsequent analysis. Both denaturation and renaturation are monitored to ensure reversibility.

Crystallization and Data Collection

To appraise changes in the interactions in the folded forms of wild-type and mutant lysozymes, determining the high resolution X-ray structure of each mutant is essential. Most mutant structures solved have contained unexpected features.

Wild-type lysozyme crystallizes in the space group $P3_221$ with unit cell dimensions $a = b = 61.1$ Å, $c = 95.3$ Å.[41] The diffraction pattern extends to at least 1.7 Å resolution. At least 80% of the mutants studied to date have crystallized in the same form as the wild-type protein. In several cases different crystal forms were obtained. Our limited experience suggests that mutants that fail to crystallize by the standard method are difficult to crystallize under any conditions.

Crystals are grown by batch methods using 4 M phosphate as the precipitant. Phosphate solutions ranging from pH 6.4 to 7.1 are usually sufficient to bracket the optimal conditions. Concentrated phosphate solutions in this pH range are obtained by mixing 4 M solutions of NaH_2PO_4 and K_2HPO_4. The precipitant is slowly added to 30 μl of 10–30 mg/ml lysozyme buffered at pH 6.7 with 10 mM $NaPO_4$, 550 mM NaCl, 14 mM 2-mercaptoethanol, and 1 mM $MgCl_2$. Precipitant is added until the solution remains slightly turbid after agitation, and the vial is capped and stored at 4°C. Crystals grow over periods ranging from 1 week to 6 months. Prior to data collection, the crystals are equilibrated with a standard mother liquor containing 1.05 M K_2HPO_4, 1.26 M NaH_2PO_4, 230 mM NaCl, 1.4 mM fresh 2-mercaptoethanol (pH 6.7).

High-resolution X-ray diffraction data are collected by oscillation photography using graphite–monochromatized copper Kα radiation from an Elliott GX-21 rotating-anode generator.[42] The data collection system is characterized by a number of unique features. These include adjustable

[41] B. W. Matthews, F. W. Dahlquist, and A. Y. Maynard, *J. Mol. Biol.* **78,** 575 (1973).

[42] M. F. Schmid, L. H. Weaver, M. A. Holmes, M. G. Grütter, D. H. Ohlendorf, R. A. Reynolds, S. J. Remington, and B. W. Matthews, *Acta Crystallogr.* **A37,** 701 (1981).

collimators, cylindrical cassettes, and film alignment pins that aid crystal alignment, improve signal to noise, reduce exposure time, simplify film processing, and generally facilitate the rapid collection of high-resolution data. For most protein crystals the exposure time for a typical high-resolution oscillation photograph is 4 hr. The films are processed by a program based on that of Rossmann.[43]

A high-resolution T4 lysozyme data set can be measured from two medium-sized crystals or one large crystal. The precision of the data from large crystals (at least 0.5 mm in each direction) is superior to that from smaller crystals (0.3–0.4 mm in each direction). Although a crystal rotation of 30° is sufficient for collection of a data set, we usually improve accuracy by measuring data for a total rotation of 40°. Because of symmetry relationships, as well as the fact that data are recorded on the top and bottom halves of the film, this data collection strategy results in an almost 3-fold redundancy in intensity measurement. Typically, the agreement between repeated intensity measurements made on different films is 5–8% for a 1.7 Å resolution data set.

For every crystal that is used to collect three-dimensional data we conclude the data collection process by taking a 15° precession photograph of the (h0l) zone. This photograph serves as a permanent reference identifying the particular crystal that was used. In several cases electron density maps have contained unanticipated features that raised questions about the identification of the amino acid substitution. In such cases it is useful to have reassurance that the measured X-ray data set reflects the designated mutant protein. The (h0l) photograph provides a "fingerprint" that can be easily compared with other reference films. Although the intensity differences between wild-type lysozyme and mutant lysozymes are small, they provide a surprisingly sensitive way of identifying each protein.

Structure Refinement

In order to understand the molecular basis of differences in thermostability, it is essential that the structures of the parent and mutant proteins be determined accurately. To improve the agreement between the protein model and the X-ray data within the bounds of stereochemical constraints, we use the refinement program TNT.[44] It should be emphasized that, no matter what refinement procedure is used, the accuracy of a protein crystal structure is ultimately limited by the resolution of the observed X-ray data. As a general guide, a preliminary (unrefined) struc-

[43] M. G. Rossmann, *J. Appl. Crystallogr.* **12,** 225 (1979).

[44] D. E. Tronrud, L. F. Ten Eyck, and B. W. Matthews, *Acta Crystallogr.* **A43,** 489 (1987).

ture determination at 2.8 Å resolution might have an overall coordinate accuracy of 0.5–1.0 Å; following refinement, still using 2.8 Å resolution data, the accuracy of the structure might be improved to about 0.4 Å. By extending the resolution of the data to 1.7 Å, careful refinement can improve coordinate accuracy to about 0.15 Å.

There are several problems in protein refinement, the first being the large size of the computational problem. Phage T4 lysozyme has 1429 nonhydrogen atoms, each of which is specified by three coordinates and a thermal motion parameter. These coordinates and thermal motion parameters have to be optimized to satisfy approximately 16,000 X-ray measurements (for a data set to 1.7 Å resolution). At the same time the refined protein structure should retain acceptable stereochemistry. Another potential difficulty is that the starting model for the protein structure may have serious errors. In such cases it will take many cycles of refinement, interspersed with manual rebuilding of the model, to achieve a satisfactory refined model. Therefore the computing time per cycle of refinement must be kept as short as possible.

The TNT refinement package is designed to solve these problems. The TNT programs carry out efficient, restrained, least-squares refinement of macromolecular crystal structures. The package was designed to be as flexible and general purpose as possible. Restraints are used to keep the protein stereochemistry within specified limits. Guide values for bond lengths and angles are specified in a straightforward, direct manner. Designated groups of atoms can be held constant or constrained to behave as a rigid body during refinement. In order to make the package as efficient as possible, the fast Fourier transform algorithm is used for all the crystallographic transformations. As an example, a cycle of phage lysozyme refinement at 1.7 Å resolution takes 45 min on a VAX 11/780.

The program package has a number of features to assist the user and to detect potential errors. For example, the user can list the atoms that have the worst bond lengths and angles, or are most at variance with the crystallographic observations. It is also possible to check that protein and solvent atoms do not sterically clash with symmetry-related neighbors.

The mode of structure refinement used by TNT is similar, in principle, to that of PROLSQ, the commonly-used program of Hendrickson and Konnert.[45] A major advantage of TNT over PROLSQ is its greater speed. In tests with thermolysin the time per cycle was reduced from 21 CPU hr per cycle to 1.5 hr.[44] Also, the way in which TNT specifies stereochemical

[45] W. A. Hendrickson and J. H. Konnert, *in* "Computing in Crystallography" (R. Diamond, S. Ramaseshan, and K. Venkatesan, eds.), p. 13.01. Indian Academy of Sciences, Bangalore, India, 1980.

restraints is superior to PROLSQ. Unusual chemical cofactors, prosthetic groups, or inhibitors can be handled with ease by TNT but can cause problems with PROLSQ.

Refinement of a Mutant Structure

As an example, we describe the refinement of the temperature-sensitive mutant lysozyme Thr 157 → Ile.[46] The mutant protein crystallizes isomorphously with wild-type lysozyme, and diffraction data were measured to 1.7 Å resolution. Structural differences between the mutant and wild-type proteins were first visualized by a Fourier map with coefficients of the form $(F_{\text{mutant}} - F_{\text{wt}})$. Such a map (Fig. 2a) shows the difference between the mutant and wild-type structures. The positive and negative peaks straddling the side chain of Asp 159 indicate that this group moves away from the site of amino acid substitution. The difference density at residue 157 is more complicated. It reflects the additional electron density of isoleucine and a rotation of the isoleucine about its C^{α}—C^{β} bond.

The structure of the mutant lysozyme was visualized directly by calculating a map with coefficients $(2F_{\text{mutant}} - F_{\text{wt,calc}})$ (Fig. 2b). This map was used for initial model building of the mutant structure with a PS300 graphics system using the program FRODO.[47] In the case of Thr 157 → Ile, the coordinates of wild-type lysozyme were used as the starting coordinates for refinement of the mutant structure because the structural shifts are small. However, the altered amino acid was first positioned in a manner consistent with its electron density (Fig. 2b). Also, the side chain of Asp 159 has moved approximately 1 Å in the mutant structure, and this residue was repositioned with the graphics system.

For the refined structure of wild-type lysozyme, the root-mean-square discrepancies between bond lengths and bond angles and the expected ideal stereochemistry are 0.02 Å and 3°, respectively. In the initial stages of refining the mutant protein these stereochemical restraints were relaxed to allow the model structure to adjust to satisfy the crystallographic observations. The weighting of the stereochemical restraints was reduced so that the discrepancies in bond lengths and bond angles increased to 0.08 Å and 10°, respectively. When the refinement with relaxed stereochemistry converged, the geometrical restraints were given additional weight to restore acceptable values for bond lengths and angles. (Conver-

[46] M. G. Grütter, T. M. Gray, L. H. Weaver, T. Alber, K. Wilson, and B. W. Matthews, *J. Mol. Biol.*, in press (1987).

[47] T. A. Jones, *in* "Crystallographic Computing" (D. Sayre, ed.), p. 303. Oxford University Press, Oxford, England, 1982.

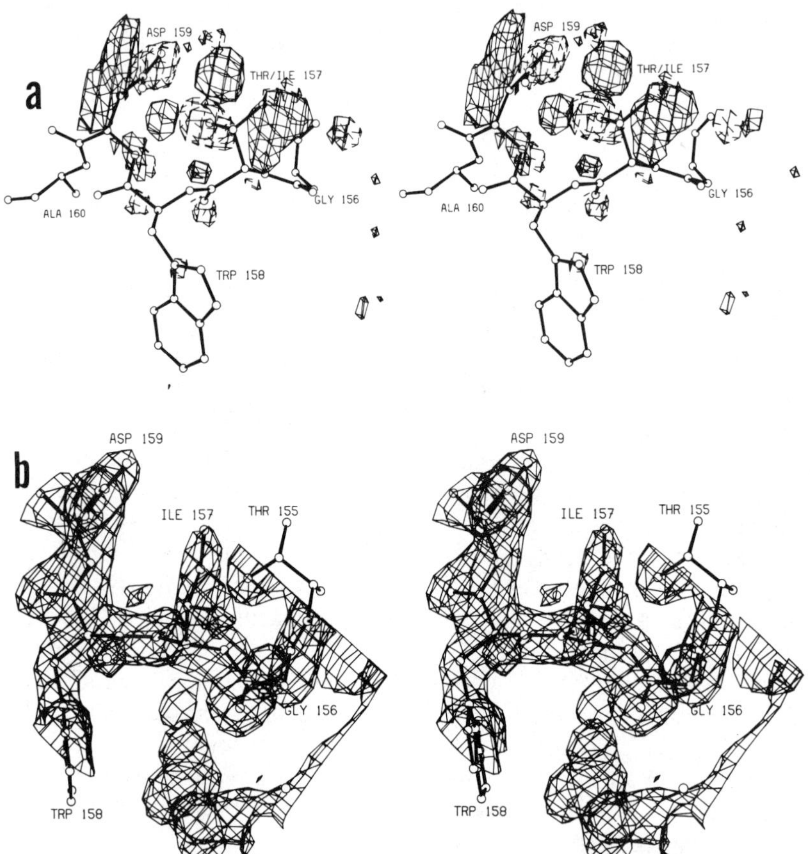

FIG. 2. Stereo views of electron density maps showing the difference between the mutant Thr 157 → Ile and wild-type lysozyme. (a) Map with amplitudes ($F_{T157I} - F_{wt}$) to 1.9 Å resolution and phases calculated from the refined model of wild-type lysozyme, superimposed on the structure of wild-type lysozyme. Contours are drawn at $+4\sigma$ (solid) and -4σ (dashed), where σ is the root-mean-square density throughout the unit cell. (b) Map with amplitudes ($2F_{T157I} - F_{wt,calc}$) to 1.9 Å resolution and phases calculated from the refined model of wild-type lysozyme, superimposed on the refined structure of T157I lysozyme.

gence with loose and tight stereochemical restraints usually require 10–20 cycles of refinement each.)

At intermediate stages, lists of bond lengths and angles with the worst stereochemistry were monitored to check for potential errors in the refined model. Difference electron density maps with coefficients ($F_{T157I,obs} - F_{T157I,calc}$) were inspected on the graphics system to detect inadequacies in the model. Such maps indicated differences in solvent structure between mutant and wild type; the mutant protein binds a new water molecule near Thr 155.

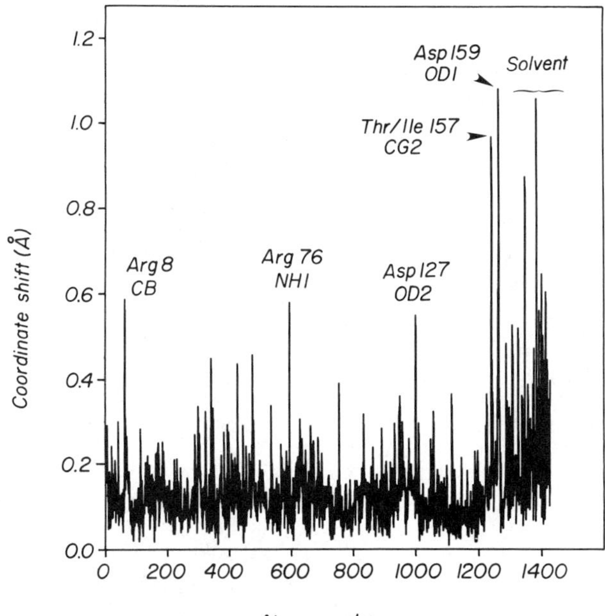

FIG. 3. "Shift plot" showing the difference in coordinates of all atoms common to the refined structures of Thr 157 → Ile and wild-type lysozymes, including solvent. Atoms with the largest apparent shift in coordinates are labeled.

Figure 3 is a "shift plot" showing the distance between each atom in the refined structure of Thr 157 → Ile and the corresponding atom in wild-type lysozyme.[46] The root-mean-square difference between all pairs of atoms, including solvent, is 0.19 Å. This provides an estimate of the overall accuracy of the coordinates. Well-defined atoms (i.e., atoms with low thermal factors) will be more accurately determined than this overall value (estimated error 0.1–0.15 Å). On the other hand, errors in the coordinates of mobile surface side chains will be much larger. The atoms of phage lysozyme (and proteins in general) that have thermal factors above 60 $Å^2$ are very poorly defined, and their coordinates should be regarded with the greatest skepticism.

Results and Discussion

Three-Dimensional Structure of Lysozyme

The X-ray crystal structure of phage T4 lysozyme provides a detailed view of the interactions in the folded protein. The structure was deter-

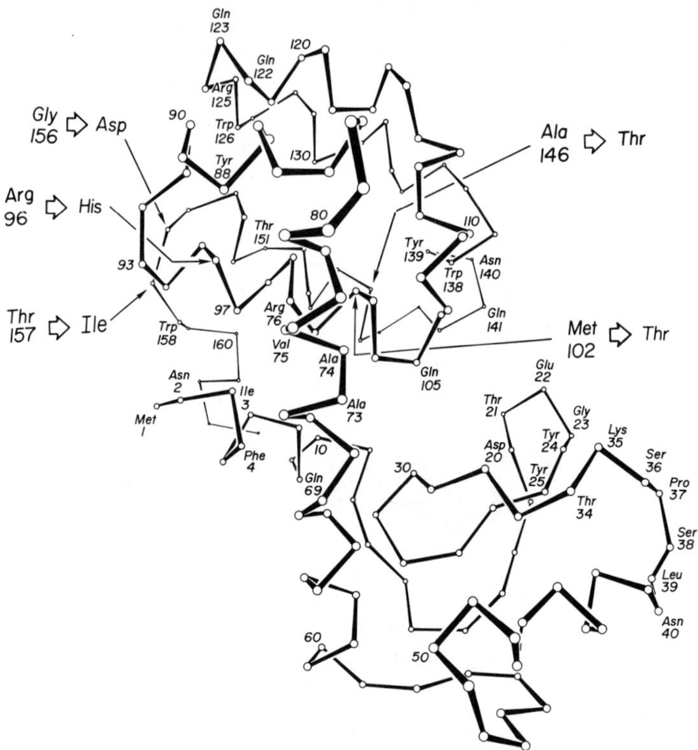

FIG. 4. Schematic drawing of the backbone of T4 lysozyme indicating the location of some of the temperature-sensitive mutations that have been analyzed crystallographically.

mined by the method of multiple isomorphous replacement[9,48] and refined to a high degree of accuracy ($R = 16.7\%$) at high resolution (1.7 Å).[49] The average error in atomic positions is of the order of 0.1–0.2 Å.

Figure 4 shows a drawing of the α-carbon backbone. Even though it is a small protein (164 amino acids), T4 lysozyme contains two distinct domains. The N-terminal domain (residues 1–60) contains all of the β-sheet in the structure, as well as two α-helices. The C-terminal domain (residues 80–164) is like a barrel whose bottom and staves are composed of seven α-helices. The domains are joined by a long α-helix (residues 60–80) that traverses the length of the molecule. Amino acids 162–164 are not well ordered in the crystals.

[48] B. W. Matthews and S. J. Remington, *Proc. Natl. Acad. Sci. U.S.A.* **71,** 4178 (1974).

[49] L. H. Weaver and B. W. Matthews, *J. Mol. Biol.* **193,** 189 (1987).

The active site is in the deep cleft between the domains. Substrate binding requires a breathing motion of the domains. Analysis of the atomic motions in the crystal structure suggests that the "hinge" for this breathing motion is centered around residue 67 in the long connecting α-helix.[49]

Random Mutants

To obtain a collection of lysozyme mutants with altered thermal stability or enzyme activity, randomly mutagenized T4 phage were screened with the plate assay. A partial list of mutants is given in Table II. The screen for leaky ts mutants (Fig. 1) identifies lysozyme variants that have suffered modest reductions in thermal stability. These proteins provide especially tractable subjects for thermodynamic and structural studies, because the lysozymes are stable enough *in vitro* to be purified in large quantities.

TABLE II
SOME MUTANT LYSOZYMES ISOLATED AFTER MUTAGENESIS OF PHAGE T4 WITH BASE ANALOGS AND HYDROXYLAMINE[a]

Tight ts mutants	Low activity mutants
Leu 66 → Pro	Glu 128 → Lys
Leu 91 → Pro	Heat resistant, low activity
	Cys 54 → Tyr
Leaky ts mutants	
Met 6 → Ile	
Leu 33 → Pro	
Arg 96 → His	
Ala 98 → Val	
Ala 98 → Thr	
Met 102 → Thr	
Val 103 → Ala	
Trp 126 → Arg	
Trp 138 → Tyr	
Arg 145 → His	
Ala 146 → Thr	
Ala 146 → Val	
Val 149 → Ala	
Gly 156 → Asp	
Thr 157 → Ile	
Trp 158 → Tyr	
Ala 160 → Thr	

[a] Mutant phenotypes are described in the text. Data are from Refs. 7, 11, 12, 46, and 50.

Most of the amino acid substitutions in this group occur in two clusters in the C-thermal domain of the protein (Fig. 4). One patch includes residues in the hydrophobic core of this domain, and the other is near the C-terminus of the molecule. This clustering is likely to be a consequence of the specificity of the mutagens used, the modest size of the collection, and the peculiarity of the screening procedure. This view is consistent with the finding that ts lesions generated by directed mutagenesis can occur outside the clusters (not shown).

The collection of mutations includes changes in many types of physical characteristics including charge, hydrophobicity, hydrogen bonding ability, and side chain volume. This suggests that no single class of interaction dominates protein stability; all types of interactions are potentially important. The thermal denaturation of several of these mutants has been investigated in detail.[7] Reductions in the enthalpy of stabilization of the folded protein are largely compensated by simultaneous reductions in the entropy change. Decreases in the temperature-dependent free energies of stabilization are modest, ranging from 2.4 to 3.3 kcal/mol at 47°C.

To investigate changes in interactions in the native conformation, the high-resolution X-ray crystal structures of six ts mutants were determined (Table III, Fig. 4). Three mutations, Arg 96 → His, Met 102 → Thr, and Thr 157 → Ile (Fig. 5), cause only localized structural adjustments. In contrast, conformational changes propagate away from the altered sites in the other three mutant proteins: Ala 98 → Val, Ala 146 → Thr (Fig. 6), and

TABLE III
TEMPERATURE-SENSITIVE MUTANT LYSOZYMES STUDIED CRYSTALLOGRAPHICALLY[12]

Lysozyme	Activity[a]	ΔT_m[b]	Resolution[c]	R[d]
Wild type	100	—	1.7	16.7
Arg 96 → His	100	−14	1.9	19.5
Ala 98 → Val	—	−13	1.8	16.8
Met 102 → Thr	60	−13	2.1	18.1
Ala 146 → Thr	55	−9	2.1	19.0
Gly 156 → Asp	50	−5	1.7	17.8
Thr 157 → Ile	90	−11	1.9	18.9

[a] Rate of hydrolysis of cell walls relative to wild type.[38]
[b] Change in melting temperature (°C) relative to wild type measured at pH 3[7] (and S. Cook, T. Alber, and B. W. Matthews, unpublished).
[c] Resolution of crystallographic data (Å).
[d] Crystallographic residual ($\Sigma|F_{obs} - F_{calc}|/\Sigma|F_{calc}|$) at the present state of refinement.

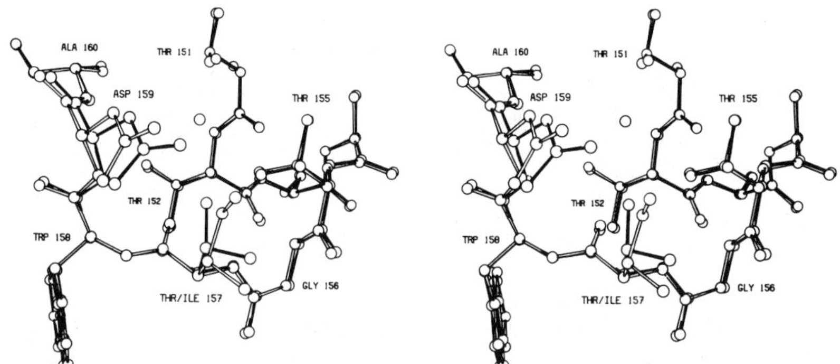

FIG. 5. Stereo drawing showing the differences in the structures of the wild-type and Thr 157 → Ile lysozymes in the vicinity of the altered amino acid. The refined model of the mutant protein (open bonds) is superimposed on the refined wild-type structure (solid bonds). An additional solvent molecule seen only in the mutant structure is also shown.

Gly 156 → Asp. The most extensive shifts result from the substitution of Ala 98 by Va in the protein interior. The bulkier valine side chain pushes apart two close-packed helices, and conformational changes are apparent in a 20 × 20 × 10 Å region of the C-terminal domain (T. Alber and B. W. Matthews, unpublished observations).

It has recently been shown that all of the temperature-sensitive mutations alter amino acid side chains that have lower than average crystallographic thermal factors and reduced solvent accessibility in the folded protein.[50] This suggests that the amino acids with well-defined conformations can form specific intramolecular interactions that make relatively large contributions to the thermal stability of the protein. Residues with high mobility or high solvent accessibility are much less susceptible to destabilizing substitutions, suggesting that, in general, such amino acids contribute less to protein stability.

Site-Directed Mutants

One interesting feature of the collection of random mutants is that destabilizing substitutions in α-helices often correlate with a reduction in statistical measures of helical propensity.[12] To investigate the relationship between protein stability and the predicted stabilities of individual elements of secondary structure, a surface residue in a helix, Pro 86, was replaced with ten other amino acids (Table I, Part a; T. Alber, J. A. Bell,

[50] T. Alber, S. Dao-pin, J. A. Nye, D. C. Muchmore, and B. W. Matthews, *Biochemistry* **26,** 3754 (1987).

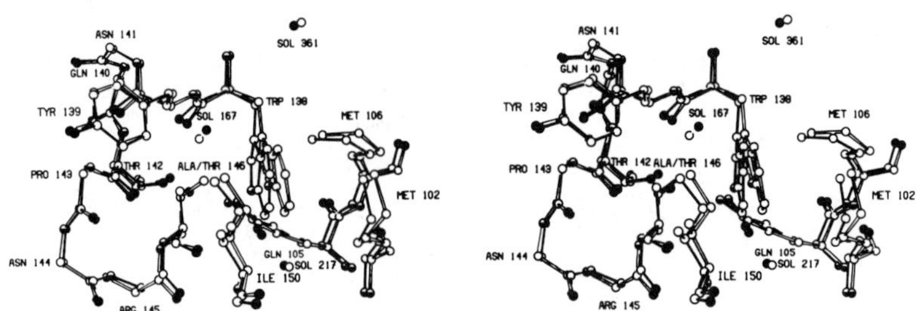

FIG. 6. Superposition of the structure of the mutant lysozyme Ala 146 → Thr (solid bonds) on wild-type lysozyme (open bonds). In this instance, structural changes propagate away from the site of substitution toward the protein interior. Thr 146 makes contact with Trp 138 which, in turn, causes repacking of Met 102 and Met 106.

and B. W. Matthews, unpublished results). These amino acids—Thr, Ser, Cys, Ala, Cys, Arg, His, Leu, Ile, and Gly—include strong helix breakers (Gly, Ser) and helix formers (Ala, Leu).[51,52] The X-ray crystal structures of seven of the mutant proteins were solved and refined at high resolution. In each case, replacement of the Pro 86 side chain causes a substantial conformational change in the vicinity of residues 82 and 83. An additional turn of distorted α-helix involving residues 82–86 is formed. Despite this large structural change, all ten amino acid substitutions cause only a slight decrease in the thermodynamic stability of the protein! Statistical analyses of secondary structures and studies of model α-helices do not provide a simple explanation for this result.

In contrast to the substitutions of Pro 86, the collection of ts mutants identifies side chain interactions that do enhance thermal stability. The best understood ts mutation is Thr 157 → Ile. Thr 157 is located on the surface of the protein in a loop between two α-helices. The ts lesion reduces T_m by 11°C at pH 2. This corresponds to a change in ΔG of approximately 2.9 kcal/mol at the T_m of the mutant. Comparison of the refined crystal structures of the wild-type and mutant proteins (Fig. 5) showed that three types of interactions are altered. The hydrogen bonds to the Thr 157 hydroxyl group are eliminated, the van der Waals contacts of the side chain γ-methyl group change, and there are differences in the solvation of Thr 155. Analysis of these changes suggested that the loss of the hydrogen bond between Thr 157 and the buried amide of Asp 159 causes the largest decrease in stability.[46]

[51] M. Levitt, *Biochemistry* **17,** 4277 (1978).
[52] H. A. Scheraga, *Pure Appl. Chem.* **50,** 315 (1978).

Site-directed mutagenesis provides a powerful method for analyzing the relative contributions of hydrogen bonding, van der Waals contacts, and solvent structure at this site because several of the natural amino acids are structurally related to the wild-type residue, Thr. For example, a measure of the importance of hydrogen bonding is provided by Ser, since it contains the hydroxyl group but not the methyl group of Thr. The role of van der Waals forces is tested by Val, since it is similar in shape to Thr but cannot hydrogen bond. The Thr side chain can be further whittled down to Ala, which lacks both the methyl and hydroxyl groups, and Gly, which lacks a side chain altogether. In addition, charged amino acids such as Arg and Asp can be introduced to see if electrostatic forces can substitute for the interactions in the wild-type protein.

Degenerate oligonucleotide primers (e.g., Table I, Part b) were used to make 12 additional amino acid substitutions at position 157. Proteins with Ser, Cys, Ala, Gly, Val, Leu, Phe, Asn, Asp, Glu, His, and Arg at this site were isolated and subjected to crystallographic and thermodynamic analysis. The changes in T_m caused by these mutations are listed in Table IV.

Several conclusions can be drawn from these studies. (1) Since all of the mutants are less stable than the wild-type protein, both the methyl and hydroxyl groups of Thr 157 contribute to thermal stability. (2) The hydrogen bonds to the hydroxyl group contribute more than the van der Waals

TABLE IV

RELATIVE STABILITIES OF LYSOZYMES WITH DIFFERENT SUBSTITUTIONS AT POSITION 157

Amino acid at position 157	Melting temperature relative to wild type ($\Delta T_m \pm 1°C$) at pH 2
Threonine (wild type)	—
Asparagine	−1.7
Serine	−2.5
Glycine	−4.2
Aspartic acid	−4.2
Cysteine	−4.9
Leucine	−5.0
Arginine	−5.1
Alanine	−5.4
Glutamic acid	−5.8
Valine	−6.0
Histidine	−7.9
Phenylalanine	−9.2
Isoleucine	−11.0

contacts of the methyl group. The X-ray crystal structures of the most stable mutants—Asn, Ser, Gly, and Asp 157—all contain hydrogen bonds analogous to those in the wild-type structure. Asn 157 causes adjustments in the conformation of the peptide backbone that allow its longer polar side chain to replace Thr. (3) Bound water can stabilize proteins. With Gly at position 157, a new bound water molecule restores the network of hydrogen bonds that is disrupted by the Thr 157 → Ile mutation! These hydrogen bonds are apparently sufficient to confer the bulk of the stability provided by Thr at this site. (4) A new ion pair involving Arg 157 can compensate for the loss of the hydrogen bonds to Thr 157. An unexpected hydrogen-bonded ion pair between Arg 157 and Asp 159 was seen in the X-ray crystal structure of the mutant protein. As the pH is raised from 2 to 6, Asp 159 becomes more ionized and the stability of Thr 157 → Arg progressively approaches that of wild-type (data not shown).

These conclusions are stated in terms of effects on the folded conformations of the mutant and wild-type lysozymes. There are several reasons to think that this is a valid way to think about the substitutions at position 157. (1) The increase in hydrophobicity due to the Thr 157 → Ile substitution would be expected to stabilize the folded state. Since the opposite effect is observed, important interactions in the folded conformation must be disrupted. (2) The stabilities of mutant lysozymes with Leu, Val, Phe, and Ile at position 157 (Table IV) do not rank according to the hydrophobicities of these side chains.[51,53,54] This implicates specific structural interactions in determining the contributions of these side chains to stability. (3) A positive correlation was found between thermal stability and a structural pattern (hydrogen bonds with residue 157) in the folded state. Nonetheless, it is possible that some of the observed effects arise from changes in the free energies of the denatured states of the proteins.

A clear example of this is provided by the studies of Perry and Wetzel.[18,55] Using site-directed mutagenesis, these workers engineered a disulfide bond in phage T4 lysozyme by replacing Ile 3 with Cys. Peptide mapping of the purified protein showed that a disulfide bond forms spontaneously between Cys 3 and the preexisting unpaired Cys 97. The disulfide bond makes the Ile 3 → Cys mutant more stable than the wild-type protein to both reversible and irreversible thermal denaturation. This is consistent with the view that covalent cross-links that are compatible with

[53] C. Tanford, *J. Am. Chem. Soc.* **84,** 4240 (1962).

[54] R. Wolfenden, L. Andersson, P. M. Cullis, and C. C. B. Southgate, *Biochemistry* **20,** 849 (1981).

[55] L. J. Perry and R. Wetzel, *Biochemistry* **25,** 733 (1986).

the native structure can make a stabilizing contribution by reducing the number of accessible conformations in the unfolded state.[56–58]

General Conclusions

Phage T4 lysozyme provides a powerful model system for studying the structural basis of protein thermal stability. By comparing the properties of mutant and wild-type proteins, the contributions of individual chemical groups can be estimated. The diversity of substitutions in the collection of mutant lysozymes indicates that many types of interactions contribute to stability. Comparison of the X-ray crystal structures of mutant lysozymes shows that the protein conformation adjusts in response to seemingly conservative amino acid substitutions. Such conformational changes can be localized or extensive. There is no correlation between the magnitude of the change in stability and the size of the conformational change. Consequently, the X-ray crystal structure of each mutant must be determined to identify the interactions in the folded conformation. This process is greatly facilitated by the fact that most of the mutant lysozymes crystallize in the same form as the wild-type protein.

Despite great advances in the technology of generating and studying mutants, the fundamental problem of assigning the observed effects to the energy of the folded or unfolded state remains unsolved. One approach to overcoming this limitation is to look for structural patterns in the native conformation that are associated with thermal stability. Identifying such patterns will deepen our understanding of nature's principles of protein design.

Acknowledgments

The results described here would not have been possible without the help of a number of talented co-workers including J. Bell, S. Cook, S. Dao-pin, T. M. Gray, M. G. Grütter, J. Nye, L. H. Weaver, K. Wilson, and J. Wozniak. We have also benefitted from help and advice from a number of other individuals involved with the lysozyme project including W. Baase, W. Becktel, F. W. Dahlquist, L. McIntosh, D. Muchmore, C. Schellman, and J. A. Schellman.

T.A. was supported by a postdoctoral fellowship from the Helen Hay Whitney Foundation. Work was supported in part by grants from the National Institutes of Health and the National Science Foundation.

[56] J. A. Schellman, *C. R. Trav. Lab. Carlsberg, Ser. Chim.* **29,** 230 (1955).

[57] P. J. Flory, *J. Am. Chem. Soc.* **78,** 5222 (1956).

[58] D. P. Goldenberg, *J. Cell. Biochem.* **29,** 321 (1985).

Author Index

Numbers in parentheses are footnote reference numbers and indicate that an author's work is referred to although the name is not cited in the text.

A

B

C

D

E

F

H

I

J

L

M

N

S

T

U

V

Subject Index

A

B

C

D

E

F

L

M

P

R